D1824317

NATURAL ENVIRONMENT RESEARCH COUNCIL

INSTITUTE OF GEOLOGICAL SCIENCES

MEMOIRS OF THE GEOLOGICAL SURVEY OF GREAT BRITAIN
ENGLAND AND WALES

Geology of the Country around Barnard Castle

Explanation of One-Inch Geological Sheet 32, (New Series)

by

D. A. C. MILLS, B.SC. and J. H. HULL, M.SC.

with contributions by

C. R. Burch, B.Sc., E. A. Francis, B.Sc., W. H. C. Ramsbottom, M.A., Ph.D., M. A. Calver, M.A., Ph.D., J. Pattison, M.Sc., G. A. L. Johnson, D.Sc., B. Owens, B.Sc., Ph.D., R. K. Harrison, M.Sc. and T. S. Tomlinson, B.Sc.

LONDON
HER MAJESTY'S STATIONERY OFFICE
1976

*The Institute of Geological Sciences
was formed by the
incorporation of the Geological Survey of Great Britain
and the Museum of Practical Geology
with Overseas Geological Surveys
and is a constituent body of the
Natural Environment Research Council*

PLATE I (*Frontispiece*)

Barnard Castle and the River Tees; the castle stands on a steep bluff of Top and Bottom Crag limestones and interbedded shales (L 392).

ISBN 0 11 880742 0*

PREFACE

THE BARNARD CASTLE district was originally surveyed on the six inch to one mile scale between the years 1870 and 1881 by W. Gunn, and by H. H. Howell who also supervised the work. The map was published, hand coloured, on the one-inch scale on Old Series Sheet 103 SW, the Solid edition in 1881 and the Drift edition in 1883. Some revisions were inserted in 1889. Some, but not all, of the six inch to one mile maps were published between 1875 and 1879 and Vertical Sections Sheet 65, published in 1878, included some sequences in the Coal Measures of the district. No explanatory memoir was published.

The district was resurveyed between 1958 and 1963, the bulk of the work being done by Mr. D. A. C. Mills in north-central, central, eastern and south-eastern areas and by Mr. J. H. Hull in north-western and south central parts. Mr. C. R. Burch resurveyed parts of the west and south, and Mr E. A. Francis the extreme north-east corner of the district. The resurvey was supervised by Mr. W. Anderson, Dr. E. H. Francis and myself. A list of six-inch maps with the names of the surveyors is given on page xi. Solid and Drift editions of the new one-inch map were published in 1968.

The task of compiling this memoir has fallen to Mr. Mills, who also wrote most of the text, though a substantial part was written by Mr. Hull. Contributions from Messrs Burch and Francis are incorporated in the structural account, and in some of the stratigraphical chapters. Viséan and Namurian macrofaunas have been identified by Dr. W. H. C. Ramsbottom and the microflora by Dr. B. Owens; Westphalian fossils were determined by Dr. M. A. Calver and those from the Permian by Mr. J. Pattison. Mr. R. K. Harrison wrote the sections on petrography. The Economic Geology chapter contains a contribution on coal rank and quality by Mr. T. S. Tomlinson of the National Coal Board. Appendix 3, an account of the Mount Pleasant Bore, was contributed by Dr. G. A. L. Johnson of Durham University. The memoir has been edited by Mr. B. J. Taylor.

Willing assistance, during the resurvey and the writing of the memoir, from officials in the National Coal Board, the various water authorities, local authorities and quarry companies, and from landowners, is gratefully acknowledged.

A. W. WOODLAND
Director

Institute of Geological Sciences
Exhibition Road
South Kensington
London SW7 2DE
15th June 1976

PREFACE

[illegible]

[illegible]

[illegible]

[illegible]

[illegible]

CONTENTS

(References are listed at the end of each chapter)

ILLUSTRATIONS

TEXT-FIGURES

EXPLANATION OF PLATES

FACING PAGE

LIST OF SIX-INCH MAPS

Six-inch geological maps covering the one-inch geological sheet 32 are listed below with the date of issue (in brackets), the initials of the surveyors and date of survey for each map. The surveyors were C. R. Burch, E. A. Francis, J. H. Hull, D. H. Land, D. A. C. Mills and R. H. Price.

Printed maps are marked P with the date of publication. Dyeline copies of the other six-inch geological maps are available, and copies of all the maps are held for public reference in the libraries of the London and Leeds offices of the Institute of Geological Sciences.

NY 91 N.E.	Hury, Baldersdale, Deepdale (1969)	C.R.B.	1961
NY 91 S.E.	Bowes, Gilmonby, Gilmonby Moor (1968)	C.R.B., D.A.C.M.	1958–63
NY 92 N.E.	Monk's Moor, Eggleston Burn (1968)	J.H.H.	1958–60
NY 92 S.E.	Mickleton, Eggleston, Romaldkirk (1969)	J.H.H.	1959–60
NZ 01 N.W.	Cotherstone, Barnard Castle (west) (1968)	C.R.B.	1963
NZ 01 N.E.	Barnard Castle (east), Streatlam (1968)	J.H.H.	1962
NZ 01 S.W.	Boldron, Greta Valley (central) (1965)	C.R.B.	1963
NZ 01 S.E.	Rokeby, Brignall, Barningham (1965)	J.H.H.	1962
NZ 02 N.W.	Eggleston Common (P 1966)	J.H.H.	1960–61
NZ 02 N.E.	Woodland, Copley (P 1966)	J.H.H.	1961
NZ 02 S.W.	Barnley, Marwood (P 1968)	J.H.H.	1961
NZ 02 S.E.	Langleydale, Shotton (P 1967)	J.H.H.	1961–62
NZ 11 N.W.	Cleatlam, Winston (1967)	D.A.C.M.	1960–62
NZ 11 N.E.	Killerby, Headlam, Gainford (1969)	D.A.C.M.	1960–62
NZ 11 S.W.	Ovington, Hutton Magna (1966)	J.H.H.	1963
NZ 11 S.E.	Caldwell, Eppleby, Fawcett (1967)	J.H.H.	1963
NZ 12 N.W.	Butterknowle, High Lands (P 1967)	D.A.C.M.	1959–60
NZ 12 N.E.	West Auckland, Etherley (P 1967)	D.A.C.M.	1959
NZ 12 S.W.	Cockfield, Staindrop (P 1966)	D.A.C.M.	1959–60
NZ 12 S.E.	Bolam, Ingleton (P 1967)	D.A.C.M.	1959–60
NZ 21 N.W.	Denton, Piercebridge, High Coniscliffe (1968)	D.A.C.M.	1961
NZ 21 N.E.	Faverdale (1976)	D.A.C.M., D.H.L.	1961–73
NZ 21 S.W.	Aldbrough, Low Coniscliffe (1967)	D.A.C.M.	1962
NZ 21 S.E.	Darlington (1976)	D.A.C.M., D.H.L.	1962–73
NZ 22 N.W.	Bishop Auckland (part), Eldon, Shildon (P 1967)	E.A.F., J.H.H.	1959–65
NZ 22 N.E.	Windlestone (P 1976)	R.H.P., E.A.F., D.H.L.	1950–73
NZ 22 S.W.	Redworth, Houghton le Side (1968)	D.A.C.M.	1961
NZ 22 S.E.	Aycliffe (1976)	D.A.C.M., D.H.L.	1961–73

A narrow strip of ground along the southern margin of the sheet is covered by six-inch geological maps which are available in manuscript form only.

EXPLANATORY NOTES

In this book, the word 'District' means the area included in the one-inch geological sheet 32 (Barnard Castle).

Numbers in square brackets are National Grid References. The prefaces NY and NZ are not used since the numerical references are unique in the district. The 100-km square NY includes a narrow strip along the western margin of the district; the rest of the sheet falls within the 100-km square NZ.

Numbers preceded by the letter E refer to the Sliced Rock Collection of the Institute of Geological Sciences.

Numbers preceded by the letter L are serial numbers in the Institute's collection of Geological Survey Photographs.

Natural sections or Boreholes marked with an asterisk (*) are detailed in the relevant appendices.

Other explanatory notes, relevant to particular parts of the book, are given in the appropriate chapters.

Authors of fossil names are given in the Index.

Chapter 1

INTRODUCTION

GEOGRAPHICAL SETTING AND INDUSTRIES

THIS MEMOIR describes the geology of the district covered by the Barnard Castle (32) Sheet of the One-Inch Geological New Series Map[1] of England and Wales. Part of the district, north of the River Tees, lies within County Durham, the remainder forms part of the North Riding of Yorkshire. The location of the district with reference to northern England is shown in Fig. 1, while Fig. 2 shows the outlines of the solid geology.

The district comprises two contrasted areas. The smaller area, in the north, is underlain by Coal Measures and is partly industrialized: the rest of the district is occupied by Carboniferous Limestone, Millstone Grit and Permian strata and is chiefly rural.

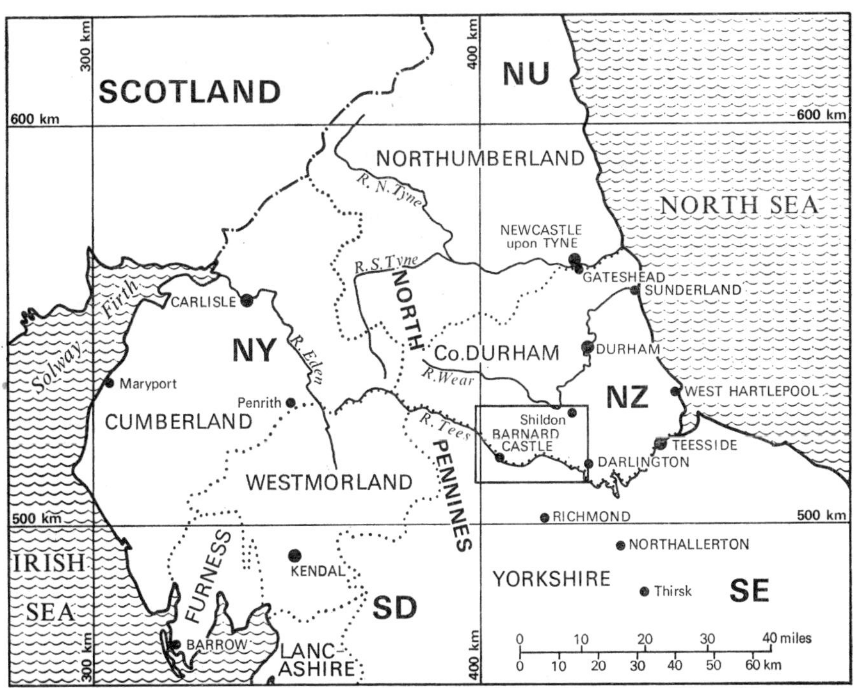

FIG. 1. *Sketch-map showing location of the Barnard Castle district. The county boundaries shown are those before the local government reorganization in 1974. Teesside refers jointly to Stockton and Middlesbrough.*

[1] Referred to as 'the district' throughout this memoir.

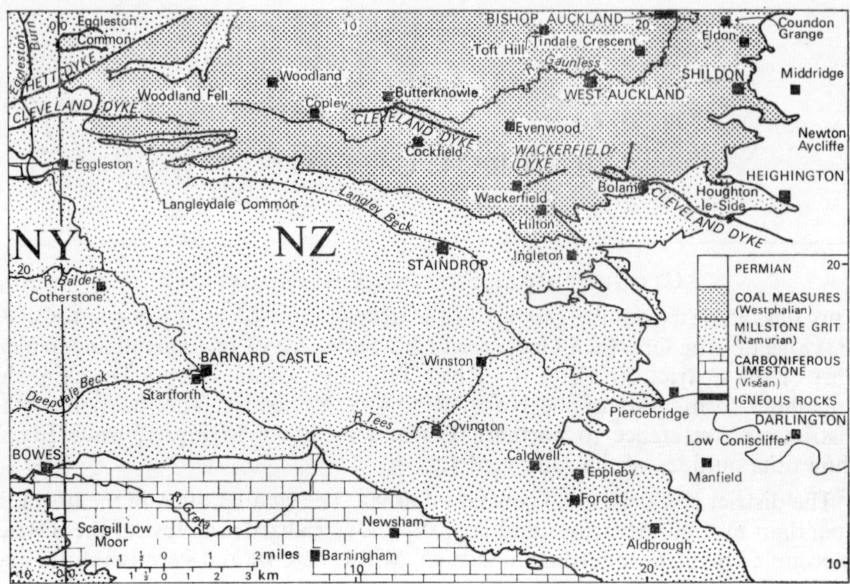

FIG. 2. *Sketch-map of the solid geology of the Barnard Castle district*

The northern area, which forms the south-western part of the Northumberland and Durham Coalfield, became industrialized in the latter half of the 19th century with the advent of large-scale coal mining. The history of coal mining, however, extends back to the year 1375 when Henry Vavasour's mine was working on Cockfield Fell—one of the first inland collieries on record. Centres around which the early coal industry grew include Hindon, Woodland, Copley, Butterknowle, Cockfield, Evenwood, Toft Hill, West Auckland, St Helen Auckland, Bishop Auckland and Shildon. The coal reserves of this area are now largely exhausted, though small areas of coal are still being mined at one locality by a private operator under licence from the National Coal Board, and small private opencast sites in areas formerly worked by underground methods are being operated between Woodland and Toft Hill. Quarrying was never a major industry in this northern part of the district, and it has, like mining, steadily declined. The Cleveland Dyke, once extensively worked for roadstone on Cockfield Fell, is now quarried only to the north-west of Bolam and even here operations have virtually ceased. Magnesian Limestone is now being removed from only two quarries near Middridge. Despite the long history of mining this part of the district retains a semi-rural aspect. Mixed agriculture is extensively carried out, though to the west on the higher ground stock rearing and sheep farming are more important. Other employment in the area is provided by light engineering, chemical and coke works at Evenwood, a coal washing plant at West Auckland, and light industrial estates at West Auckland, St Helen Auckland, Tindale Crescent and at Shildon, where there is also a railway wagon works. There are brickworks and a coal washing plant at Eldon.

In the north-western, central southern and eastern parts of the district farming is particularly important being mainly of mixed type, sheep and stock cattle becoming more important on the higher ground in the west. Minor industries in this area include the quarrying of Carboniferous limestone and sandstone, and sand and gravel. A chemical plant at Barnard Castle is the largest single source of employment in the western part of the area, while in the east many people are employed in a light industrial estate at Newton Aycliffe and in Darlington, whose western suburbs extend into the area. The main administrative centre of the Durham part of this area is Barnard Castle—a market town of about 6000 population—while those parts that are in the North Riding are administered from Barnard Castle and Startforth, and from Richmond, which lies in the (Sheet 41) district to the south.

Roads are the main means of communication within the district, the most notable being the A66 route from Scotch Corner to Penrith which traverses the southern part of the district. Railway services have been discontinued except for the line from Darlington to Bishop Auckland.

Water for parts of the district is obtained from storage reservoirs to the west in the adjacent Brough-under-Stainmore (Sheet 31) district. Some supplies, however, are obtained locally from springs and boreholes, while a large part of Darlington's water is obtained by pumping from the River Tees near Low Coniscliffe.

Physical Geography

A large part of the district falls within the Mid-Tees valley which may be taken to extend eastward from the Pennine slopes of Stainmore to the broad dale around Darlington. In the south-western part of the district the River Tees is joined by three major south-bank tributaries, namely the River Balder, Deepdale Beck and the River Greta. Elsewhere, however, the streams draining into the River Tees are, except for Eggleston Burn and Langley Beck, small. An area up to four miles wide, which extends along most of the northern margin of the district, falls within the catchment of the River Wear, the chief tributary here being the River Gaunless. On the southern margin of the district a small area east of Newsham is drained by the River Gilling, a tributary of the River Swale.

Most of the high ground lies in the west and north where it forms part of a dissected plateau tilted to the east. In the north-west it includes the wide moorland expanse of Eggleston Common (up to 1590 ft OD), Woodland Fell (1448 ft OD) and Langleydale Common (1408 ft OD) all forming part of the Tees–Wear watershed. The south side of the divide between the Wear and Tees is marked by a strong south-facing scarp composed mainly of massive sandstones of late Namurian and early Westphalian age. It extends from Eggleston in the west to Houghton le Side in the east, though it is ill-defined between Wackerfield and Hilton because of faulting. To the east and north-east of Houghton le Side the scarp dies out as Magnesian Limestone caps the sandstone sequence. In the south-west, beyond Scargill and Barningham, is Scargill Low Moor rising to over 1450 ft and leading to the divide between Teesdale and Arkengarthdale. The major south-bank Tees tributaries have cut deep gorges for some distance above their respective confluences; these gorges are well-wooded and contrast with the bleak open moorlands farther upstream. Below Barnard Castle the Tees Valley broadens out progressively, though locally the river flows through a series of gorges, the broad ridges and valleys of the Carboniferous country giving way farther east to the gently rolling area formed by the Magnesian Limestone.

Geological Sequence

The formations represented on the One-Inch New Series Geological Map and sections of this district are summarized below.

Superficial Formations (Drift)

Recent and Pleistocene

Landslip	Fluvio-glacial Terrace
Peat	Glacial Sand and Gravel
Alluvium	Boulder Clay
River Terrace Deposits	Glacial Lake Deposits, including Brick Clays and Laminated Clays
Fluvio-glacial Sand and Gravel	
Older River Gravel	Morainic Drift

Solid Formations

	Generalized thickness in ft
Permian	
Magnesian Limestone: Limestones, dolomitic limestones and dolomites with interbedded marl and ?anhydrite; Marl Slate at base	up to 430
Basal Permian Sandstone and Breccia, with subordinate Basal Permian Sands (Yellow Sands) in north-east	up to 44
Carboniferous	
UPPER CARBONIFEROUS (NAMURIAN AND WESTPHALIAN)	
Middle Coal Measures (*Westphalian B*)	
Mudstones and shales; sandstones; numerous coal seams and seatearths; Harvey Marine Band at base	up to 860
Lower Coal Measures (*Westphalian A*)	
Mudstones and shales; sandstones, especially in lower part; numerous coal seams in upper part, becoming thin and subordinate to base; seatearths; marine bands in basal 140 ft; Quarterburn Marine Band at base	up to 660
Millstone Grit Series (*Namurian*)	
Sandstones, commonly coarse-grained; thin seatearths, coals, mudstones and shales; *Reticuloceras* cf. *stubblefieldi* at base	up to 280
Mudstones, shales; sandstones (locally thick); occasional thin coals and seatearths; limestones, including Great Limestone at base	up to 1100
LOWER CARBONIFEROUS (VISÉAN)	
Limestones, mudstones and shales with subordinate sandstones including channel sandstones	up to 500

Intrusive Igneous Rocks

Tertiary

Tholeiitic dolerites and allied tholeiite, e.g. Cleveland Dyke

Carboniferous–Permian

Quartz-dolerite and allied tholeiite, e.g. Hett Dyke

Classification

The Carboniferous rocks of England and Wales have traditionally been divided on a lithological basis into Carboniferous Limestone at the base overlain successively by Millstone Grit and Coal Measures. It has been known for many years that the divisions so called in northern England are not equivalent to their counterparts farther south. Faunal correlations have become sufficiently precise to allow for the reclassification of the northern England sequence. The standard classification now applied to the rocks of the Barnard Castle district is shown in Table 1 together with its relationship to older classifications now abandoned.

Table 1

Original Survey of District (1881)	Survey Practice, Northumberland and Durham (1926-1955)			Standard Classification: Northern England	Standard Classification: N.W. Europe		
MIDDLE COAL MEASURES Brockwell Coal	Upper Carboniferous	COAL MEASURES	UPPER GROUP Shield Row or High Main Coal MIDDLE GROUP Brockwell Coal	MIDDLE COAL MEASURES *Harvey Marine Band*	*A. vanderbeckei*	WESTPHALIAN B	UPPER CARBONIFEROUS OR SILESIAN
LOWER COAL MEASURES Top of Third Grit			LOWER GROUP Ganister Clay Coal	LOWER COAL MEASURES *Quarterburn Marine Band*	*G. subcrenatum*	WESTPHALIAN A	
MILLSTONE GRIT Base of First Grit		MILLSTONE GRIT Base of First Grit		Top of Second Grit MILLSTONE GRIT SERIES		NAMURIAN	
CARBONIFEROUS LIMESTONE SERIES	Lower Carboniferous	CARBONIFEROUS LIMESTONE SERIES	UPPER LIMESTONE GROUP Base of Gt. Limestone	Base of Gt. Limestone	*C. leion*		
			MIDDLE LIMESTONE GROUP	CARBONIFEROUS LIMESTONE SERIES		VISÉAN	LOWER CARBONIFEROUS OR DINANTIAN

The classification of the Permian rocks, in terms of formational names, is a modified form of that adopted in the Durham and West Hartlepool district (Smith and Francis 1967, pp. 91, 95). It is essentially based on the Durham sequence, while taking into account adjoining parts of Yorkshire. Recent work has enabled further refinements, which are not applied here, to be made in terms of beds and evaporitic cycles, and this work has been summarized by Smith (1970, p. 67).

Geological History

Nothing is known of the pre-Carboniferous rocks of the district, for the oldest known strata are of high Viséan (P_2) age. It is probable that earlier Viséan sediments are present at depth resting on Skiddaw Slates or granite as they do in neighbouring districts (Woolacott 1923, p. 60; Dunham and others 1965, p. 383). These sediments were probably deposited in a gradually subsiding shallow sea in which conditions particularly favoured the formation of limestone and shale, though locally 'channel' and 'sheet' sandstones were introduced.

Interpretation of the depositional history owes much to the concept of rigid and buoyant crustal blocks with intervening subsiding troughs (for criteria see Kent 1966, p. 325). The south-eastern part of the Alston Block occupies the north-west of the district with the Cotherstone (Stainmore) Syncline to the south. The greater part of the district would appear, therefore, to be marginal between block and trough.

There was little change in conditions of deposition in early Namurian times; the sediments are similar to those of the Viséan except that 'channel' sandstones are more common and limestones are generally thinner. Towards the end of Namurian times, however, there was a marked change: where, hitherto, sandstones, other than 'channel' sandstones, had been only component members of cyclic units they now became dominant. Limestones are absent from these cyclothems and marine shales are subordinate. These conditions continued into the early Westphalian. They were followed by a phase of sedimentation in a near-shore, estuarine and deltaic environment where local intermittent subsidences were followed by relatively stable periods conducive to coal formation.

During late Carboniferous and early Permian times the district was affected by the main Hercynian orogeny, which resulted in the folding and faulting of the Carboniferous rocks and the intrusion of quartz-dolerite dykes and the Great and Little Whin sills. These earth movements led to uplift and erosion, so that by the end of Lower Permian times a low-lying desert fringed by low hills had been produced. The axis of this shallow depression probably extended east-south-east from the Piercebridge area. The first deposits laid on this surface were sporadic breccias probably representing the result of ephemeral sheet floods, gully deposits and residual talus accumulations. During Upper Permian times the Zechstein Sea transgressed into much of the district, locally depositing the Marl Slate, followed upwards by carbonate rocks, limestones and dolomites, laid down for the most part under shallow water with restricted access to the open sea. The carbonates were followed by a series of interbedded siltstones, mudstones and evaporites forming the Middle Permian Marl succeeded by further limestones of the Upper Magnesian Limestone and in turn by the Upper Permian Marls. It is not known what thickness of higher Permian and younger measures may have been laid down and subsequently removed by erosion.

Both Carboniferous and Permian formations were faulted by Tertiary earth movements. During this time Hercynian faults affecting Carboniferous rocks were reactivated and their lines can be traced in the overlying Permian beds in the eastern part of the district. Associated with the orogeny was the intrusion of the Cleveland Dyke, which is thought to be of late Eocene to early Oligocene age. The uplifting and eastward tilting of the region, which commenced at this time, formed a surface on which the present pattern of erosion was initiated. The drainage system first established on this surface was substantially modified by the later Pleistocene glaciation. After the retreat of the ice the present regime was established, and while over much of the eastern part of the district streams are fairly well graded, in the west erosion is still a significant factor in the higher reaches of the Tees and Gaunless drainage systems.

Summary of Previous Research

The earliest geological work relevant to this district was that of Forster (1809) who named lithological units in the Carboniferous. In 1836 Phillips described the Carboniferous Limestone Series strata of the district in his classic *Geology of*

Yorkshire. The district was first surveyed between 1870 and 1881 by H. H. Howell and W. Gunn for the Geological Survey; most of their six inches to one mile geological maps were published, though a few sheets in County Durham for the area east of Piercebridge and Bolam were made available only as library reference copies. The one-inch map (Old Series 103 SW) was published in solid edition in June 1881 and in drift edition in July 1883. Further editions, with minor amendments, were published in December 1889. The maps were not accompanied by an explanatory memoir though an account by Gunn relating to the west of the district survives in manuscript. The Carboniferous strata of the north-western part of the district are marginal to the Northern Pennine Orefield and were discussed in general terms by Dunham (1948); this memoir was the only stratigraphical account of the Pennines north of the Tees–Swale watershed until work on the Carboniferous Limestone Series and Millstone Grit Series was carried out in the south-western and southern parts of the district respectively by Reading (1957) and Wells (1955, 1957) and more recently by Johnson and others (1962), Mills and Hull (1968), and Hull (1968).

The only work relating to the coalfield is by Kirkby and Duff (1872, pp. 150–98) who gave a short account of part of the area between Bishop Auckland, Woolly Hill and Arnghyll. This account incorporated some shaft and borehole sections and brief notes on adjacent areas to south and east, including the geology of the area between Staindrop and Heighington (Millstone Grit Series and Coal Measures) the Cleveland (or Cockfield) Dyke, the Magnesian Limestone and Drift.

The Permian rocks of the district have been mentioned by Winch (1817), Sedgwick (1829) and Browell and Kirkby (1866). They have also been referred to in the wider context of the descriptions of the Permo-Triassic rocks of north-eastern England by Trechmann (1913, 1914, 1921 and 1925). The same author (1921) refers specifically to the Permian exposures around Thickley.

The igneous rocks have been dealt with by Teall (1884), who gave an account of the Cleveland Dyke, and by Holmes and Smith (1921), who undertook research on the Wackerfield Dyke.

Detailed work on the Pleistocene deposits of the western part of the district was carried out by Dwerryhouse (1902), who described the glaciation of Teesdale following earlier mention by Goodchild (1875) and Dakyns (*in* Dakyns and others 1891). The only published work on the eastern part of the district is by Fawcett (1916), who researched on the pre- and post-glacial drainage system of the Mid-Tees valley. Allusion to this work and that of Dwerryhouse (1902) was made by Trotter (1929a, b) in his description of the glaciation and drainage of the country to the west of this district. More recently the Quaternary of Durham county and the north of England has been summarized, respectively, by Francis (1970) and Smith (*in* Taylor and others 1971). D.A.C.M., J.H.H.

References

Borings and Sinkings. 1878–1910. An account of the Strata of Northumberland and Durham as proved by Boring and Sinkings. The Council of the North of England Institute of Mining and Mechanical Engineers (usually bound in 4 books, but originally issued as 7 volumes. Generally referred to as "Borings and Sinkings").

Browell, E. J. and Kirkby, J. W. 1866. On the chemical composition of various beds of the Magnesian Limestone and associated Permian rocks of Durham. *Nat. Hist. Trans. Northumb.*, **1**, 204–30.

Dakyns, J. R., Tiddeman, R. H., Russell, R., Clough, C. T. and Strahan, A. 1891. The Geology of the country around Mallerstang. *Mem. geol. Surv. Gt Br.*

Dunham, K. C. 1948. The Geology of the Northern Pennine Orefield; Vol. 1, Tyne to Stainmore. *Mem. geol. Surv. Gt Br.*

——Dunham, A. C., Hodge, B. L. and Johnson, G. A. L. 1965. Granite beneath Viséan sediments with mineralization at Rookhope, northern Pennines. *Q. Jnl geol. Soc. Lond.*, **121,** 383–417.

Dwerryhouse, A. R. 1902. The Glaciation of Teesdale, Weardale and the Tyne Valley, and their Tributary Valleys. *Q. Jnl geol. Soc. Lond.*, **58,** 572–608.

Fawcett, C. B. 1916. The Middle Tees and its tributaries: a study in river development. *Geogrl. Jnl,* **48,** 310–25.

Forster, W. 1809. *A treatise on a Section of the Strata from Newcastle-upon-Tyne to the Mountain of Cross Fell, in Cumberland, with remarks on Mineral Veins in general.* 1st Edit. Alston.

——1821. 2nd Edit. Alston.

——1883. 3rd Edit., revised by W. Nall. Newcastle.

Francis, E. A. 1970. Quaternary (*in* Geology of Durham County). *Trans. nat. Hist. Soc. Northumb.*, **41,** 134–52.

Goodchild, J. G. 1875. The Glacial Phenomena of the Eden Valley and the Yorkshire-Dale District. *Q. Jnl geol. Soc. Lond.*, **31,** 55–99.

Holmes, A. and Smith, S. 1921. The Wackerfield Dyke, Co. Durham. *Geol. Mag.*, **58,** 440–54.

Hull, J. H. 1968. The Namurian stages of north-eastern England. *Proc. Yorks. geol. Soc.*, **36,** 297–308.

Johnson, G. A. L., Hodge, B. L. and Fairbairn, R. A. 1962. The base of the Namurian and of the Millstone Grit in north-eastern England. *Proc. Yorks. geol. Soc.*, **33,** 341–62.

Kirkby, J. W. and Duff, J. 1872. Notes on the geology of part of South Durham. *Nat. Hist. Trans. Northumb.*, **4,** 150–98.

Kent, P. E. 1966. The structure of the concealed Carboniferous rocks of north-eastern England. *Proc. Yorks. geol. Soc.*, **35,** 323–52.

Mills, D. A. C. and Hull, J. H. 1968. The Geological Survey borehole at Woodland, Co. Durham (1962). *Bull geol. Surv. Gt Br.*, **28,** 1–37.

Phillips, J. 1836. *Illustrations of the Geology of Yorkshire. Part II: The Mountain Limestone District.* London.

Reading, H. G. 1957. The stratigraphy and structure of the Cotherstone Syncline. *Q. Jnl geol. Soc. Lond.*, **113,** 27–56.

Sedgwick, A. 1829. On the geological relations and internal structure of the Magnesian Limestone, and the lower portion of the New Red Sandstone Series in their range through Nottinghamshire, Derbyshire, Yorkshire and Durham to the southern extremity of Northumberland. *Trans. geol. Soc. Lond.*, Ser. **2, 3,** 37–124.

Smith, D. B. 1970. Permian and Trias (*in* Geology of Durham County). *Trans. nat. Hist. Soc. Northumb.*, **41,** 66–91.

Smith, D. B. and Francis, E. A. 1967. Geology of the country between Durham and West Hartlepool. *Mem. geol. Surv. Gt Br.*

Taylor, B. J., Burgess, I. C., Land, D. H., Mills, D. A. C., Smith, D. B., and Warren, P. T. 1971. Northern England. *Br. reg. Geol.*

Teall, J. J. H. 1884. Petrological Notes on some north of England Dykes. *Q. Jnl geol. Soc. Lond.*, **40,** 209–27.

TRECHMANN, C. T. 1913. On a mass of anhydrite in the Magnesian Limestone at Hartlepool, and on the Permian of south-eastern Durham. *Q. Jnl geol. Soc. Lond.*, **69,** 184–218.

——1914. On the lithology and composition of Durham Magnesian Limestones. *Q. Jnl geol. Soc. Lond.*, **70,** 232–65.

——1921. Some remarkably preserved brachiopods from the Lower Magnesian Limestone of Durham. *Geol. Mag.*, **58,** 538–43.

——1925. The Permian Formation in Durham. *Proc. Geol. Ass.*, **36,** 135–45.

TROTTER, F. M. 1929a. The Tertiary uplift and Resultant Drainage of the Alston Block and Adjacent Areas. *Proc. Yorks. geol. Soc.*, **20,** 161–80.

——1929b. On the Glaciation of Eastern Edenside, the Alston Block and the Carlisle Plain. *Q. Jnl geol. Soc. Lond.*, **85,** 549–612.

WELLS, A. J. 1955. The development of chert between the Main and Crow Limestones in North Yorkshire. *Proc. Yorks. geol. Soc.*, **30,** 177–96.

——1957. The Stratigraphy and Structure of the Middleton Tyas—Sleightholme Anticline, North Yorkshire. *Proc. Geol. Ass.*, **68,** 231–54.

WINCH, N. J. 1817. Observations on the geology of Northumberland and Durham. *Trans. geol. Soc. Lond.*, **4,** 3–7.

WOOLACOTT, D. 1923. A Boring at Roddymoor Colliery, near Crook, Co. Durham. *Geol. Mag.*, **60,** 50–62.

Chapter 2

LOWER CARBONIFEROUS
VISÉAN (CARBONIFEROUS LIMESTONE SERIES)

INTRODUCTION

THE Lower Carboniferous[1] crops out in a small area north of the Lunedale–Staindrop Fault near Eggleston, but the main outcrop extends across the south-western part of the district from Bowes to East Layton. Elsewhere these measures are buried beneath younger rocks and have been proved only in a few boreholes.

Only the late Viséan strata are known within the district, the sequence extending from close above the inferred position of the Scar Limestone to the base of the Namurian, which is taken in this account, following Johnson and others (1962), at the base of the Great Limestone. This succession is about 500 ft thick, and includes representatives of the Five-Yard, Three-Yard and Four-Fathom limestones (Fig. 3). The Rookhope and Roddymoor bores provide well-documented sections through upper Viséan measures in districts to the north and north-west and are included for comparison in Fig. 4.

CLASSIFICATION

Within the Lower Carboniferous of the district a fauna diagnostic of precise horizon has been obtained only from above the Four-Fathom Limestone, where *Girtyoceras*? *costatum* indicates a P_{2c} age (Johnson and others 1962, p. 349). Correlation with the measures in other districts suggests that the whole sequence here lies within the P_2 goniatite stage and the D_2 coral-brachiopod zone (Rayner 1953, pp. 285–95) and thus forms part of the upper Viséan.

These measures also fall within the Middle Limestone Group, as defined in Northumberland by Fowler (1926, p. 18), a classification later extended into Cumberland by Trotter and Hollingworth (1932, p. 16) and into Durham by Dunham (1948, p. 14). However, there has been some confusion about the lower limit of the group (Johnson 1959, p. 95; Frost 1969, p. 279) and this, together with the proposed abandonment of the term 'Upper Limestone Group' (Mills and Hull 1968, p. 1) for the overlying strata, has led to the discontinuance of this classification.

By still older terminology these measures form part of the 'Yoredale Series'—a term originally proposed by Phillips (1836, p. 37) and used by him as a stratigraphical unit. The usage is now obsolete, though the term 'Yoredale' has survived, after Hickling (1931, p. 222), to describe a type of sedimentary facies.

[1]Shown as Carboniferous Limestone Series d^3 on the One-Inch geological map.

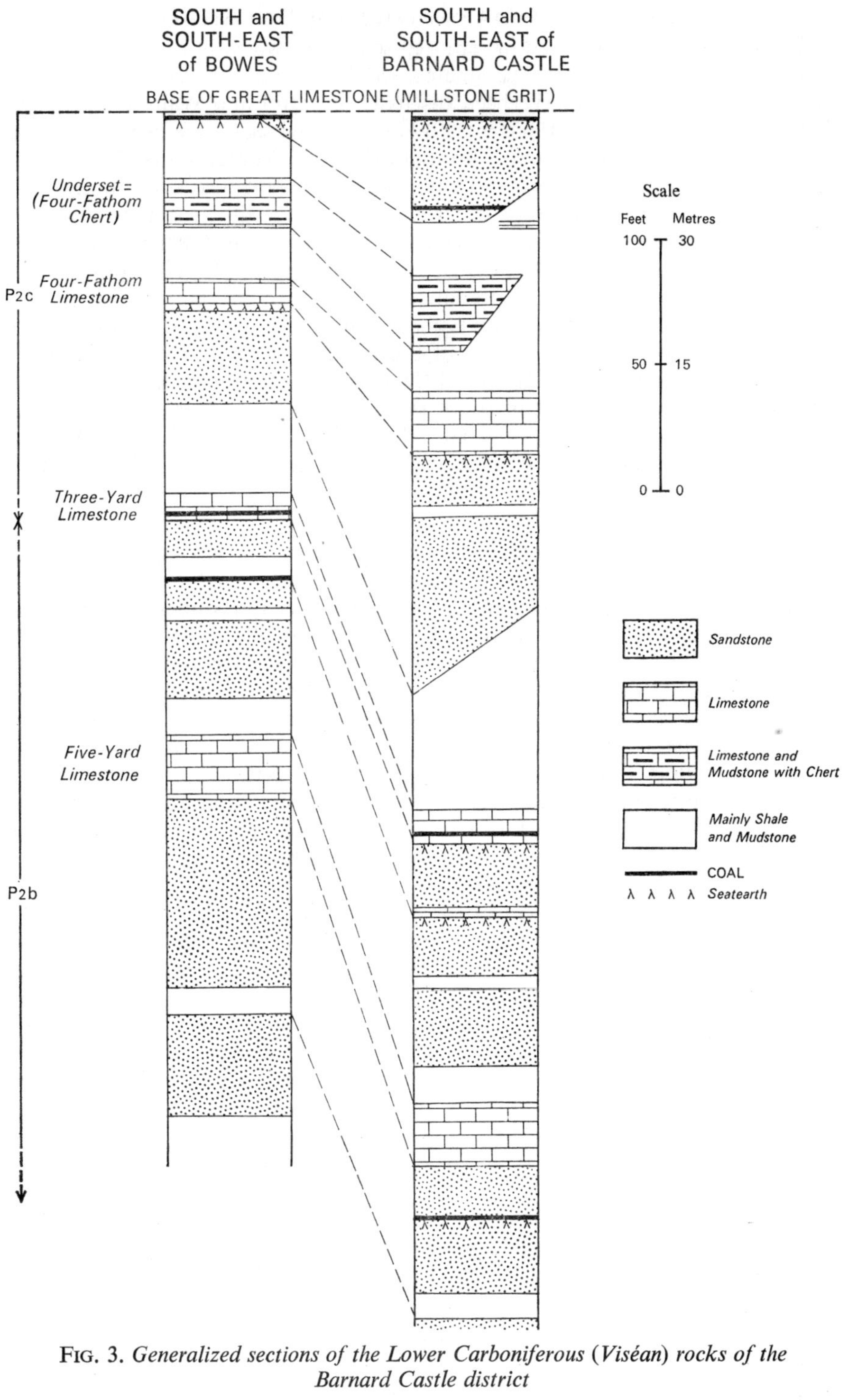

SOUTH and SOUTH-EAST of BOWES

SOUTH and SOUTH-EAST of BARNARD CASTLE

BASE OF GREAT LIMESTONE (MILLSTONE GRIT)

Underset = (Four-Fathom Chert)

Four-Fathom Limestone

P2c

Three-Yard Limestone

Five-Yard Limestone

P2b

Scale

Feet Metres

100 30

50 15

0 0

Sandstone

Limestone

Limestone and Mudstone with Chert

Mainly Shale and Mudstone

COAL

Seatearth

FIG. 3. *Generalized sections of the Lower Carboniferous (Viséan) rocks of the Barnard Castle district*

Lithology and Conditions of Deposition

The sediments are arranged in repeated sequences which have been variously called, by different authors, cyclothems, rhythms or cycles. Variations in the meanings of these terms have been discussed by Duff and Walton (1962, pp. 235–55), though as Wells (1960, p. 390) has indicated they have become almost synonymous. The 'rhythmic unit' (Brough 1929, p. 117) or 'cyclothem' (Dunham 1948, p. 14) typical of the northern Pennines is, in upward succession: limestone, shale with marine fossils, unfossiliferous dark shale with ironstone nodules, sandy shale or shaly sandstone, sandstone, ganister or fireclay, coal. The sequences in the Barnard Castle district generally conform to this arrangement, but additional minor cycles, represented by seatearth or coal and/or marine strata, are recorded in the sandstone phase. The local pattern of sedimentation is thus more comparable with that of the Northumberland Trough (Johnson 1959, p. 89; 1962, p. 328) than that of the rest of the northern Pennines. Other observed differences are that many of the sandstones grade up into limestone with no intervening seatearth or coal. Alternatively where coal is present it is commonly separated from the base of the limestone by a few inches of shale with abundant shell debris.

It is generally accepted that the deposition of the Lower Carboniferous in northern England was influenced by the presence of a relatively stable block flanked by more rapidly subsiding sedimentary troughs or gulfs. The concept of a stable block in the northern Pennines during Lower Carboniferous times was first postulated by Marr (1921, pp. 63–72). Subsequently the northern half of the structure became known as the Alston Block (Trotter and Hollingworth 1928, p. 433) and the southern half the Askrigg Block (Hudson 1938, p. 309), the two being separated by the Cotherstone (Stainmore) Syncline (Reading 1957, p. 27). The south-eastern margin of the Alston Block was later more closely delineated by Bott (1961, fig. 4), who showed it to comprise a 'hinge' approximating to the line of the Butterknowle Fault; this interpretation was subsequently followed by Kent (1966, fig. 1). In terms of these structural elements, as defined by previous authors, the Barnard Castle district includes the south-eastern part of the Alston Block, the north-eastern portion of the Askrigg Block and an eastward continuation of the Cotherstone Syncline, in the form of a gentle monocline (see Fig. 17). The criteria, based on the nature and thickness of the sediments, for distinguishing between blocks and troughs (gulfs) have been listed by Kent (1966, p. 325). With these in mind it would appear that in Carboniferous times the depositional environment over most of the district was intermediate between block and trough. Furthermore, the evidence from the Woodland Bore (Mills and Hull 1968, p. 11) seems to indicate that margins may have migrated with time and also that they are not necessarily coincident with faults.

As with the overlying Namurian sediments, the upper Viséan beds generally show evidence of deposition in relatively shallow water, irrespective of whether or not the environment was one of block or trough. Mechanisms by which shallow-water environments can be maintained in spite of differential subsidence, and in which cyclic sedimentation can occur are adequately reviewed by Dunham (1950, pp. 46–63), Wells (1960, pp. 389–403), Bott (1964, p. 1082), and Bott and Johnson (1967, pp. 421–41).

Nomenclature and Correlation

Beds in the Lower Carboniferous in this district have in the past been named in terms of local usage on either the Askrigg or the Alston block. In Swaledale it was formerly the practice of miners to name limestones by counting down from the Main or Great Limestone (Namurian, see p. 28) and calling them respectively Underset, Third Set, Fourth Set, etc. This classification was seen by the primary surveyors (Dakyns and others 1891, p. 107) to break down when the Third Set was locally absent; thus the Fourth Set of one area became the Third of another. They resolved this problem by using the current Alston Block terms Three-Yard and Five-Yard limestones for the upper and lower beds respectively, a practice subsequently followed by other workers and summarized by Dunham (1959, fig. 3). In this account it is proposed to discontinue the use of the term 'set' completely and to use Alston Block names (Dunham 1948, fig. 3); accordingly in addition to the names mentioned the Underset Limestone is called Four-Fathom Limestone.

Though the marker beds in the Lower Carboniferous do not contain any unique palaeontological characteristics their exposure is sufficiently continuous, and their lithologies and thicknesses distinctive enough, for correlation to be relatively easy.

General Stratigraphy

Strata below the Five-Yard Limestone (Fig. 4) crop out in the River Greta and also farther south on Barningham Moor south-west of Barningham. The lowest exposed measures comprise 18 ft of dark grey silty mudstone with thin sandstone laminae and are inferred to occur close above the Scar Limestone and therefore, by comparison with other areas (Johnson and others 1962, p. 347), to be of early P_{2b} age. They are overlain by 100 to 125 ft of sandstones, including a mudstone band, 7 to 10 ft thick, in the lower 50 ft and a thin coal about 20 ft from the top; the mudstone has yielded a marine fauna in the River Greta, near Rutherford Bridge, but is apparently unfossiliferous elsewhere. The uppermost 1 to 1½ ft of sandstone, immediately beneath the Five-Yard Limestone, contains shell casts and the sandstone–limestone junction is sharp at some localities and gradational at others.

The Five-Yard Limestone (Fig. 4) is a grey, fine-grained, thin- to thick-bedded[1], bioclastic limestone with a fauna including corals and brachiopods (p. 18).

Locally, as on the north side of Barningham Moor, the limestone includes bands of shale containing an abundant fauna. The thickness of the limestone ranges from 25 to 30 ft, compared with the 20 to 30 ft recorded farther south in Swaledale (Dakyns and others 1891, p. 107) and with the 28 ft proved in the Roddymoor Bore, near Crook, County Durham (Woolacott 1923, p. 51). On the Alston Block the comparable range is from 7 to 23 ft (Dunham 1948, p. 21) with a thickness of almost 16 ft recorded in the Rookhope Bore some 4 miles NW of Stanhope, County Durham (Dunham and others 1965, pl. 34).

Strata between the Five-Yard and Three-Yard limestones (Fig. 4) are discontinuously exposed at two localities in the River Greta. Near Thackholme Farm they are about 79 ft thick, and 3½ miles farther east, 98 ft. At both localities the lowest beds are similar, consisting, in upward sequence, of 10 to 14 ft of mudstone, fossiliferous in the lower part, 30 ft of sandstone and 3 ft of silty shale. The higher beds, however, are different at the two localities. In the west the silty shale is overlain successively by 12½ ft of sandstone with a shale parting, 14 in of coal with dirt bands, 7 ft of shale with *Lingula* at the base, and 11½ ft of

[1]In this account bedding is described as follows: 'massive' if in units over 3 ft thick; 'thick' if in units between 1 and 3 ft.; and 'thin' if in units between a quarter of an inch and a foot. Regular beds less than a quarter of an inch thick are termed 'laminated'.

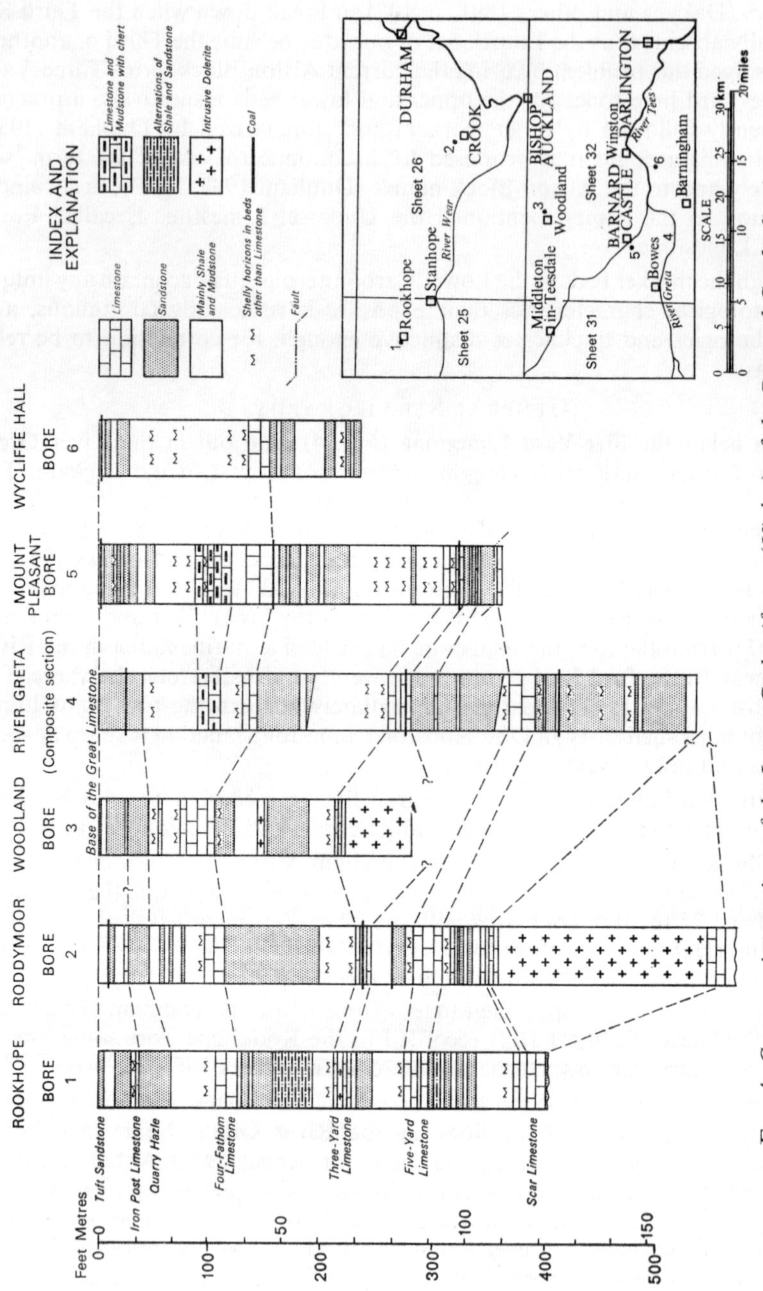

FIG. 4. *Comparative sections of the Lower Carboniferous (Viséan) of the Barnard Castle district and adjacent districts to the north*

sandstone; east of Brignall in the bed of the River Greta near Scotchman's Stone, the equivalent beds comprise 51 ft of sandstones with a 1-ft limestone, which is correlated with the *Lingula* band of the western sequence.

These measures were also proved in the Mount Pleasant Bore* (see Appendix 3) where the sequence, though lithologically similar to that near Scotchman's Stone, was only 31 ft thick. As this thickness is even less than the minimum recorded on the Alston Block (Dunham 1948, p. 21) it is thought that 40 to 60 ft of strata may have been cut out by faulting. In the northern part of the district the Woodland Bore* penetrated below the Three-Yard limestone but then entered and ended in the Great Whin Sill.

The only completely exposed section of the **Three-Yard Limestone** (Fig. 4) is in the River Greta, near Scotchman's Stone, where it is 12 ft thick and includes shale partings and a 1½-in coal. A similar section 14¼ ft thick, was proved in the Mount Pleasant Bore*. Other partial sections are exposed in the River Greta and on Barningham Moor, where the limestone is estimated to have an overall thickness of at least 16 ft, including a 1-ft seatearth-sandstone which locally cuts out some of the shales in the lower part of the limestone. In the Woodland Bore* 6 ft 3½ in of limestone occurs close above the Great Whin Sill; it is uncertain whether this bed represents the total thickness of the Three-Yard Limestone or whether more limestone may be present beneath the sill. With this exception thicknesses in the Barnard Castle district are rather greater on average than those recorded to the north and south on the Alston and Askrigg blocks respectively (Dunham 1948, p. 22; Dakyns and others 1891, p. 108).

Strata between the Three-Yard and Four-Fathom limestones (Fig. 4) have their best outcrops in the River Greta. They are variable in thickness and fall broadly into a lower argillaceous and an upper arenaceous subdivision. These subdivisions are respectively 36 and 40 ft thick in the area south-east of Bowes increasing to 79 and 62 ft respectively 5½ miles to the east, south of Greta Bridge. Erosion is locally apparent at the base of the upper sandstone division in the eastern area, near Eastwood Hall, while west-south-west of Brignall, near Moor House Farm, the base of the sandstones extends down to the horizon of the Three-Yard Limestone.

Elsewhere in the district good sections through these measures are recorded only in boreholes. In the Woodland Bore,* where they are almost 112 ft thick, the lower argillaceous division is relatively thin and the upper is split by the intrusion of the Little Whin Sill. In the Mount Pleasant Bore* these beds are 150 ft thick.

The upper, mainly arenaceous, division is correlated with the Nattrass Gill Hazle of the Alston Block (Forster 1809, p. 58; Dunham 1948, p. 21) and at the Mount Pleasant Bore*, as in the Wycliffe Hall Bore*, it consists of four discrete sandstones separated by shale bands, the middle one of which contains marine fossils.

Throughout the district the contact between the Four-Fathom Limestone and the underlying strata is relatively sharp, limestone or a thin marine shale overlying either a coal or seatearth resting on a sandstone of the upper arenaceous division of the measures. The ranges of thickness and lithology exhibited by the measures in the district are broadly comparable with those in adjacent areas to the north and south (Dunham 1948, p. 22; Dakyns and others 1891, p. 110). A P_{2c} age may be inferred (Johnson and others 1962, p. 348) for the shales above the Three-Yard Limestone.

The Four-Fathom Limestone (Fig. 4) generally consists of a grey, fine-grained, bioclastic limestone; massive at the base, it becomes more thinly bedded upwards. Like the underlying strata, it crops out not only in the southern part of the district but also farther north in the River Tees, near Eggleston Hall, on the upthrow side of the Lunedale–Staindrop Fault. Its thickness is broadly in accord with the range of 20 to 40 ft recorded in Swaledale (Dakyns and others 1891, p. 108).

At the Browson Bank section east of Newsham (p. 21) and at the exposure in the River Tees (p. 21) clisiophylloid corals, including *Aulophyllum fungites pachyendothecum* and *Dibunophyllum bipartitum bipartitum*, have been collected from near the base of the limestone. This coral band, originally noted in Wensleydale by Hudson (1924, p. 131) and more recently described elsewhere by Turner (1956, p. 414) and Reading (1957, p. 33), is thus more laterally persistent than is suggested by Wells (1957, p. 241). The coral band is associated with a 14-in bed containing nodules of pale grey fine-grained siliceous decalcified limestone, with shell casts. The shape of the nodules (Plate II) and their disposition suggests that they are concretions formed in place even though the limestone in which they occur is current-bedded. This nodular bed is known to thicken from 9 to 16 ft in an east-south-easterly direction (Hull 1964, p. 49). At Browson Bank the limestone also contains an 8-in band rich in *Saccamminopsis fusulinaformis*, which is common at this horizon elsewhere in northern England (Dunham 1948, p. 22; Wells 1957, p. 241).

The strata between the Four-Fathom and Great limestones (Fig. 4) are considered by some authors to comprise two cyclothems because of the presence of the Iron Post Limestone and the underlying Quarry Hazle sandstone (Dunham 1948, p. 23) midway between the two limestones. In this district the only possible representative of the Iron Post is the 5-ft limestone recorded about 70 ft above the Four-Fathom Limestone in the area south of Ovington (p. 22). This limestone, however, may equally well be correlated with the Underset Chert, because the Quarry Hazle, which elsewhere intervenes, appears to be absent over the greater part of the district.

Though the measures between the Four-Fathom and Great limestones crop out extensively in the southern part of the district, and to a more limited extent south-west of Eggleston, exposures are generally poor because of thick drift deposits. The measures, which in much of the district comprise a lower argillaceous and an upper arenaceous sequence, range in thickness from 60 to 138 ft. In exposures along the River Greta at Bowes the argillaceous sequence includes about 20 ft of interbedded calcareous mudstones, cherty mudstones and chert, which lie some 20 ft above the top of the Four-Fathom Limestone. These cherty beds, which locally contain thin bands of siliceous limestone, are the equivalents of the Underset (= Four-Fathom) Chert of other authors (Wells 1957, p. 242; Johnson and others 1962, fig. 2). In the Mount Pleasant Bore* they are 32 ft thick, the base being separated by 13 ft of shales from the Four-Fathom Limestone, and it is from these intervening shales that *Girtyoceras*? *costatum*, which indicates a P_{2c} age, has been collected (Johnson and others 1962, pp. 349–50). The cherty beds are absent farther east in the Wycliffe Hall Bore* and this accords with the statement by Wells (1957, p. 242) that they are known to be absent north and east of a line drawn roughly from Barnard Castle to Barningham Moor, along the south side of the Gilling Valley and across to Middleton Tyas in the Richmond (41) district.

(L 342)

A, B. Siliceous concretions; base of the Four-Fathom Limestone, Browson Bank Quarry, Newsham

PLATE II

(L 343)

The upper arenaceous part of these measures consists mainly of sandstone and it is the variation in thickness of these beds which accounts for the overall variation between the Four-Fathom and the Great. The uppermost sandstone post of this sequence, which lies close below the base of the Great Limestone, ranges from 2 to 9 ft in thickness. This bed is the probable equivalent of the Tuft or Water Sill of the Alston Block, where it ranges in thickness from 3 to 29 ft (Dunham 1948, p. 25).

A section of special note immediately beneath the Great Limestone was formerly exposed in the floor of East Layton Quarry (p. 23) where the limestone is separated from the top of the local equivalent of the Tuft Sill by 2 to 4 in of silty sand with numerous nodules of bornite and covellite. J.H.H.

DETAILS

(Comparative sections of these measures are shown in Figs. 3 and 4)

Measures below the Five-Yard Limestone crop out in the River Greta from above Rutherford Bridge [0348 1218] down to Tebb Wood [0737 1179]. At the base are 18 ft of dark grey silty mudstone with numerous sandstone laminae; these beds are presumed to lie close above the Scar Limestone and are exposed at the foot of Black Scar [0584 1119], about 750 yd NE of Scargill Castle. Above them are 40 ft of white and pale grey fine-grained limonitic sandstones, with worm tubes, possible shell casts and carbonaceous partings near the base. The upper part of these sandstones forms the lowest part of the sequence exposed on a cliff [0451 1137], about ¾ mile SE of Rutherford Bridge, as follows:

	ft	in
FIVE-YARD LIMESTONE	—	—
Not exposed	15	0
Sandstone, white massive false-bedded fine-grained siliceous ..	60	0
Mudstone, grey fossiliferous; silty at top and base; sandstone laminae	9	10
Sandstone, white and brown thin-bedded fine-grained siliceous; calcareous, with shell fragments in top 6 in	2	0
Shale, grey; sandstone laminae ..	2	0
Sandstone, white thin- and thick-bedded fine-grained siliceous; carbonaceous micaceous partings ..	22	6

About 7 ft of the upper mudstone member of this sequence are further exposed in the east bank of Gill Beck [0622 1107], and much of the upper sandstone can be seen in Brignall Banks [0541 1128], as follows:

	ft	in
Sandstone, grey massive false-bedded medium- to coarse-grained limonitic; shell casts, coal scars; undulating base	39	0
Shale, dark grey micaceous; ironstone nodules and sandstone lenses .. .. .. 9 in to	1	4
Sandstone, grey fine-grained siliceous; chert nodules; sporadic shell casts	1	0
Shale, black carbonaceous micaceous	0	2
Sandstone, dark grey fine-grained; chert nodules; shell casts ..	0	2
Sandstone, grey thin-bedded fine-grained siliceous	1	0
Sandstone, grey thick-bedded fine-grained carbonaceous micaceous	1	1
Sandstone, grey thick-bedded fine-grained siliceous	0	9

The top of the upper sandstone of this last section, together with the overlying beds which are within 20 ft of the base of the Five-Yard Limestone, are again exposed [0671 1161], near Brignall Quarry as follows:

	ft	in
Shale, dark grey micaceous ..	1	6
Coal	0	1
Mudstone-seatearth..	0	6
Sandstone, grey shaly	1	0
Sandstone, grey to white massive fine-grained siliceous, carbonaceous micaceous..	8	2
Sandstone, grey to brown massive fine-grained limonitic	3	6
Sandstone, grey thin-bedded fine-grained; carbonaceous micaceous streaks	3	6

The measures immediately beneath the Five-Yard Limestone are exposed in the

River Greta [0366 1210], 220 yd ESE of Rutherford Bridge, where the sequence is:

	ft	in
FIVE-YARD LIMESTONE	—	—
Sandstone, white and brown massive fine-grained siliceous; shell fragments in the top 1½ ft	8	6
Shale, grey sandy	0	10
Sandstone, grey thin-bedded fine-grained micaceous siliceous ..	7	0

At this locality the junction between the limestone and the underlying sandstone is sharp, whereas in an exposure [0393 1187], about 350 yd to the south-east, it is gradational.

South of the River Greta the measures below the Five-Yard Limestone are seen only on the north side of Barningham Moor, where the sequence exposed [0643 0949] near Cowclose House Farm, is

	ft	in
FIVE-YARD LIMESTONE	—	—
Not exposed	0	6
Sandstone, grey and brown fine-grained calcareous; micaceous and shaly in basal 14 in; shell fragments, roots and plants	2	5
Sandstone, grey shaly micaceous fine-grained	0	6
Mudstone, grey silty	0	2

The Five-Yard Limestone is exposed in Eller Beck [0298 1160], about 430 yd SW of Rutherford Farm, and also at the confluence of the beck with the River Greta. It is best seen, however, in the Greta valley, especially on the north bank between Rutherford Bridge and Tebb Wood, as exemplified by the following section [0366 1210], about 220 yd ESE of the bridge:

	ft	in
Limestone, bluish grey thin- and thick-bedded fine-grained nodular bioclastic; *Dibunophyllum bipartitum*, *Quasiavonia* cf. *aculeata*, rhynchonelloid indet., trilobite hypostome	7	6
Limestone, as above; *Cleiothyridina sp.*, *Quasiavonia* cf. *aculeata*, *Semiplanus* cf. *latissimus*, *Spirifer sp.* and *Parallelodon sp.* in lower 5 ft.	10	0
Limestone: *Lithostrotion junceum*, *Clisiophyllum keyserlingi* ..	1	0
Limestone, bluish grey thin-bedded, fine-grained	3	0
Sandstone	—	—

About 350 yd to the south-east, 22 ft of limestone are exposed, while in a quarry [0480 1146], about 320 yd NE of Brignall Mill, the bed is 20½ ft thick with solitary corals, crinoids and brachiopods. The limestone is continuously exposed and about 25 ft thick in the north bank of the River Greta between a locality [0557 1146] 500 yd E of Moor House Farm and Tebb Wood, where it dips into the river. Here [0756 1187] the upper 15 ft of limestone contain *Dibunophyllum sp.*, *Avonia youngiana*, *Eomarginifera sp.* and a smooth indeterminate spiriferoid. On the south bank small exposures of limestone can be seen near an old lime kiln [0672 1148] and in a quarry [0562 1123] in an outlier, about 750 yd NE of Scargill.

The limestone is well exposed between Cowclose Hall Farm and Moorcock Hall Farm. In Nor Beck, some 100 yd SSE of Cowclose Hall Farm [0643 0949], for example, it comprises at least 25 ft of grey thin-bedded fine-grained shelly limestone. In another exposure [0701 1012] lower down Nor Beck the limestone is particularly bioclastic, and together with a shale band contains *Amplexizaphrentis sp.*, *Diphyphyllum sp.*, *Fenestella sp.*, bryozoa [indet.], *Alitaria sp.*, *Antiquatonia sulcata*, *Cleiothyridina sp.*, *Dielasma sp.*, *Eomarginifera* cf. *setosa*, *Gigantoproductus sp.* [fragment], *Pleuropugnoides sp.*, *Rhipidomella michelini*, and smooth indeterminate spiriferoids.

The only other record of the limestone in the district is from the Mount Pleasant Bore* [0328 1508] near Barnard Castle, where the upper 8½ ft were recorded.

Measures between the Five-Yard and Three-Yard limestones are patchily exposed in the Greta valley and on Barningham Moor. An exposure [0361 1221] in the north bank of the river, about 150 yd ENE of Rutherford Bridge, shows the limestone to be succeeded by 10 to 14 ft of mudstone, the basal 3 ft of which contains foraminifera (including *Endothyra*, *Stacheoides* and *Tetrataxis*), Holothuroidea, *Fenestella frutex*, *F.* aff. *occulata*, *Penniretepora sp.*, *Polypora verrucosa*, *Rhabdomeson sp.*, *Streblotrypa sp.*, *Alitaria panderi*, *Antiquatonia sp.*, *Craniops sp. nov.*, *Cleiothyridina fimbriata*, *Pleuropugnoides sp.*, *Plicochonetes sp.*, *Pugnoides* aff. *triplex*, *Quasiavonia* cf. *aculeata*, *Rhipidomella michelini*, *Rugosochonetes sp.*, *Semiplanus* cf. *latissimus*, *Spiriferellina* aff. *etheridgei*,

Hypergonia sp., Aviculopecten sp., Leiopteria sp., Limipecten dissimilis, Streblochondria sp.

Above the mudstone are about 30 ft of grey and brown, thin- to thick-bedded, false-bedded fine-grained, micaceous sandstone with shaly partings and thin coals. These beds are intermittently exposed between 600 and 800 yd upstream from Rutherford Bridge. This sandstone is succeeded by 3 ft of dark grey silty shale, with thin sandstone ribs, exposed [0765 1190] in the south bank of the river, about 485 yd NNE of Crook's House.

The measures between this shale and the base of the Three-Yard Limestone are exposed at two localities, about 3½ miles apart, in the Greta valley. The westernmost section, on the north bank [0222 1281], about 250 yd W of Thrackholme Farm, is as follows:

	ft	in
THREE-YARD LIMESTONE	—	—
Sandstone, white thin- and thick-bedded fine-grained micaceous siliceous; calcareous at top, grading up into the limestone; shaly partings and coal films in basal 8 ft	11	6
Shale, grey sandy; *Lingula* at base	7	0
Coal; dirt bands and plants ..	1	2
Sandstone-seatearth, grey fine-grained siliceous 1 to	3	0
Shale, grey sandy; plants and roots	2	6
Sandstone, white and brown fine-grained siliceous	7	0

The eastern exposure extends from the remains of St Mary's Church [0772 1220] down river to Scotchman's Stone, and consists of

	ft	in
THREE-YARD LIMESTONE	—	—
Sandstone, grey and white siliceous thin-bedded to massive mainly fine-grained but medium-grained at base; shell casts and roots in top 1 ft; coal fragments, carbonaceous partings; *4½ ft about middle not exposed*	28	0
Limestone, dark grey fine-grained siliceous	1	0
Sandstone, grey and brown thin- and thick-bedded fine- to medium-grained limonitic; jointed; *incompletely exposed*	23	0

The 1-ft limestone here is correlated with the *Lingula* band of the western section.

Small sections in the measures below the Three-Yard Limestone are also exposed to the south on Barningham Moor. In a beck [0728 0997], about 500 yd W of Barningham Park, 1½ ft of white, fine-grained, calcareous sandstone with shell casts rest on 3 ft of grey thin-bedded fine-grained carbonaceous, micaceous sandstone. In Gordale Gill [0673 0936], just beyond the southern margin of the district, 17½ ft of sandstone are exposed below the limestone.

In the Mount Pleasant Bore* these measures appear to be lithologically similar to those upstream of Scotchman's Stone, but they total only 31¼ ft in thickness, part of the sequence being cut out by faulting.

The Three-Yard Limestone is proved in the north of the district only in the Woodland Bore* where 6¼ ft of limestone resting on 2¼ ft of mudstone were recorded. The limestone lies close above the Great Whin Sill and it is probably only the upper part which is represented. A complete section through the limestone was proved farther south in the Mount Pleasant Bore* where it is 14¼ ft thick including shale bands and a 7-in coal near the base, and where it resembles exposed sections farther south.

In an exposure [0218 1282] on the north side of the Greta valley, about 250 yd W of Thackholme Farm, the lower part of the limestone is seen resting on calcareous sandstone as follows:

	ft	in
Limestone, bluish grey fine-grained crinoidal	3	6
Shale, grey	0	2
Coal, contorted .. .. 1 in to	0	2
Sandstone-seatearth, white fine-grained siliceous; abundant roots	1	0
Limestone, greyish blue fine-grained slightly sandy; grades down to underlying sandstone	1	0

The sandy limestone post and the underlying sandstone are also partially exposed in the Greta valley as follows: [0262 1265], about 230 yd SE of Thackholme Farm, limestone 1 ft; [0763 1231], about 150 yd NNW of St Mary's Church, limestone 2 ft; [0772 1191], about 600 yd W of Wilson House, limestone 4½ ft (? not *in situ*). The following complete section in the limestone is exposed in the river at Scotchman's Stone [0807 1245]:

	ft	in
Shale	—	—
THREE-YARD LIMESTONE		
Limestone, grey massive fine-grained siliceous; *Caninia cornucopiae*, *Cyathaxonia cornu rushianum* group, *Fasiculophyllum densum*, *Zaphrentites enniskilleni derbiensis*, *Avonia youngiana*	5	2
Shale, dark grey calcareous; endothyrid foraminifera, indeterminate clisiophylloid corals, *Fenestella sp.*, *Penniretepora sp.*, *Tabulipora* cf. *scotica*, *Avonia youngiana*, '*Camarotoechia*' *sp.*, *Cleiothyridina* cf. *fimbriata*, *Craniops sp.*, *Crurithyris sp.*, *Echinoconchus* cf. *elegans*, *Eomarginifera setosa*, *Plicochonetes* aff. *interstriatus*, *Rugosochonetes sp.*, *Semiplanus sp.* [latissimoid], *Spirifer* cf. *bisulcatus*, *Spiriferellina octoplicata*, *Tornquistia polita*, indeterminate turreted gastropods, *Acanthopecten* cf. *nobilis*, *Aviculopecten sp.*, *Limipecten sp.*, *Weberides sp.*, and ostracods including *Hollinella sp.*, and *Kirkbya sp.*	4	0
Limestone, dark grey fine-grained argillaceous; foraminifera, *Heterophyllia sp.*, *Aulophyllum fungites*, *Dibunophyllum bipartitum bipartitum*	1	8
Shale, dark grey	0	1½
Coal	0	1½
Mudstone-seatearth, pale grey silty rooty	0	2
Limestone, dark grey thin-bedded fine-grained sandy; grading down to	0	9
Sandstone	—	—

Farther south in a stream [0775 0986], about 500 yd W of Barningham Park, the following comparable section is exposed:

	ft	in
THREE-YARD LIMESTONE		
Limestone, grey thin-bedded fine-grained; plants at top	5	9
Not exposed	1	0
Mudstone, grey calcareous ..	1	0
Limestone, grey fine-grained; shelly	1	0
Not exposed (probably including thin coal)	0	6
Sandstone, grey and white thin-bedded fine-grained siliceous; rooty	1	0
Mudstone, dark grey;	1	6
Shale, black papery ..	1	6
Limestone, grey thin-bedded fine-grained rooty; becoming sandy upwards	2	7
Sandstone, grey thin-bedded fine-grained carbonaceous; shelly at top, worm borings at base ..	4	6

About 100 yd upstream the 1-ft sandstone of this sequence rests on limestone, the 3 ft of fossiliferous mudstone and shale having being cut out.

The measures between the Three-Yard and Four-Fathom limestones are not well exposed in the northern part of the district. In the north bank of the River Tees, however, the upper part of the sequence is exposed [9949 2342], about 250 yd WNW of Eggleston Hall, as follows:

	ft	in
FOUR-FATHOM LIMESTONE	—	—
Mudstone-seatearth, dark grey carbonaceous micaceous; thin coal laminae at base	2	0
Sandstone, grey fine-grained siliceous	0	8
Sandstone-seatearth, grey thin-bedded fine-grained silty micaceous	7	6

Between this area and the main outcrop to the south, which extends from Bowes to East Layton, these measures were proved in the Woodland* and Mount Pleasant* bores, where thicknesses of 111½ and 150 ft respectively were recorded. The sequence in the Woodland Bore* is notable for including a representative of the Little Whin Sill (Mills and Hull 1968, p. 9; Harrison 1968, p. 38). The upper part of the sequence was also proved in the Wycliffe Hall Bore*[1228 1328]

In the area south-south-east of Bowes these measures are partially exposed in the western reaches of the River Greta, where the succession east of Robin Hood's Scar [0026 1322] is:

	ft	in
FOUR-FATHOM LIMESTONE	—	—
Sandstone, grey massive fine-grained siliceous; roots at top	40	0
Shale, grey sandy; sandstone ribs ..	6	0
Not exposed	20	0
Shale, grey; ironstone nodules ..	5	0
Not exposed	5	0
THREE-YARD LIMESTONE	—	—

Farther east a more complete section is exposed in the River Greta between Scotchman's Stone and a locality [0851 1300], about 200 yd upstream from Greta Bridge:

	ft	in
Sandstone, grey and brown thick-bedded and massive jointed medium- to coarse-grained with carbonaceous scars; undulating base	38	0
Shale, grey, silty fine-grained; sandstone ribs; plant debris	7	0
Sandstone, grey fine-grained; carbonaceous micaceous partings 6 in to	2	6
Shale, dark grey silty micaceous; shaly sandstone ribs and lenses at top	18	6
Shale, dark grey; calcareous at base; *Fenestella sp.*, *Rhombopora sp. nov.*, *Crurithyris sp.*, *Dielasma sp.*, *Eomarginifera setosa*, indeterminate gigantoproductids [latissimoid, juv.]., *Lingula sp.*, *Linoprotonia sp.*, *Martinia* cf. *glabra*, *Phricodothyris sp.*, *Rugosochonetes sp.*, *Spirifer sp.*, *Tornquistia polita*, indeterminate pleurotomarian, *Straparollus carbonarius*, indeterminate pectinoid, coiled nautiloid fragments, *Cornulitella carbonaria* and ostracods including *Kirkbya sp.* (*not continuously exposed*) ..	51	0
THREE-YARD LIMESTONE	—	—

The top of this section is thought to lie about 24 ft below the Four-Fathom Limestone.

The thick sandstone below the Four-Fathom Limestone is sporadically exposed to the south, in Barningham Park. It is also equated with the 30 ft of grey massive jointed medium- to coarse-grained sandstone, with an undulating base, which is exposed in a quarry [1324 0944], 90 yd E of Dunsa Manor, on the southern margin of the district. In a quarry [0162 1028], about 150 yd N of Spanham Farm, the upper part of this sandstone is rooty and is separated from the overlying limestone by 2 in of black, sandy micaceous shale, with roots and slickensides.

The Four-Fathom Limestone crops out [9944 2342] in the River Tees west-north-west of Eggleston Hall, where it is 18½ ft thick and grey, fine-grained and massive, with sporadic solitary corals at the base.

The only other record in the northern part of the district is from the Woodland Bore* where a thickness of 32 ft 1 in was recorded. Farther south the limestone was also proved in the Mount Pleasant* and Wycliffe Hall* bores (Fig. 4) with thicknesses of 25 and 18 ft respectively.

In the southern half of the district the Four-Fathom Limestone crops out extensively between Bowes and East Layton. In an exposure [9959 1321], near Gilmonby Bridge, it comprises 12 ft of earthy, fine-grained limestone, with crinoid debris. Other small exposures occur south-west of Plover Hall on Scargill Low Moor, but the best section in the south-western part of the district is seen at a locality [0162 1028], about 150 yd N of Spanham Farm, where the sequence is:

	ft	in
FOUR-FATHOM LIMESTONE		
Limestone, greyish brown; crinoid and brachiopod debris	6	0
Limestone, pale grey or brown thin-bedded bioclastic slightly dolomitic	5	0
Limestone, grey and brown thin-bedded crinoidal; arenaceous especially at base	3	6
Shale, black silty	0	2
Sandstone	—	—

Farther east 5 ft of limestone are exposed [1222 1064] on the south side of the A66 trunk road, and it can be followed south-eastwards along the outcrop in a number of small quarries. The best section is in a quarry [1297 1012] north-east of Browson Bank, where the limestone notably contains some siliceous concretions (Plate II).

	ft	in
Limestone, pale grey medium-grained bioclastic rubbly crinoidal	2	6
Limestone, grey and brown thick-bedded fine-grained; calcite on joints; *Rhopalolasma sp.*, *Avonia ?*, *Echinoconchus punctatus*, *Phricodothyris* ? [juv.], *Pugnax pugnus*, *Spirifer* cf. *bisulcatus*	3	6
Limestone, grey and brown massive jointed fine- to medium-grained crinoidal; *Avonia ?*, *Eomarginifera* cf. *setosa*, orthotetoid [costae like *Schellwienella aspis*], *Spirifer* cf. *bisulcatus*	6	6
Limestone; abundant *Saccamminopsis fusulinaformis*	0	8

	ft	in
Limestone, grey thin-bedded fine- to coarse-grained bioclastic; calcite veins and siliceous decalcified concretions in basal 1 ft; *Aulophyllum fungites pachyendothecum, Dibunophyllum bipartitum bipartitum, Avonia sp.*, indeterminate gigantoproductid fragments [thin shell], *Phricodothyris sp.* [small].	2	10
Limestone, dark grey thin-bedded fine-grained crinoidal; sporadic zaphrentoids	2	6
Limestone, dark grey fine-grained argillaceous	1	0

Still farther south-eastwards other small exposures can be seen along the outcrop, the most easterly being in a quarry [1314 0998], about 550 yd N of Dunsa Manor, where 8½ ft of limestone are exposed.

The measures between the Four-Fathom and Great limestones are only poorly exposed in the district. North of the Lunedale–Staindrop Fault they can be seen in the north bank [9954 2342] of the River Tees, near Eggleston Hall, where 10 ft of dark grey, slightly silty mudstone, with ironstone nodules is seen to overlie the Four-Fathom Limestone; in Stobgreen Sike [0016 2356], where a dark grey mudstone is exposed about 15 ft below the Great Limestone, and in the adjacent stream [9977 2368] to the west, where a similar horizon is exposed.

About 2½ miles to the north in WT/4 Bore* [9986 2750] on Eggleston Common some 20 ft of strata underlying the Great Limestone consisting of alternating mudstones and sandstones were proved; two thin calcareous bioturbated shelly sandstone ribs were recorded 17 ft 1 in and 11 ft 5 in respectively below the Great Limestone. Nearly 6 ft of thin to thick-bedded grey and dark grey fine-grained siliceous sandstone immediately below the Limestone are regarded as the equivalent of the Tuft Sill.

In this northern area these measures were proved to be 69 ft 2 in thick in the Woodland Bore* where they comprise an upper arenaceous sequence overlying some 30 ft of mudstones and shales with thin bands of limestone and calcite mudstone.

In the south-western part of the district near Bowes 14 ft of hard calcareous mudstone overlying 6 ft of cherty mudstone, with chert bands, are seen in the River Greta [9915 1326], about 470 yd W of Gilmonby Bridge; these beds, which lie about 20 ft above the Four-Fathom Limestone, probably represent in part the Underset (=Four-Fathom) Chert (Wells 1957, p. 242; Johnson and others 1962, fig. 2). Part of the Underset Chert is further partially exposed in a road cutting [9961 1324] due north of Gilmonby Bridge as follows:

	ft	in
Limestone, grey arenaceous crinoidal	1	6
Mudstone, hard; alternating bands of calcareous shale	10	0
Mudstone, calcareous; cherty bands	6	0

The 5 ft of grey and brown thin-bedded fine-grained limonitic and dolomitic limestones exposed in a small quarry [1299 1075] south of Tefit Hall and occurring about 70 ft above the Four-Fathom Limestone may be ascribed to the chert horizon. A complete sequence, 125 ft thick, was proved in the Mount Pleasant Bore* which includes 32 ft of Underset Chert separated by some 13 ft of shale from the top of the Four-Fathom Limestone. *Girtyoceras? costatum* was found 4 ft above the Four-Fathom Limestone. A broadly comparable sequence was proved in the Wycliffe Hall Bore*, but no Underset Chert was present.

The measures between a horizon about 10 ft above the Underset Chert and the Great Limestone crop out in Chert Gill [9889 1240], about 170 yd WNW of West Plantation, where the sequence is:

	ft	in
GREAT LIMESTONE	—	—
Shale, grey sandy	0	5
Coal	0	6
Mudstone-seatearth, grey; sandy in basal 1 ft	5	6
Shale, grey; sandy at top; ironstone nodules in lower 9 ft	19	0

A somewhat different sequence underlies the Great Limestone about 6 miles to the east-north-east, where the section [0788 1427] in the south bank of the River Tees, about 400 yd W of Rokeby Hall, is as follows:

	ft	in
GREAT LIMESTONE	—	—
Sandstone, grey and white thin-bedded massive fine-grained; current-bedded at base; carbonaceous micaceous siltstone partings and laminae; rooty at top, worm tubes at base	22	6

	ft	in
Coal .. .. 6 in to	0	10
Sandstone-seatearth, grey medium-grained carbonaceous; rooty at top	1	0
Sandstone, grey thick-bedded fine-grained ripple-marked; worm tubes at top and silty near base; carbonaceous micaceous laminae	14	6
Mudstone, dark grey; ironstone nodules and sandstone ribs at top; indeterminate productoid fragments 18 ft from top	21	0

The sandstone at the top of this section is again exposed in the River Greta upstream of a locality [0849 1423], about 150 yd W of Mortham Tower. About 4¾ miles SE of Mortham Tower the following mineralized sequence can be seen in East Layton Quarry [1555 1066]:

	ft	in
GREAT LIMESTONE	—	—
Sand, black soft; sporadic shell fragments and numerous nodules of bornite and covellite covered with malachite .. .. 2 in to	0	4
Sandstone, grey thick-bedded hard fine-grained siliceous calcareous; malachite and chalcocite	0	9
Sandstone, pale grey finely laminated fine-grained micaceous; carbonaceous streaks and scars; malachite on joints and bedding planes	1	0
Shale, dark grey silty carbonaceous micaceous; numerous sandstone ribs and a little malachite ..	5	0

J.H.H., C.R.B., D.A.C.M.

References

Bott, M. H. P. 1961. A gravity survey off the coast of north-east England. *Proc. Yorks. geol. Soc.*, **33**, 1–20.

——1964. Formation of sedimentary basins by ductile flow of isostatic origin in the upper mantle. *Nature. Lond.*, **201**, 1082–4.

——and Johnson, G. A. L. 1967. The controlling mechanism of Carboniferous cyclic sedimentation. *Quart. Jnl geol. Soc. Lond.*, **122**, 421–41.

Brough, J. 1929. On rhythmic deposition in the Yoredale Series. *Proc. Univ. Durham phil. Soc.*, **8**, 116–26.

Dakyns, J. R., Tiddeman, R. H., Russell, R., Clough, C. T. and Strahan, A. 1891. The geology of the country around Mallerstang. *Mem. geol. Surv. Gt Br.*

Duff, P. McL. D. and Walton, E. K. 1962. Statistical basis for cyclothems: quantitative study of the sedimentary succession in the east Pennine coalfield. *Sedimentology*, **1**, 235–55.

Dunham, K. C. 1948. The geology of the Northern Pennine Orefield; Vol. 1, Tyne to Stainmore. *Mem. geol. Surv. Gt Br.*

——1950. Lower Carboniferous sedimentation in the northern Pennines (England). *Rept XVIIIth Int. geol. Congr.*, Pt. 4, 46–63.

——1959. Epigenetic mineralization in Yorkshire. *Proc. Yorks. geol. Soc.*, **32**, 1–29.

——Dunham, A. C., Hodge, B. L. and Johnson, G. A. L. 1965. Granite beneath Viséan sediments with mineralization at Rookhope, northern Pennines. *Quart. Jnl geol. Soc. Lond.*, **121**, 383–417.

Forster, Westgarth. 1809. *A treatise on a section of the strata from Newcastle-on-Tyne to the mountain of Cross Fell, in Cumberland; with remarks on mineral veins in general.* 1st edit. Alston.

——1821. 2nd edit. Alston.

——1883. 3rd edit., revised by W. Nall. Newcastle.

Fowler, A. 1926. The geology of Berwick-on-Tweed, Norham and Scremerston. *Mem. geol. Surv. Gt Br.*

FROST, D. V. 1969. The Lower Limestone Group (Viséan) of the Otterburn district, Northumberland. *Proc. Yorks. geol. Soc.*, **37,** 277–309.

HARRISON, R. K. 1968. The petrology of the Little and Great Whin Sills in the Woodland Borehole, Co. Durham. *Bull. geol. Surv. Gt Br.*, No. 28, 38–54.

HICKLING, H. G. A. 1931. Contributions to the geology of Northumberland. *Proc. Geol. Ass.*, **42,** 219–28.

HUDSON, R. G. S. 1924. On the rhythmic succession of the Yoredale Series in Wensleydale. *Proc. Yorks. geol. Soc.*, **20,** 125–35.

——1938. *In* The geology of the country around Harrogate. *Proc. Geol. Ass.*, **59,** 295–330.

HULL. J. H. 1964. In *Summ. Prog. geol. Surv. Gt Br. for 1963*, p. 49.

JOHNSON, G. A. L. 1959. The Carboniferous stratigraphy of the Roman Wall district in western Northumberland. *Proc. Yorks. geol. Soc.*, **32,** 83–130.

——1960. Palaeogeography of the northern Pennines and part of north-eastern England during the deposition of Carboniferous cyclothemic deposits. *Rept XXIst Int. geol. Cong.*, Pt. 12, 118–28.

——1962. Lateral variation of marine and deltaic sediments in cyclothemic deposits with particular reference to the Viséan and Namurian of northern England. *C. r. Congrès IV Avanc. Etud. Stratigr. Geol. carbonif. Heerlen*, 1958, **2,** 323–30.

——HODGE, B. L. and FAIRBAIRN, R. A. 1962. The base of the Namurian and of the Millstone Grit in north-eastern England. *Proc. Yorks. geol. Soc.*, **33,** 341–62.

KENT, P. E. 1966. The structure of the concealed Carboniferous rocks of north-eastern England. *Proc. Yorks. geol. Soc.*, **35,** 323–52.

MARR, J. E. 1921. The rigidity of north-west Yorkshire. *The Naturalist*, No. 769, 63–72.

MILLS, D. A. C. and HULL, J. H. 1968. The Geological Survey borehole at Woodland, Co. Durham (1962). *Bull. geol. Surv. Gt Br.*, No. 28., 1–37.

PHILLIPS, J. 1836. *Illustrations of the geology of Yorkshire. Part II. The Mountain Limestone district*. London.

RAYNER, D. H. 1953. The Lower Carboniferous rocks in the north of England: a review. *Proc. Yorks. geol. Soc.*, **28,** 231–315.

READING, H. G. 1957. The stratigraphy and structure of the Cotherstone Syncline. *Quart. Jnl geol. Soc. Lond.*, **113,** 27–56.

TROTTER, F. M. and HOLLINGWORTH, S. E. 1928. The Alston Block. *Geol. Mag.*, **65,** 433–48.

—— ——1932. The geology of the Brampton district. *Mem. geol. Surv. Gt Br.*

TURNER, J. S. 1956. Some faunal bands in the Upper Viséan and early Namurian of the Askrigg Block. *Lpool Manchr geol Jnl*, **1,** 410–19.

WELLS, A. J. 1957. The stratigraphy and structure of the Middleton Tyas-Sleightholme Anticline, North Yorkshire. *Proc. Geol. Ass.*, **68,** 231–54.

——1960. Cyclic sedimentation: a review. *Geol. Mag.*, **97,** 389–403.

WOOLACOTT, D. 1923. A boring at Roddymoor Colliery, near Crook, Co. Durham. *Geol. Mag.*, **60,** 50–62.

Chapter 3

UPPER CARBONIFEROUS
NAMURIAN (MILLSTONE GRIT SERIES)

INTRODUCTION

NAMURIAN rocks, shown as Millstone Grit Series (d^4) on the One-Inch map, crop out in a belt 6 to 8 miles wide, crossing the district in an east-south-easterly direction. Towards the north and north-east these beds pass down dip beneath the Coal Measures; to the east they are overlain unconformably by Permian rocks.

In accordance with current practice in other parts of England and Wales, the Namurian is defined as extending from the base of the band containing *Cravenoceras leion* Bisat to the base of the band containing *Gastrioceras subcrenatum*. The lower boundary is close to the base of the Great Limestone (Johnson and others 1962); the upper is correlated with the base of the Quarterburn Marine Band (p. 76) though the diagnostic goniatites have yet to be found in north-eastern England.

As thus defined the Namurian ranges in thickness from 1000 to 1750 ft and comprises a lower Yoredale-type facies and an upper arenaceous facies. In terms of the former classification in Durham and Northumberland (Hull 1968, table 1) these facies are respectively equivalent to the 'Upper Limestone Group' and the lower two-thirds of the Durham 'Millstone Grit'. The variation in the succession is shown in Fig. 5, together with the approximate limits of the goniatite stages, established farther south primarily by Bisat (1924, 1928) and Hudson (1945). These have been tentatively fixed in this district partly on evidence of newly discovered goniatites and partly on palynological grounds (Mills and Hull 1968, p. 9; Neves 1968; Hull 1968).

The rocks of Yoredale facies forming the lower part of the Namurian are broadly similar to the cyclic repetitions of limestone, shale, sandstone, seatearth and coal in the underlying Viséan. They differ slightly, however, for between any two major limestones there are a number of incomplete cycles in which the sandstone or seatearth is succeeded by either a subordinate thin limestone or marine shale and then by a further sandstone and seatearth. Superimposed on these generalized cycles are a number of 'channel' sandstones, which are generally false-bedded and much more coarse-grained than the usual sandstone members of the cycles. Locally 'channel' sandstones cut out underlying strata and thus complicate any correlation based solely on lithology.

The upper part of the Namurian corresponding with the lower two-thirds of the Durham 'Millstone Grit' consists of a number of coarse-grained, massive, false-bedded sandstones separated by relatively thin argillaceous measures, which locally include marine bands and seatearths, some of the latter being overlain by thin coals. These measures too are cyclic and, typically, consist of repetitions of the following in upward sequence: siltstone, sandstone, seatearth and coal; a few cycles have an additional marine mudstone unit beneath the siltstone. In detail, however, many of the cycles are incomplete and consist only of alternations of sandstone and seatearth.

Conditions of Deposition

In early Namurian times the factors controlling sedimentation were probably similar to those operating during the Viséan (p. 12). The Alston and Askrigg blocks, separated by the Cotherstone Syncline, continued to function as structural units, and over them were deposited the sediments of Yoredale type characteristic of these areas. In general those beds north of the Lunedale–Staindrop Fault are similar in thickness and in nature to those of the Alston Block (Fig. 5), whereas those south of the fault are similar to the Cotherstone Syncline sequence (Reading 1957, fig. 2). This southern sequence differs only slightly in thickness from that of the Askrigg Block and lithologically the sequences are similar.

Superimposed on this division of the sediments of the Barnard Castle district there is a further regional variation, which is independent of the effect of the Lunedale–Staindrop Fault. These changes occur across a line which trends approximately north-north-east to south-south-west and which appears to have migrated progressively with Namurian time between Barnard Castle and Greta Bridge. To the west of this line, 'channel' sandstones are common throughout the lower part of the Namurian, whereas to the east they are absent and even the usual sandstone members of the cycles are relatively subordinate to the thicker developments of limestones and shales.

The upper, arenaceous, part of the Namurian was probably deposited nearer to the contemporary shore-line than the underlying strata. As evidence of this the sandstones are false-bedded and variable in thickness and grain size, though they are relatively persistent over that part of the district where they crop out when compared to the 'channel' sandstones of the underlying measures. There is evidence to suggest that the sandstones of this facies thin to the south-east. This, together with an increase in thickness of the marine parts of cycles, suggests that sedimentation may have been controlled along a similar migratory north-north-east to south-south-west line to that in the underlying measures.

In the eastern part of the district reddened measures of Namurian age are recorded from a number of localities adjacent to the sub-Permian unconformity and from boreholes which have penetrated Permian strata. Some of these localities are up to half a mile away from the Permian outcrop; they include quarries near High Hulam and Headlam. The basal Permian and the underlying reddened measures are exposed south-west of Cleasby, while reddening has also been recorded in boreholes near Aycliffe, Burtree Gate, Thornton Hall*, Newton Morrell and south of Manfield. Because strata of Coal Measures age are also reddened the phenomenon is described in more detail in Chapter 4, p. 75.

Nomenclature and Correlation

Within this district there are extensions of a number of well-known structural elements, each of which has its own well-documented stratigraphical nomenclature. The nomenclature prevalent on the Alston Block (Dunham 1948) has been used throughout this memoir, chiefly because mapping commenced in an area of 'block' facies. The names employed here were subsequently carried into areas

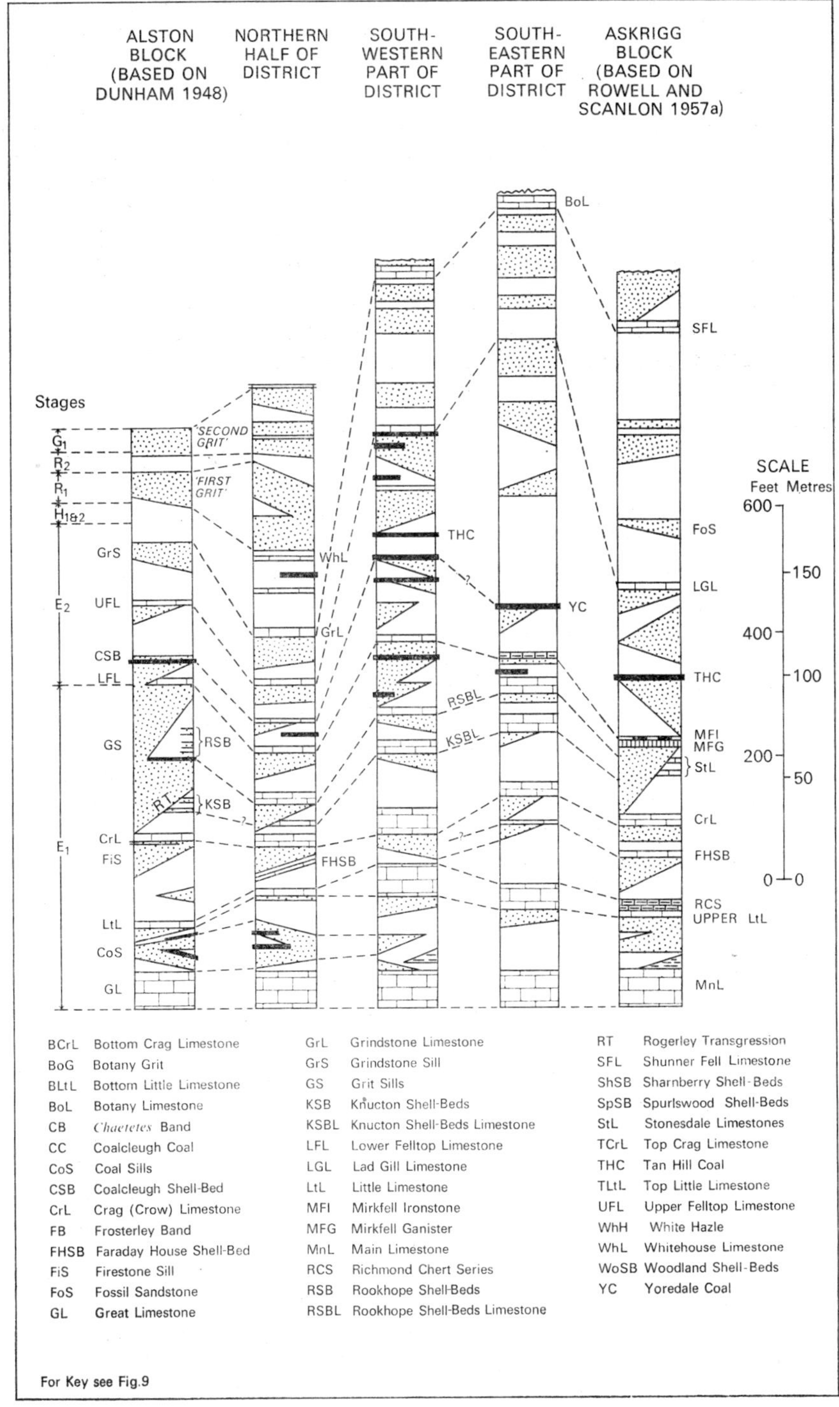

FIG. 5. *Comparative generalized sections of the Namurian strata of the Barnard Castle and adjacent districts*

which have affinities with the Cotherstone Syncline (Reading 1957), the Askrigg Block (Rowell and Scanlon 1957a) and the Middleton Tyas–Sleightholme Anticline (Wells 1957) and consequently the various nomenclatures can be compared (Fig. 5).

Lithological correlations, both within the district and with neighbouring areas, have been made by tracing marker bands from area to area or by a comparison with established sequences. These marker bands, which are generally distinctive marine horizons, form the basis of the subdivision of the following stratigraphical account. Where palaeontological or palynological evidence exists it has been used to confirm or modify the lithological correlations.

The lower part of the Namurian can be correlated with reasonable certainty irrespective of area. Thus the Great, Little and Crag limestones of the north have for some time been correlated with the Main, Upper Little and Crow limestones to the south.

The Knucton Shell-Beds (Carruthers 1938, p. 238) constitute the next marker-band above the Crag Limestone on the Alston Block. These have been recognized in the Cotherstone Syncline by Reading (1957, p. 40 and fig. 4) who considers them to lie below the Lower Stonesdale Limestone; this correlation was initially followed by Hull (1961, p. 41). In its type area, however, the Lower Stonesdale Limestone is the next characteristic marine bed above the Crag Limestone (Turner 1955, p. 350). Furthermore, in the Brough-under-Stainmore (31) district there is evidence that this limestone occurs close above the Crag Limestone and their relationship is similar to that between the Knucton Shell-Beds and the Crag Limestone in the Woodland Bore* (Mills and Hull 1968, p. 8) and in more recent boreholes in the north-west part of the district, including, for example WT/4 Bore*. Accordingly the Knucton Shell-Beds of the Alston Block are here correlated with the Lower Stonesdale Limestone of the Askrigg Block. This correlation further implies that the Upper Stonesdale Limestone of the type area is equivalent, at least in part, to the Rookhope Shell-Beds of the Alston Block (Carruthers 1938, pl. xiii; Dunham 1948, p. 39). On the Askrigg Block the lower part of these shell-beds may well be represented by limestones that are known to be developed between the Lower and Upper Stonesdale limestones in the Brough-under-Stainmore district.

The Mirkfell Ironstones of the Askrigg Block have been shown to be of low E_2 age (Hudson 1941, pp. 279–83; Wilson and Thompson 1959, p. 54). These ironstones are underlain by the Mirkfell Ganister which, it is thought, can be mapped into the top of the sandstones below the Lower Felltop Limestone. Spore evidence from this district (Neves 1968, pl. i) suggests that the E_1–E_2 boundary may lie close above the Lower Felltop Limestone and approximate closely to the horizon of the overlying Coalcleugh Shell-Bed (Carruthers 1938, p. 237). The Coalcleugh Coal (Dunham 1948, p. 40), however, occurs beneath the shell-bed and it is correlated with the Yoredale Coal of this district, which has yielded spores thought to be low E_2 in age (p. 59). This somewhat conflicting evidence suggests that the E_1–E_2 boundary lies between the Lower Felltop Limestone and the overlying Coalcleugh Shell-Bed, but because the mapping suggests a correlation between the Mirkfell Ironstones and the limestone and because the latter is a more persistent marker horizon it is considered that the Pendleian–Arnsbergian boundary is best taken at the base of the limestone.

The correlation of the Upper Felltop Limestone with the Hearne Beck Limestone of Swaledale and the Lad Gill Limestone, proposed by Reading

(1957, p. 46), is supported by the work in this district. It should be noted, however, that the limestone is not everywhere present in the southern part of the district and the correlation hinges in part on the recognition and persistence of the overlying marker-bed, the Fossil Sandstone, which is also well developed in neighbouring areas (Reading 1957, p. 47; Rowell and Scanlon 1957a, fig. 12). This sandstone has no obvious correlative in the north on the Alston Block, the next marker above the Upper Felltop Limestone being the Grindstone Limestone, here correlated with the Botany Limestone of the area west-south-west of Romaldkirk (Carruthers 1938, p. 244) which in turn has been correlated by Reading (1957, p. 49) with the Shunner Fell Limestone. The Botany Limestone in the type area is lithologically, and faunally very similar (p. 63) to the Grindstone Limestone in the Woodland Bore* (Mills and Hull 1968, p. 6). The record of *Cravenoceras sp.* from below the former (Richardson 1961, p. 41) is not thought to conflict with the spore evidence from above the Grindstone Limestone, which suggests a probable E_{2b} age (Neves 1968, pl. i).

The succeeding measures in this district are comparable with those to the north proved in the Roddymoor Bore (Woolacott 1923; Dunham 1948, figs. 4 and 16). Lithological correlations within these beds are of only local significance and the most effective correlations with other parts of northern England are those based on the recognition of the standard Namurian goniatite stages. Various correlations, incorporating evidence from this district, have been attempted on this basis, including those of Wilson and Thompson (1965, p. 221), Ramsbottom (1966a, pl. 4) and Hull (1968, p. 297).

Between the marker bands described above many beds have been given local names and these are indicated in the following stratigraphical account.

General Stratigraphy

The Great Limestone (Fig. 6) (Plate IIIA) crops out across the southern half of the district and is generally a grey, fine- to medium-grained massive bioclastic rock except for the top 15 to 20 ft where it is commonly argillaceous and thin-bedded.

The thickness of the limestone ranges from 55 ft in the Mount Pleasant Bore* (Johnson and others 1962, fig. 2) to 72 ft about two miles to the east in the River Tees, though in most other parts of the district the thicknesses approximate closely to 55 ft. This is slightly less than the 60-ft average recorded on the Alston Block (Dunham 1948, p. 26) and the Askrigg Block (Wells 1957, p. 244) though in the Cotherstone Syncline thicknesses of up to 102 ft are known (Reading 1957, p. 36).

The Great Limestone is known to contain three biostromes (Johnson 1958) named in ascending order the *Chaetetes* Band, Brunton Band and Frosterley Band. In this district the *Chaetetes* Band has been recognized only in quarries near West and East Layton, where it occurs 1 to 3½ ft above the base of the limestone and is 1½ to 2½ ft thick. In addition to *Chaetetes depressus*, the band also contains *Syringopora sp.* and productoids. Generally the band is strongly mineralized, being dolomitic, ferruginous and siliceous, with calcite veinlets and laminae. The Brunton Band has been recognized only in the River Tees, near Egglestone Abbey, where the alga *Calcifolium bruntonense* has been identified 40 ft from the top of the limestone. At this locality the *Chaetetes* Band cannot be certainly recognized, though it may be represented by a band containing zaphren-

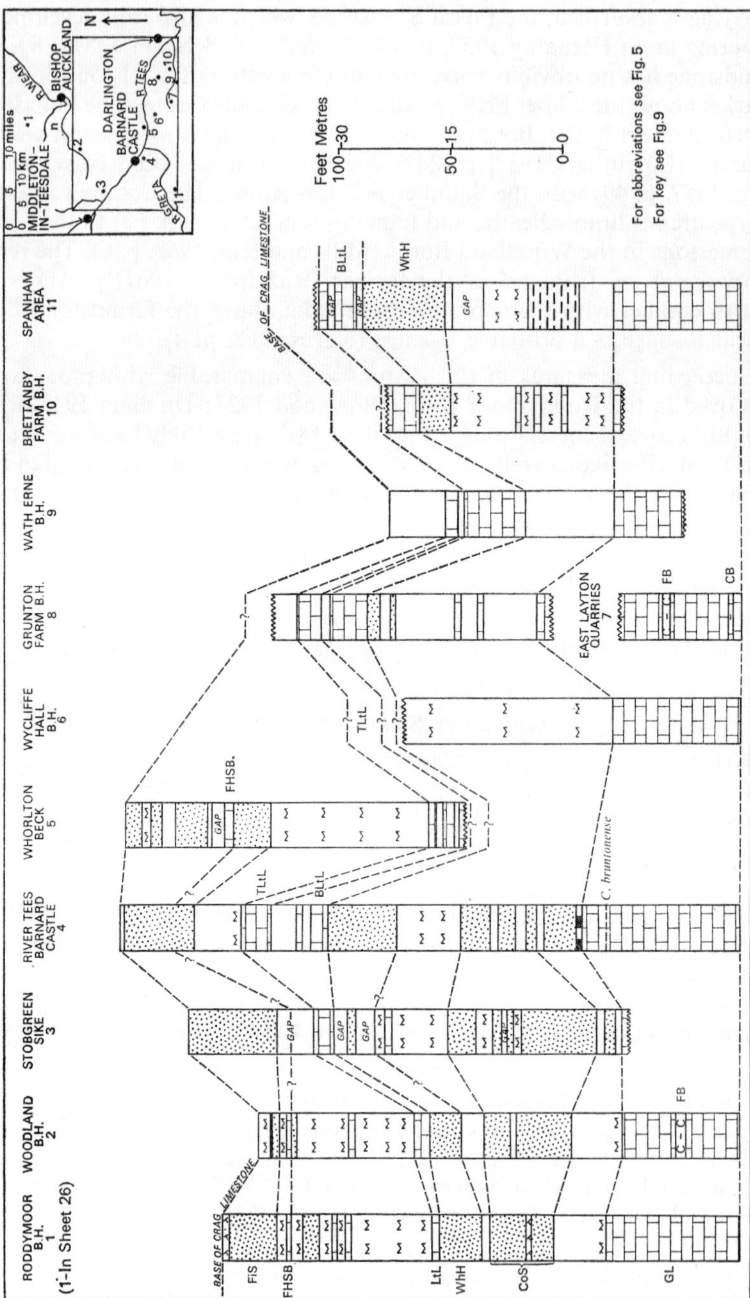

FIG. 6. *Comparative sections of the Namurian. Base of Great Limestone to Crag (Bottom Crag) Limestone*

toids and productoids 4 ft above the base of the limestone. The Frosterley Band, typified by *Dibunophyllum bipartitum bipartitum*, has been more widely recognized throughout the district. It is relatively thin compared to areas farther west (Johnson 1958, p. 152; Hull 1967, p. 75, pl. ii), ranging in thickness from 2 to nearly 7 ft and lying 17 to 33½ ft below the top of the limestone. The fauna collected from the band in this district includes *D. bipartitum bipartitum* and *Koninckophyllum magnificum*, together with a variety of brachiopods (Mills and Hull 1968, pp. 9, 29). It is notable that the solitary corals in both the Frosterley and *Chaetetes* bands are associated with stylolites and with undulating argillaceous partings containing shell debris, suggesting the action of strong currents or waves during deposition.

The measures between the Great and Little limestones (Fig. 6) range in thickness from 40 to 140 ft, though on the eastern margin of the adjacent Brough-under-Stainmore (31) district a thickness of 203 ft is recorded.

This range of thickness, which is slightly greater than that recorded on the Alston Block, on the Askrigg Block or in the Cotherstone Syncline, is directly related to the presence or absence of a lower group of sandstones known as the Coal Sills, and an upper sandstone, the 'White Hazle'. The Coal Sills, collectively up to 66 ft thick, are generally present north of the Lunedale–Staindrop Fault and extend as far south as the River Tees between Barnard Castle and Greta Bridge. West of Barnard Castle they are absent within the district, though they reappear on the margin of the Brough-under-Stainmore (31) district where they are 68 ft thick; they are also absent east of Greta Bridge where the measures are correspondingly thin. The 'White Hazle' has a maximum thickness of 41 ft, in the south-west, but is absent in the south-eastern part of the district. It is generally shelly and calcareous at the top locally grading up into the overlying Little Limestone.

A thin limestone or calcareous sandstone locally overlies the Coal Sills. It now seems possible that Wells (1955, p. 178) miscorrelated this limestone with the Little and was consequently led to postulate an unconformity at this level here and farther south in Wensleydale. No support for an unconformity has since been found in Wensleydale (Wilson 1960a, p. 300), nor has the resurvey of the present district produced evidence to suggest that the south-eastward attenuation of the measures between the Great and Little limestones is due to anything other than a slight change in the depositional environment.

The measures between the Great and Little limestones also include banded cherts and siliceous shales up to 25 ft thick. They are generally separated from the underlying Great Limestone by about 10 ft of mudstone and shale and form the northward extension of the Main Chert of Swaledale (Wells 1955, p. 177). Their generally unfossiliferous character, their petrography and their field relationship with the associated measures suggest that they are a syngenetic deposit, as was concluded by previous workers (Sargent 1929; Wells 1955, p. 191). The chert, however, is commonly shattered, especially in the top 12 ft, and the fractures and cavities produced are often infilled with secondary silica in the form either of crystalline quartz or chalcedony. It may be of significance that the cherts occur in the south-west part of the district in areas where the Coal Sills are generally not developed.

No diagnostic faunas have been obtained from the measures in this district to supplement the records in Northumberland of *Cravenoceras leion* (E_{1a})

(Johnson and others 1962, p. 352) and *Eumorphoceras medusa* (E_{1a}) (Ramsbottom 1966b, p. 56).

The Little Limestone (Fig. 6), sometimes called the Upper Little Limestone to distinguish it from the Lower Little Limestone of the Viséan beds below, crops out parallel to, and to the north of, the Great Limestone across the whole district. North of the Lunedale–Staindrop Fault the limestone is 2½ to 10 ft thick and is fine-grained, thickly bedded, compact, argillaceous and locally siliceous, with brachiopod fragments and sporadic pipe structures in the base (cf. Wells 1955, p. 180). The range of thickness is generally comparable with that of 3 to 21 ft recorded on the Alston Block (Dunham 1948, p. 33). South of the fault the limestone contains appreciable shale partings and Top and Bottom Little Limestones can generally be recognized, though in the area south of Forcett a single argillaceous, bioclastic limestone up to 20 ft thick is mapped.

The Bottom Little Limestone (Fig. 6) ranges in thickness from 7½ to 27 ft being thickest in the south-east of the district. The limestone is often fine-grained, thinly bedded and siliceous, but near Hutton Magna it is almost completely silicified. It is separated from the overlying Top Little Limestone by 3 to 32½ ft of siliceous shales and mudstones, these measures being thickest around Barnard Castle and thinnest in the south-eastern part of the district.

The Top Little Limestone (Fig. 6) has a thickness ranging from 4 ft in the south-east to almost 16 ft near Barnard Castle. Generally the limestone is fine-grained, thinly bedded, siliceous and shelly, but in the Bowes area it is cherty. Farther south, on the Askrigg Block, the measures equivalent to this limestone and the underlying shales are so siliceous (some of the silica being secondary) that Wells (1955, p. 180) called them the Richmond Chert Series, a practice subsequently followed by Rowell and Scanlon (1957a, p. 9 and fig. 4) who recorded a thickness range from 11 to 24 ft.

The measures between the Little or Top Little and Crag limestones (Fig. 6) comprise fossiliferous shales, often with thin limestones, overlain by sandy beds. North of the Lunedale–Staindrop Fault the thickness varies from 50 to almost 80 ft, the upper arenaceous part of the sequence, which here consists of the Firestone Sill (Dunham 1948, p. 32) ranging from 4 to 40 ft. Below the Firestone Sill and underlying shale a thin shelly sandstone or limestone, which provides a useful marker throughout the district, is termed the Faraday House Shell-Bed (Plate IIIB) to indicate its position at the base of the Faraday House Marine Band (Turner 1955, p. 350).

In the northern part of the district there is generally very little sandstone between the Faraday House Shell-Bed and the Little Limestone. This contrasts with areas still farther to the north, where sandstones termed the White and Pattinson sills give rise to an increase in the overall thickness (Dunham 1948, p. 33).

South of the Lunedale–Staindrop Fault, these measures range in thickness from 24 to 237 ft, being generally thickest in the south-western part of the district. The upper arenaceous measures were called the Ten Fathom Grit by Dakyns and others (1891, p. 10) who recognized that it comprised two sandstones separated by coal, shales and a thin 'crow' limestone, which is taken to be the Faraday House Shell-Bed. The lower sandstone has since been termed the Faraday House Sill (Rowell and Scanlon 1957a, p. 10) and the upper sandstone the Uldale Sill (Turner 1955, p. 350). As the Uldale Sill lies between the Faraday

(L 328)

A. Working face (1963) in Great Limestone, Lamb Hill Quarry near Bowes

PLATE III

B. Faraday House Shell-Bed, Eggleston Burn, near Eggleston

(L 314)

House Shell-Bed and the Crag Limestone it is equivalent to the Firestone Sill of the Alston Block. Only in the River Tees at Barnard Castle, is the Faraday House Shell-Bed not recognized in sections through these measures. Here the Ten Fathom Grit, almost 33 ft thick, is only 20 ft above the Top Little Limestone. Other examples of a unified Ten Fathom Grit occur to the west of the district, in the Cotherstone Syncline, where it is 15 to 65 ft thick and is considered to have cut out the Faraday House Shell-Bed by washout (Reading 1957, p. 39).

The Faraday House Sill which ranges in thickness from 0 to 140 ft on the Askrigg Block (Rowell and Scanlon 1957a, p. 11) and from less than 10 ft up to 60 ft in the Cotherstone Syncline (Reading 1957, p. 39) is up to 18½ ft thick in this district.

West of Barnard Castle the Firestone Sill (Uldale Sill) is almost 18 ft thick, whereas to the east it is up to 39½ ft. In both areas the sandstone contains shale and mudstone bands up to 6 ft thick, resembling the sections in the Woodland Bore* (Mills and Hull 1968, p. 8) and the Roddymoor Bore (Woolacott 1923). They contrast with the massive sandstones of the north-western part of the district and the Askrigg Block, where thicknesses of more than 60 ft are recorded (Rowell and Scanlon 1957a, fig. 6).

In the south-eastern part of the district the measures between the Little and Crag limestones are notable for the absence of sandstones, and they are accordingly thinner here than anywhere else in the district or the surrounding country.

The only record of a diagnostic fauna from these measures is that of *Tylonautilus sp. nov.* [= *T. nodiferus* 'early mut.' of Stubblefield (see, for example, Fowler and Robbie 1961, p. 83)] from 31 ft above the Little Limestone in the Woodland Bore* (Mills and Hull 1968, p. 8). This indicates a late E_1 age and supports the goniatite evidence from districts to the north-west, where *Cravenoceras spp.* and a number of specimens of *Cravenoceras* aff. *malhamense* (E_{1c}) have been collected from above the Little Limestone (Dunham and Johnson 1962, p. 243; Johnson and others 1962, p. 354).

The Crag Limestone (Fig. 7) is known to occur as a single bed only at Stobgreen Sike, near Eggleston, where it is 2½ ft thick, fine- to medium-grained, argillaceous, and contains crinoids and brachiopods. Its lithology here is similar to that on the Alston Block, where it ranges from 1 to 6 ft (Dunham 1948, p. 34; Dunham and Johnson 1962, p. 246), but over the rest of the district it comprises two leaves known as Top and Bottom Crag limestones. The two leaves and intercalated shales together form a sequence up to 58 ft thick as compared with 30 ft and 34 ft, respectively, in the Cotherstone Syncline and the Askrigg Block (Reading 1957, p. 40; Rowell and Scanlon 1957a, p. 13). The measures are generally siliceous, but locally, as near Little Hutton, they have suffered complete secondary silicification. Hey (1956, p. 297) working to the south of the district, concluded that the silicification was secondary, though virtually penecontemporaneous. Thus the origin of the silica in these beds may be similar to that of the 'Richmond Chert Series' (p. 32) but different from that of the Main Chert (p. 31) which was of primary origin.

The Bottom Crag Limestone (Fig. 7, Plate I) ranges in thickness from 9½ to 24½ ft and is generally grey, fine-grained, siliceous, argillaceous and shelly, with siliceous shale partings. It crops out, as does the Top Crag Limestone, in Deepdale Beck, Bessy Sike, Smart Gill and in the River Tees and its tributaries at, and below, Barnard Castle. The Top Crag Limestone (Plate IVA) is also grey,

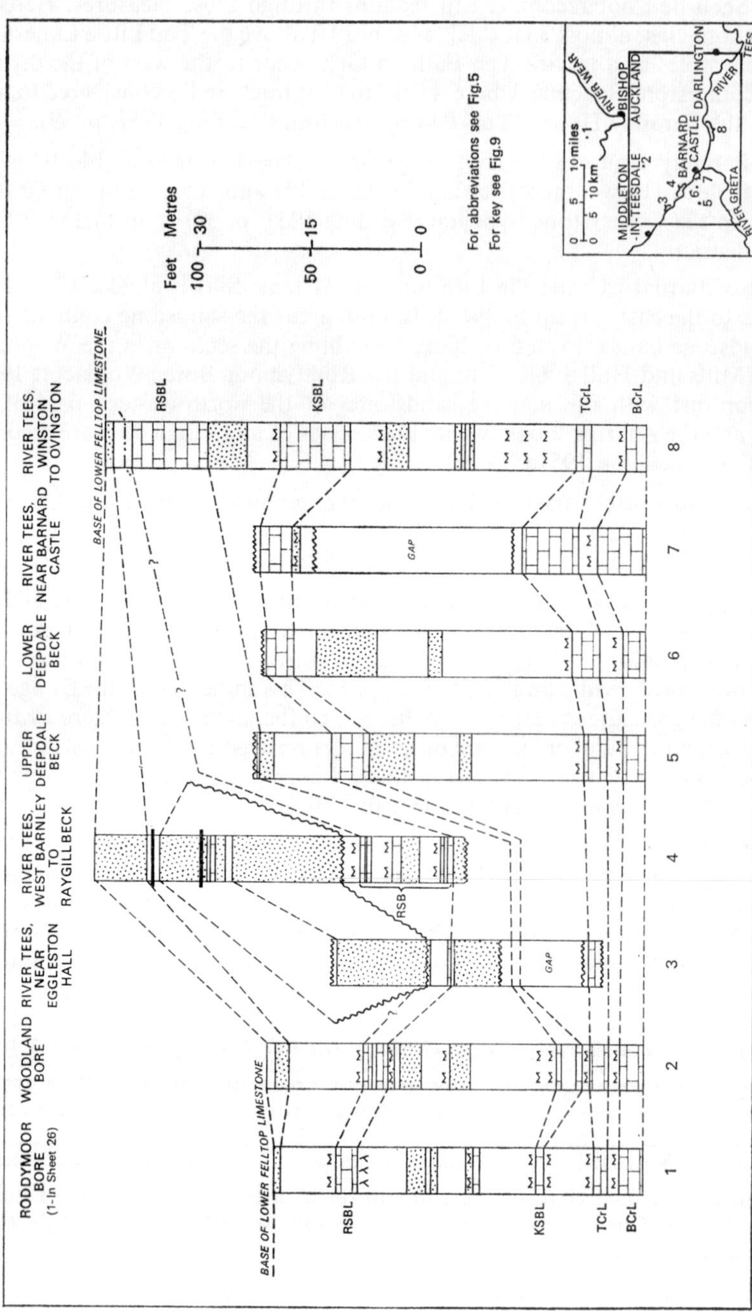

FIG. 7. *Comparative sections of the Namurian. Base of Bottom Crag Limestone to base of Lower Felltop Limestone*

fine-grained, thin-bedded, siliceous and shelly, but it commonly contains cauda-galli structures; it ranges in thickness from 7½ to 25 ft.

No faunas of diagnostic merit have been collected from the Crag Limestone.

The measures between the Top Crag Limestone and the Knucton Shell-Beds (Fig. 7) are not well documented in the north of the district, where they comprise only about 5 ft of strata; atypically, in WT/4 Bore* on Eggleston Moor the Knucton Shell-Beds form the roof of the Top Crag Limestone. In the south of the district these beds are better known; they crop out at a number of localities including the River Tees below Ovington. In these areas they range in thickness from 84 to 120 ft, being thickest in the west. They generally comprise a lower argillaceous sequence up to 52 ft thick overlain by a succession of interbedded sandstones and shales (Fig. 7). The argillaceous beds are often marine at the base with a fauna including sponge spicules, brachiopods, gastropods and bivalves.

In the north-western part of the district these measures are thin compared with similar sequences on the adjacent Alston Block (Dunham 1948, p. 35) where, locally, the upper part may be cut out by the Rogerley transgression (op. cit., p. 36). On the Askrigg Block the equivalent strata range in thickness from 8 to 72 ft, while in the Cotherstone Syncline a section measured in Shield's Beck shows them to be 32½ ft thick. This range shows that the trend of increasing thickness from the south to the north-east of the Askrigg Block (Rowell and Scanlon 1957a, p. 15) is continued into the southern part of this district.

The Knucton Shell-Beds (Fig. 7) as defined by Carruthers (1938, p. 238), comprise thin sandstones with brachiopod casts and crinoids. They have been recognized in this form only on the eastern margin of the adjacent Brough-under-Stainmore (31) district in Snaisgill Sike, north of Middleton-in-Teesdale. Elsewhere they are represented by shelly limestones with interbedded shales, which have been named the Knucton Shell-Beds Limestone in the southern part of the district. In the northern half of the district, the horizon is represented by 10 ft 1 in of beds including a shale parting in the Woodland Bore* and by 12 ft 2 in of muddy limestone and calcareous mudstone in a borehole north of Eggleston. In the south, the limestone is intermittently exposed in Deepdale and the River Tees above Barnard Castle, and again in the River Tees below Winston. In these areas it increases from 17½ to 32½ ft in thickness from west to east and is argillaceous, shelly and locally silicified. The fauna of the limestone includes corals, brachiopods, bivalves, nautiloids and occasional trilobites.

The equivalent limestone in areas to the south-west of the district is the Lower Stonesdale Limestone, which is about 6 ft thick (Rowell and Scanlon 1957a, fig. 8) and thus accords with the general trend of east-to-west thinning in this region.

In the northern half of the district the **measures between the Knucton Shell-Beds and the Rookhope Shell-Beds** (Fig. 7) are exposed only in the River Tees, downstream from Eggleston Hall, where the uppermost 21 ft, comprising sandstones, are seen. In the Woodland Bore* the measures are 76 ft thick, being argillaceous in the lower part and mainly sandstone with marine shale partings above. In most of the northern part of the district this sandstone-on-shale sequence persists, but the measures are reduced in thickness, ranging from 32½ to 46½ ft between Deepdale and the River Tees, near Winston. Recent boreholes in the extreme north-west of the district have proved thick sandstones with subordinate mudstones extending up to 220 ft above the Knucton Shell-Beds.

Some sandstones are in part 'channel' sandstones like their approximate correlatives farther north-west, the Grit Sills, whose base is known as the Rogerley Transgression (Dunham 1948, p. 36); the others are essentially similar to those found elsewhere on the Alston Block, for example at Coalcleugh (Carruthers 1938, pl. xiii). Recent mapping in the adjacent Brough-under-Stainmore (31) district suggests that the equivalent measures to the west may be at least 50 ft thick in Lunedale, thinning westwards to about 35 ft in Stainmore (Owens and Burgess 1965, p. 21).

The Rookhope Shell-Beds (Carruthers 1938, pl. xiii) consist, in the type area, of three minor cyclic units each consisting of shale, shaly sandstone and sandstone (Fig. 7). In this district equivalent measures with broadly similar cyclic pattern have been proved in boreholes, and are seen at outcrop only in the River Tees near West Barnley where three limestones, each 3 to 4 ft thick, are separated by 10 to 20 ft of mudstone and sandstone. Mapping in the adjacent district to the west has shown the upper two limestones of this sequence to be equivalent to the Upper Stonesdale Limestone of the Askrigg Block and Cotherstone Syncline (Rowell and Scanlon 1957a, p. 17; Reading 1957, p. 42). Locally, in the north-west of the district, there is evidence that these beds are cut out by the transgression of the overlying sandstones. Elsewhere the shell-beds are represented by thin limestones and interbedded shales and they have been mapped as the Rookhope Shell-Beds Limestone. In the Woodland Bore*, the limestone is almost 13 ft thick including a shale parting. Farther south the lower posts of the limestone are exposed west of Barnard Castle in Deepdale and Percy becks and also in the River Tees. The limestone is thickest in the River Tees, above Winston Bridge, where together with subordinate shale partings it amounts to 32 ft 1 in.

The measures between the Rookhope Shell-Beds and the Lower Felltop Limestone (Fig. 7) vary greatly in thickness and lithology. They are thickest in the River Tees, near West Barnley, where they consist of up to 9 ft of shale succeeded by about 115 ft of sandstone with a median 15-ft band of shale including a thin limestone. In an adjacent section near the confluence of the Tees and Wilden Beck the upper part of the sandstone overlies a 26-in coal, with interbedded shale, and contains a 2-in coal about 31 ft from the top. The latter is correlated with a seam, 10 to 12 in thick, recorded by W. Gunn (in manuscript) at Barnard Castle, where it is estimated to be only 17 ft below the Lower Felltop Limestone. To the west of Barnard Castle, in Percy Beck, the measures are about 50 ft thick, including a 1-in coal near the top. They include only a little sandstone in the Woodland Bore*, where a thickness of about 45½ ft is recorded.

In the Eggleston area the sandstone sequence is demonstrably locally transgressive. It is in part equivalent to the Mirkfell Ganister and the Lower Howgate Edge Grit, which rests unconformably on the underlying strata in certain areas of the Askrigg Block (Rowell and Scanlon 1957a, p. 18; 1957b, fig. 1). This unconformity is further correlated with that at the base of the Grassington Grit in the Greenhow area (Dunham and Stubblefield 1945, p. 234; Rowell and Scanlon 1957b, p. 89; Wilson 1960a, p. 304). To the north, this unconformity has no strict equivalent. It could be represented, however, by either the Rogerley and Coalcleugh transgressions (Dunham 1948, pp. 36, 40), or by an unrecognized intermediate transgression. Of these transgressions, which are essentially the relatively restricted results of the 'channeling' of sandstones, a correlation with the Rogerley Transgression is thought to be more likely.

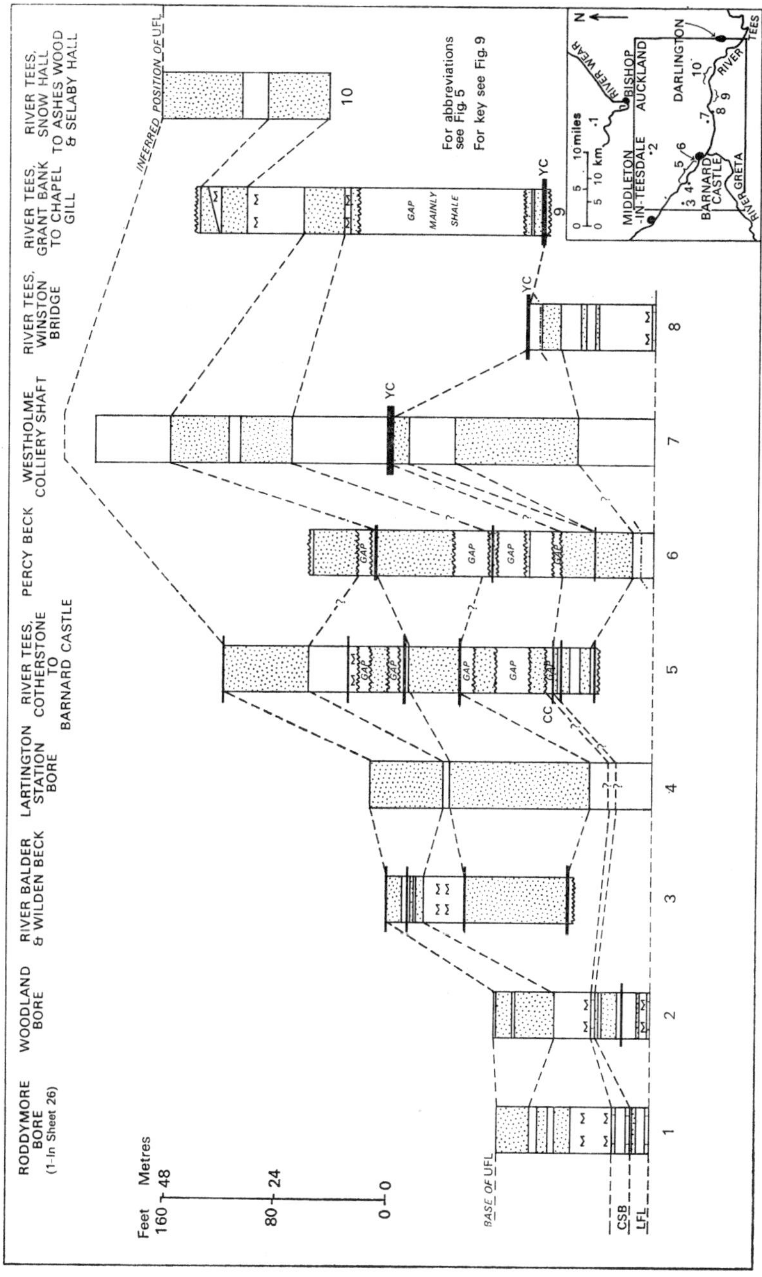

FIG. 8. *Comparative sections of the Namurian. Base of Lower Felltop Limestone to base of Upper Felltop Limestone*

The Lower Felltop Limestone (Fig. 8) is 6 to 8 in thick north of the Lunedale–Staindrop Fault and 1½ to 3 ft farther south. It is locally silicified, as east of Barnard Castle, but is otherwise fine-grained and shelly with shale partings. The base of this limestone is thought to be the best position for the boundary between the Pendleian and Arnsbergian stages (p. 28).

The measures between the Lower and Upper Felltop limestones (Fig. 8) can be divided naturally at the Coalcleugh Shell-Bed (Carruthers 1938, p. 237). In the north-western part of the district the strata below the shell-bed comprise 25 to 30 ft of shale overlain by 36 to 60 ft of sandstone, which is here equated with the Coalcleugh Transgression Beds of the Alston Block (Dunham 1948, p. 40). Farther east and south, however, these measures are very variable; moreover, east of a line from Staindrop to Gainford, where the measures are thought to be thickest, they are difficult to correlate because of the extensive cover of Permian strata and drift. The shales overlying the Lower Felltop Limestone are fossiliferous at the base and range in thickness from 16 ft in the River Tees near Cotherstone to 50 ft near Barnard Castle. In the same area the overlying equivalent of the Coalcleugh Transgression Beds comprise two sandstones separated by up to 26 ft of silty shales with sandstone laminae. The lower sandstone, which has a transgressive base, thins eastwards from the River Tees, above Barnard Castle, where it is up to 15 ft thick, and is apparently absent east of Barnard Castle. The upper sandstone is almost 53 ft thick in Percy Beck where it includes a 1- to 2-in coal and a 3-in shale band. It thins westwards to less than 25 ft in the River Tees, where it contains a 4-in shaly coal and more siltstone partings, and it also appears to be thin near Whorlton. Still farther east, however, near Winston Station, a borehole proved a 130-ft sequence consisting mainly of sandstone, which may represent either the reappearance of the lower sandstone or the re-establishment and splitting of the upper sandstone. The top of the Coalcleugh Transgression Beds is usually marked by the Coalcleugh Coal (Carruthers 1938, pl. xiii) or its seatearth. In the south-eastern part of this district, around Winston, this coal is sufficiently thick to have been mined under the name of Yoredale or Winston Coal. It was once worked about 1 mile west of Winston in the North Tees Drift, where it ranges in thickness, including bands, from 33 in to 66 in. A spore analysis of the seam by the Scientific Department of the National Coal Board suggested a low E_2 age.

The Coalcleugh Shell-Bed (Fig. 8) was recorded in the Woodland Bore* as 1 ft 5 in of dark grey, fine-grained shelly limestone. Elsewhere in the district the position of the bed is inferred from the presence of the underlying Coalcleugh Coal or its seatearth.

The beds succeeding the Coalcleugh Shell-Bed in the north-western part of the district consist of about 30 ft of shale and mudstone overlain by a 30-ft sandstone; elsewhere, however, they are much more variable and comparable to the Askrigg Block sequences (Fig. 8). In general they comprise a lower mainly argillaceous sequence and an upper arenaceous one, which includes thin coals and marine bands and which overlies locally a prominent coal, the probable equivalent of the Tan Hill or Kettlepot Coal (Rowell and Scanlon 1957a, p. 22). The lower measures range in thickness from 40 to 80 ft and locally include a sandstone equivalent to the Upper Howgate Edge Grit and the Kettlepot Ganister of the Askrigg Block (op. cit., p. 21). The upper sandy beds range in thickness from 85 to 250 ft and for the most part represent the Tan Hill Grit and overlying Lad Gill Sill of the Askrigg Block (op. cit., p. 23).

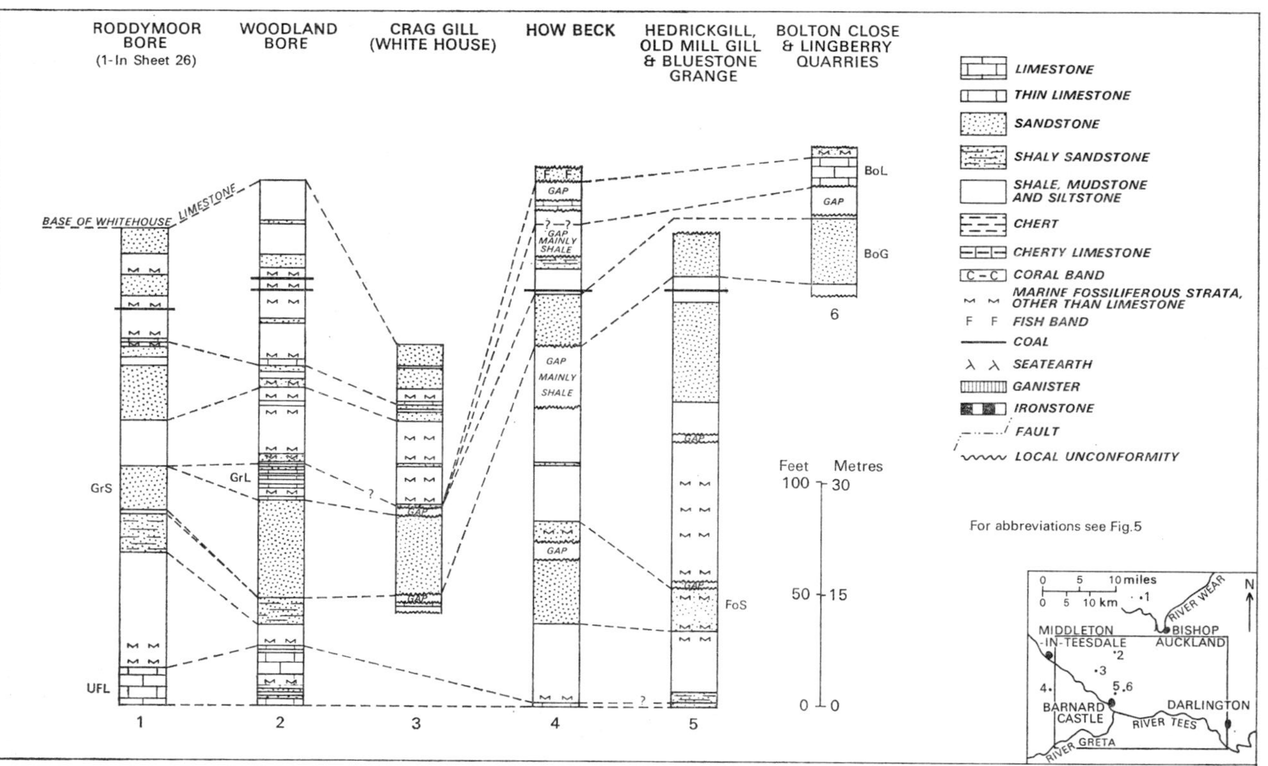

FIG. 9. *Comparative sections of the Namurian. Base of Upper Felltop Limestone to base of Whitehouse Limestone*

The Upper Felltop Limestone (Fig. 9) comprises alternations of limestone and shale up to 25½ ft thick in the area north of the Lunedale–Staindrop Fault. South of the fault, however, the limestone is only 1 ft thick on the western margin of the district and is only locally present to the east, where its position is generally inferred from the overlying strata, especially the Fossil Sandstone. The thick development of the limestone in the northern part of the district is similar to that in the Roddymoor Bore where it was termed the Felltop Limestone (Woolacott 1923, p. 51) but it differs in thickness from that in any other neighbouring area (Dunham 1948, p. 41; Reading 1957, p. 46; Rowell and Scanlon 1957a, p. 23).

The measures between the Upper Felltop and Grindstone (Botany) limestones (Fig. 9) are markedly different in thickness and lithology on either side of the Lunedale–Staindrop Fault. North of the fault they are up to 67 ft thick comprising a lower argillaceous and an upper arenaceous succession, of which the Grindstone Sill forms the major part. The Grindstone Sill ranges in thickness from 36 to 43 ft and resembles its equivalent on the Alston Block (Dunham 1948, p. 42), except that in Knotts Plantation, near Eggleston, and Crag Gill to the east, it includes a shale parting. South of the fault the beds are much thicker (Fig. 9), with a range from 225 ft in How Beck to the west, to 260 ft north-east of Barnard Castle. They include at least three sandstones of which the lowest is the Fossil Sandstone of Carruthers (1938, p. 250) or the 'Fossil Grit' of Dakyns and others (1891, p. 156). This is 12 to 20 ft thick, and lies 20 to 33 ft above the Upper Felltop Limestone. The next sandstone above is generally 40 to 45 ft thick and has been extensively quarried (p. 62 and Plate IVB), though it is absent on the western margin of the district in How Beck. The uppermost sandstone, which is the probable equivalent of the Botany Grit (Reading 1957, p. 47) is up to about 30 ft thick and is separated from the overlying Botany Limestone by 14 to 26 ft of mudstone. On the eastern margin of the adjacent Brough-under-Stainmore (31) district *Cravenoceras sp.*, indicating a low E_2 or earlier age, has been collected (Richardson 1961, p. 41) from the top of these argillaceous measures.

The Grindstone Limestone (Fig. 9) was defined in the Woodland Bore* as being the group of limestones with intercalated shales which lie immediately above the Grindstone Sill (Mills and Hull 1968, p. 20). It reaches a maximum known thickness of just over 14 ft in the borehole, but elsewhere it is thought to be impersistent and locally represented by a shelly quartzitic sandstone. The limestone is the probable equivalent of the Botany Limestone, which crops out near Botany Farm (Carruthers 1938, p. 244) and which was subsequently recognized near Barnard Castle (Reading 1957, p. 49), where it is 13¼ ft thick with a fauna including *Dibunophyllum sp.*, *Fenestella sp.*, *Lingula sp.*, *Schellwienella sp.* and other fossils that are also common in the type area.

The measures between the Grindstone and Whitehouse limestones (Fig. 9) are the youngest Namurian strata to have been preserved south of the Lunedale–Staindrop Fault, and they are there represented by a few feet only of fossiliferous sandstone. North of the fault these strata crop out extensively north and north-east of Eggleston, where they range in thickness from 73 to 127 ft and comprise a number of minor cyclothems, one of which includes a thin unnamed limestone (Fig. 9). In the Woodland Bore* *Homoceras* cf. *henkei* (indicative of the R_{1a} Zone) was recorded (Mills and Hull 1968, p. 6) about 83½ ft above the Grindstone Limestone.

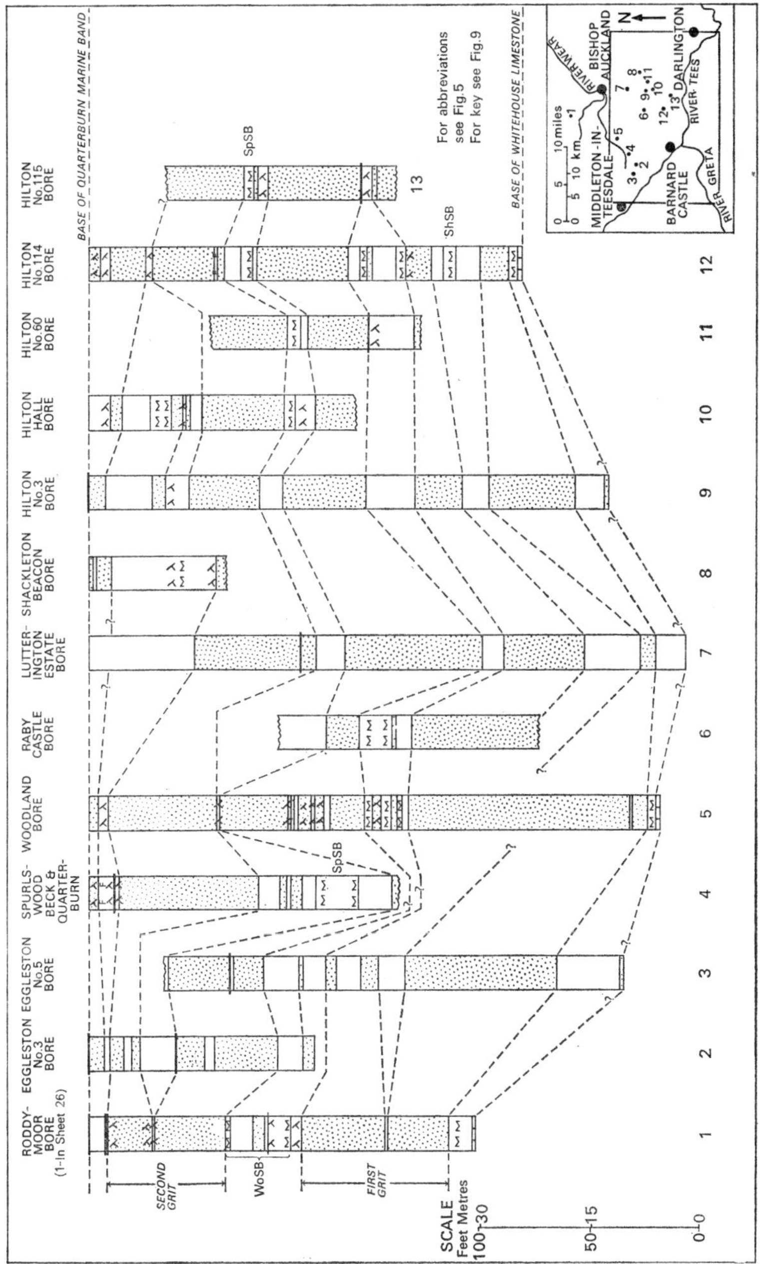

FIG. 10. *Comparative sections of the Namurian. Base of Whitehouse Limestone to Quarterburn Marine Band*

The Whitehouse Limestone (Fig. 10) was defined in the Woodland Bore* (op. cit., p. 6); it crops out near, and north-east of Eggleston. It generally ranges in thickness from 1 in to 1½ ft, but locally it may be up to 5 ft thick (p. 64). The limestone was also recorded in the Roddymoor Bore but it has not been described elsewhere in County Durham.

The measures above the Whitehouse Limestone to the top of the Namurian (Fig. 10) are 199 to 330 ft thick and consist mainly of sandstones, with subordinate thin coals, seatearths and shales which are sporadically fossiliferous. This part of the succession includes the First Grit and the Second Grit of the Primary Survey, which together constituted the lower two-thirds of the 'Millstone Grit' of former classifications.

The 'grits' give rise to well-defined features, and it is clear from the mapping and from a comparison of stream and borehole sections that they each split into two or more sandstone units which are only locally persistent. The 'First and Second grits' are thus of doubtful stratigraphical significance and the names are likely to be misleading if extrapolated over a wide area. For this reason an attempt has been made to correlate the sequence by means of marine marker bands, some of which are named here for the first time, others having been defined in the Woodland Bore* (Mills and Hull 1968, p. 6). Each of these markers contains a restricted brachiopod, bivalve and gastropod fauna and each is liable to be cut out by overlying sandstones. The most notable of such markers in these measures lies close above the Whitehouse Limestone, where *Reticuloceras stubblefieldi* (R_{1b}) has been recorded (Hull 1963, p. 45). Goniatites of the *R. nodosum* group were found at a similar horizon in the Woodland Bore*.

The lowest marker, the Sharnberry Shell-Beds, takes its name from Sharnberry Beck in the adjacent Wolsingham (26) district, where it is 4¼ ft thick and where the fauna includes *Reticuloceras sp.*, probably indicating a Lower *Reticuloceras* (R_1) age. In terms of the former classification it lies midway in the 'First Grit', overlying a sandstone, locally called the 'Basement Grit'.

The Woodland Shell-Beds which were defined in the Woodland Bore*, consist of 11 to 28 ft of mudstones, generally with thin sandstones, and with a fauna commonly including *Serpuloides sp.*, *Lingula mytilloides*, *Orbiculoidea* cf. *nitida*, productoids and *Retispira sp.* These shell-beds occur in the measures between the 'First' and 'Second' grits. The lower leaf of the 'Second Grit' separates the Woodland Shell-Beds from the next marker band above. This sandstone contains a number of shale partings and mapping suggests that one of these crops out at a locality where Kirkby and Duff (1872, p. 186) described shales and an inferior limestone containing marine shells. An unnamed marine horizon may exist therefore between the Woodland and Spurlswood Shell-Beds.

The Spurlswood Shell-Beds are named after a locality in Spurlswood Beck, where they consist of two thin bands of fossiliferous shale separated by 16 ft of apparently unfossiliferous mudstone and siltstone. At other localities the beds are represented by an argillaceous sequence from 10¼ to 12¾ ft thick. The fauna from these measures generally includes *Lingula* and productoids. The shell-beds generally occur midway in the 'Second Grit', but the overlying upper leaf commonly cuts them out. The Spurlswood Shell-Beds are separated from the base of the Coal Measures by 72 to 104 ft of strata consisting of the upper sandstone leaf of the 'Second Grit' and by up to 103 ft of overlying variable measures, which locally contain, where they are thickest, an unnamed marine band. J.H.H., D.A.C.M.

DETAILS

Comparative sections of the Namurian sequence are shown in Figs. 6–10; Fig. 11 shows most of the localities referred to in the text.

North of the Lunedale–Staindrop Fault the **Great Limestone** (Fig. 6) is exposed only in Eggleston Burn [9898 2427 and 9883 2393] where the Frosterley Band (Johnson 1958, p. 151), with corals and brachiopods, is present about 18 ft from the top.

Elsewhere it is proved in boreholes, notably WT/4 Bore* [9986 2750] on Eggleston Common, WT/6 Bore [0030 2423] north of Eggleston and the Woodland Bore* [0910 2770]. In these boreholes the thickness ranges from 52 ft 11 in up to at least 64 ft 6 in in Bore WT/6. In WT/4 Bore* a pronounced coral band 6 ft 9 in thick and a coral-brachiopod band 1 ft 10 in thick are present 25 ft and 33 ft 6 in respectively from the top of the limestone. In the WT/6 and Woodland bores the Frosterley Band is 3 ft 5 in and 3 ft 10 in thick respectively.

In the southern half of the district good sections in the limestone can be seen around Lamb Hill (Plate IIIA) where its average thickness is 55 ft, and in the River Tees below Barnard Castle downstream from Abbey Mill [0634 1513], where it is 72 ft. Throughout this area the limestone is generally grey or dark grey, massively bedded, fine- to medium-grained, bioclastic and crinoidal; in the top 15 ft and the basal 8 ft it is generally more thinly bedded and argillaceous. Sporadic clisiophylloid corals and brachiopods are found but no distinctive Frosterley Band has been recognized. The *Chaetetes* Band (Johnson 1958, p. 149) is similarly difficult to discern in this area, though its horizon may be indicated by sporadic zaphrentoids and brachiopods about 4 ft above the base of the limestone. A specimen 40 ft from the top of the limestone (loc. 1c) [1] near Abbey Bridge [0663 1496] contains the alga *Calcifolium bruntonense* indicative of the Brunton Band (Johnson 1958, p. 150). At an old quarry [0229 1339] on the south side of the A66 road at Lamb Hill the section in the Great Limestone is as follows (loc. 1a):

	ft
Limestone, grey thin-bedded crinoidal; *Avonia youngiana*; *Spirifer* cf. *trigonalis*	7
Limestone, grey-brown massive; *Dibunophyllum bipartitum bipartitum*, zaphrentoid indet., *Plicochonetes sp.*	23
Limestone, dark grey hard finely crystalline	4
Limestone; *Semiplanus* cf. *latissimus*	2

In the Mount Pleasant Bore* the Great Limestone was 55 ft thick. Sections occur in the River Greta below Greta Bridge [0862 1317] and also 2½ miles to the east-south-east in Lanehead Old Quarry [1260 1177] where 37 ft of limestone are exposed. In the Wycliffe Hall Bore* [1228 1328] a full thickness of 56 ft 10 in was proved.

Of 32 ft of Great Limestone exposed in a quarry [1433 1047] north of West Layton (loc. 1b) the lowest 2½ ft represent the *Chaetetes* Band, consisting of thin-bedded limonitic limestone with abundant laminae of *Chaetetes depressus* together with *Lithostrotion pauciradiale*. About 1 ft 2 in above occurs a 6-in band of fine-grained dolomitic limestone with coarsely bioclastic patches containing sporadic *Koninckophyllum sp.* and *Semiplanus sp.*, [latissimoid]. A fauna collected from the overlying 27 ft 6 in of limestone included *Aulophyllum fungites*, *Caninia sp.*, *Dibunophyllum bipartitum bipartitum*, *Echinoconchus* cf. *elegans*, *Etheridgina complectans*, *Plicochonetes* aff. *interstriatus*, *Schizophoria sp.* and *Semiplanus sp.* [latissimoid].

The *Chaetetes* Band is also exposed in the floor of East Layton Quarry [1556 1066], where it is 18 in thick and lies 14 to 20 in above the base of the limestone. The Great Limestone has a total thickness of at least 52 ft and a composite section of the northern and eastern faces reads:

	ft	in
Limestone, grey massive fine-grained; jointed; earthy partings	11	10
Limestone, grey thin-bedded rubbly fine-grained crinoidal	4	8
Limestone, dark grey rubbly earthy fine-grained; abundant shell debris	0	6

[1]For list of localities quoted thus see pp. 68–69 and Fig. 11

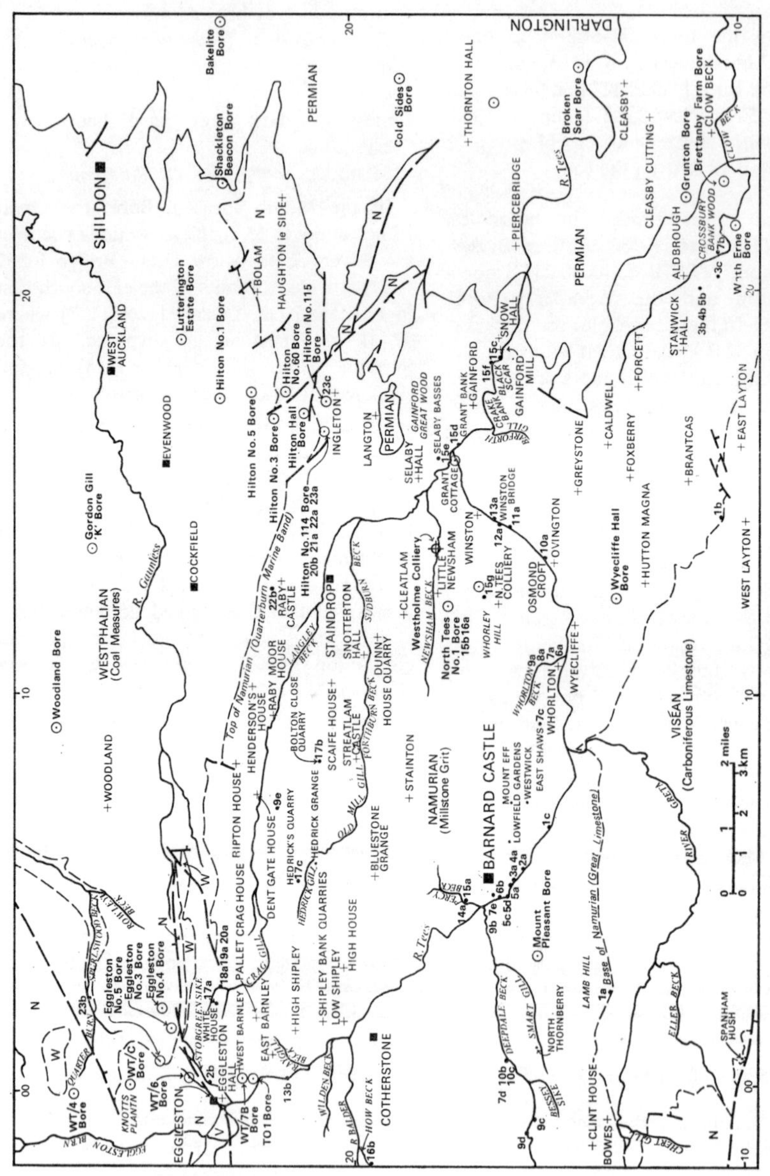

FIG. 11. *Distribution of the Namurian rocks, showing sites of boreholes and fossil localities in the Barnard Castle district.*

V = Viséan, N = Namurian, W = Westphalian. Small black dots are surface localities, ringed dots are boreholes.

FROSTERLEY BAND	ft	in
Limestone, grey well-bedded fine-grained bioclastic: clisiophylloid corals, *Dibunophyllum sp.*; about	2	6
Limestone, grey thin-bedded fine-grained bioclastic; thin undulating argillaceous bands	7	0
Limestone, dark grey fine-grained argillaceous; abundant shell debris	0	2
Limestone, grey massive fine-grained; calcite on joints ..	5	3
Limestone, dark grey argillaceous; shell debris	0	2
Limestone, grey massive jointed fine-grained crinoidal; some dolomitization and silicification at base	11	8
Limestone, brown fine-grained siliceous; dolomite and ?ankerite; limonite where weathered ..	3	6
CHAETETES BAND		
Limestone, grey to black fine-grained cherty; silicified *Chaetetes* laminae and sporadic solitary corals at top	1	8
Limestone, red and brown fine-grained decalcified ferruginous ..	0	4
Not exposed (probably limestone) ..	3	0
Shale	—	—

The *Chaetetes* Band is further exposed, with a thickness of 26 to 28 in, in the floor of Forcett Old Quarry [1674 1100], where the basal 10 ft of limestone are seen to be dolomitic and ferruginous. There are other exposures of the Great Limestone at East Layton [1606 1000], in a quarry [1679 1026] SSW of Carkin Field, and in Gill Wells Plantation [1702 1013]. Still farther east, the top of the Great Limestone was proved in the Grunton* [2239 1148] and Brettanby Farm* [2279 1038] boreholes, and in a borehole* [2160 1001] at Wath Erne.

J.H.H.

The measures between the Great and Little Limestones (Fig. 6) are partially exposed north of the Lunedale Fault in Eggleston Burn [9893 2471]. The Little Limestone is seen to rest on a 3-ft siliceous sandstone—the White Hazle—overlying 10 ft of shale and 8 ft of thin-bedded sandstone with shale partings, one of which, about 3 ft from the top, yielded orthotetoid fragments. A more complete section is exposed, a mile south-east, in Stobgreen Sike (loc. 2b) [0019 2378]:

	ft	in
LITTLE LIMESTONE	—	—
Not exposed (including White Hazle)	7	0
Sandstone	2	0
Not exposed	10	0
Mudstone, with limestone ribs; trepostomatous bryozoa, *Dictyoclostus sp.*, *Eomarginifera sp.*, *Spirifer* cf. *trigonalis*, gastropods, and bivalves	33	1
Not exposed	7	0
Mudstone; '*Camarotoechia*' *sp.*, *Lingula sp.* [large], *Spirifer sp.* ..	5	4
Coal Sills		
Sandstone	12	6
Mudstone, silty; ironstone nodules	5	6
Mudstone; *Fenestella sp.*, *Eomarginifera sp.*, *Euphemites sp.* ..	2	0
Mudstone-seatearth	0	3
Sandstone	3	9
Not exposed	2	0
Sandstone (ganister)	2	6
Not exposed	2	0
Mudstone; *Eomarginifera sp.* ..	0	6
Sandstone, partly obscured ..	35	0
Mudstone, silty; *Paleyoldia macgregori*	4	0
Sandstone	4	6
Shale with sandstone ribs	2	0
GREAT LIMESTONE	—	—

Further detailed sections of these beds are recorded in WT/4 Bore* and the Woodland Bore* where they are 82 and 92 ft thick respectively and include prominent sandstones forming the White Hazle and Coal Sills. In WT/C Bore [0041 2596] at Blackton Head and WT/6 Bore the White Hazle is respectively 21 and 8¼ ft thick and is separated from the top of the Coal Sills sandstone by strata up to 2¼ ft thick containing a thin muddy sandy limestone and calcareous sandstone. It is possible that this marine horizon may be equated with the 1-ft limestone overlying the Coal Sills in the Sleightholme Beck section, which is also locally recorded elsewhere.

J.H.H., D.A.C.M.

In the southern half of the district complete sequences are not seen, but a short distance to the west, in the Brough-under-Stainmore (31) district, the following sequence is exposed in Sleightholme Beck:

	ft	in
LITTLE LIMESTONE [9565 1068] ..	—	—
White Hazle		
Sandstone, greyish-brown thick-bedded fine-grained calcareous; shell fragments; grades up into limestone	3	0
Sandstone, grey thick-bedded to massive fine-grained feldspathic siliceous..	16	6
Sandstone, white thin-bedded fine-grained micaceous carbonaceous patchily calcareous; worm tubes and shell moulds..	22	0
Siltstone, grey sandy; calcareous nodules	22	0
Shale, grey silty; marine fossils ..	6	0
Limestone, grey sandy; abundant shell fragments	1	0
Siltstone, grey sandy; fossiliferous	2	0
Coal Sills		
Sandstone, white thin-bedded fine-grained feldspathic micaceous; sporadic false-bedding and shale partings	68	0
Siltstone, grey; sandstone ribs ..	20	0
Shale, grey silty; black and fossiliferous near base	30	0
Shale, dark grey siliceous, interbedded with argillaceous limestones; *Lingula* band 4 ft from base	13	0
GREAT LIMESTONE [9648 1123] ..	—	—

Lateral variation is apparent by comparing the above with the following section, exposed about 3½ miles to the south-east in Spanham Hush [0087 0989]:

	ft	in
LITTLE LIMESTONE	—	—
Not exposed	2	0
White Hazle		
Sandstone, white and brown massive shattered fine-grained limonitic; calcareous at top ..	14	0
Sandstone, grey and brown thin-bedded false-bedded fine-grained micaceous; shaly and rooty at base	23	0
Siltstone, grey sandy; thin sandstone ribs	3	0
Not exposed	16	0
Shale, black; ironstone nodules; shell fragments	18	6
Chert, grey or black, banded; brecciated in part	9	0

The base of this section is estimated to lie about 30 ft above the top of the Great Limestone. The 9 ft of chert are taken to represent the top of the Main Chert Series in the Richmond (41) district (Wells 1955, p. 177). C.R.B.

These cherts are exposed in the streams which drain Scargill High Moor northwards into Eller Beck and in Chert Gill [9867 1193], where they are 23 ft thick, but they do not persist as far north as the Mount Pleasant Bore*, nor are they significantly represented in the River Tees between Abbey Mill and Demesne Mill*, Barnard Castle. In this Tees section the Coal Sills are in three posts totalling 26½ ft of sandstone in 34 ft of strata separated by almost 16 ft from the top of the Great Limestone. Here, the Coal Sills are overlain by 27½ ft of mudstone, the basal 18 ft yielding an exceptionally rich fauna of brachiopods as well as bryozoa, bivalves, ostracods and a fish tooth (loc. 2a). This mudstone is succeeded by the White Hazle, which is 33½ ft thick. The top 8 in of the White Hazle is calcareous and shelly and grades up into the overlying limestone. J.H.H., C.R.B.

A similar sequence is exposed in the River Tees between its confluence with the Greta and a locality three-quarters of a mile downstream near Thorpe Scar [0961 1443]. About 1¾ miles SE of Thorpe Scar, in the Wycliffe Hall Bore* the Coal Sills are absent, the Great Limestone being overlain by at least 94 ft of dark grey fossiliferous shale with sporadic ironstone nodules. East of the borehole, these measures are not now well exposed, but Gunn records (in manuscript) 5 ft of blue shale on siliceous flags immediately beneath the Little Limestone in a quarry [1779 1068] south-west of Stanwick Hall. J.H.H.

East and south-east of Aldbrough the measures were proved in two water boreholes at Grunton* and Brettanby Farm* to be 75 ft and 89 ft thick respectively, in both cases including a representative of the White Hazle. At another borehole* at Wath Erne the equivalent measures consisted only of 40 ft of shale. D.A.C.M.

The Little Limestone (Fig. 6) crops out at two localities north of the Lunedale Fault. In Eggleston Burn [9893 2471] it is approximately 5 ft thick, the basal beds comprising 3 in of medium-grained sandy ferruginous limestone overlying, and piping down into, 2¼ ft of dark grey, fine-grained bioclastic limestone. In Stobgreen Sike* the limestone is 10 ft thick; it is grey and fine-grained, with brachiopod fragments. The only other records of this limestone in the north of the district are in WT/4 Bore*, WT/C Bore, WT/6 Bore and the Woodland Bore*, where it is almost 4½ ft thick, the proved thicknesses ranging between 4 and 5 ft. J.H.H., D.A.C.M.

In the southern half of the district shale partings are common in the Little Limestone; and both a Top and a Bottom Little Limestone, separated by a sequence of shale and mudstone, can be recognized in most areas. J.H.H.

The Bottom Little Limestone is seen in Spanham Hush [0087 0989]:

	ft	in
Limestone, grey thin-bedded fine-grained siliceous decalcified; inter-bedded with calcareous, siliceous shale; 'cauda-galli' markings on bedding planes	1	0
Not exposed	3	0
Limestone, pale grey fine-grained decalcified; sporadic chert streaks and shells	3	0
Limestone, grey thin-bedded fine-grained siliceous; interbedded shales; shelly; 'cauda-galli' on bedding planes	2	6
Limestone, black thin-bedded fine-grained cherty decalcified; limonitized shell casts	5	0
Not exposed	2	0
Sandstone	—	—

A similar sequence is exposed in West Spanham Gill [0058 0995]. C.R.B.

Near Demesne Mill* [0535 1582] in the River Tees, the Bottom Little Limestone consists of almost 7½ ft of dark grey thin-bedded fine-grained earthy siliceous limestone, with brachiopods and ostracods (loc. 3a). It is continuously exposed east-south-east of Demesne Mill* at the top of the scarp behind the sewage works and also in the scarp in Cross Berry Plantation [0900 1466]. The limestone passes gradually down into the underlying sandstone and there are vestigial pipe-like structures near the base.

In a beck [0930 1463], 550 yd SSE of East Shaws, the Bottom Little Limestone is 10¼ ft thick and a similar thickness is seen in the south bank of the River Tees [0957 1434] near West Thorpe Wood.

South of the Tees, in a quarry [1273 1290] north of Hutton Magna, the limestone is at least 12 ft thick and is, like the overlying 3½ ft of shale, completely silicified: only vestigial patches of the original grey siliceous limestone remain. The limestone is also silicified in an outlier 2 miles SE of Hutton Magna, where the uppermost 5½ ft are exposed [1602 1187]. About a mile north-east of this the limestone is relatively unaltered, and the following section is exposed in a railway cutting [1689 1259]:

	ft	in
Limestone, pale grey thin-bedded fine-grained siliceous	2	2
Limestone, orange-weathered; abundant brachiopods	1	2
Limestone, pale grey fine-grained silicified; sporadic brachiopods	3	6
Limestone, pale grey fine-grained; passes laterally into silicified limestone	3	6
Limestone, grey fine-grained; abundant brachiopods..	1	6
Shale, grey; shell fragments ..	0	9
Limestone, grey bioclastic; crinoidal	0	6

About 8 ft of grey and fawn siliceous dolomitic limestone exposed in Cat Wood [1790 1234] represents part of the same sequence.

Between Stanwick Hall [1843 1145] and Aldbrough, the Top and Bottom Little limestones are thought to come together into one bed, the lower part of which is exposed in Hergill Quarry [1759 1101]:

	ft	in
Limestone, pale grey massive fine-grained with undulating argillaceous partings; brachiopods, including gigantoproductids, and rare solitary corals at top ..	10	6
Limestone, bluish grey thin- and thick-bedded fine-grained bioclastic siliceous; sporadic shaly partings with abundant shell debris; many brachiopods in top 3 in	9	2

Comparable sections are exposed at two other localities, 150 yd SSW and 700 yd SE respectively, of Stanwick Hall. J.H.H.

A partial section is also seen in the quarry [1989 1103] south-west of Aldbrough. East of this quarry the Little Limestone is again split, but the parting is too thin to warrant the mapping of two limestones. In Harthorn Quarry* [2060 1100] about 350 yd SW of Aldbrough Church, the upper part of the Bottom Little Limestone consists of 2 ft 8 in of dark grey fine-grained impure limestone with *Alitaria* cf. *panderi* and *Martinia sp.* (loc. 3b). Beds at about the same horizon exposed [2034 1056] in Melsonby Lane, Aldbrough, consist of about 10¾ ft of argillaceous shelly limestone with a calcareous mudstone parting 5 to 8 in thick; the fauna which is abundant in the top 2 ft includes *Alitaria* cf. *panderi*, *Avonia davidsoni*, *Pleuropugnoides sp.*, *Quasiavonia* cf. *aculeata*, *Spiriferellina sp.* and *Limipecten dissimilis* (loc. 3c).

Farther east, where the limestone is concealed by thick drift or Permian strata, the Bottom Little Limestone was 17 ft thick in the Grunton Bore*, at least 14 ft thick in the Brettanby Farm Bore* and 27 ft thick in the Wath Erne Bore*. D.A.C.M.

The measures between the Bottom Little and Top Little Limestones (Fig. 6) consist of siliceous shales with thin limestone ribs and sporadic silty bands. These beds are continuously exposed at Demesne Mill*, Barnard Castle, where they are 30 ft 4 in thick and contain *Avonia sp.*, *Crurithyris sp.*, *Echinoconchus* cf. *elegans*, *Rugosochonetes sp.* and *Spirifer* cf. *trigonalis* (loc. 4a). At Harthorn Quarry* [2060 1100] the equivalent strata are 16 ft 8 in thick and contain *Hyalostelia smithii*, *Fenestella frutex*, *Fistulipora sp.* and *Cleiothyridina sp.* (loc. 4b). East of Aldbrough these measures are represented by only 4 ft of black shale with banded chert in the Grunton Bore* and 3 ft of shale in the Wath Erne Bore*. J.H., D.A.C.M.

The Top Little Limestone which generally crops out at the same localities as the Bottom Little Limestone is represented around Bowes by 6 ft of dark grey cherty limestone with irregular nodules of purer chert; it can be seen, for example, in a quarry [9821 1429] about 750 yd NW of Clint House. In the River Tees the top 10 ft of the limestone can be seen downstream from the Woollen Mill [0484 1617] at Barnard Castle. Collections obtained from various exposures in the limestone hereabouts include *Avonia youngiana*, *Buxtonia sp.*, *Cleiothyridina* cf. *fimbriata*, *C.* aff *glabristria*, *Eomarginifera* cf. *setosa*, *Phricodothyris sp.* and *Spirifer* cf. *trigonalis* (locs. 5c, 5d). C.R.B.

The most complete section of the Top Little Limestone is near Demesne Mill* where 17½ ft of siliceous shelly limestone, including a 2-in mudstone band, crops out; it has yielded *Avonia* cf. *youngiana*, '*Camarotoechia*' *sp.*, *Crurithyris sp.*, *Eomarginifera* cf. *setosa* and *Spirifer* cf. *trigonalis* (loc. 5a). The limestone is well exposed in a half-mile stretch of the river [1035 1473 to 1187 1434] between Whorlton and Wycliffe, with a typical section:

	ft
Limestone, dark grey fine-grained bioclastic; argillaceous bands	2½
Shale, grey calcareous; brachiopods ..	3
Limestone, grey massive fine-grained siliceous limonitic; shelly	3½
Shale, dark grey; brachiopods ..	1½
Limestone, dark grey thin-bedded fine-grained argillaceous shelly ..	1½

Three miles to the south-east the lowest 8 ft 2 in of the limestone are exposed in a quarry [2153 1146], 200 yd NE of Brantcas Farm; it is pale grey, massive and fine-grained, with a few bioclastic argillaceous partings. A section exposed in Caldwell Beck [1684 1281], ½ mile SE of Caldwell Church shows:

	ft	in
Limestone, dark grey thin-bedded fine-grained siliceous; abundant small brachiopods	0	6
Shale, grey calcareous; shelly ..	1	8
Limestone, grey thin-bedded fine-grained siliceous	1	0
Shale, grey calcareous	0	3
Limestone, grey thin-bedded fine-grained siliceous; sporadic brachiopods	1	6

J.H.H.

At Harthorn Quarry* [2060 1100], Aldbrough, the lower part of the Top Little Limestone consists of 12¼ ft of brownish grey flaggy argillaceous siliceous crinoidal limestone with calcareous shale partings at the top. *Hyalostelia smithii*, *Alitaria panderi* and an indeterminate turreted gastropod were collected from the basal 2¼ ft (loc. 5b). Farther east in the Grunton Bore* the limestone is 10 ft thick, being reddened at the top, while in the Wath Erne Bore* it is only 4 ft thick. D.A.C.M.

The measures between the Top Little and Crag Limestones (Fig. 6) are almost continuously exposed in Eggleston Burn as follows:

	ft	in
Horizon of CRAG LIMESTONE [9892 2515]	—	—
Coal and seatearth-mudstone in thin alternations	3	10
Firestone Sill		
Seatearth-sandstone	8	0
Sandstone, pale grey massive false-bedded medium- to coarse-grained; jointed ..	30	0
Sandstone, grey thick-bedded medium-grained with highly micaceous shale partings ..	1	6
Shale, dark grey silty	0	1½
Sandstone, grey thin-bedded fine-grained micaceous	1	6
Shale, dark grey silty micaceous; sandstone ribs	7	10
Faraday House Shell-Bed		
Sandstone, grey thick-bedded fine-grained calcareous; crystals of mica, pyrite and calcite; shells including productoids, spiriferoids and gastropods (Plate IIIB)	2	4
Sandstone, grey thin-bedded fine-grained calcareous shaly micaceous partings; vugs of calcite; sporadic shells	0	9
Shale, dark grey silty	2	4
Not exposed	9	0
Shale, dark grey and black, slightly micaceous; ironstone nodules and ribs; ten fossiliferous horizons with alternating rhynchonelloid-productoid and nuculoid-bellerophontoid assemblages	9	0
Not exposed	8	9
Limestone, dark grey argillaceous	2	6
Shale, dark grey	3	0
Limestone, grey sandy; productoids and spiriferoids	0	9
Shale, grey; sparsely fossiliferous..	0	5¾
LITTLE LIMESTONE [9893 2478] ..	—	—

The Firestone Sill is also exposed [0037 2390] in Stobgreen Sike* where its thickness of 40 ft contrasts with 7 ft 5 in, including 1 ft of shale and coal, proved in the Woodland Bore* to the north-east. The Faraday House Shell-Bed is faulted out in the Stobgreen Sike section, but is represented in the bore by an 11-in limestone rich in brachiopods.

J.H.H.

In recent Bores WT/4*, WT/C and WT/6 in the north-west of the district these measures are between 50 and 70 ft thick. A thick sequence of mudstones and shales also containing calcareous bands overlies the Little Limestone; in Bore WT/C one of these bands expands to form a limestone 3 ft 3 in thick lying 11 ft above the Little Limestone. The Faraday House Shell-Bed can be recognized in all cases either as a limestone or calcareous shelly sandstone. The Firestone Sill is thin and does not exceed 7 ft. In WT/4 Bore* it is separated from the base of the Crag Limestone by 1 ft 7 in of mudstone containing the Crag Coal, one inch thick.

D.A.C.M.

South of the Lunedale Fault, and north of Bowes, these measures are incompletely exposed. The best section is in the north bank of Deepdale Beck [9953 1538]:

	ft	in
CRAG LIMESTONE	—	—
Firestone Sill		
Sandstone, grey fine-grained; shelly, with worm tubes ..	1	0
Not exposed: shale debris ..	6	0
Coal	0	9
Sandstone, white thick-bedded false-bedded fine-grained feldspathic; roots at top	2	0
Sandstone, white thin-bedded false-bedded fine-grained micaceous carbonaceous; calcareous	8	0
Not exposed	12	0
Faraday House Shell-Bed		
Limestone, grey hard thin-bedded jointed sandy; brachiopods	3	0
Shale, dark grey fossiliferous ..	1	0
Sandstone, white and brown thin-bedded fine-grained	3	0

The base of this section is about 105 ft above the top of the Top Little Limestone.

In Bessy Sike [0018 1535] the sequence underlying the Crag Limestone is similar to that in Deepdale Beck, but 3 miles to the east-north-east, at Barnard Castle, it is much thinner, includes loc. 6b, and the sandstones in the upper part of the sequence form a single post that has been called the Ten Fathom Grit (see p. 32). The following section can be seen below the castle [0482 1640]:

	ft	in
CRAG LIMESTONE	—	—
Shale, black	0	5
Not exposed (probably including a coal)	0	6
Ten Fathom Grit		
Sandstone-seatearth, white siliceous rooty; shale partings towards the base	0	11
Shale, dark grey micaceous ..	0	2
Sandstone-seatearth, white thin-bedded fine-grained micaceous rooty	1	7
Sandstone, white to brown thin-bedded and thick-bedded false-bedded fine-grained feldspathic micaceous; sporadic pyrite nodules	30	0
Shale, grey; crinoids and brachiopods towards base	20	0
TOP LITTLE LIMESTONE	—	—

C.R.B.

The equivalent measures are continuously exposed in Whorlton Beck* below the bridge [1066 1499]. They are 137 ft 1 in thick, including the Faraday House Shell-Bed (3 ft 2 in) and the Firestone Sill (37½ ft). The fauna at several horizons is abundant and consists of bryozoa, brachiopods, nautiloids and bivalves, the most important forms present being *Fenestella* cf. *frutex*, *Cleiothyridina* cf. *fimbriata*, *Dielasma sp.*, *Dictyoclostus sp.*, *Echinoconchus* cf. *punctatus*, *Pleuropugnoides* cf. *pleurodon*, *Spirifer* cf. *bisulcatus*, *Rugosochonetes sp.*, *Limipecten dissimillis*, *Parallelodon* cf. *semicostatus* and *Pernopecten sowerbii* (loc. 6a). Good sections are also exposed in the banks of the River Tees between a locality [1009 1484] west of Whorlton suspension bridge and the weir [1314 1497] at Ovington. J.H.H.

In the Grunton Bore* the Top Little Limestone is succeeded by 10 ft of shale, reddened beneath the Permian unconformity, while in the Wath Erne Bore* the complete sequence between the Top Little and Crag Limestones is represented by only 24 ft of shales. D.A.C.M.

The Crag Limestone (Fig. 7) appears to be split into Top and Bottom leaves throughout most of the district, although it is present as a single bed, 2½ ft thick in Stobgreen Sike*. Farther west in Eggleston Burn the limestone is apparently absent, its position being indicated by the Crag coal. In Bores WT/4*, WT/C and WT/6 the Bottom Crag Limestone ranges between 9 ft 3 in and 19 ft 1 in. In Bore WT/4* where it is at its thickest it includes 1 ft 7 in of conglomerate consisting of limestone fragments set in a matrix of calcareous mudstone. In this borehole also the limestone is separated by only 5 in of calcareous mudstone from the Top Crag Limestone. To the north-east in the Woodland Bore* it is represented by 10 ft 4 in of limestone. D.A.C.M., J.H.H.

The Bottom Crag Limestone (Fig. 7) crops out in Bessy Sike [0018 1535], on the western margin of the district, where 12 ft of greyish brown thin-bedded sandy limestone with siliceous laminae rest on shelly sandstone; a similar section is seen in Deepdale Beck [0071 1604] where on the south side of the Raven's Nest fault 10½ ft of limestone contains *Caninia juddi*, *Chaetetes depressus*, *Martinia sp.*, *Rugosochonetes sp.* and *Aviculopecten* cf. *plicatus* (loc. 7d). The best of several exposures near Thornberry Farm is in a quarry [0107 1541] 65 yd N of the farmhouse, where the limestone is 10 ft thick and is grey thin-bedded fine-grained and decalcified with siliceous wisps; it contains crinoids, brachiopods and fish scales. In Smart Gill [0296 1586] 17 ft of siliceous limestone with chert lenses are ascribed to this horizon, though correlation with the Top Crag cannot be ruled out. The Bottom Crag is more certainly recognized beneath the castle in the north bank of the River Tees at Barnard Castle [0482 1640] (Plate I) where the section is

	ft	in
Limestone, grey thin-bedded crinoidal	1	0
Limestone, grey thin-bedded fine-grained; siliceous streaks and nodules and sporadic argillaceous partings	14	8
Limestone, grey fine-grained; sporadic nodules of limestone and chert; shelly	3	3
Shale, grey hard siliceous	0	11
Limestone, grey thin-bedded sandy	3	9
Limestone, grey thin-bedded fine-grained argillaceous	0	10

The fauna, ranged throughout the limestone, consists of bryozoa, *Caninia sp.*, *Koninckophyllum* cf. *echinatum*, *Cleiothyridina sp.*, *Quasiavonia* cf. *aculeata* and bivalves (loc. 7e). C.R.B.

A comparable section is exposed in Sledwick Hall quarry [0914 1525] near East Shaws where the fauna comprises *Chaetetes depressus*, *Koninckophyllum magnificum*, *Eomarginifera* cf. *setosa*, *Limipecten dissimilis* (loc. 7c). In Whorlton Beck* (loc. 7a) [1068 1501] the Bottom Crag Limestone has thinned to 9 ft; in the north bank of the River Tees below Wycliffe it is at least 10 ft thick. Other exposures are to be seen in the south bank of the Tees at, and to the west of, Ovington Weir* [1314 1497]; here it is particularly siliceous with well marked 'cauda-galli' structures and has a thickness of 6¼ ft including shale partings. East of Little Hutton the lower part of the Bottom Crag Limestone can be seen [1455 1269] to consist of at least 5 ft of grey, massive fine-grained siliceous limestone almost completely silicified; the top of the limestone, however, is dark grey thin-bedded fine-grained siliceous and relatively unaltered at outcrop some 300 yd farther east [1481 1268]. In the quarries [1541 1293] north-east of Foxberry the limestone is at least 5 ft thick and grey thin-bedded fine-grained bioclastic, crinoidal and dolomitic. J.H.H.

In the Wath Erne Bore* 21 ft of hard yellow limestone immediately underlying the drift is taken to be the equivalent of at least the Bottom Crag Limestone. In this same area small exposures of the limestone can be seen in Crossbury Bank Wood [2094 1056] (loc. 7b) where a typical section is:

	ft	in
Limestone, bluish grey hard siliceous; some chert nodules; brachiopods including *Alitaria* cf. *panderi* and ostracods	2	6
Limestone, bluish grey to yellow siliceous; brachiopods and ostracods	0	3
Mudstone, dark grey; calcareous with shale partings	4	0

Other exposures can be found in the beck [2122 1027] immediately north of Micklow Farm and in the two quarries east of Micklow Hill [2128 1002]. D.A.C.M.

The measures between the Bottom and Top Crag Limestones (Fig. 7) consist of dark grey calcareous, locally siliceous, fossiliferous shale or mudstone with thin limestone ribs. Thickness ranges from 5 in at WT/4 Bore* in the north to 16 ft in the River Tees near Ovington*. In Whorlton Beck* [1068 1510], for example, they are 12 ft thick and contain bryozoa, *Antiquatonia* cf. *insculpta*, *Linoprotonia sp.*, *Tornquistia* cf. *polita*, *Cypricardella sp.* and ostracods (loc. 8a). J.H.H., D.A.C.M.

The Top Crag Limestone (Fig. 7) was recognized in several boreholes in the northern part of the district but is exposed at only two localities. One is in the River Tees [9905 2308], about 800 yd SW of Eggleston Hall, where 5½ ft of dark grey thick-bedded jointed fine-grained crinoidal limestone are seen; the other [0706 2200] is in a railway cutting approximately 700 yd SW of Dent Gate Farm:

	ft	in
Limestone, dark grey thin-bedded fine-grained; *Fenestella sp.* ..	3	6
Shale, grey calcareous	1	0
Limestone, dark grey argillaceous; shells	3	0

These beds contain an abundant fauna of brachiopods and bivalves including '*Camarotoechia*' *sp.*, cf. *Schellwienella rotundata*, *Spirifer* cf. *trigonalis*, *Nuculopsis gibbosa* and *Polidevcia attenuata;* gastropods and trilobites are also present (loc. 9e). Immediately beneath this section Gunn (in manuscript) recorded 5½ ft of shale on 2½ ft of limestone, the latter being the possible local equivalent of the Bottom Crag Limestone. J.H.H.

In Bores WT/4*, WT/C and WT/6 in the north-west part of the district the Top Crag Limestone ranges between 2 ft 2 in and 9 ft 8 in, being thickest in Bore WT/4 where it is a dark grey muddy limestone with calcareous mudstone bands. In the Woodland Bore* it is represented by 7 ft 5 in of limestone with shale bands. D.A.C.M.

In the south-western part of the district the Top Crag Limestone is exposed in Deepdale Beck [9920 1546], about 80 yd E of Crag Bridge (Plate IVA), where it consists of at least 9½ ft of bioclastic limestones interbedded with fossiliferous calcareous shales and impure limestones, all of which are decalcified (locs. 9c, 9d) and again [0070 1602] south of the Raven's Nest Fault, where it comprises 12 ft 10 in of grey thin-bedded sandy shelly limestone with siliceous wisps. Still farther down the beck [0281 1627] the upper part of the limestone is seen as follows:

	ft	in
Limestone, grey to brown fine-grained crinoidal	0	4
Shale, dark grey; shell fragments ..	0	3

	ft	in
Limestone, grey to brown fine-grained; crinoidal	0	8
Limestone, grey siliceous; shelly ..	0	7
Shale, grey calcareous; limestone laminae	0	6
Limestone, grey fine-grained argillaceous; shelly	1	3
Shale, grey calcareous; partings containing abundant shell debris including brachiopods and crinoids	1	2
Limestone, grey hard siliceous crinoidal	0	11
Shale, grey calcareous; shell fragments	0	5
Limestone, grey thin-bedded argillaceous shelly	0	6

In the north bank of the River Tees [0483 1642] at Barnard Castle (Plate I) the Top Crag Limestone consists of at least 25 ft of grey thin-bedded fine-grained sandy limestone with 'cauda-galli' structures and siliceous wisps. The fauna consists of bryozoa, crinoids (in top 4½ ft), *Cleiothyridina sp.*, *Quasiavonia sp.* [juv.], bivalves and ostracods (loc. 9b). About 1 mile ESE in a quarry [0631 1600] near Lowfield Garden the basal members of the limestone are exposed:

	ft
Limestone, grey thin-bedded fine-grained, becoming coarser upwards, siliceous; mudstone bands; crinoids and sporadic brachiopods	6
Mudstone, grey siliceous calcareous; abundant crinoid debris and bryozoa, sporadic brachiopods and bivalves	4
Limestone, grey coarse-grained argillaceous crinoidal	1

J.H.H.

In Whorlton Beck* [1068 1515] a continuous section of the Top Crag Limestone is exposed, comprising limestone 2 ft 8 in on mudstone 2 ft and limestone 5 ft. The fauna consists mainly of brachiopods including *Linoprotonia sp.* and *Martinia sp.* (loc 9a). The Top Crag Limestone is further exposed in the south bank of the River Tees near Ovington*, where it has a total thickness of 6 ft 7 in including 1½ ft of mudstone.

D.A.C.M.

About 2 miles SE of Ovington in a small quarry [1598 1349] in Caldwell the section exposed is:

	ft	in
Shale	1	0
TOP CRAG LIMESTONE		
Limestone, dark grey thin-bedded fine-grained siliceous	0	9
Shale, grey siliceous	1	6
Limestone, pale grey thin-bedded fine-grained silicified and decalcified; carbonaceous streaks	1	0
Limestone, pale grey massive fine-grained siliceous	6	0
Limestone, dark grey fine-grained argillaceous siliceous; sporadic crinoids and brachiopods ..	2	6
Mudstone, dark grey to black; *Fenestella sp.*, brachiopods ..	2	0

J.H.H.

The measures between the Top Crag Limestone and the Knucton Shell-Beds (Fig. 7) are poorly exposed in the northern part of the district; they are recorded however in the WT/4*, WT/6 and Woodland* bores. Atypically, in WT/4 Bore the Knucton Shell-Beds are formed by 3 ft 5 in of calcareous mudstone and siltstone resting directly on the Top Crag Limestone. In WT/6 Bore and the Woodland Bore 5 and 3 ft respectively of calcareous mudstones separate the two horizons. D.A.C.M. J.H.H.

In the south-western part of the district, these beds are well exposed in Deepdale Beck between Crag Force and the River Tees at Barnard Castle. Between Crag Force and the Raven's Nest Fault, for instance, the Top Crag Limestone is succeeded by 68 ft of shale (loc. 10b) containing a fauna including *Hyalostelia sp.*, bryozoa, *Rugosochonetes sp.*, *Euphemites sp.* and *Palaeoneilo* cf. *mansoni;* this in turn is overlain by an 8-ft sandstone. Immediately north of the fault [0075 1610] the following more complete section is exposed.

	ft
KNUCTON SHELL-BEDS LIMESTONE ..	—
Sandstone, white to brown thin-bedded fine-grained	25
Shale, dark grey silty; sandstone ribs	20
Sandstone, brown thin-bedded false-bedded fine-grained	8
Shale, dark grey; becoming silty upwards; basal 2 ft contains *Fenestella sp.*, *Echinoconchus* cf. *elegans*, *Aviculopecten* cf. *plicatus* (loc. 10c)	40
Not exposed (probably shale) ..	10
TOP CRAG LIMESTONE	—

(L 311)

A. Top Crag Limestone, Crag Bridge, Deepdale Beck

PLATE IV

B. Working face (1963) in massive siliceous sandstone, Dunn House Quarry, Staindrop

(L 336)

Lateral variation in the upper part of these measures is demonstrated by comparing the above section with that in a beck near Rigg Farm*, where the Knucton Shell-Beds Limestone is separated from an underlying 28-ft sandstone by 10 ft of sandy shale. About 300 yd due east a similar sequence is seen in a southern tributary stream of Deepdale Beck. Beneath the lowest sandstone, however, a further 52 ft of beds are exposed consisting of 46 ft of silty shales with sandstone ribs split by a 6-ft median sandstone. The base of the lowest shales is 30 to 40 ft above the Top Crag Limestone. Part, if not all, of this interval consists of argillaceous beds, locally containing ironstone nodules and fossils; they are poorly exposed in Deepdale Ravine and nearby in Smart Gill [0312 1621], where the lowest 12 ft of siltstone and shale rests directly on the Top Crag Limestone. Still farther east, in Deepdale Beck, the Knucton Shell-Beds Limestone is underlain in the north bank [0444 1667] by 4 ft of grey fine-grained micaceous shelly sandstone with argillaceous wisps, 'cauda-galli' structures and brachiopods below which lie 5 ft of grey pyritous siltstone with sandstone ribs. The siltstone is also partially exposed at the mouth of Deepdale Beck, where it contains lenses and nodules of white, fine-grained, siliceous sandstone: the overlying sandstone can be seen in the north bank of the River Tees [0477 1659] beneath the gasworks at Barnard Castle. C.R.B.

These measures are next seen nearly 6 miles to the east, in the Tees below Ovington*, where they consist of 101½ ft of mudstone, with sandstone ribs in the upper 40 ft, and ironstone nodules, sporadic bryozoa, brachiopods and bivalves in the basal 30 ft (loc. 10a). A borehole [1503 1427] near Greystone penetrated these measures and proved the Knucton Shell-Beds Limestone to rest on 15 ft of sandstone, 4 in of coal and 1 ft of sandstone. J.H.H.

In the south-eastern part of the district a borehole [2404 1064] at Clow Beck Farm recorded the following section in the lower part of these measures:

	ft
PERMIAN strata	—
Shale, greenish sandy	10
Sandstone, green and purple shaly	2
Shales, sandy; purplish tinge	10
Limestone, impure; rare crinoids ..	2
Shales, dark grey massive; purple ironstone rib in top foot and a 3-in rib 24 ft from top; rare brachiopods	45

The base of this section is considered to be close above the top of the Crag Limestone. The measures exposed in Cleasby cutting* [2458 1231] are thought to be above those proved at Clow Beck Farm. D.A.C.M.

The Knucton Shell-Beds (Fig. 7) are not exposed in the northern part of the district but were proved in three boreholes. In WT/4 Bore* they are represented by 3 ft 5 in of calcareous mudstone and siltstone containing ribbed shelly fossils at the top; atypically the bed rests directly on top of the Top Crag Limestone. In WT/6 Bore, these beds are represented by 12 ft 2 in of dark grey muddy thin-bedded shelly limestone interbedded with hard calcareous laminated mudstones. In the Woodland Bore* they were proved to consist of a lower 13-in and an upper 2-in limestone separated by 8 ft 10 in of calcareous mudstone. J.H.H., D.A.C.M.

In the south-western part of the district the beds are represented by the Knucton Shell-Beds Limestone. This crops out in a beck near Rigg Farm* where it is a grey thin-bedded fine-grained siliceous arenaceous shelly limestone 17½ ft thick. It is exposed intermittently farther east in both banks of Deepdale Beck, being at least 20 ft thick in an exposure [0170 1624] about 390 yd NNE of Raven's Nest Farm. It is completely silicified with some chert at another locality [0444 1667], 130 yd W of Deepdale Bridge. Farther east, in the River Tees the limestone is 12 to 14 ft thick; the lower part, exposed in the south bank [0453 1674] 45 yd N of Deepdale Aqueduct, is as follows:

	ft	in
Limestone, grey hard thin-bedded fine-grained siliceous; layers of siliceous shale containing limestone lenses	5	6
Limestone, grey thin-bedded fine-grained part sandy with siliceous wisps; shelly	5	0

In the north bank of the river between the mouth of Percy Beck [0456 1679] and Barnard Castle gasworks [0477 1659] the limestone is so intensely silicified that only small remnants of the original rock remain. C.R.B.

The Limestone is next exposed in the River Tees, over 5 miles farther east, upstream from Winston Bridge*, where it is 32½ ft thick and is grey, fine-grained, thin-bedded, rubbly and shelly. An abundant fauna is present and includes *Hyalostelia smithii*, *Caninia sp*, *Cleiothyridina* cf. *fimbriata*, *Echinoconchus punctatus*, *Aviculopecten* aff. *interstitialis*, *Sulcatopinna costata* and *Rayonnoceras sp*. (loc. 11a).

According to Gunn (in manuscript) a bore [1503 1427] near Greystone proved the basal 9 ft of the limestone to be siliceous and impure. J.H.H.

The measures between the Knucton Shell-Beds and the Rookhope Shell-Beds (Fig. 7) are exposed in the northern half of the district only in the River Tees [9956 2338] downstream from Eggleston Hall, where a grey, medium- to coarse-grained, thin- and thick-bedded sandstone, at least 20 ft thick, passes up into a 9-in fine-grained rooty sandstone lying immediately beneath the Rookhope Shell-Beds. This sandstone, which forms a pavement in the river bed hereabouts, has been proved in TO 1 Bore [0016 2277] to be 26 ft thick and is underlain by 6 in of coal on 2 ft 9 in of sandstone-seatearth. The upper part of the sandstone bears root impressions, is bioturbated and very hard. In recent boreholes in the north-western part of the district, notably WT/4* and WT/C bores, the greater part of the sequence is composed of sandstones, the probable correlatives of the Grit Sills of the district to the north and north-west. In WT/C Bore, of 237 ft of strata overlying the Top Crag Limestone, 220 ft were sandstones with subordinate bands of mudstone, the Knucton and Rookhope Shell-Beds being cut out. A broadly similar sequence is recorded in WT/4 Bore* although not all of it is proved. In the Woodland Bore* these measures are 76 ft thick, consisting of 37 ft 7 in of sandstone with shale layers, overlying 38 ft 5 in of shales with thin sandstone ribs.

In a borehole [0027 2299] near West Barnley Farm and on the south side of the Lunedale–Staindrop Fault, the Rookhope Shell-Beds are underlain by 76 ft of grey moderately bedded, variably textured locally coarse-grained pebbly sandstone. J.H.H., D.A.C.M.

In the south-western part of the district these measures are completely exposed in the tributary of Deepdale Beck south of Rigg Farm, where they consist of 7½ ft of sandstone overlying 26 ft of shales. In Percy Beck, west of Barnard Castle, the following exposed section is comparable, despite a small fault:

	ft	in
ROOKHOPE SHELL-BEDS LIMESTONE [0470 1695]	—	—
Sandstone, brown fine-grained massive siliceous feldspathic; limonitized shell casts at base . .	2	4
Sandstone, brown fine-grained thick-bedded micaceous feldspathic	3	8
Not exposed	5	6
Sandstone, white fine-grained thin-bedded micaceous; sporadic shaly partings	16	10
Shale, grey sandy micaceous; fine-grained sandstone ribs . .	3	0
Not exposed	15	0
KNUCTON SHELL-BEDS LIMESTONE [0456 1679]	—	—

C.R.B.

These measures are again completely exposed in the River Tees, upstream from Winston Bridge*, where 17½ ft of sandstone overlie 15¼ ft of shale containing sponge spicules, bryozoa, brachiopods, bivalves including nuculoids and nautiloids (loc. 12a). D.A.C.M.

The Rookhope Shell-Beds (Fig. 7) are continuously exposed in a bluff [0033 2187] of the River Tees, near West Barnley, about 1 mile SE of Eggleston Hall (loc. 13b). The section, together with the overlying beds, is:

	ft	in
Sandstone, white to yellow coarse-grained massive	2	0
Mudstone, grey silty micaceous; ironstone nodules; fauna near base including crinoid debris, *Eomarginifera* cf. *setosa*, *Rugosochonetes sp. nov.* [tumid], *Posidonia corrugata*	8	8
ROOKHOPE SHELL-BEDS		
Limestone, grey argillaceous; crinoid fragments, coral fragment, shell fragments	2	2
Mudstone, grey calcareous; productid fragments	0	6
Limestone, grey argillaceous; quartz grains; *Buxtonia sp.*, *Lingula sp.*	0	6

	ft	in
Mudstone, grey; ironstone ribs; '*Chonetes*' *sp.* [juv.], *Spirifer sp.*, gastropods indet. and *Posidonia corrugata* in the lower 3 ft	10	4
Limestone, grey fine-grained argillaceous; rare shell fragments	1	5
Limestone, grey to green; quartz grains; nodules of dark grey limestone	0	3
Limestone, grey to brown argillaceous; abundant quartz grains; rare shell fragments ..	1	3
Sandstone, grey medium- to coarse-grained thin-bedded calcareous; micaceous carbonaceous laminae..	0	4
Sandstone, pale grey fine-grained micaceous	0	4
Sandstone, grey fine-grained thin-bedded; siltstone ribs ..	9	0
Mudstone, grey silty micaceous; sandstone lenses, ironstone nodules; rare shell fragments at base	10	0
Ironstone	0	4
Limestone, dark grey fine-grained cherty arenaceous	0	9
Mudstone, grey	1	2
Limestone, dark grey fine-grained cherty	0	5
Mudstone, grey	0	5
Limestone, dark grey fine-grained arenaceous argillaceous ..	1	2
Sandstone-seatearth	—	—

The lowest 3 ft 11 in of strata contain a fauna which includes sponge indet., productoid fragments indet., *Martinia sp.*, *Spirifer* cf. *trigonalis*, *Aviculopecten sp.* and *Weberides sp.* [pygidium] (loc. 13b).

J.H.H.

The sequence proved in the Rookhope Shell-Beds in T0 1 Bore was as follows:

	ft	in
Limestone, dark grey sandy muddy thin-bedded very fine; shell debris	0	8
Sandstone, grey and greenish grey thin-bedded fine- to medium-grained	0	9
Siltstone, calcareous dark grey laminated micaceous and carbonaceous	6	0
Mudstone, dark grey finely micaceous; occasional shell debris; calcareous siltstone and mudstone ribs	12	1
Siltstone, calcareous and limestone, muddy; shell debris	3	8

D.A.C.M.

The sandstone above the Rookhope Shell-Beds is transgressive and locally cuts out much of the upper part of the shell-beds. In the River Tees, 10 yd upstream from Eggleston Bridge [9966 2325], for instance, the section seen is:

	ft	in
Sandstone, grey medium- to coarse-grained massive	40	0
ROOKHOPE SHELL-BEDS		
Sandstone, grey thin-bedded silty rooty	1	9
Mudstone, grey silty calcareous; pyrite and ironstone nodules	1	0
Not exposed	2	0
Mudstone, grey silty calcareous slightly micaceous	6	0
Limestone, grey fine-grained hard argillaceous	0	8
Mudstone, grey calcareous; limestone ribs and chert	1	9
Sandstone-seatearth	—	—

In TO 1 Bore the sandstone above the Rookhope Shell-Beds is again transgressive and more than 37 ft thick. A limestone one foot thick is present near its base. Another record of the shell-beds in the northern part of the district is in the Woodland Bore*, where they are represented by the Rookhope Shell-Beds Limestone, nearly 13 ft thick, which includes a 2½-ft median bed of mudstone. This limestone persists throughout the southern part of the district. J.H.H.

The lower leaf of the limestone is seen in Deepdale Beck [0079 1619] and Percy Beck [0470 1695], where it is 1 ft 7 in and 2 ft 4 in thick respectively and grey, fine-grained, thin-bedded, sandy and crinoidal, with hematite grains near the top. C.R.B.

Farther east in the River Tees [1429 1619] above Winston Bridge* the Rookhope Shell-Beds Limestone consists of at least 33 ft of grey fine- to medium-grained shelly limestone, with numerous bands of calcareous mudstone. Brachiopods are locally abundant and include notably *Semiplanus* aff. *latissimus* and *Tornquistia* cf. *polita* (loc. 13a). D.A.C.M.

The measures between the Rookhope Shell-Beds and the Lower Felltop Limestone (Fig. 7) consist of mudstone below and a variable thickness of sandstone above. On the eastern margin of the Brough-under-Stainmore (31) Sheet the sandstone is 50 ft thick. This thickness is more than doubled farther east near West Barnley, where in the River Tees and its tributaries, between Eggleston Hall [9977 2334] and the mouth of Raygill Beck [0053 2133] the following composite section is exposed:

	ft	in
Horizon of the LOWER FELLTOP LIMESTONE	—	—
Sandstone, grey massive false-bedded jointed coarse-grained feldspathic pebbly	50	0
Shale, dark grey	1	3
Limestone, dark grey fine-grained cherty [0017 2274]	1	3
Sandstone, dark grey fine-grained shaly	4	8
Shale, dark grey silty	5	0
Shale, *partly obscured*	3	0
Sandstone, grey coarse-grained pebbly false-bedded massive ..	50	0
Mudstone; crinoids, brachiopods and bivalves in basal 2 ft, including *Eomarginifera* cf. *setosa*, *Rugosochonetes sp. nov.* [tumid] and *Posidonia corrugata*	8	8
ROOKHOPE SHELL-BEDS	—	—

Because of faulting the upper 50-ft sandstone is exposed again in the River Tees [0077 2117] below the confluence with Raygill Beck. At this locality it contains a median bed of mudstone and two coals; the section is as follows:

	ft	in
Sandstone, grey medium- to coarse-grained thin-bedded	25	0
Mudstone, grey micaceous rooty ..	2	2
Coal, with numerous shale bands ..	4	0
Seatearth-mudstone, pale grey silty rooty	0	6
Sandstone, grey fine-grained thin-bedded	5	6
Sandstone, grey fine-grained thick-bedded	10	0
Sandstone, grey fine-grained thin-bedded micaceous carbonaceous	2	2
Coal, with numerous shale bands	2	2

The coal at the base of this section was recorded during the original geological survey both at this locality and in the stream 200 yd E of Eggleston Hall, where it is not now exposed. It is correlated with a 10- to 12-in coal which lies beneath Cambridge House [0548 1691], Barnard Castle, and which is calculated to be approximately 17 ft below the Lower Felltop Limestone and also with the 12-in coal proved at 472 ft 9 inches in Westholme Colliery No. 1 Shaft* [1347 1785]. The only other continuous record of these measures in the northern half of the district was obtained in the Woodland Bore* and is generalized as follows:

	ft	in
LOWER FELLTOP LIMESTONE ..	—	—
Mudstone	1	4
Sandstone	7	0
Shale and mudstone	36	1
Rookhope Shell-Beds Limestone	—	—

J.H.H.

In the south-western part of the district these measures are exposed in Percy Beck where 20 ft of fossiliferous shale (loc. 15a) are overlain by 34 ft of silty shale with ironstone nodules and thin sandstones. C.R.B.

The measures beneath the Lower Felltop Limestone are also seen in a quarry [1295 1064] near Osmondcroft Farm on the north side of the River Tees near Ovington, where they consist of 4 ft of white, fine- to medium-grained, quartzitic sandstone. The equivalent sandstone is also exposed in the River Tees, downstream from Winston Bridge*. It there contains a 12-in cherty bed rich in sponge spicules. D.A.C.M.

In thin section [E 32023] this bed is seen to be composed of large discrete areas of uniformly extinguishing carbonate (ω=1.766$\pm$ ·005) identified as ferrodolomite, in a chert matrix strewn with spicules of carbonate. The spicules are 0.02 to 0.14 mm long with generally circular cross-sections. They are accompanied by subangular to subrounded quartz grains of 0.05 mm average diameter and show marginal carbonate replacement. R.K.H.

The Lower Felltop Limestone (Fig. 8) is 8 in thick in the Woodland Bore* and is represented by 6 in of grey fine-grained shelly limestone at outcrop [0698 2244] in Langley Beck, about 530 yd W of Dent Gate Farm. It is not otherwise exposed or proved in the northern half of the district. Farther south, however, it crops out at two localities. In a section [0790 1636], near Westwick, the sequence is as follows:

	ft	in
Shale, dark grey; ironstone nodules	3	0
Chert	0	6
LOWER FELLTOP LIMESTONE		
Limestone, dark grey fine-grained siliceous..	1	6
Silica-rock, pale grey fine-grained well bedded	2	0

At the second locality in the River Tees below Winston Bridge* the limestone is 3 ft thick and is grey, fine- to medium-grained and argillaceous with sporadic crinoid and brachiopod fragments. J.H.H., D.A.C.M.

The measures between the Lower and Upper Felltop limestones (Fig. 8) have the following generalized sequence in the area north-west of Eggleston:

		ft	in
Horizon of UPPER FELLTOP LIMESTONE		—	—
Sandstone		30	0
Shale		30	0
Inferred position of Coalcleugh Shell-Bed		—	—
Sandstone	36 to	60	0
Shale	25 to	30	0
Horizon of LOWER FELLTOP LIMESTONE		—	—

The variable lower sandstone (Coalcleugh Transgression Beds), exposed in the bed of Eggleston Burn [9867 2629], north of the confluence with Druvy Burn, is pale grey, medium- to coarse-grained, pebbly, massive and false-bedded, with sporadic micaceous carbonaceous scars. The upper sandstone, exposed to a thickness of 10 ft farther north-east in two quarries [9894 2605 and 9893 2722], is white, fine- to medium-grained, thin- and thick-bedded and jointed. To the east the Woodland Bore* proved these measures to be 112½ ft thick, including the Coalcleugh Shell-Bed represented by a 17-in limestone. Farther south, in Langley Beck [0736 2238], the following sequence was proved in the upper part of these measures:

	ft	in
UPPER FELLTOP LIMESTONE	—	—
Sandstone, pale grey fine-grained thin- to thick-bedded; siliceous at top; sporadic shell casts at base..	13	6
Sandstone, grey silty micaceous rooty; carbonaceous laminae and siltstone bands	2	0
Siltstone, dark grey micaceous; interbedded silty mudstone bands	8	6
Sandstone, pale grey fine- to medium-grained thin-bedded; carbonaceous laminae	0	10
Not exposed	6	0
Sandstone, grey fine-grained thin-bedded flaggy; cross-bedded with ripple marks; sporadic worm tubes, carbonized plant fragments and mudstone pellets	37	0
Siltstone, grey; numerous sandstone ribs..	11	0

The base of this section is taken to be approximately 75 ft above the Lower Felltop Limestone.

South of the Lunedale–Staindrop Fault and east of the River Tees, near East Barnley, the following generalized sequence is applicable:

		ft	in
HORIZON OF UPPER FELLTOP LIMESTONE		—	—
Sandstone	15 to	20	0
Shale		30	0
Sandstone		40	0
Shale (including Coalcleugh Shell-Bed equivalent)		30	0
Horizon of LOWER FELLTOP LIMESTONE		—	—

Some 8 ft of the 40-ft sandstone is exposed in a quarry [0083 2225], 120 yd N of East Barnley, where it is fine- to medium-grained, thin-bedded and flaggy. Minor exposures also occur in Raygill Beck, while in the top of the Tees Bank [0088 2117] about 20 ft of grey fine-grained, massive, jointed micaceous sandstone is exposed at this horizon. In the River Tees [0114 2074 to 0161 2003], north of Cotherstone this bed is seen to consist of at least 30 ft of grey, medium- to coarse-grained, thick-bedded to massive false-bedded sandstone, which becomes more fine-grained and thin-bedded towards the top. J.H.H.

In the River Balder, between the western margin of the district and the River Tees, these measures are similar in sequence to those near East Barnley, but the sandstones are thicker as illustrated by the following section:

	ft	in
UPPER FELLTOP LIMESTONE	—	—
Coal [9739 1908]	0	3
Sandstone-seatearth, white fine-grained siliceous micaceous ..	1	6
Sandstone, white to brown fine-grained thin-bedded micaceous limonitic	10	0
Shale, grey silty	1	0
Sandstone, white to brown fine-grained thin-bedded carbonaceous micaceous	1	6
Shale, grey soft silty; passing down into a structureless clay	1	0
Coal, dirty	0	3
Mudstone-seatearth, grey silty ..	2	0
Sandstone, grey fine-grained thin-bedded shaly micaceous ..	2	0
Shale, grey silty; passing down to siltstone	1	0
Sandstone, white fine-grained micaceous carbonaceous limonitic	2	0
Not exposed	36	6
Sandstone, white to brown fine- to coarse-grained pebbly thin- to thick-bedded and false-bedded, limonitic feldspathic micaceous..	75	0
Coal, dirty (?Tanhill Coal) 1 ft to	3	0
Shale, grey silty micaceous ..	3	0
Sandstone, white to brown; fine-grained and thin-bedded in top 2 to 3 in, coarse-grained, locally pebbly and massive below [9889 1989]	4	3

C.R.B.

In Osmond Gill [9909 1984] 12 in of coal with shale partings immediately overlies the 75-ft sandstone of this section, while in Wilden Beck [9768 2105], just beyond the western margin of the district, the remainder of the 36½ ft of unexposed measures are seen to consist of 6 to 7 ft of sandstone overlying 30 ft of grey silty mudstone, with at least two *Lingula* bands about 20 ft above the base. At the confluence of the rivers Balder and Tees the measures below the Upper Felltop Limestone are further intermittently exposed and are generally thinner; the 75-ft sandstone of upper Baldersdale is about 40 ft thick and in an exposure [0315 2015] in the River Tees it is separated from the underlying coal (?Tanhill Coal) by 6 ft of silty shale. J.H.H.

The sandstone exposed above the north bank of Deepdale Beck, between Crag Hill [9967 1618] and Cat Castle Quarries [0115 1636] is thought to be part of this sequence. It is 56 ft thick and is white and brown, coarse-grained, pebbly, massive, locally false-bedded and feldspathic.

In the River Tees between Cotherstone and Barnard Castle there are good sections in these measures. A composite section* shows them to be up to 314 ft thick and to include four good coals in the upper part and three coaly shales below. The lowest good coal, tentatively correlated with the Tanhill Coal, is 12 to 24 in thick and is overlain by a sandstone which ranges in thickness from 37 to 50 ft. Another sandstone of variable thickness occurs about 16 ft above the Lower Felltop Limestone and contains posts which, for example, increase in thickness from 2 in to 2 ft 9 in over a lateral distance of 13 ft. C.R.B.

North-east of Barnard Castle the only exposed measures within this sequence are those close below the Upper Felltop Limestone. Thus at Old Mill Gill* [0714 1971] 10 ft of silty mudstone overlie 45 to 49 ft of shaly sandstone on 15 ft of silty mudstone beneath the horizon of the limestone and up to 6 ft of the same sandstone can be seen in two streams [0955 1827 and 0958 1830] west of High Barford and in Newsham Beck [1286 1788]. In these three exposures the sandstone is greyish brown, fine-grained, flaggy and limonitic, with 'cauda-galli' structures in the latter section.

In a stream [0790 1636], about ½ mile NE of Mount Eff, the Lower Felltop Limestone is overlain by 5 ft of dark grey shale with ironstone nodules; 10 ft above this section is the base of a poorly exposed sandstone estimated to be 26 ft thick, and which, in an adjacent borehole [0747 1679] is succeeded by a 6-in coal seam taken to be the equivalent of the Yoredale Coal. J.H.H.

This 6-in coal was proved at a depth of 139 ft in a borehole [about 0555 1746] at Barnard Castle where it overlies a shale which contains a diverse well-preserved miospore assemblage. The assemblage is dominated by the genus *Lycospora* and contains a large number of stratigraphically useful accessory spores. Several of the accessory spores, i.e. *Punctatisporites aerarius*, *Acanthotriletes castanea*, *Convolutispora ampla*, *Microreticulatisporites concavus*, *M. microreticulatus*, *Rotaspora fracta*, *R.*

knoxi, *Crassispora maculosa* and *Rugospora corporata* have been previously recorded from both the Upper Viséan and lower Namurian whilst others, i.e. *Convolutispora cerebra*, *Mooreisporites trigallerus*, *Knoxisporites seniradiatus*, *Proprisporites laevigatus* and *Schulzospora ocellata* have been previously recorded only from the lower Namurian of Britain. On the basis of this evidence together with the presence of several undescribed forms previously recorded only between the Little Limestone and the Mirk Fell Ironstones in the Stainmore Outlier, Westmorland (Brough under Stainmore (31) district), an upper E_1 or low E_2 age is suggested for this assemblage. B.O.

Farther east these measures are very variable and they are best discussed by separately considering two geographical areas; that between Whorlton and Winston and that around Gainford.

Between Whorlton and Winston the following generalized sequence obtains:

		ft	in
Horizon of UPPER FELLTOP LIMESTONE (position of)		—	—
Sandstone	30 to	40	0
Shale	20 to	30	0
Sandstone		20	0
Shale	0 to	30	0
Sandstone		30	0
Shale	36 to	96	0
Yoredale Coal	3½ to	5	0
Sandstone	0 to	26	0
Shale, with thin sandstone ribs	35 to	88	0
LOWER FELLTOP LIMESTONE		—	—

These measures have been proved in the North Tees No. 1 Bore* [1212 1762] (loc. 15b) at Little Newsham, and those beds between the Lower Felltop Limestone and the Yoredale Coal are best exposed in the west bank of the River Tees, 50 to 150 yd below Winston Bridge (Fig. 8). In this section the limestone is overlain by 65 ft 2 in of shale with two thin sandstones, 13½ ft of sandstone, 10 to 15 ft of unexposed measures and then the Yoredale Coal. Broadly similar sequences were proved in a borehole [1266 1674] near Walker Hall (Borings and Sinkings No. 3075), in which a 2½ ft Yoredale Coal lay about 158 ft above the estimated position of the Lower Felltop Limestone, and in Westholme Colliery No. 1 Shaft*.

The Yoredale Coal was mined until recently from the North Tees Colliery [1220 1626] and was formerly worked by opencast immediately to the south and half a mile east of the mine. The seam section is variable. In the south-western part of the workings the section is: coal 30 in, on band 7 in, and coal 29 in, while in the north-eastern workings the seam section is locally reduced to: coal 6 in, on band ½ in, coal 5 in, band ½ in and coal 21 in. Records from old workings (1897) around Osmondcroft show the coal to have been about 46 to 50 in thick, brassy at the top and inferior at the base, while south of Winston the worked seam was 36 to 48 in thick including an impersistent 1-in band. The seam was wrought in the south bank of the River Tees [1465 1680] across from Winston, but no records have been preserved apart from a boring [1463 1679], about 220 yd N of Hedgeholme in the delivery drift of an old pit, which is said to have proved 4 ft of coal (Kirkby and Duff 1872, p. 184).

An analysis by the National Coal Board of the spore assemblage in specimens of the coal from the North Tees Colliery Drift show it to contain *Lycospora spp.*, *Rotaspora knoxi* and *Crassispora kosankei*, which are taken to indicate an age not greater than E_2.

In the Whorlton–Winston area the remaining part of these measures, that is beds between the Yoredale Coal and the Upper Felltop Limestone, are incompletely exposed.

A section in the lower part was recorded in the North Tees Colliery Back Drift [1226 1657], as follows:

	ft	in
Boulder clay	15	0
Sandstone, dull yellow to brown fine- to medium-grained hard; a little mica	24	0
Sandstone, grey fine-grained carbonaceous micaceous; grading down to shaly sandstone	3	6
Sandstone, flaggy carbonaceous	4	0
Mudstone, flaggy at top; fine-grained sandstone laminae; grades down to dark grey hard micaceous carbonaceous mudstone	32	0
Shale, micaceous blocky	6	6
Mudstone, dark grey blocky hard micaceous carbonaceous; *Myalina sp.* (loc. 15g)	30	0
Yoredale Coal	4	0

Sections have also been proved in these measures in several boreholes between Whorley Hill and Winston. Two of these boreholes are recorded in Borings and Sinkings (Nos. 2884 and 3075) while a third, Westholme Colliery Shaft* is shown in Fig. 8. In North Tees No. 1 Bore* near Little Newsham, 255½ ft of measures were proved underlying the Upper Felltop Limestone, the Lower Felltop Limestone having not been proved or reached, although a 6-in limestone containing spiriferoids was proved at 270½ ft. The measures predominantly consist of siltstones and mudstones although two sandstones are present near the top. A fauna (loc. 15b) consisting of crinoid debris, brachiopods, ostracods and a trilobite, is present at various horizons. Among the most interesting forms present are cf. *Planolites montanus*, *Productus carbonarius*, *Rugosochonetes sp.*, *Tornquistia* cf. *polita*, and the trilobite *Weberides* cf. *mucronatus*.

Around Gainford the measures between the Lower and Upper Felltop limestones are not less than 400 ft thick, increasing towards the east, especially those beds above the Yoredale Coal. The main exposures are in the River Tees and its tributaries and the following is a generalized sequence:

	ft	in
Horizon of UPPER FELLTOP LIMESTONE	—	—
Sandstone (of Selaby Park and West Scar)	60	0
Shale with sandstone bands ..	40	0
Sandstone with shale bands ..	44	0
Sandstone (of Black Scar) .. 8 ft to	12	0
Shale .. .. 0 ft to	12	0
Sandstone (of Black Scar)	20	0
Shale	30	0
Sandstone (of Grant Bank and Boat Scar)	30	0
Shale .. .. about	140	0
Sandstone	8	0
Yoredale Coal	2	0
Shale	80	0
Horizon of LOWER FELLTOP LIMESTONE	—	—

The measures below the Yoredale Coal in the Gainford area are exposed only at Crake Bank [1634 1635], where the seam is underlain by 4 ft of shale with ironstone nodules. The coal was formerly dug at outcrop at the base of Crake Bank, near the mediaeval village of Old Richmond, and cropped out in an adit near the bottom of Barforth Gill [1632 1637], where it ranged in thickness from 15 to 20 in. It was 2 ft 2 in thick in a borehole [1563 1754] near Grant Cottage, north of the Tees (Borings and Sinkings No. 3076). Kirkby and Duff (1872, p. 184) refer to a further sinking near Gainford, where the coal was 2 ft 6 in thick, including 1½ ft of black band, at a depth of 102 ft. In the Barforth Gill adit the coal is overlain by 12 ft of yellow, fine-grained, massive sandstone which includes an 18-in bed of carbonaceous silty shale. The measures overlying the sandstone are indifferently exposed farther up the gill [1636 1610], and consist of grey, sandy or silty shales with ironstone nodules and ribs and with shaly sandstone bands. Still higher measures are exposed in the north-east bank of the River Tees at Grant Bank [1618 1736] as follows:

	ft	in
Sandstone, yellow grey and brown fine- to medium-grained; massive at top; silty with micaceous carbonaceous laminae and plant fragments below	30	0
Mudstone, grey silty micaceous; sparse crinoid debris; brachiopod fragments and *Actinopteria regularis* (loc. 15d)	2	10
Mudstone-seatearth, grey and white; abundant sphaerosiderite	0	7
Sandstone, grey silty medium-grained laminated; abundant carbonaceous laminae .. 4 ft to	6	6

At the base of Selaby Basses Wood [1582 1772] 1 ft 10 in of impure, arenaceous ochreous limestone, with brachiopods including *Rugosochonetes sp.* (loc. 15e) overlies 4 ft 2 in of brown, fine- to medium-grained massive sandstone with carbonaceous micaceous blebs. It is inferred that these beds are older than those exposed at Grant Bank and that they lie approximately 100 ft above the Yoredale Coal. The 30-ft sandstone of Grant Bank is also partially exposed in the south bank of the River Tees at Boat Scar [1678 1637], where it consists of 20 ft of grey and white mainly massive and locally false-bedded sandstone with ripple-marks, resting with an irregular base on unfossiliferous shale. This sandstone further crops out widely to the north and east of Winston, and in Westholme Colliery Shaft* its equivalent is 80 ft thick, comprising two posts separated by only 5 ft of mudstone. The sandstone of Grant Bank is succeeded by 160 ft of variable strata equivalent to only 30 or 40 ft of measures farther west. They are exposed in the River Tees between Black Scar [1759 1626] and Chapel Gill [1871 1594], the section at the former being as follows:

	ft	in
Sandstone, buff to white fine- to medium-grained; sporadic carbonaceous bands; worm tracks and plant fragments .. 8 ft to	10	0
Sandstone, grey silty; massive at top, thin-bedded at base; locally current bedded; micaceous carbonaceous bands; worm markings	20	0
Mudstone and shale, micaceous carbonaceous; thin sandstone and siltstone ribs	40	0
Sandstone (of Grant Bank and Boat Scar)	—	—

The mudstone and shale are exposed in a bluff [1756 1637] opposite Black Scar where they are at least 17 ft thick and contain sporadic ironstone nodules. At a high level in the same sequence a 5-in band rich in crinoid columnals occurs [1782 1639] about 290 yd upstream. East of Black Scar the two sandstone members at the top of the sequence are separated by 4 to 8 ft of grey silty micaceous mudstone with siltstone and sandstone laminae. Some 200 yd east the mudstone [1827 1638] contains two thin calcareous bands which themselves unite and thicken eastward. At an exposure [1828 1639] the upper calcareous rib is 2 ft thick and contains *Lingula* cf. *squamiformis*, *Naticopsis sp.*, *Pseudozygopleura sp.*, *Straparollus sp.*, *Myalina* cf. *verneuilii* and *Parallelodon* cf. *squamifer* (loc. 15f). The upper member of the sandstone overlying this sequence crops out impersistently in the River Tees between the old railway bridge and Chapel Gill. Above Gainford Mill it shows a great eastwards thickening and possesses the features of a 'washout' or 'channel' sandstone.

At Gainford Mill the 'channel' sandstone cuts out the mudstone parting but it reappears downstream, and at the Chapel Gill–Tees confluence [1871 1594] the sequence is:

	ft	in
Sandstone, pale grey (weathering brown) massive fine-grained; bivalve fragments, plants and worm markings .. .. 4½ ft to	12	0
Sandstone, fine-grained compact thin-bedded; quartz pebbles ..	2	6
Mudstone, grey micaceous.. ..	2	6
Sandstone, fine-grained compact calcareous	2	3
Mudstone, grey silty; bivalve fragments	5	0
Sandstone, pale grey; massive at top, finely cross-bedded with abundant siltstone bands below; worm markings and plant fragments	18	0

The whole sequence is equivalent to the 30-ft sandstone at the top of Black Scar.

This sandstone and the succeeding measures are exposed in the Durham Bank of the river south of Snow Hall and at the base of the section between West Scar and Ashes Wood* where it is 43 ft 4 in thick, including mudstone bands and a 6-in shelly sandstone rib, and is succeeded by 40 ft of siltstone and 17 ft of sandstone. The upper sandstone forms the extensive bed assumed to underly the Upper Felltop Limestone. It is 50 to 60 ft thick at outcrop around Selaby Hall [1530 1830], but it is split by a shale parting up to 40 ft thick east of Gainford Great Wood [1650 1795]. Outside the valley of the Tees, the sandstone is rarely seen but the upper leaf is exposed at Balmer's Hill Quarry [1722 1804], where it consists of at least 13 ft of dull yellow or pale grey massive fine-grained iron-speckled sandstone.

In the country east of Gainford, the only records are from an old borehole* [2480 1634], south-east of Thornton Hall, where over 250 ft of Carboniferous beds were proved between the presumed horizon of the Upper Felltop Limestone and the Yoredale Coal, and in a further borehole* [about 254 188], 1½ miles to the north-north-east near Cold Sides. In both cases the correlation is tentative. D.A.C.M.

The Upper Felltop Limestone (Fig. 9) is exposed in West Rake Hush [9741 2972], about 600 yd NW of the district boundary, in the Alston (25) district, where the section is:

	ft	in
Shale, dark grey	1	0
UPPER FELLTOP LIMESTONE		
Limestone, brown hard limonitic sandy	2	6
Sandstone, grey fine- to medium-grained well-bedded; siliceous in part	3	0

In the Woodland Bore* the limestone comprises 25 ft 5 in of interbanded limestone and shale with two calcareous sandstone ribs, but it is not otherwise proved north of the Lunedale–Butterknowle fault complex. In Langley Beck [0736 2238], between the Butterknowle and Wigglesworth fault-systems 180 yd S of Dent Gate Farm the section is as follows:

	ft	in
Limestone, grey thin-bedded argillaceous fine-grained; solitary corals, brachiopods and crinoids	10	0
Limestone, grey rubbly argillaceous fine-grained	0	10
Limestone, grey thick-bedded fine-grained; sparsely crinoidal . .	1	6
Shale, grey calcareous	0	2
Limestone, grey thick-bedded argillaceous pyritous; brachiopods, trilobites and rare crinoids . .	3	0
Not exposed	1	6
Limestone, grey thin- and thick-bedded fine-grained; bands of calcareous shale	9	0

J.H.H.

South of the Lunedale–Staindrop Fault the limestone is exposed only in How Beck [9740 1907] on the eastern margin of the adjacent Brough-under-Stainmore (31) Sheet, where the succession is:

	ft	in
Shale, dark grey; fossiliferous . .	2	0
UPPER FELLTOP LIMESTONE		
Limestone, grey fine-grained sandy; small shell fragments . .	1	3
Coal	0	3

C.R.B.

Beds at the same horizon are completely exposed north-east of Barnard Castle in Old Mill Gill [0714 1971], but the limestone is apparently absent, though it may be represented by 1½ ft of calcareous sandstone. Farther east, however, in North Tees No. 1 Bore* near Little Newsham, 2 ft of pale grey, fine-grained, crinoidal limestone is thought to represent the Upper Felltop Limestone.

J.H.H.

The measures between the Upper Felltop and Grindstone (= Botany) limestones (Fig. 9) range from 67 ft in the Woodland Bore* to about 225 ft in and around How Beck. In the borehole the section is relatively simple, consisting of 55 ft 11 in of sandstone on 11 ft 1 in of shale. Most of the upper part of the sandstone represents the Grindstone Sill, the probable equivalent of the Botany Grit. The section in and around How Beck* (loc. 16b) includes several beds of shale and sandstone. One of the latter is the Fossil Sandstone, 7½ ft thick and some 73 ft above the Felltop Limestone, which contains crinoid ossicles and shell casts.

The thinner sequence as proved by the borehole is best seen at outcrop in Crag Gill*, below White House [0268 2362], where the Grindstone Sill is over 36 ft thick and is underlain by an unnamed 2-ft limestone containing *Spirifer* cf. *trigonalis*, estimated to be 26½ ft above the Upper Felltop Limestone.

South-east of Eggleston the measures are comparable to those of How Beck. In Raygill Beck [0110 2192] the horizon of the Upper Felltop Limestone is succeeded first by 20 ft of sandy shale with numerous sandstone ribs, then by about 12 ft of grey fine-grained, thin-bedded, carbonaceous sandstone. This sandstone is also exposed on the northern edge of Shipley Wood [0116 2092] where it is calcareous with worm tubes at the base. Contained shell fragments suggest that it is the equivalent of the Fossil Sandstone. The fossiliferous part of the sandstone is represented by 3½ ft of arenaceous limestone [0156 2031] about 170 yd downstream from Low Shipley. The Fossil Sandstone is locally succeeded by up to 85 ft of dark grey, silty, apparently unfossiliferous shales which are partially exposed in the three streams west of High House [0308 2029]. Above the shales are 40 ft of pale grey, fine-grained, thick-bedded sandstone which is being worked in Shipley Bank Quarries [0182 2075] and also at Stainton [1699 1884] and Dunn House [1136 1938] (see p. 63). This sandstone is further exposed in Howegill Quarries [0284 2244], where at least 20 ft are exposed, and also in Baxtongill Quarry [0257 2078].

North-east of Barnard Castle there are good exposures in Hedrick Gill*, Forthburn Beck* and Old Mill Gill* which together range upwards from the horizon of the Upper Felltop Limestone to the base of the Botany Grit. In these sections the Fossil Sandstone is up to 20⅓ ft thick, lying about 32 to 33½ ft above the horizon of the Upper Felltop Limestone. It is succeeded in Hedrick Gill by almost 83 ft of shale and siltstone which is particularly fossiliferous in the basal 47½ ft. Above these shales are 45 ft of sandstone, which are separated by 11 ft of unexposed measures from the lower 20 ft of the Botany Grit. Small exposures in the Fossil Sandstone also occur in Old Mill Gill east of a locality [0817 1974], about 150 yd W of Streatlam Castle and in Nelly's Gill, north of the castle. J.H.H.

Farther east it is indifferently exposed in Sudburn Beck between Snotterton Hall and Sudburn Bridge, though in a quarry [1094 1967] about 120 yd SSE of Snotterton Hall, the following section is seen:

	ft	in
Sandstone, fine-grained rubbly 4 ft to	5	0
Shale, pale grey and brown; thin sandstone ribs	1	6
Sandstone, white fine-grained thick-bedded .. .. .. 3 ft to	4	0

The North Tees No. 1 Bore* near Little Newsham proved 79 ft 6 in of measures above the Upper Felltop Limestone, consisting predominantly of mudstone and siltstone containing crinoid debris, brachiopods including *Crania sp.*, bivalves, and ostracods at various horizons (loc. 16a). D.A.C.M.

The upper part of the thick mudstone sequence which succeeds the Fossil Sandstone and the overlying sandstone, are best exposed in working quarries. In the north-west face [0699 1884] of Stainton Quarries the sandstone, of which 15 ft is exposed, contains a 6-in shaly coal 6 ft from the base of the exposure. In the south-east corner [0712 1877] of the quarry, abundant 'cauda-galli' structures can be seen. At Dunn House Quarry [1136 1938] (Plate IVB) the sandstone, at least 40 ft thick, is pale grey, fine-grained, predominantly massive or thick-bedded, with sporadic lenses of highly micaceous carbonaceous shale and siltstone; it is said to contain 95 per cent free silica together with mica, feldspar and carbonaceous fragments. To the south-east in Cleatlam Old Quarry [1171 1832] 2 to 3 ft of shaly micaceous sandstone overlie 14 ft of grey and brown fine-grained thick-bedded to massive sandstone ascribed to this horizon.

J.H.H., D.A.C.M.

Above the Stainton–Dunn House sandstone are some 10 to 15 ft of shales which are not now exposed, though 12 ft were proved in a well at Pallet Crag House [0181 2257] and 10 ft of shale with a 6-in coal were recorded during the original geological survey in railway cuttings [0750 1891 and 0556 1974] near Mount Pleasant and Bluestone Grange.

The sandstone which succeeds this shale is probably the lateral equivalent of the Botany Grit of the How Beck section*. It is about 30 ft thick and crops out extensively east of the River Tees and south of the Lunedale–Staindrop Fault. About a mile south-east of Eggleston exposures in Roger Moor Quarry [0194 2247] and Pallet Crag [0273 2261] show the sandstone to be pale grey, coarse-grained, thin-bedded to massive and feldspathic. North of Barnard Castle small exposures can be seen in a railway cutting [0554 1968] near Bluestone Grange, while in Lingberry Quarry [0859 1057] it is at least 15 ft thick lying about 14 ft below the base of the Botany Limestone. Farther east its probable equivalent at Langton Grange Quarry [1531 1995] comprises 9 ft of yellow and grey medium- to coarse-grained current-bedded sandstone.

The 14 ft of measures separating the sandstone from the Botany Limestone are seen in this part of the district only in a beck [1044 2021], near Scaife House, where a thin unnamed limestone with associated shale is exposed at the base of the sequence. These measures are taken to be the equivalent of the 26 ft of mudstone which underlie the Botany Limestone in the type area and which are exposed in How Gill [9551 2045] in the Brough-under-Stainmore (Sheet 31) district.

In the Woodland Bore* the **Grindstone Limestone** (Fig. 9) consists of 14 ft 1 in of interbedded impure shelly limestone, shale and mudstone with a 7-in calcareous sandstone rib. At a small quarry [0216 2356], about 500 yd W of White House the following sequence is seen at the position of the limestone (loc. 17a):

	ft	in
Mudstone, grey, silty; fauna of brachiopods and bivalves including *Orbiculoidea* cf. *nitida*, *Euphemites sp.*, *Myalina mitchelli*, *Nuculopsis gibbosa* and *Polidevcia attenuata*	6	0
Sandstone, yellowish brown and pale grey thick-bedded fine-grained quartzitic decalcified; brachiopod casts and bivalve fragments	5	0
Grindstone Sill		
Sandstone, grey thin-bedded fine-grained rooty; carbonaceous micaceous partings (base not seen)	4	0

The 5-ft sandstone has the characteristic appearance of a weathered limestone and is taken to be the siliceous lateral equivalent of the Grindstone Limestone. Similarly in neighbouring Crag Gill* a ganister is taken to indicate the position of the limestone. The equivalent Botany Limestone is exposed in Hedrick's Quarry [0518 2144] (loc. 17c), about 2½ miles north of Barnard Castle, where the sequence is:

	ft	in
Boulder clay	5	6
Limestone, grey to brown thin-bedded rubbly fine-grained; *Caninia sp.*, *Dibunophyllum bipartitum bipartitum*, *D. bipartitum konincki*, *Koninckophyllum sp.*, *Lithostrotion portlocki*, *Cleiothyridina* aff. *fimbriata*, *Quasiavonia* aff. *aculeata*, cf. *Sinuatella sinuata*, *Spirifer sp.*, *Weberides sp.* [pygidium]	3	6
Mudstone, grey calcareous; *Composita*?	0	4
Limestone, grey fine-grained argillaceous; fauna including *Fasciculophyllum sp.*, *Composita sp.*, trilobite free cheek indet... ..	0	8
Mudstone, grey calcareous ..	0	5
Limestone, grey to brown fine-grained; brachiopods	0	10
Limestone, dark grey argillaceous bioclastic	0	3
Mudstone, grey calcareous ..	0	5
Limestone, grey thick bedded fine-grained; crinoidal	1	6
Mudstone, grey calcareous; fauna including *Fenestella sp.*, *Pugilis sp.* and *Rugosochonetes sp.* ..	1	6
Limestone, grey thick-bedded to massive fine-grained siliceous; fauna of crinoids, corals and brachiopods including *Dibunophyllum bipartitum bipartitum*, *Echinoconchus* cf. *punctatus*, *Phricodothyris sp.*, *Plicochonetes sp.*, productoid *nov.* (with diaphragm), *Spirifer* cf. *bisulcatus*	4	4

Almost 2 miles farther east the following contrasting section with the overlying beds, is exposed in Bolton Close Quarry [0822 2090] (loc. 17b):

	ft	in
Sandstone, brown fine-grained decalcified argillaceous; brachiopods, crinoids and 3-in bryozoa band	2	0+
BOTANY LIMESTONE		
Limestone, grey massive fine-grained; honeycombed appearance; crinoids, *Fenestella* and brachiopods	7	2
Limestone, dark grey fine-grained argillaceous; irregular thin bedding and micaceous shale partings; brachiopods including *Lingula*	4	4

Strata between the Grindstone (= Botany) Limestone and the Whitehouse Limestone (Fig. 9) are continuously exposed in Crag Gill* (loc. 18a) and were proved in the Woodland Bore* being respectively about 73 and 127 ft thick and comprising several cycles, one of which includes a thin limestone. In the borehole a specimen of *Homoceras* cf. *henkei* (R_{1a}) was collected about 83 ft above the Grindstone Limestone.

South of the Lunedale–Staindrop Fault only the lowest few feet of this sequence are present. At Bolton Close Quarry [0822 2090] they include 2 ft of thin-bedded fine-grained decalcified sandstone, which in addition to crinoids and brachiopods, carries a 3- to 6-in band rich in bryozoa at the base.

The Whitehouse Limestone (Fig. 10) in the type area of Crag Gill* consists of 11 in of impure limestone containing crinoid debris and fragments of *Productus sp.* (loc. 19a), while in the Woodland Bore* it is 1 ft 1 in thick. The Whitehouse Limestone may further be represented by 5 ft of crinoidal limestone occurring in Stobgreen Sike*, though faulting in the vicinity of the section may make correlation with a stratigraphically lower limestone more feasible. J.H.H.

To the south-east a further possible equivalent is 1½ ft of grey, fine-grained limestone recorded below the First Grit in Hilton No. 114 Bore* [1654 2082] west-north-west of Ingleton. D.A.C.M.

Strata between the Whitehouse Limestone and the Sharnberry Shell-Beds (Fig. 10) consist of a few feet of shale and shaly sandstone and an overlying sandstone containing coarse, and in places, pebbly layers, which is locally termed the Basement Grit and which comprises the lower leaf of the 'First Grit'. The most complete section in these measures is exposed in Crag Gill* where the Whitehouse Limestone is succeeded by 7 ft of mudstone, 9 ft of thin-bedded silty sandstone and the lower 20 ft of the 'First Grit'. A particularly rich fauna (loc. 20a) was obtained from the mudstone about 10 in above the Whitehouse Limestone and included an indeterminate zaphrentoid coral, *Productus carbonarius*, *Catastroboceras* cf. *neilsoni* and *Reticuloceras stubblefieldi*, the latter indicative of an R_{1b} horizon. The horizon is particularly high for the occurrence of a coral. A comparable section, but in which no mudstone is present, is exposed in a quarry [9908 2608] at Knotts Plantation, about 1½ miles NNW of Eggleston Parish Church, where the sequence is:

	ft	in
Sandstone, yellow to grey thin-bedded medium- to coarse-grained	19	10
Sandstone, grey massive coarse-grained	3	2
Sandstone, grey thin-bedded fine-grained; base undulating ..	5	8

Other exposures of Basement Grit occur in Cloudlam Beck [0092 2871] on the northern margin of the district between Stobgreen Sike* [0055 2407] and Crag Gill*, in a former railway cutting [0814 2318] 100 yd N of Ripton House, and in Langley Beck between the Old Lodge [0902 2218] and Raby Moor House [0944 2206]. J.H.H.

A further exposure in Basement Grit is seen in a quarry [1141 2167] at Raby Plantation, where 3 ft of red, thin-bedded, shaly sandstone overlie 4 ft of red, coarse-grained, massive, soft sandstone. D.A.C.M.

In the Eggleston No. 5 Bore* [0073 2518] on Eggleston Common the grit was proved to be 66 ft thick while north of Ingleton it ranges in thickness from 7 ft at the Lutterington Estate Bore* [1875 2447] Bildershaw to 38½ ft at Hilton No. 3 Bore* [1679 2203] near Hilton. Hilton No. 114 Bore* (loc. 20b) records that locally the top of the Basement Grit is separated from the Sharnberry Shell-Beds by about 11 ft of siltstone.

The Sharnberry Shell-Beds (Fig. 10) take their name from a beck [0158 3087] immediately north of the district where they consist of 3 ft 9 in of bluish grey shale on 6 in of 'bullion'; the fauna from these beds includes *Orbiculoidea*, *Euphemites* and *Reticuloceras*. Within the district the shell-beds have been proved only in Hilton No. 114 Bore*, where they comprise 5 ft of shale with *Retispira* on a 7-in ironstone with shell debris (loc. 21a). Elsewhere, though the beds are not exposed, they can, together with the shales in which they occur, be identified lithologically in several old boreholes. North of Eggleston 7 ft of 'dark metal' were recorded in Eggleston No. 5 Bore* and 14¼ ft in Eggleston No. 4 Bore* [0202 2518] on Woodland Fell.

Near Ingleton these measures range from 3 ft 2 in at Hilton No. 1 Bore* [1730 2342] near Bolton Garths to 25 ft at the Lutterington Estate Bore*.

The measures between the Sharnberry Shell-Beds and the Woodland Shell-Beds (Fig. 10) consist mainly of sandstone, representing the upper leaf of the 'First Grit'. In Sharnberry Beck the shell-beds are succeeded immediately by the sandstone, whereas in Hilton No. 114 Bore* they are separated by 5 ft of dark grey siltstone with plant debris. In the Woodland Bore* the Sharnberry Shell-Beds and associated argillaceous beds have been cut out by the overlying sandstone, which joins with the one beneath to form a united First Grit, 108⅓ ft thick, broken only by a few shale bands. J.H.H., D.A.C.M.

In the north-western part of the district the upper leaf of the First Grit is a poorly exposed grey, coarse-grained, pebbly, thin-bedded sandstone, which ranges in thickness from 66 to 72 ft along the outcrop between Stobgreen Plantation and the Wigglesworth Fault [0350 2361]. In Eggleston No. 5 Bore* it consisted of a lower 8-ft and upper 5-ft post separated by 12 ft of 'grey metal', while to the east in Gordon Gill 'K' Bore*

[1380 2687] at Snape Foot on the upthrow side of the Butterknowle Fault, the lowest 52 ft of mainly sandstone measures are ascribed to this horizon.

South and east of the Wigglesworth Fault the upper leaf of the First Grit lies close above the lower leaf. It is exposed at Jagger Hill [0402 2343] and Pinstone Crag [0710 2364], where it is 35 ft thick and is grey, coarse-grained, pebbly, massive and false-bedded. J.H.H.

The Raby Castle Water Bore* [1288 2218] proved at least 57 ft of grey and red sandstone at this horizon. Near Ingleton the sandstone ranges in thickness from 35¾ ft in Hilton No. 1 Bore* to 11¼ ft in Hilton No. 114 Bore* and is pale grey, fine-grained, flaggy and micaceous, with shaly partings at the base. D.A.C.M.

The Woodland Shell-Beds (Fig. 10) comprise 11 to 28 ft of strata lying between the First and Second Grits. They consist of up to three beds of shale or mudstone with marine shells separated by thin beds of sandstone. In the Woodland Bore* they contain three fossiliferous beds, whereas in the Raby Castle* (loc. 22b) Hilton No. 114* (loc. 22a) and Gordon Gill 'K'* bores, only two such beds were proved or can be inferred from the logs. Other boreholes giving sections through these measures, and listed in Appendix 2, are Eggleston Nos. 3 and 5, Hilton No. 3 and Lutterington Estate bores.

J.H.H., D.A.C.M.

Strata between the Woodland Shell-Beds and the Spurlswood Shell-Beds (Fig. 10) consist mainly of the sandstone forming the lower leaf of the 'Second Grit' together with the overlying few feet of silty shale. The sandstone is at least 60 ft thick in a quarry [0608 2891] on the north bank of Spurlswood Beck and is grey, coarse-grained, flaggy and false-bedded. In the Woodland Bore* these measures are over 64 ft thick consisting mainly of sandstone and in Gordon Gill 'K' Bore* the sandstone assumed to be equivalent is 20 ft 3 in thick.

North of Eggleston the sandstone is exposed at the head of Stobgreen Sike* [0109 2430], where it is grey, medium- to coarse-grained, thin-bedded and false-bedded. In the Eggleston Nos. 3, 4 and 5 bores* the sandstone ranges in thickness from 43⅓ ft to 46½ ft; it includes 4 in of metal and coal 15 ft from the base of the sandstone in No. 5 Bore. At a small quarry [0417 2400], about 2½ miles ENE of Eggleston, the sandstone contains baryte; baryte and calcite are also present on joints and in thin veins in the sandstone in Penny Hill Plantation [0781 2371]. There are other exposures at a quarry [1004 2274] near Henderson House and West Broombank Quarry [1088 2246], in Raby Park. J.H.H.

Around Ingleton the sandstone is exposed only in Holywell House Quarry [1747 2152], where 15 ft of grey, medium- to coarse-grained, false-bedded, micaceous sandstone are seen. The variable nature of the bed in this area is recorded in the Lutterington Estate* and Hilton No. 1* bores. East of Hilton the lower leaf of the Second Grit is split into two posts. Part of the lower post is represented in a quarry [2213 2185], about 350 yd W of Houghton-le-Side, by 15 ft of medium- to coarse-grained, massive, pebbly, feldspathic sandstone, The upper post was formerly seen in an old quarry [2221 2195] about 150 yd to the north-east, and described by Kirkby and Duff (1872, p. 186) as a yellow, fine-grained, thick-bedded sandstone. On the south side of this quarry and passing beneath the sandstone, Kirkby and Duff also saw a dark grey limestone overlying a "limestone shale" which together yielded: "*Productus semireticulatus*, *var. martini*, Sow., *Spirifera lineata*, Martin, *Athyris ambigua* Sow., *Rhynchonella pleurodon* (?) Phillips. Remains of crinoids"; and they recorded the same marine band farther east of the road to Houghton-le-Side. A marine band at this locality would be below the Second Grit and thus be part of the Woodland Shell-Beds. D.A.C.M.

The measures between the top of the lower leaf of the 'Second Grit' and the Spurlswood Shell-Beds (Fig. 10) are exposed [0226 2686] in Spurlswood Beck* where they comprise about 28 ft of shale with sandstone ribs. In the Hilton area they are mainly argillaceous and range in thickness from 1½ to 8½ ft. J.H.H., D.A.C.M.

The Spurlswood Shell-Beds (Fig. 10) take their name from Spurlswood Beck* [0226 2686] where they consist of 17½ ft of mudstone including a 2½ ft siltstone rib. The basal 6 in of mudstone contains a fauna of brachiopods, bivalves, plants and fish which includes *Orbiculoidea* cf. *nitida*, *Productus carbonarius*, *Rugosochonetes* cf. *hindi*, *Actinopteria* *regularis*, *Serpuloides sp.* and *Paraconularia sp.*, while the upper 6 in of mudstone contain brachiopods, gastropods and plants including *Rugosochonetes* cf. *hindi*, *Retispira sp.* and *Serpuloides sp.* (loc. 23b). About 2¼ miles ENE in Greenless Beck [0576 2775] the beds are cut out by the overlying sandstone and only 15 ft of argillaceous strata remain between the top and bottom leaves of the Second Grit; neither are they present in the Woodland Bore*, where the measures between the two leaves of the Second Grit are represented by only 8 in of shale and mudstone. J.H.H.

South-east of Hilton about 6 ft of shale in this position were proved in Hilton Hall* [1698 2143] and Hilton No. 60* [1754 2175] bores near Hilton and in both cases the fossils included *Lingula sp.* Near Ingleton at the Hilton No. 114 Bore* (loc. 23a) the same shale yielded *Orbiculoidea nitida* and productoids and at Hilton No. 115 Bore* [1721 2072] Ingleton (loc. 23c), *Serpuloides sp.*, *Lingula sp.*, *Productus sp.*, *Donaldina sp.* and crinoids were collected. Farther east in the area north of Houghton-le-Side the shell beds are not present, the upper and lower leaves of the Second Grit uniting to form a single massive sandstone. D.A.C.M.

Strata between the Spurlswood Shell-Beds and the base of the Coal Measures (Quarterburn Marine Band) (Fig. 10) are exposed between a bluff [0226 2686] in the Spurlswood Beck* and the east bank [0170 2676] of Quarter Burn*, where they consist of almost 104 ft of strata including the upper leaf of the Second Grit 63 ft thick. This sandstone locally cuts out the underlying strata and its thickness increases east-north-eastwards to about 95 ft at Greenless Beck [0583 2757]. At the northern margin of the district, on the western side of Spurlswood Beck, the sandstone splits into two posts estimated to be 55 and 45 ft thick, separated by 5 ft of mainly argillaceous strata; the thinner lower post is exposed [0448 2689] on the southern edge of Hamsterley Forest and in the lower reaches of Rowley Gill [0495 2683]. About 2 miles to the east in the Woodland Bore* a single sandstone 49 ft thick is recorded, and still farther east, in Gordon Gill 'K' Bore*, the thickness is 25 ft.

In Knott Hole Quarry [9969 2629], about 1½ miles N of Eggleston Parish Church, part of the sequence exposed is:

	ft	in
Sandstone	17	9
Mudstone-seatearth, grey and brown porcellanous siliceous ferruginous rooty	2	10
Sandstone-seatearth, grey fine-grained fissile carbonaceous rooty	3	6
Sandstone, grey and white medium- to coarse-grained pebbly massive; sporadic silty bands in top 6 ft ..	22	9

The seatearths of this section are equated with those below the Quarterburn Marine Band of Quarter Burn*. The sandstone at the top of this section is therefore assumed to be in the Lower Coal Measures and to cut out the marine band in that area. J.H.H.

Near Ingleton continuous sections between the Spurlswood Shell-Beds and the Quarterburn Marine Band were proved in the Hilton No. 114* and Hilton Hall* bores. The Hilton Hall Bore recorded *Lingula* bands between the 'Second Grit' and the Quarterburn Marine Band, whereas farther east in the Shackleton Beacon Bore* [2281 2336], 1½ miles WNW of Heighington productoids were recorded from an analogous horizon. The measures overlying the 'Second Grit' are exposed in Brownside Quarry [2035 2283], near Bolam, where 4 ft of dark grey shale, containing ironstone nodules and sporadic plant fragments, overlie 6 ft of fine-grained, carbonaceous, micaceous, baked, shattered sandstone.

Although a number of boreholes have penetrated Namurian strata underneath the Permian in the eastern part of the district, there is insufficient information for firm correlation. Four of these boreholes, notably that near Thornton Hall, Bakelite Water Bore at Newton Aycliffe, Cold Sides, and Broken Scar Pumping Station Bore, Darlington, are quoted in Appendix 2. D.A.C.M.

Namurian Palaeontology

The fauna, which comprises many forms similar to those found in the Viséan rocks below, are those of a well aerated shelf sea with abundance especially of brachiopods and molluscs. There is no repeated change of faunal phase, and fossil assemblages found in the limestones, mudstones and more rarely in sandstones, are each distinctive of the lithological type. Notable features in the district are:

1. Corals are abundant only in the Great, Bottom Crag and the Botany/Grindstone limestones; at other levels they are rare.

2. At many horizons there are nuculoid phases with *Nuculopsis*, *Palaeoneilo*, and *Paleyoldia*, often associated with bellerophontoid gastropods especially *Euphemites* and *Retispira* [formerly called *Bucanopsis*].

3. As Wilson (1967) noted in Scotland there are few forms with restricted stratigraphical range. Amongst the brachiopods the latissimoid gigantoproductids, here referred to *Semiplanus*, occur above the Great Limestone only in the Rookhope Shell-Beds Limestone; and *Antiquatonia* cf. *costata* only in the Bottom Crag Limestone.

In the following lists of the fossils collected during the resurvey, the numbers and letters following the fossil names indicate respectively the horizon (according to the list below), and the locality (as in the list and below Fig. 11). The horizons are as follows:

1. Great Limestone
2. Between Great Limestone and Bottom Little Limestone
3. Bottom Little Limestone
4. Between Bottom Little and Top Little Limestones
5. Top Little Limestone
6. Between Top Little and Bottom Crag Limestones
7. Bottom Crag Limestone
8. Between Bottom Crag and Top Crag Limestones
9. Top Crag Limestone
10. Between Top Crag and Knucton Shell-Beds Limestone
11. Knucton Shell-Beds Limestone
12. Between Knucton Shell-Beds Limestone and Rookhope Shell-Beds Limestone
13. Rookhope Shell-Beds Limestone and beds immediately above
14. Lower Felltop Limestone
15. Between Lower and Upper Felltop Limestone
16. Between Upper Felltop and Botany/Grindstone Limestone
17. Botany/Grindstone Limestone
18. Between Botany/Grindstone and Whitehouse Limestone
19. Whitehouse Limestone
20. Between Whitehouse Limestone and Sharnberry Shell-Beds
21. Sharnberry Shell-Beds
22. Woodland Shell-Beds
23. Spurlswood Shell-Beds

Localities

1a. Old quarry [0229 1339] at Lamb Hill
1b. Quarry north of West Layton [1433 1047]
1c. *R. Tees [0663 1496] near Abbey Bridge
2a. *R. Tees [0552 1554] near Lendings Mill
2b. Stobgreen Sike [0019 2378]
3a. *R. Tees near Demesne Mill [0535 1582]
3b. Harthorn Quarry, Aldbrough [2060 1100]
3c. Both sides of Melsonby Lane, ¾ mile S. of Aldbrough [2034 1056]
4a. R. Tees near Desmesne Mill [see 3a, 5a]
4b. Harthorn Quarry, Aldbrough [2060 1100]
5a. R. Tees [0512 1598] near Desmesne Mill
5b. Harthorn Quarry, Aldbrough [2060 1100]
5c. R. Tees [0489 1593] at end of Thorngate, Barnard Castle
5d. R. Tees [0479 1605] south of road bridge, Barnard Castle
6a. Whorlton Beck [1067 1498 to 1088 1457]
6b. R. Tees [0480 1615] south of road bridge, Barnard Castle
7a. *Whorlton Beck [1068 1501]
7b. Crossbury Bank Wood [2094 1056]

Localities—continued

7c. Sledwick Hall quarry, near East Shaws [0914 1525]
7d. Deepdale Beck [0071 1604]
7e. Cliff [0481 1641] north of road bridge, Barnard Castle
8a. Whorlton Beck [1068 1510]
9a. Whorlton Beck [1068 1515]
9b. Cliff [0483 1642] north of road bridge, Barnard Castle
9c. Deepdale Beck [9920 1546]
9d. Deepdale Beck [9878 1563]
9e. Railway cutting [0706 2200] south-west of Dent Gate Farm
10a. Deep gully entering R. Tees at Ovington [1328 1500]
10b. Deepdale Beck [0073 1607]
10c. Deepdale Beck [0075 1610]
11a. *R. Tees [1408 1591] above Winston Bridge
12a. *R. Tees [1425 1617] above Winston Bridge
13a. *R. Tees [1429 1619] above Winston Bridge
13b. R. Tees [0033 2187] near West Barnley
14a. *Percy Beck [0472 1701], Barnard Castle
15a. *Percy Beck [0474 1706], Barnard Castle
15b. *North Tees No. 1 Bore [1212 1762]
15c. R. Tees [1838 1628] near Snow Hall
15d. Grant Bank [1618 1736]
15e. Selaby Basses Wood [1582 1772] near Gainford
15f. R. Tees [1828 1639] near Greystone Hall
15g. N. Tees Colliery Drift [1226 1657]
16a. *North Tees No. 1 Bore [1212 1762]
16b. *How Beck [9757 1915]
17a. Quarry [0216 2356] 500 yd W of White House
17b. Bolton Close Quarry [0822 2090]
17c. Hedrick Grange Quarry [0518 2144]
18a. *Crag Gill [0268 2362] near White House
19a. *Crag Gill [0268 2362] near White House
20a. *Crag Gill [0268 2362] near White House
20b. *Hilton No. 114 Bore [1654 2082]
21a. *Hilton No. 114 Bore [1654 2082]
22a. *Hilton No. 114 Bore [1654 2082]
22b. *Raby Castle Water Bore [1288 2218]
23a. *Hilton No. 114 Bore [1654 2082]
23b. Spurlswood Beck [0226 2686]
23c. *Hilton No. 115 Bore [1721 2072]

Fossils

Calcifolium bruntonense 1*c*
Hyalostelia parallela 4*a*
H. smithii 4*b*, 5*b*, 11*a*
Hyalostelia sp. 9*c*, 12*a*
Sponge indet. 13*b*
?Sponge spicules 5*a*
Paraconularia sp. 17*c*, 23*b*
Aulophyllum fungites 1*b*
Caninia juddi 7*d*
Caninia sp. 1*b*, 7*e*, 11*a*, 17*c*
Chaetetes depressus 1*b*, 7*c*, 7*d*
Dibunophyllum bipartitum bipartitum 1*a*, 1*b*, 17*c*
D. bipartitum craigianum 17*c*
D. bipartitum konincki 17*c*
Fasciculophyllum sp. 17*c*
Koninckophyllum cf. *echinatum* 7*e*
K. magnificum 7*c*
Koninckophyllum sp. 1*b*, 17*c*
Lithostrotion pauciradiale 1*b*
L. portlocki 17*c*
Rhopalolasma? 17*c*
Zaphrentoid indet. 1*a*, 1*b*, 20*a*
Fenestella frutex 2*a*, 4*b*, 6*a*, 12*a*
F. cf. *oblongata* 2*a*
Fenestella sp. 1*a*, 2*b*, 4*b*, 6*a*, 9*b*, 9*c*, 10*a*. 10*b*, 11*a*, 13*a*, 17*b*, 17*c*, 20*a*
Fistulipora sp. 2*a*, 4*b*
Penniretepora sp. 2*a*
Rhombopora sp. 1*a*, 18*a*, 20*a*
Trepostomatous bryozoa 2*b*, 6*a*, 17*c*
Bryozoa indet. 8*a*
Chondrites? 15*b*
Cornulitella sp. 18*a*
Cf. *Planolites montanus* 15*b*
Serpuloides sp. 22*a*, 23*b*, 23*c*
Actinoconchus? 1*a*
Alitaria panderi 3*b*, 3*c*, 5*b*, 7*b*
Antiquatonia cf. *insculpta* 8*a*
A. muricata 1*a*
A. cf. *costata* 7*c*
Antiquatonia sp. 3*b*, 3*c*, 5*d*, 6*a*, 7*e*, 9*a*, 9*e*, 11*a*, 15*b*, 17*c*
Avonia youngiana 1*a*, 3*c*, 5*a*, 5*d*
Buxtonia sp. 2*a*, 3*c*, 5*d*, 6*a*, 6*b*, 9*e*, 13*b*, 17*b*, 17*c*
'*Camarotoechia*' *sp.* 2*a*, 2*b*, 4*a*, 5*a*, 5*d*, 6*b*, 7*b*, 9*e*, 13*b*

Fossils—continued

Cleiothyridina fimbriata 1*a*, 5*a*, 5*d*, 6*a*, 6*b*, 11*a*, 17*c*
C. aff. *glabristria* 5*d*
Avonia sp. 3*a*, 4*a*, 5*a*
Brachythyris sp. 1*a*, 5*d*, 6*a*, 9*a*
Cleiothyridina sp. 4*b*, 6*a*, 6*b*, 7*a*, 9*b*, 10*b*
Composita sp. 7*e*, 9*e*, 17*b*, 17*c*
Crania sp. 16*a*
Crurithyris sp. 3*c*, 4*a*, 5*a*, 9*a*, 9*d*, 10*b*, 16*b*, 20*a*, 22*b*
Dictyoclostus sp. 2*b*, 6*a*, 18*b*, 20*b*
Dielasma sp. 4*a*, 5*a*, 6*a*, 7*b*
Echinoconchus cf. *elegans* 1*b*, 2*a*, 4*a*, 6*a*, 6*b*, 10*b*
E. punctatus 6*a*, 11*a*, 17*c*
Echinoconchus sp. 3*c*
Eomarginifera lobata 1*a*, 2*a*
E. cf. *setosa* 2*a*, 5*a*, 5*d*, 7*c*, 13*b*
Eomarginifera sp. 1*a*, 2*b*, 3*a*, 4*a*, 6*a*, 6*b*, 7*d*, 9*a*, 13*a*, 15*a*, 18*a*, 22*b*
Etheridgina complectans 1*b*
Lingula mytilloides 9*e*, 17*c*, 20*b*, 22*a*, 23*b*
L. cf. *squamiformis* 6*a*, 15*f*
Lingula sp. 2*b*, 3*a*, 6*a*, 13*b*, 14*a*, 18*a*, 22*a*, 23*c*
Linoprotonia sp. 8*a*, 9*a*
Martinia sp. 1*a*, 1*b*, 2*b*, 3*b*, 3*c*, 4*a*, 6*b*, 7*d*, 9*a*, 13*a*, 13*b*, 16*b*
Martinothyris cf. *lineata* 1*a*, 5*a*, 7*b*
Orbiculoidea nitida 3*a*, 5*d*, 9*a*, 9*e*, 15*a*, 16*a*, 17*a*, 18*a*, 22*a*, 22*b*, 23*a*, 23*b*
Orthotetoids indet. 2*b*, 4*a*, 5*a*, 6*a*, 7*b*, 7*c*, 7*e*, 13*b*, 14*a*, 18*a*, 20*a*, 20*b*, 22*a*, 23*b*
Phricodothyris sp. 1*b*, 3*b*, 3*c*, 4*a*, 5*c*, 17*b*, 17*c*
Pleuropugnoides cf. *pleurodon* 6*a*
Pleuropugnoides sp. 3*c*, 6*a*, 12*a*
Plicochonetes aff. *interstriatus* 1*b*
Plicochonetes sp. 1*a*, 13*a*, 17*d*
Productoid *nov.* [with diaphragm] 17*b*, 17*c*
Productus carbonarius 15*b*, 18*a*, 20*a*, 22*a*, 23*b*, 23*c*
Productus sp. 2*a*, 19*a*, 22*b*
Pugilis sp. 17*c*
Pugilis? 4*a*, 5*a*
Pugnoides triplex 1*a*
Quasiavonia cf. *aculeata* 3*c*, 7*e*, 17*c*
Quasiavonia sp. 7*d*, 9*b*
Rhipidomella michelini 3*b*
Rugosochonetes sp. 2*b*, 4*a*, 6*a*, 7*c*, 7*d*, 9*c*, 9*d*, 10*a*, 11*a*, 12*a*, 15*a*, 15*b*, 15*e*, 17*c*, 20*a*, 20*b*, 22*a*, 23*b*
R. sp. nov. [tumid] 13*b*, 18*a*
cf. *Schellwienella rotundata* 9*e*
Schellwienella sp. 15*b*
Schizophoria resupinata 6*a*
Schizophoria sp. 1*a*, 1*b*, 3*a*, 9*e*, 18*a*
Schuchertella sp. 1*a*, 3*b*, 9*a*
Semiplanus cf. *latissimus* 1*a*, 1*b*, 13*a*
Semiplanus sp. [latissimoid, broad folds on trail] 1*b*
cf. *Sinuatella sinuata* 17*c*
Smooth spiriferoids indet. 3*a*, 12*a*, 14*a*, 15*b*, 16*a*, 17*c*, 23*b*
Spirifer cf. *bisulcatus* 6*a*, 7*d*, 10*b*, 11*a*, 17*c*
S. cf. *trigonalis* 1*a*, 2*b*, 3*a*, 3*b*, 4*a*, 5*a*, 5*c*, 6*a*, 7*a*, 7*b*, 7*c*, 9*a*, 9*e*, 13*b*, 15*a*
Spirifer sp. 1*a*, 1*b*, 2*b*, 3*c*, 4*a*, 5*c*, 5*d*, 6*a*, 6*b*, 7*e*, 9*b*, 13*a*, 13*b*, 14*a*, 15*b*, 17*c*, 20*a*
Spiriferellina sp. 3*c*
Temnocheilus cf. *coronatus* 6*b*
Tornquistia cf. *polita* 3*a*, 7*b*, 8*a*, 13*a*, 15*b*
Aclisina cf. *elongata* 2*a*
Bellerophontoid indet. 2*b*
Donaldina sp. 18*b*, 23*c*
Euphemites sp. 2*b*, 9*d*, 17*a*, 20*a*, 22*a*
Glabrocingulum sp. 6*b*
Ianthinopsis sp. 2*a*
Naticopsis sp. 1*b*, 15*f*
Platyceras neritoides 9*e*
Platyceras sp. 18*a*
Pseudozygopleura cf. *rugifera* 2*a*
Pseudozygopleura sp. 15*f*
Retispira cf. *striata* 9*e*
Retispira sp. 2*a*, 18*a*, 21*a*, 22*a*, 23*b*
Shansiella globosa 22*b*
Straparollus cf. *carbonarius* 9*a*
Straparollus sp. 10*a*, 15*a*, 15*f*
Turreted gastropods indet. 1*a*, 5*b*, 20*a*
Coleolus cf. *namurcensis* 18*a*, 20*a*
Coleolus sp. 2*b*, 18*a*
Actinopteria regularis 15*d*, 23*b*
Aviculopecten cf. *clathratus* 2*a*
A. aff. *interstitialis* 11*a*
A. cf. *plicatus* 2*a*, 7*d*, 10*b*
Aviculopecten sp. 1*a*, 6*a*, 6*b*, 9*e*, 12*a*, 13*b*, 15*a*, 20*a*, 22*a*
Aviculopinna mutica 9*e*
Cypricardella rectangularis 6*b*
Cypricardella sp. 8*a*, 11*a*
cf. *Edmondia expansa* 6*b*
E. cf. *gigantea* 5*d*
E. laminata 2*a*, 4*a*, 6*a*
Edmondia sp. 7*e*, 9*b*, 9*e*, 14*a*, 23*b*
Euchondria sp. nov. 20*a*
Euchondria? 16*b*
Limipecten dissimilis 3*c*, 4*a*, 6*a*, 7*c*
Myalina mitchelli 17*a*
M. cf. *verneuilii* 15*f*

Fossils—continued

Myalina sp. 15*c*, 15*g*
Nuculoids indet. 12*a*
Nuculopsis gibbosa 9*e*, 13*b*, 17*a*
Palaeoneilo cf. *laevirostrum* 2*a*
P. cf. *mansoni* 9*d*
Palaeoneilo sp. 2*b*, 6*a*, 15*a*, 17*a*, 20*a*, 23*b*
Paleyoldia macgregori 2*a*, 2*b*, 20*a*
Parallelodon cf. *obtusus* 22*a*
P. cf. *semicostatus* 6*a*
P. cf. *squamifer* 6*b*, 15*f*
Parallelodon sp. 11*a*
Pectinoid fragments indet. 4*b*, 16*a*, 17*a*, 23*b*
Pernopecten sowerbii 6*a*
Pernopecten sp. 9*e*
Polidevcia attenuata 9*e*, 15*a*, 17*a*
Posidonia corrugata 13*b*
Sanguinolites aff. *v–scriptus* 18*a*
Sanguinolites sp. 10*a*, 10*b*, 18*a*, 23*b*
Schizodus sp. 6*a*, 15*c*
Streblochondria cf. *anisota* 2*a*
Streblochondria sp. 15*b*, 18*a*
Sulcatopinna costata 11*a*
Catastroboceras cf. *neilsoni* 20*a*
Catastroboceras sp. 6*a*, 9*a*, 11*a*
'*Cyrtoceras*' *rugosum* 6*b*
Metacoceras? 20*a*
Orthocone indet. 6*a*
Rayonnoceras sp. 11*a*
Nautiloids indet. 12*a*, 20*a*
Reticuloceras stubblefieldi 20*a*
Weberides cf. *mucronatus* 15*b*
Weberides sp. [pygidia] 13*b*, 15*a*, 17*c*
Trilobite pygidia fragments 1*a*, 9*e*, 17*c*
Hollinella sp. 16*a*, 20*a*
Paraparchites sp. 3*a*, 11*a*
Ostracods indet. 2*a*, 3*a*, 3*c*, 7*b*, 8*a*, 9*b*, 16*a*, 20*a*
Dithyrocaris sp. [includes buccal tooth] 20*a*
Archaeocidaris sp. 1*a*, 20*a*
Fish teeth 2*a*, 11*a*, 20*a*

W.H.C.R.

REFERENCES

BISAT, W. S. 1924. The Carboniferous goniatites of the north of England and their zones. *Proc. Yorks. geol. Soc.*, **20**, 40–12.

——1928. The Carboniferous goniatite zones of England and their continental equivalents. *C. r. Congr. Avanc. Etud. Stratigr. Geol. carbonif.*, 117–33.

CARRUTHERS, R. G. 1938. Alston Moor to Botany and Tan Hill: an adventure in stratigraphy. *Proc. Yorks. geol. Soc.*, **23**, 236–53.

DAKYNS, J. R., TIDDEMAN, R. H., RUSSELL, R., CLOUGH, C. T. and STRAHAN, A. 1891. The geology of the country around Mallerstang. *Mem. geol. Surv. Gt Br.*

DUNHAM, K. C. 1948. The geology of the Northern Pennine Orefield; Vol. 1, Tyne to Stainmore. *Mem. geol. Surv. Gt Br.*

——and JOHNSON, G. A. L. 1962. Sub-surface data on the Namurian strata of Allenheads, south Northumberland. *Proc. Yorks. geol. Soc.*, **33**, 235–54.

——and STUBBLEFIELD, C. J. 1945. The stratigraphy, structure and mineralization of the Greenhow mining area, Yorkshire. *Q. Jnl geol. Soc. Lond.*, **100**, 209–68.

FOWLER, A. and ROBBIE, J. A. 1961. The geology of the country around Dungannon. *Mem. geol. Surv. N. Ireland.*

HEY, R. W. 1956. Cherts and limestones from the Crow Series near Richmond, Yorkshire. *Proc. Yorks. geol. Soc.*, **30**, 289–99.

HUDSON, R. G. S. 1941. The Mirk Fell Beds (Namurian E_2) of Tan Hill, Yorkshire. *Proc. Yorks. geol. Soc.*, **24**, 259–89.

——1945. The goniatite zones of the Namurian. *Geol. Mag.*, **82**, 1–9.

HULL, J. H. 1961. In *Summ. Prog. geol. Surv. Gt Br. for 1960*, 41.

——1963. *Ibid. for 1962*, 45.

——1967. *Ibid. for 1966*, 75, pl. II.

——1968. The Namurian stages of north-eastern England. *Proc. Yorks. geol. Soc.*, **36**, 297–308.

JOHNSON, G. A. L. 1958. Biostromes in the Namurian Great Limestone of northern England. *Palaeont.*, **1,** 147–57.

——(Ed.) 1970. Geology of Durham County. *Trans. nat. Hist. Soc. Northumb.*, **41.**

——HODGE, B. L. and FAIRBAIRN, R. A. 1962. The base of the Namurian and of the Millstone Grit in north-eastern England. *Proc. Yorks. geol. Soc.*, **33,** 341–61.

KIRKBY, J. W. and DUFF, J. 1872. Notes on the geology of part of south Durham. *Nat. Hist. Trans. Northumb.*, **4,** 150–98.

MILLS, D. A. C. and HULL, J. H. 1968. The Geological Survey borehole at Woodland, Co. Durham (1962). *Bull. geol. Surv. Gt Br.*, No. 28, 1–34.

NEVES, R. 1968. The palynology of the Woodland Borehole, Co. Durham. *Bull. geol. Surv. Gt Br.*, No. 28, 55–60.

OWENS, B. and BURGESS, I. C. 1965. The stratigraphy and palynology of the Upper Carboniferous outlier of Stainmore, Westmorland. *Bull. geol. Surv. Gt Br.*, No. 23, 17–44.

RAMSBOTTOM, W. H. C. 1966a. A pictorial diagram of the Namurian rocks of the Pennines. *Leeds Geol. Ass.*, **7,** 181–4.

——1966b. In *Summ. Prog. geol. Surv. for 1965*, 56.

READING, H. G. 1957. The stratigraphy and structure of the Cotherstone Syncline. *Q. Jnl geol. Soc. Lond.*, **113,** 27–56.

RICHARDSON, G. 1961. In *Summ. Prog. geol. Surv. Gt Br. for 1960*, 41.

ROWELL, A. J. and SCANLON, J. E. 1957a. The Namurian of the north-west quarter of the Askrigg Block. *Proc. Yorks. geol. Soc.*, **31,** 1–38.

—— ——1957b. The relation between the Yoredale Series and the Millstone Grit on the Askrigg Block. *Proc. Yorks. geol. Soc.*, **31,** 79–90.

SARGENT, H. C. 1929. Further studies in chert. *Geol. Mag.*, **66,** 399–413.

TURNER, J. S. 1955. Upper Yoredales and Millstone Grit relations in the Stainmore coalfield. *Geol. Mag.*, **92,** 350.

WELLS, A. J. 1955. The development of chert between the Main and Crow limestones in north Yorkshire. *Proc. Yorks. geol. Soc.*, **30,** 177–96.

——1957. The stratigraphy and structure of the Middleton Tyas–Sleightholme anticline, north Yorkshire. *Proc. Geol. Ass.*, **68,** 231–54.

WILSON, A. A. 1960a. The Carboniferous rocks of Coverdale and adjacent valleys in the Yorkshire Pennines. *Proc. Yorks. geol. Soc.*, **32,** 285–316.

——1960b. The Millstone Grit Series of Colsterdale and neighbourhood, Yorkshire. *Proc. Yorks. geol. Soc.*, **32,** 429–52.

——and THOMPSON, A. T. 1959. Marine bands of Arnsbergian age (Namurian) in the south-eastern portion of the Askrigg Block, Yorkshire. *Proc. Yorks. geol. Soc.*, **32,** 45–67.

—— ——1965. The Carboniferous succession in the Kirkby Malzeard area, Yorkshire. *Proc. Yorks. geol. Soc.*, **35,** 203–27.

WILSON, R. B. 1967. A study of some Namurian faunas of central Scotland. *Trans. R. Soc. Edinb.*, **66,** 445–90.

WOOLACOTT, D. 1923. A boring at Roddymoor Colliery, near Crook, Co. Durham. *Geol. Mag.*, **60,** 50–62.

Chapter 4

UPPER CARBONIFEROUS WESTPHALIAN (COAL MEASURES)

INTRODUCTION

COAL MEASURES occupy a broad, complex and much faulted structural depression trending east-north-east across the northern part of the district and extending into the Wolsingham (26) district to the north. They are conformable on the Namurian rocks, which form a rim to the syncline on its western and southern sides. To the east of Shildon, Coal Measures are concealed beneath the unconformable Permian which progressively oversteps southwards on to lower horizons.

A full sequence of Lower Coal Measures is present, as well as a great part of the Middle Coal Measures, to a maximum total thickness of 1520 ft.

Little research has been done on the Coal Measures of this district specifically, although various workers (Kirkby and Duff 1872; Calvert 1884; Anderson and Dunham 1953) have referred to the area in general geological accounts of the Coal Measures of Northumberland and Durham or of special aspects of the Carboniferous. The Coal Measures of the Durham and West Hartlepool (27) district to the north-east have been described by Smith and Francis (1967) and their account contains much general information relevant to the Coal Measures in this district.

CLASSIFICATION

The classification adopted in this memoir is shown in Fig. 12, which also indicates the distribution of the main coal seams, the thicker sandstones and more persistent faunal horizons in the Coal Measures.

Traditionally, the base of the Lower Coal Measures in Durham was taken at the Ganister Clay coal which is the lowest workable coal of the sequence and overlies the highest sandstone, sometimes referred to as the Third Grit, of the old 'Durham Millstone Grit' sequence. This lithological classification no longer accords with palaeontological evidence from other coalfields in Britain and north-west Europe, where the base of the Coal Measures (Westphalian) is defined by a marine band containing *Gastrioceras subcrenatum* (Frech). Although this fossil has yet to be found in north-east England, other palaeontological evidence indicates that the equivalent horizon in this district is the Quarterburn Marine Band (Mills and Hull 1968, pp. 4–5) between the Second and Third Grits of the old 'Durham Millstone Grit' Series.

The division between the Lower and Middle Coal Measures is taken at the base of the Harvey Marine Band, represented in other coalfields by the marine band containing *Anthracoceratites vanderbeckei*.

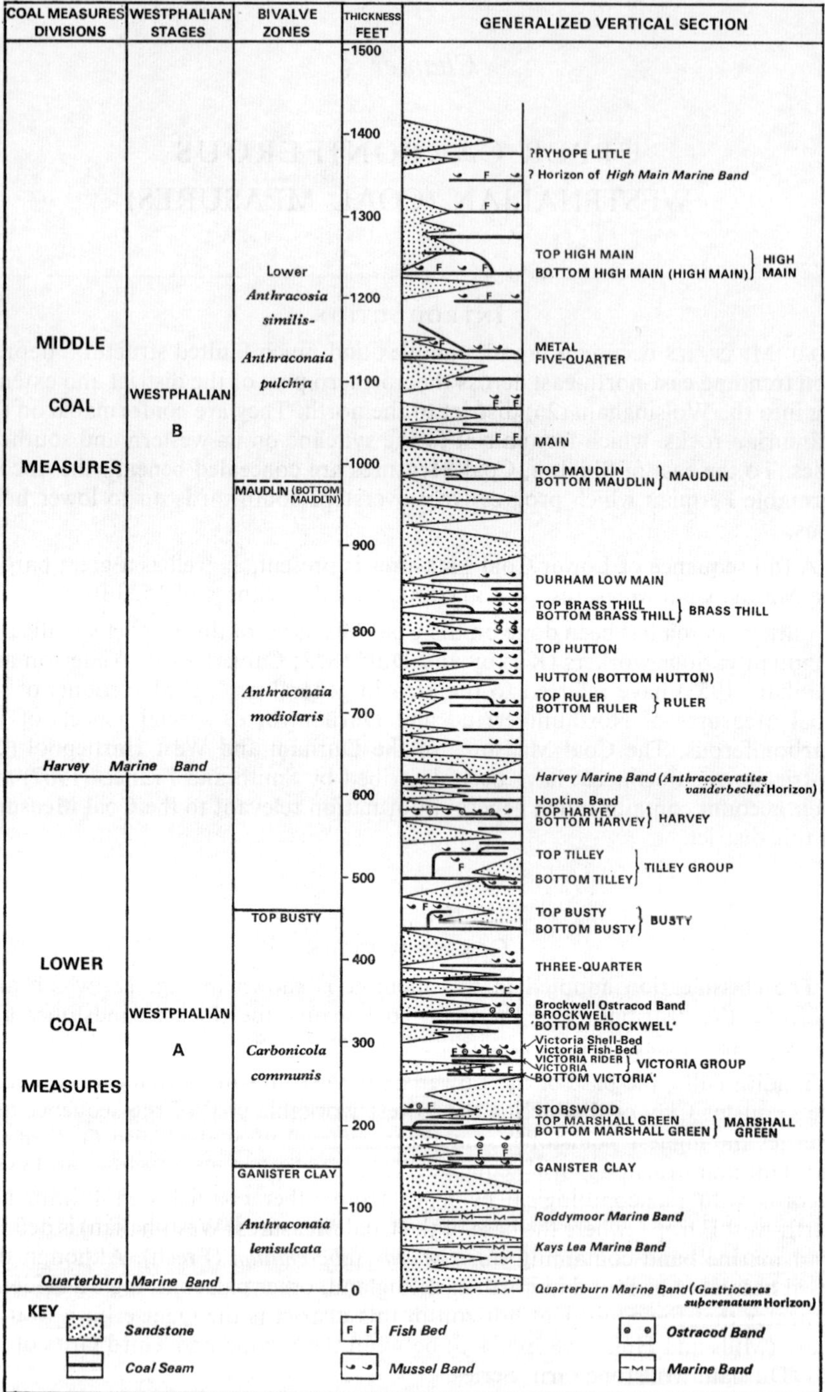

FIG. 12. *Generalized vertical section of the Westphalian (Coal Measures) showing the principal coal seams, sandstones and fossil bands*

Lithology

As in the Durham and West Hartlepool (27) district the Coal Measures sequence here consists of alternations of shales, mudstones, siltstones and sandstones together with coals and associated seatearths. Although the sequence may be regarded as rhythmic, wide variations in local conditions gave rise to modifications of the standard cyclothem.

For convenience, the sediments may be divided into five main groups: argillaceous sediments, 'mixed beds', arenaceous sediments, seatearths and coals. Argillaceous sediments, generally grey or dark grey in colour, range from highly fissile micaceous carbonaceous shales to blocky mudstones with a variable sand and silt content. The shade of grey is usually directly related to the contained proportion of finely disseminated carbonaceous material, which is often abundant. More rarely, disseminated sulphides occur, usually in the form of pyrite. Ironstone bands and nodules are common at various horizons and in these 'mussels' are commonly preserved.

'Mixed beds' is a convenient term for describing a rapid alternation of mudstones, siltstones and fine sandstones in a sequence in which individual beds nowhere predominate. Small-scale false-bedding and turbulence is often present.

Arenaceous sediments range from dark grey, muddy, micaceous and carbonaceous, silty sandstones to paler grey or off-white bedded and massive sandstones. At outcrop they often weather to a dull yellow or rust colour. Most of the sandstones are fine- to medium-grained. Quartz is the dominant mineral and feldspars are subordinate. Near the base of the sequence the sandstones are commonly coarser, with conglomerate or pebbly layers in places. Conglomerate bands often contain subrounded to subangular elongate fragments of argillaceous sediments. The cement in most of the sandstones is kaolinitic, but in the harder sandstones it is predominantly of silica or ankerite.

Seatearths range from dull grey rooty mudstone in which ironstone nodules are common to hard dull grey siliceous ganister. Fireclay of good commercial quality is not common.

Coals are bituminous, generally have low ash and sulphur content and are suitable for coking. They are more fully described in Chapter 10.

Seam Nomenclature

The coal seam names adopted in this district, and which appear on the six inches and one inch to the mile maps, are those used by the National Coal Board. Local names which have been applied in the past are however quoted on the first mention of the current coal seam name. In addition, a glossary of terms applicable to rock types is given in Appendix 2.

Reddened Beds in the Coal Measures

By the end of Lower Permian times, the Carboniferous rocks had been uplifted, folded and eroded, and the resulting topography was of generally low relief. Anderson and Dunham (1953) concluded from examination of the Coal Measures of Durham and South Northumberland, including those at Eldon Hill Opencast Quarry in this district, that reddening took place before local Permian sedimentation. These authors noted that reddening was generally absent directly beneath the Permian rocks, but that it usually appeared a few feet below the unconformity. Most of the boreholes which have penetrated the unconformity and subjacent Coal Measures rocks in the north-eastern part of the district have proved an upper grey zone, 2 to 8 ft thick, and two have recorded as much as 20 ft. About half the boreholes proved the reddened zone to extend to between 30 and 50 ft below the unconformity. In one case the base of the reddened zone lay at only 14 ft, and a few boreholes proved reddened beds at over 70 ft, the maximum of 76 ft being recorded in the extreme north-east of the district near Eldon Blue House. Where sandstone lies directly under the plane of unconformity it is typically leached to a light grey colour, and commonly contains a high proportion of pyrite. This pyrite, and the grey colour are thought to be the result of reduction of secondarily oxidised iron by downward-percolating water from the Marl Slate sea (Smith and Francis 1967, pp. 18–19). The petrographical evidence adduced by Anderson and Dunham (1953) from Eldon Hill suggests that reddening is due to *in situ* oxidation of pyrite and siderite, and to the introduction of red iron oxide along joints and in the pore spaces of sandstones.

General Stratigraphy

LOWER COAL MEASURES

Lower Coal Measures account for more than two-thirds of the total area of Coal Measures outcrop in the district. The beds are extensively concealed by glacial drift and exposure is generally confined to short sections in valleys in the western part of the area and to isolated sandstones cropping out on steeper or higher ground. Details of the stratigraphy have been derived from shafts, boreholes and mining operations.

The maximum recorded thickness of Lower Coal Measures in this district is 600 ft. The lowest part, that between the Quarterburn Marine Band and the Ganister Clay Coal, is a predominantly arenaceous or 'mixed bed' sequence, and includes the Third Grit of the old 'Durham Millstone Grit'. Only a few thin coals are present, none of which has been worked. Of the five marine bands found in the Lower Coal Measures of Durham, four are in this part of the sequence. A similar close grouping of marine bands is noted in the '*Anthraconaia*' *lenisulcata* Zone of the Cumberland, Lancashire and Yorkshire coalfields.

Between the Ganister Clay and the base of the Brockwell Coal there are up to nine thin coals, three of which have been worked locally. Again, the sequence is predominantly arenaceous, particularly above the Marshall Green. The upper part of the Lower Coal Measures includes the three seams, Brockwell, Busty and Harvey, which have been of leading importance in the economy of the district. The Tilley group of coals, also included in these measures, has only been worked sporadically on account of the variable quality and thickness of the seams.

The Quarterburn Marine Band (Mills and Hull 1968, p. 4) consists generally of between 1 and 10 ft of shale and mudstone containing brachiopods, bivalves, gastropods and fish debris. Locally the marine band is replaced or cut out by sandstone. Correlations based on comparative lithologies and the presence of productoid fragments, shown on Plate VI, include records of this marine band in the Roddymoor Bore near Crook, at 442 ft 7 in (Woolacott 1923; Mills and Hull 1968, p. 4), and in the Middle Stotfold Bore between West Hartlepool and Fishburn at about 942 ft (Magraw and others 1963, p. 157).

The measures between the Quarterburn and Kays Lea marine bands range in thickness (Plate VI) from 26 ft at Quarterburn to 90 ft in boreholes between Cockfield and Toft Hill. In general, the thickness is directly related to the amount of sandstone in the succession. Two unnamed marine bands occur in the upper part of these measures in a borehole near Keverstone, the upper of which has also been proved in two other boreholes near Hilton. The fauna collected from these bands includes foraminifera, brachiopods, a conodont, and fish scales; plant fragments also occur. In the Durham and West Hartlepool (27) district, these measures are only 17 ft thick if the mudstone with *Planolites sp.* recorded at and above 920 ft in the Middle Stotfold Bore (Magraw and others 1963, p. 196) is the assumed equivalent of the Kays Lea Marine Band. The absence of foraminifera, however, makes correlation uncertain. The Kays Lea Marine Band has not been recognized in the Roddymoor Bore but it is thought that about 42 ft of strata there separate the horizons of the Quarterburn and Kays Lea marine bands (Mills and Hull 1968, p. 4).

The Kays Lea Marine Band (Plate VI) is of interest in that it is the lowest marine band containing foraminifera in the Coal Measures. It was defined in the Woodland Bore* (Mills and Hull 1968, p. 4) and has been proved in boreholes elsewhere in this district, though it is apparently absent to north and north-east of Woodland. In the Roddymoor Bore it is apparently cut out by the overlying sandstone (Third Grit).

Most of the interval between the Kays Lea and Roddymoor marine bands (Plate VI) generally consists of sandstone—the Third Grit of the primary surveyors. The Third Grit thins generally towards the south-east from a recorded maximum of 56 ft in Spurlswood Beck to about 6 ft in a borehole near Hilton. Its top is in places separated from the Roddymoor Marine Band by a seatearth and a few inches of coal which has been called the Gubeon Coal in the Durham (27) district (Smith and Francis 1967, p. 13). In the Roddymoor Bore the Third Grit is thought to be about 38 ft thick at about 395 ft (Mills and Hull 1968, p. 4), while in the Durham (27) district it has been shown to range between about 30 and 40 ft (Smith and Francis 1967, p. 35) although in that area a single thick sandstone at this position is unusual.

The Roddymoor Marine Band (Plate VI) is defined as the *Lingula* band which is present at the base of 6 ft of shale at 350 ft 1 in in the Roddymoor Bore (Mills and Hull 1968, p. 4). In the north-western part of this district it is unrecorded, though near-marine conditions are indicated by the occurrence of *Planolites sp.* and fish at this horizon in Spurlswood Beck and in the Woodland Bore* (Mills and Hull 1968, p. 4). It is however proved in boreholes around Hilton where its thickness ranges from about 2 to 5 ft. It contains a fauna which includes worm trace fossils, horny brachiopods, gastropods, fish scales and foraminifera. Plant debris is also present.

The Roddymoor Marine Band appears to correlate with the Gubeon Marine Band which overlies the Third Grit and Gubeon Coal in the Durham (27) district (Smith and Francis 1967, p. 13), and in the Offshore No. 1 Bore north of Hartlepools (Magraw and others 1963, p. 157). Some confusion has arisen in this nomenclature, since the Gubeon seam in its type locality in Northumberland, as shown on the Morpeth (14) Sheet, has been shown to equate with the Marshall Green of Durham. It is here recommended, therefore, that the term Roddymoor Marine Band should be used for this horizon in Northumberland and Durham in preference to Gubeon Marine Band.

The measures between the Roddymoor Marine Band and the Ganister Clay Coal (Plate VI) show wide lateral variation in lithology and thickness. North and north-west of Woodland the succession consists of 3 to 21 ft of sandstone, overlain by 33 to 52 ft of an argillaceous or 'mixed beds' sequence, with a thin sandstone near the top in places. South-eastwards the measures thicken to between 65 and 105 ft and a thick sandstone is often present, as well as sporadic seatearths and coals.

There are several fossil bands in the mudstones, and in the Woodland Bore* these have yielded a fauna of ostracods and non-marine bivalves, the latter commonly referred to as mussels. The occurrence of *Planolites ophthalmoides* at different levels suggests that these measures may contain a number of cyclothems. The occurrence in one borehole of *Carbonicola* aff. *proxima* is of interest in that it affords a favourable comparison with the *C. proxima* fauna from the Norton mussel band of the East Midlands, which is in the upper part of the '*Anthraconaia*' *lenisulcata* Zone.

In the Roddymoor Bore equivalent measures are about 60 ft thick, and farther east in the Durham (27) district (Smith and Francis 1962, p. 35; Magraw and others 1963, p. 157) they range between 40 to 70 ft.

The Ganister Clay Coal (Plate VII) which was once taken as the base of the Coal Measures in County Durham, is a thin seam, generally ranging between 8 and 20 in. It is thickest on the southern and western margins of the coalfield area where it is often strongly banded and where it has been sporadically worked at or near the outcrop. In the Wolsingham (26) and Durham and West Hartlepool (27) districts the coal is usually thin or absent (Smith and Francis 1967, p. 36).

The beds between the Ganister Clay and Marshall Green (Plate VII) range in thickness between 30 and 65 ft. There is considerable variation in lithology, and in places the greater part of the sequence consists of sandstone. Mussels, in places associated with *Planolites* and fish remains, have been recorded in the middle and upper parts of the sequence in several boreholes.

The Marshall Green (Plate VII) is generally split into Top and Bottom coals, and only to the north-east of Woodland is it a single united seam up to 24 in thick. The thickness of the Bottom Marshall Green rarely exceeds 24 in, including bands. Locally it is absent, its assumed horizon being underlain by seatearth or fireclay. Although it has been worked sporadically in the west and south of the coalfield area and at Randolph Colliery, Evenwood, it has in general been considered too thin to warrant exploitation.

The measures between the Bottom and Top Marshall Green coals (Plate VII) generally range between 15 and 50 ft and usually consist of sandstone, the texture and bedding of which recalls the coarser 'grits' lower down in the sequence. This sandstone is particularly thick south of Cockfield and around Woodland. Locally it apparently cuts out the Top Marshall Green seam while in one locality it extends upwards to the base of the Victoria group of coals. Where the interval between the two seams is small the measures tend to be more shaly, though a thin sandstone often overlies the Bottom Marshall Green. The Top Marshall Green Coal rarely exceeds 10 in and in the east and north-east it fails locally, its assumed horizon being underlain by a thin fireclay. In places it is cut out by sandstone.

The Top Marshall Green and Stobswood coals (Plate VII) are generally 12 to 30 ft apart with local maximum and minimum of 40 ft and 2 ft. In many cases, the Stobswood Coal, which is not marked on the one-inch map of the district but is indicated and first named on the vertical section, is either absent or thin, the maximum recorded thickness being 11 in.

Above the Stobswood, up to the Victoria group of coals (Plate VII), lie 28 to 74 ft of variable strata. Generally the lower part of the sequence consists of argillaceous beds, overlain by a 'mixed bed' sequence or one in which sandstone tends to predominate. The roof of the Stobswood Coal locally contains a fauna of mussels, in places associated with *Planolites* and fish debris. This horizon is thought to be the equivalent of the Stobswood Marine Band, proved in a borehole at Stobswood, Northumberland (first published on 6-inch geological sheet NZ 38 SW, Blyth) but as yet no marine fossils have been recorded from this position in County Durham. In this sequence, also, there is an impersistent coal, sometimes split into two leaves, which has been termed the Bottom Victoria by the National Coal Board. Where the coal is absent the horizon is marked by the top of a seatearth. Although it approaches within 2 ft of the Victoria in one case, generally the separation is greater and there is no indication in this district or adjacent areas that this coal is the result of a split. A similar coal ranging from 4 to 22 ft below the Victoria is present in the Tynemouth (15) district (Land 1974). The roof measures of this coal, or where split its upper leaf, locally contain *Carbonicola*, *Naiadites*, *Carbonita*, and fish scales, while at one locality mussels occur between the two leaves of the seam. Non-marine fossils have been recorded from these horizons in the East Hetton Busty Bore in the Durham and West Hartlepool district (Smith and Francis 1967, p. 37).

Rarely, virtually the whole sequence between the Top Marshall Green and the Victoria group is occupied by sandstone.

The Victoria group of coals (Plate VII) is, typically, the Victoria Coal separated by a well developed seatearth from the Victoria Rider Coal. Three variations may occur: (1) the Victoria and Victoria Rider coals are united. Where a single seam is present it generally ranges in thickness between 3 and 13 in; (2) only a thick seatearth is present at this horizon; (3) the Victoria Rider Coal is absent, a thick seatearth being present underlain by the Victoria Coal. Where the typical sequence exists the separation usually does not exceed 5 ft, the interval being mainly seatearth with only occasional mudstone bands. Fish scales are occasionally present in the roof of the Victoria Coal.

The measures between the Victoria group of coals and the Brockwell (Plate VII) range in thickness between about 40 and 80 ft. A thick sandstone commonly forms the middle or upper part of the sequence. More rarely sandstone also occupies the lower part of the sequence. The Victoria Rider is overlain by the Victoria Fish-Bed, comprising shale with fish scales and spines, and sometimes with *Planolites ophthalmoides* and mussels. This in turn is overlain by the Victoria Shell-Bed which is correlated with the Upper Victoria Shell-Bed of the Durham and West Hartlepool district (Smith and Francis 1967, p. 39) and consists of shale with non-marine bivalves including *Carbonicola declivis*, *C. pseudorobusta*, and *Curvirimula sp.* and *Naiadites sp.* This bed either directly overlies the Victoria Fish-Bed or is separated from it by a few feet of barren mudstone. A coal is locally found close below the Brockwell with a maximum recorded thickness of 35½ in. It is thought to be the equivalent of the Bottom Brockwell seam of the Durham and West Hartlepool district (Smith and Francis 1967, p. 39).

The Brockwell (Plates IX and X), the lowest major coal in the sequence, has been extensively worked in the district. Its thickness ranges from 5 to 7 ft, and in the thicker sections the coal is frequently canneloid at the top.

The measures between the Brockwell and Busty coals (Plates IX and X) average about 100 ft, and typically consist of two sandstones and two thin coals, the lower coal being impersistent; exceptionally, up to four sandstones and three coals are present. The uppermost coal, although of no economic importance, may be equated with the Three-Quarter Coal of the Durham and West Hartlepool (27) district. Two or occasionally three fossil bands have been proved between the Brockwell and Three-Quarter coals. The lowest band, near the roof of the Brockwell, contains mussels, fish fragments and in one locality ostracods, and is the probable equivalent of the Brockwell Ostracod Band of Durham and West Hartlepool (Smith and Francis 1967, pp. 15, 24, 43); the highest band, whose top lies up to 60 ft above the Brockwell, contains fish fragments and mussels but no ostracods.

The Fishburn '*Estheria*' Band, in the roof of the Three-Quarter Coal in the Durham and West Hartlepool district (Smith and Francis 1967, pp. 15, 25, 45), is not recorded here, although mussels are locally recorded at that horizon. There is also a band containing mussels some 16 ft higher.

Over most of its area the Busty Coal (Plates IX and X) is split into two distinct sections. Only west of a line from between Evenwood Gate and Spring Gardens to Toft Hill and Woodhouses is it more often a single composite seam. Even here, however, it is often strongly banded and is locally split by up to 3 ft of seatearth or mudstone. Where regarded as a single seam it generally ranges between 3 and 5 ft and has been extensively worked. East of the line of split the Bottom Busty is usually 2½ to 4 ft in thickness. The Bottom Busty seam is prone to washout or impoverishment.

The measures between the Bottom and Top Busty (Plates IX and X) are generally between 5 and 20 ft thick, a sandstone often being present in the roof of the Bottom Busty. Near West Auckland where the Bottom Busty is impoverished the separation between the two seams is in places as much as 54 ft. The Top Busty Coal generally ranges between 1½ and 3 ft, but in some localities it is much thinner and of little economic importance.

The measures between the Top Busty seam and the Tilley group of coals (Plates IX and X) are of variable lithology and range in thickness between about 15 and 104 ft, though 40 to 50 ft is most common. A thick sandstone usually overlies the Busty in the western part of its area, and in a few places the whole sequence between Bottom Busty and the Tilley group is occupied by sandstone. To the east the sequence is mainly argillaceous, and a prominent mussel band, sometimes associated with fish remains in the immediate coal roof, commonly overlies the Top Busty or its horizon as in the Durham and West Hartlepool district (Smith and Francis 1967, pp. 15, 26 and 50). In places scattered mussels occur higher in the sequence between the Top Busty and Bottom Tilley.

The Tilley group of coals (Plates IX and X) occur in up to 48 ft of strata. Individual seams rarely exceed 2 ft and are usually less; some are locally absent. Typically, the group forms Top and Bottom sub-groups of which the Bottom is the most consistent, and gives rise at some localities to a single coal up to 3 ft thick. The measures between the two sub-groups vary considerably; a sandstone in places occupies the whole interval, but more usually the measures are mainly argillaceous. Mussels occur sporadically between the leaves of the Bottom Tilley and between the Bottom and Top Tilley sub-groups. Fish fragments are locally recorded from the roof of the top leaf of the Bottom Tilley. The Tilley coals have been little worked and are of small economic value.

The measures between the Tilley group and the Harvey (Plates IX and X) generally consist of about 60 ft of variable measures in which sandstone or alternations of sandstone and shale commonly predominates. Between St Helen Auckland and Shildon however the interval is greater and includes up to 109 ft of sandstone. A thin coal or seatearth occasionally present up to 16 ft below the Harvey Coal may be the equivalent of the Hodge seam of north-west Durham. In a few localities one or two thin coals occur immediately under the Harvey Coal and may be splits off the base of the seam. The sequence is relatively unfossiliferous.

The Harvey seam (Plates IX and X) locally prone to washout or impoverishment, ranges generally between 3 and 4 ft. Locally in the eastern part of the district it is split into Top and Bottom coals separated by up to 8 ft of seatearth or mudstone. Like the Brockwell and Bottom Busty it has been extensively worked.

The Harvey–Harvey Marine Band interval (Plates IX and X) is generally between 40 and 60 ft, and contains a variable sequence comprising roughly equal proportions of thin sandstones and shales. Two thin coals or seatearths are occasionally present midway in the sequence. The Harvey, or where present the Top Harvey, is overlain by the Hopkins Band (Smith and Francis 1967, pp. 15, 26 and 56), the most persistent ostracod-bearing marker band in the Durham Coal Measures. Mussels and fish fragments are also present at this horizon.

Middle Coal Measures

The Middle Coal Measures are preserved mainly in the north-eastern part of the district, where they are partly concealed by Permian rocks. Farther west, Middle Coal Measures occur in narrow belts along the south sides of the Butterknowle and Wigglesworth faults respectively. The base of the Middle Coal Measures is taken at the base of the Harvey Marine Band, near the middle of the *A. modiolaris* Zone. The marine band is generally thin and represented only by dark shale or mudstone; only at two localities have marine fossils been found, and these were *Lingula*. Over most of the district the horizon can only be inferred by correlation.

Between the Harvey Marine Band and the Hutton (Bottom Hutton) (Plate XI) are 100 to 130 ft of measures with up to six sandstones. The sandstones are generally thin, though rarely they form the greater part of the sequence. Several thin coals also occur, the highest of which is most persistent in the west, while the others are most persistent in the east. This upper coal, the Ruler or Jubilee, is equated with the Plessey seam of Northumberland. In the western part of the outcrop it attains a maximum recorded thickness of 38 in. Generally however it is significantly thinner than this and is locally split into Top and Bottom seams.

Fossils are recorded from several levels in the sequence. An important mussel band, sometimes with fish debris at the base, occurs close above the Harvey Marine Band, and although not widely recorded from this district it is well known as one of the more persistent marker bands of the Durham coalfield (Smith and Francis 1967, pp. 16, 59). Three impersistent faunal bands are recorded between this shell bed and the Ruler Coal. Mussels have been recorded from the roof of the Ruler Coal but there is no fossil band comparable with the Plessey Shell-Bed of Northumberland.

In this district the Hutton seam (Plate XI) is split into Top and Bottom coals as in the southern half of the Durham and West Hartlepool (27) district (Smith and Francis 1967, p. 16). The Bottom Hutton, termed the Hutton Coal except in the north-eastern part of the district, generally occurs as a single composite seam between 36 and 48 in thick. In the extreme north-east the seam is generally slightly thinner; and in one or two localities it is impoverished. The seam has been widely worked throughout the district.

The Hutton (Bottom Hutton)–Top Hutton interval (Plate XI) varies from 24 to 70 ft, the usual range being of the order of 30 to 40 ft. The measures are mainly shale, although sandstones may be present, one underlying the Top Hutton. Mussels are common in the shales above the Hutton Coal, and at one locality an abundant mussel fauna, associated with a band containing ostracods persists throughout the greater part of the sequence between the two coals.

The Top Hutton Coal is thin and impersistent over most of its area and of no economic value; occasionally it is split into two or even three seams.

The Top Hutton to Brass Thill interval (Plate XI) varies from 20 to 60 ft, the more usual thickness being about 30 to 40 ft. The beds are mainly argillaceous, although one or more sandstones occur, especially in the west. Poorly preserved mussels have been noted at several horizons, and include *Anthracosia* and *Naiadites* as the main genera.

The Brass Thill (Plate XI) throughout most of its area is a single composite seam 3 to 4 ft thick, often with a thin band near the base of the coal. Locally the seam is inferior or impoverished. In some localities, especially in the north-eastern part of the district, the coal is split into two leaves separated by up to about 11 ft of strata, and here the lower leaf usually ranges between 12 and 18 in, the upper between 29 and 44 in.

The Brass Thill (or Top Brass Thill) to Durham Low Main interval (Plate XI) varies from 20 to 60 ft, being generally between 30 and 40 ft. The beds are mainly argillaceous with thin impersistent sandstones, though locally sandstone occupies most of the sequence. Occasionally the Durham Low Main Coal has a thin underrider, while a thin coal or seatearth is sometimes present lower in the sequence. The Brass Thill Coal is commonly overlain by shales containing an abundant fauna of mussels in which *Anthracosia* and *Naiadites* are common. This bed equates with the Brass Thill Shell-Bed of the Durham and West Hartlepool district (Smith and Francis 1967, pp. 16, 27).

The Durham Low Main (locally known as the Chatham) ranges up to 63 in, but the general range of thickness is between 20 and 36 in; it is locally prone to impoverishment or small washouts, especially around and to the east of Auckland Park.

The Durham Low Main to Main Seam interval (Plate XI) has a maximum thickness of 220 ft, the more general range being 110 to 170 ft. These beds are mainly arenaceous and in most cases include a thick sandstone, the Durham Low Main Post either directly overlying, or separated by a few feet of shaly beds from, the roof of the Durham Low Main. The sandstone, either as a single bed or split into two or three leaves, has a maximum recorded thickness of 120 ft. At the few localities where shales occur in the roof of the Durham Low Main, mussels including *Anthracosia* are recorded. The Maudlin Coal, a composite seam up to a maximum thickness of 77 in and up to 70 ft below the Main, is impersistent. In the extreme north-east of the district it is split into Top and Bottom coals up to 17 ft apart. In the north-east part of the district, where both Top and Bottom Maudlin are present, the Durham Low Main post tends to be thinner, and this feature becomes even more pronounced in the Durham and West Hartlepool (27) district (Smith and Francis 1967, p. 16). Where the Maudlin is present the measures between it and the Main Coal are predominantly argillaceous.

The Main Coal (Plate XI) though of limited extent in this district, is the thickest seam and has been widely worked. Where the seam is united it ranges between 36 and 132 in; where split into two leaves by bands (which never exceed 3 ft thick) its composite thickness ranges between 80 and 171 in.

Measures above the Main Coal (Plate XII) occur only in the north-eastern part of the district. Except for a limited zone on the south side of the Butterknowle Fault north-west of Woodhouses, they are restricted to the east side of the Gaunless valley between South Church and West Thickley, and bounded to the south-east by an irregular line from West Thickley to Windlestone Park. The highest Middle Coal Measures crop out on Shawbrow Hill north-west of Shildon.

The measures between the Main and Five-Quarter coals (Plate XII) are mainly shales and range in thickness from 47 to 120 ft; sandstones sometimes occur in the thicker sequences; three thin sporadic coals are also present. Fossils are not common; fish fragments are recorded from the roof of the Main, while a band with fragmentary mussels including *Anthracosia* and sometimes underlain by shales with fish debris, has been recorded above the uppermost thin coal.

The Five-Quarter (Plate XII) locally termed the Bottom Five-Quarter, is generally a single seam ranging between 60 and 72 in. Locally, as between Fieldon Bridge and Shildon, and in places in the sub-Permian concealed coalfield, it is much thinner and in a few isolated localities it is absent. Locally it is united with the Metal seam.

The measures between the Five-Quarter and the Metal seams (Plate XII) range up to 49 ft, the widest separation being mainly confined to the synclinal area. The beds are mainly shales and mudstones sometimes with mussels and fish fragments, but they occasionally include a thin sandstone in the middle or lower part of the sequence.

The Metal Coal (Plate XII) locally termed the Jet, Dicky Dant or Top Five-Quarter, is generally a single seam or composite banded seam between 6 and 66 in thick. The name 'Metal Coal' is derived from the Newcastle upon Tyne area thereby avoiding the inference that it is an upper split of the Five-Quarter Coal although very locally it is united with this seam. This accords with practice in the Durham and West Hartlepool (27) district where Smith and Francis (1967, p. 17) refer to the confusion that has arisen over the nomenclature and correlation of this seam.

The measures between the Metal and High Main (or Bottom High Main) (Plate XII) range between 40 and 81 ft and consist mainly of shale with local thin sandstones. An impersistent coal with a maximum recorded thickness of 26 in occurs in the upper part of the sequence. Fragmentary mussels are occasionally recorded in the roof of the Metal, while fish debris and mussels are recorded in the upper part of the sequence.

The High Main (Plate XII) occurs as a single seam ranging between 18 and 24 in at several localities but is generally split and divided into Top and Bottom High Main. Because the seams are generally very close, and the Top High Main is thin or perhaps locally absent, the two seams have been mapped where they occur as 'the High Main'.

The Bottom High Main, locally known as the Willy Winter, generally ranges between 18 and 38 in, but west of Shildon thicknesses of up to 85 in inclusive of band and shale have been proved.

Where the High Main is split, the measures between the two units are shales generally between 10 and 15 ft thick, with a local maximum of 24 ft, and where separation is greatest, up to two thin sandstones sporadically occur. Fish debris and scattered mussel fragments have been reported from the roof of the Bottom High Main.

The Top High Main ranges up to a maximum recorded thickness of 33 in; more commonly between 6 and 21 in. Locally it is absent or cut out by an overlying sandstone—the High Main Post.

A maximum of 160 ft of Carboniferous strata have been proved above the Top High Main in this district (Plate XII). They are confined to the area of Auckland Park, where they are arenaceous, and around Shawbrow Hill where they are mainly shales and mudstones. The measures include four coals of which the three lowest are thin and impersistent. The lowest coal, up to 7 in thick, carries mussels in its roof. It is overlain by, or separated by a few feet of argillaceous measures from, the High Main Post, a sandstone which in the Auckland Park area ranges up to 67 ft thick, cuts out this coal and extends down to, or locally cuts out, the Top High Main. Farther south, the sandstone rarely exceeds 15 ft, and is locally absent. Both mussels and fish fragments are recorded locally from the roof of two succeeding higher thin coals. The higher band containing fish and mussels and which is not necessarily associated with coal, may be at about the same horizon as the High Main Marine Band of the Durham and West Hartlepool (27) district (Smith and Francis 1967, p. 17). Mussels are known to occur in the sequence between the Marine Band and the succeeding coal. The highest of the four coals, thought to be the Ryhope Little, is recorded as being up to 24 in thick on Shawbrow Hill. This coal is overlain by a sandstone over 30 ft thick. It is possible that a few feet of higher measures are preserved along the synclinal axis extending from Coppy Crook towards Coundon Grange.

Details

In the following details positions of outcrops, shafts and boreholes are as far as possible related to localities shown on the One-inch Geological Map but the exact position is indicated by reference to the National Grid. These localities and their grid references are shown on Plate V. Representative sections of both Lower and Middle Coal Measures are shown graphically in Plates VI, VII, IX–XII. The numbers in parentheses following the names of sections refer to sites on Plate V and to the numbered graphic sections on the relevant plates. Detailed records and synopses of some of the more important outcrops, shafts and boreholes are included in Appendices 1 and 2 and references to these in the text are marked with an asterisk (*).

Geology of the country around Barı

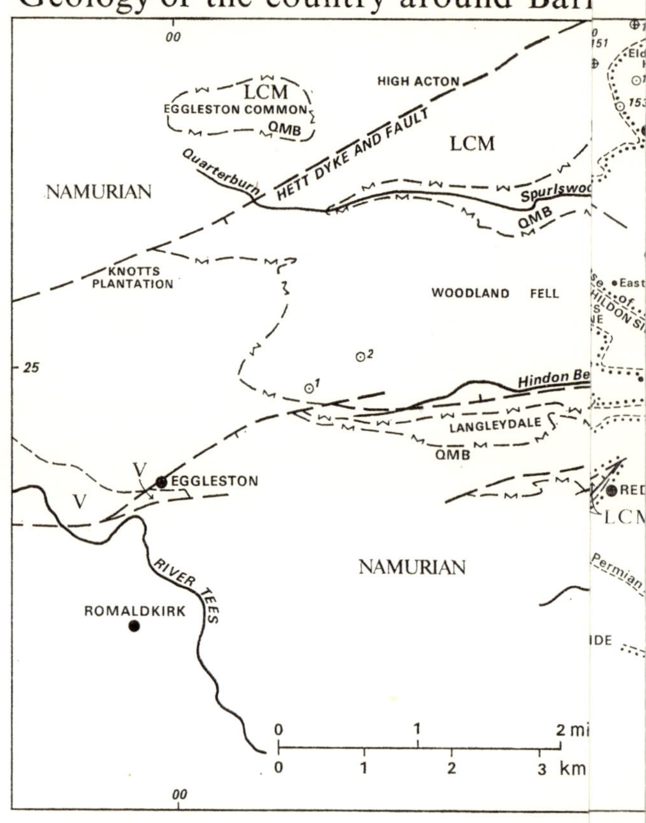

- * 1. Eggleston No.3 Bore (01532483)
- * 2. Eggleston No.4 Bore (02022518)
- • 3. Woodland Colliery No.47 Bore (05132495)
- • 4. Woodland Colliery No.10 Bore (06162522)
- * 5. Cowley Bore (06292532)
- * 6. Woodland Colliery No.29 Bore (06142702)
- * 7. Arnghyll No. 3 Bore (06672455)
- 8. Arnghyll No. 2 Bore (06672452)
- • * 9. Arnghyll No. 1 Bore (06682450)
- • 10. Woodland Colliery Shaft (06802680)
- • 11. Cowley Shaft (06762542)
- • * 12. Burfoot Leazes Bore (06902458)
- * 13. Pioneer Shaft & Bore (07922759)
- 14. Moorhill Bore (08072852)
- • 15. High Copley Bore (08712547)
- • * 16. Woodland Bore (09102770)
- 17. High Wood Farm Bore (08902381)
- • * 18. Moorhill No.118 Bore (09782817)
- • * 19. Quarry West Pit (09822566)
- 20. Moorhill No.121 Bore (09942840)
- • * 21. Moorhill No.1 Bore (10272801)
- 22. Moorhill No.2 Bore (10632818)
- * 23. High Wham No.31 Bore (10652750)
- • * 24. Diamond Pit (10942556)
- 25. Oliver's Mill Bore (11052554)
- • 26. Gordon Pit (11192540)
- • 27. High Wham No.32 Bore (11312734)
- * 28. High Wham No.33 Bore (11412694)
- • * 29. John Pit (11562397)
- • 30. Maddison's No.13 Bore (11712576)
- • * 31. Gordon House th Pit (2
- 32. High Wham No on Colli
- • 33. Gordon House dridge N
- • * 34. Keverstone N on New
- 35. High Wham No dridge N
- • 36. Moorhill No.1 Eldon '
- • 37. Gordon House t Thickl
- 38. Keverstone N on 'A' B
- • 39. High Wham No rles Pit
- 40. Keverstone N on Unde
- • * 41. Gordon House on Unde
- 42. Gordon House dridge N
- • 43. Gordon House on Moor
- • 44. Engine Pit, N Eldon '
- 45. Hall Pit (1314 dlestone
- * 46. Gordon House dlestone
- • * 47. Gordon House on Moor
- • * 48a Gordon House dlestone
- 48b Gordon House
- 49. Keverstone N
- 50. Keverstone N 752879)
- • 51. High Wham No ne (139
- • 52. Keverstone N k (15022
- • 53. High Wham No 272520)
- * 54. Gordon Gill 'F
- • * 55. Gordon Gill 'K *es and*
- • 56. Gordon Gill 'L *graphi*
- • * 57. Gordon House
- 58. Gordon Gill 'A *es and*
- • 59. Gordon House *led in d*
- • 60. Esperley Lane

MAP OF THE

The following account is derived from many more shaft and borehole sections than can be quoted in detail here, and these are kept for reference in the Northern England Office of the Institute of Geological Sciences.

LOWER COAL MEASURES

Quarterburn Marine Band to Ganister Clay seam

(shown graphically on Plate VI)

The Quarterburn Marine Band is exposed in Quarter Burn* [0170 2676] where it consists of 1 ft 7 in of mudstone with sporadic limestone nodules, containing a fauna of *Lingula mytilloides* and *Productus carbonarius*. To the east-north-east in Greenless Beck [0597 2748], the marine band itself is not present, but its position can be inferred by correlating equivalent sandstones and seatearths with those in Quarterburn.

J.H.H.

In the Woodland Bore* (16) the marine band consists of 2 ft 10 in of shale containing *Lingula mytilloides*, *Productus carbonarius*, and *Rhabdoderma sp.* About 3 miles farther east-south-east, however, in boreholes near Snape Foot [1379 2691] the horizon cannot be recognized with any certainty although in Gordon Gill 'K' Bore* (55) it may be present in 9 ft of shale at 249 ft. South of Cockfield at Burnt Houses the marine band was recognized in Keverstone No. 44 Bore* (34) and consisted of at least 1½ ft of mudstone containing *Lingula sp.*

The marine band has been proved in a number of boreholes in the vicinity of Hilton including Hilton Moor No. 2 Bore* (80), Hilton No. 113 Bore* (74), Hilton No. 114 Bore* (87), and in the Hilton Hall Bore* (91). The marine band which consists of shale, mudstone or siltstone is between 6 in and 9 ft 9 in thick and contains brachiopods, bivalves, gastropods and fish scales. The richest fauna was obtained from Hilton Moor No. 2 Bore* (80) and included *Aviculopecten* cf. *delepinei* and a turreted gastropod at the top, and *Lingula mytilloides*, *Orbiculoidea?* and *Donaldina?* at the base. In the Hilton Hall Bore* (91) productoid fragments and a palaeoniscid scale were present. The most easterly recorded occurrence of the presumed Quarterburn Marine Band is from the Shackleton Beacon Bore* (146), where '*Productus sp.*' is recorded from a thin shale at 58½ ft.

D.A.C.M.

The marine band and associated measures are thought to be absent in Knott Holes Quarry [9968 2628] 1½ miles N of Eggleston. Here the Third Grit is separated by only 2 ft 10 in of seatearth mudstone from the top of the Second Grit.

The measures between the Quarterburn and Kays Lea marine bands were proved in the Woodland Bore* (16) to consist of 47 ft 11 in of strata, 44¼ ft of which were sandstone, while in Quarter Burn* (Plate VI) [0170 2676] 4¾ miles WSW the measures are only 26 ft thick with a 12½ ft sandstone. In Greenless Beck [0592 2736] the equivalent strata consist entirely of coarse-grained sandstone overlain by an 8-in coal.

J.H.H.

An exceptional thickness of 89¾ ft of sandstone was proved farther east in Gordon Gill 'F' Bore* (54) at Snape Foot.

Sandstone is locally less abundant in these beds around Cockfield and Hilton, and in Keverstone No. 44 Bore* (34) they include two unnamed marine bands. The upper is equated with a 4 ft mudstone band containing a marine fauna at the bottom of Hilton No. 53 Bore* (97), and a marine band some 16 ft below the Kays Lea Marine Band in Hilton Moor No. 2 Bore* (80).

The unnamed marine band in Hilton Moor No. 2 Bore* (80) is underlain by a 13-in coal, and coals are recorded from several other boreholes at about the same horizon, including the Hilton Hall Bore* (91), and Hilton No. 113 Bore* (74). The upper marine band was not recorded in the Hilton Hall Bore though its position is assumed to be in shale overlying a 36-in coal at 151 ft. In Hilton No. 114 Bore* (87) the lower of the two marine bands recorded in Keverstone No. 44 Bore* (34) may be represented by 2½ ft of shale containing *Lingula mytilloides* and *Rhadinichthys sp.* only 15 ft above the Quarterburn Marine Band although here only part of the sequence between the Quarterburn and Kays Lea marine bands is recorded.

The Kays Lea Marine Band varies in thickness from 3 to 8 ft and in lithology from shale to siltstone. It is not seen anywhere at the surface and has been recorded only in boreholes. In the Woodland Bore* (16) it consists of 6 ft 5 in of shale containing *Planolites ophthalmoides* at the top, *Agathamminoides*? and *Ammodiscus sp.* in a 1 ft 5 in band below, underlain by *Lingula mytilloides* in a 3-in band at the base. The marine band has not been proved farther north-west. Other boreholes in which the marine band or its assumed horizon has been recorded include Moorhill No. 1* (21), Gordon House No. 1* (46), Keverstone No. 44* (34), Hilton Moor No. 2* (80), Hilton No. 53* (97) and Hilton No. 113* (74).

Lithologies and faunas are detailed in Appendix 2.

Measures between the Kays Lea and Roddymoor marine bands are exposed [0172 2660] in Spurlswood Beck* (Plate VI) where they consist of 1¼ ft of seatearth-mudstone on 56½ ft of predominantly medium-grained sandstone —the Third Grit—with a coal horizon some 7 ft from the base. The lower part of the sandstone is also exposed [0595 2723] about 2¾ miles to ENE in Greenless Beck. The Third Grit crops out extensively in the area of Woodland Fell and Eggleston Common but only partial sections near the base of the sandstone are seen, examples of which are as follows:

Knott Holes Quarry [9969 2629] 1½ miles N of Eggleston: 10¾ ft of pink to greenish brown hard thin- to well-bedded sandstone overlain by 7 ft of grey well-bedded medium-grained jointed sandstone.

Quarry [0494 2359], north of Billy Lane House: pink to grey, massive coarse-grained sandstone, 10 ft.

Quarry [0736 2399], about 370 yd NNE of Crag Top: grey, thin-flaggy and false-bedded, fine- to medium-grained sandstone.

J.H.H.

A number of boreholes show that the sequence is generally dominated by the Third Grit which is separated from the top of the Kays Lea Marine Band by between 1 and 12 ft of shale and/or siltstone and occasionally a coal, and from the base of the Roddymoor Marine Band by a seatearth and thin coal ranging between 1 and 18 in thick. In general these measures thin from north-west to south-east corresponding to a thinning of the Third Grit in the same direction. Typical sections are provided by the following boreholes, detailed in Appendix 2: Woodland* (16), Moorhill No. 1* (21), Gordon Gill 'F'* (54), Toft Hill Underground 'B'* (77), Gordon House No. 1* (46), Hilton Moor No. 2* (80), Hilton No. 53* (97), and Hilton No. 113* (74). Atypical sections are proved in Keverstone No. 44 Bore* (34) and Hilton Hall Bore* (91). The section in the former is exceptional in that there is a 2 ft 10 in sandstone immediately overlying the Kays Lea Marine Band, a position normally occupied by argillaceous measures. The latter is unusual in containing no apparent Third Grit.

The Roddymoor Marine Band is absent in the northern part of this district although near-marine conditions may be indicated by the presence of *Planolites* and fish fragments at this horizon in Spurlswood Beck* (Plate VI) in the Woodland Bore* (16) and in Toft Hill Underground Bore 'B'* (77). Similarly, in Keverstone No. 44 Bore* (34) its horizon may be underlying a mudstone containing mussels and fish fragments at 134 ft 11 in. The Roddymoor Marine Band has been recorded in several boreholes in the Hilton area. In Hilton Moor No. 2 Bore* (80) the fauna included *Ammodiscus sp.*, *Lingula mytilloides*, *Donaldina?* and *Elonichthys sp.* [scales]. At Hilton No. 53 Bore* (97) *Serpuloides stubblefieldi*, *Planolites ophthalmoides*, *Lingula mytilloides* and *Euphemites sp.* were recorded, and at Hilton No. 113 Bore (74) the fauna included *Planolites ophthalmoides*, *Serpuloides stubblefieldi*, *Paraconularia sp.* and *Orbiculoidea* cf. *nitida*. All these boreholes are detailed in Appendix 2.

The measures between the Roddymoor Marine Band and the Ganister Clay Coal are very variable in lithology and thickness. In Spurlswood Beck* [0173 2659] (Plate VI) they are 37 ft 4 in thick with only a thin sandstone present under the assumed position of the Ganister Clay coal.

J.H.H.

In the Woodland Bore* (16) these measures are 40 ft 3 in thick and contain two thin sandstones near the top, underlain by predominantly argillaceous beds containing two

Cast

Roddymo
Bore, Cro
In Sheet

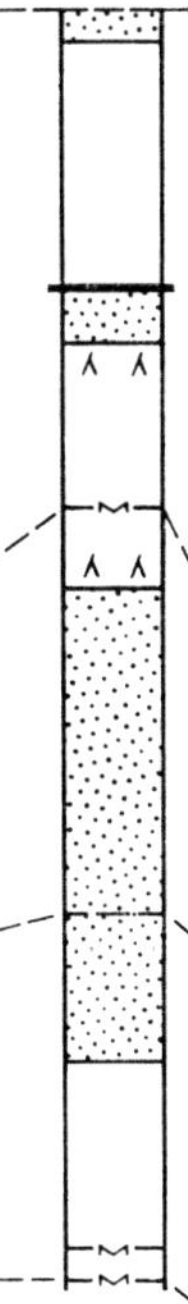

Undi
argil

Thol

Mari

R CO

fossil bands. The lower band some 15 ft 1 in above the assumed horizon of the Roddymoor Marine Band contains *Planolites ophthalmoides* and *Curvirimula sp.*, while the upper band some 4 ft 8 in higher contains *Carbonicola sp.*, *Curvirimula sp.*, *Naiadites sp.* and *Geisina arcuata*. To the east in Moorhill No. 1 Bore* (21) these measures are 73 ft thick and include an 18 ft sandstone at the top of the sequence, the remainder of the measures being predominantly argillaceous and including a 4 in coal 8 ft from the base. In the adjacent Moorhill No. 2 Bore* (22), the upper part of the sequence is broadly similar but in the shale underlying the sandstone and some 20 ft below the Ganister Clay a fauna consisting of *Carbonicola sp.*, *Naiadites sp.*, *Geisina arcuata* and fish remains has been recorded. They are partially exposed in the Gaunless Valley* [1032 2430] west of Peathrow West where they consist of 67 ft of mudstone and shale, but include a 13 ft sandstone about 10 ft below the Ganister Clay. In Gordon Gill 'F' Bore* (54) these beds include three thin sandstones in a predominantly mixed beds sequence, of which the exact thickness is doubtful on account of faulting. In Toft Hill Underground Bore 'B'* (77) the sequence is 52¼ ft thick and includes three sandstones. These measures were penetrated in Keverstone No. 44 Bore* (34) where they are at least 104 ft 11 in thick and include two sandstones separating measures of a dominantly argillaceous or mixed bed character. The lower sandstone is underlain by a coal 1 ft 4 in thick while in the silty measures below fish fragments are present at two horizons, the lowest together with a mussel band immediately overlying the assumed position of the Roddymoor Marine Band. Fossils are also recorded at other horizons in this sequence notably mussels and fish fragments in 8 ft of shale and mudstone overlying the lower sandstone, fish scales at three horizons in silty shaly laminae in the lower part of the upper sandstone, and fish scales about 17 ft below the assumed position of the Ganister Clay. The sequence is somewhat thinner around Hilton although in most boreholes there is a high proportion of sandstone and sandy mudstone. In Hilton Moor No. 2 Bore* (80) the beds measure 75 ft 4 in, and include a sandstone nearly 35 ft thick at the top. In Hilton No. 53 Bore* (97) there are three separate sandstones; two argillaceous beds containing *Curvirimula sp.* are present 40 ft 5 in and 57 ft 11 in below the Ganister Clay. These forms also occur in the Woodland Bore* (16) at about the same position. In Hilton Moor No. 1 Bore* (81) near Hilton Moor Farm a mudstone containing *Spirorbis sp.* and *Carbonicola* aff. *proxima* is present 40½ ft below the Ganister Clay. Although only one specimen of *C.* aff. *proxima* is available it is comparable with *C. proxima* from the Norton mussel band of the East Midlands Coalfield, in the upper '*A*'. *lenisulcata* Zone. The band is possibly represented by the upper of the two faunal bands already referred to in the Woodland Bore. In the Hilton Hall Bore* (91) at least 66 ft of strata separate the Roddymoor Marine Band from the Ganister Clay. A 7-in coal 7 ft above the Roddymoor Marine Band is possibly the equivalent of the 1 ft 4 in coal at 114 ft in Keverstone No. 44 Bore* (34). The measures however are complicated by faulting and correlation is dubious. At the top of the sequence a second thin seam—6 in of inferior coal—lies near the middle of a 17-ft sandstone immediately below the Ganister Clay Coal. Further sections or synopses of sections of boreholes which show all or part of the measures between the base of the Quarterburn Marine Band and the Ganister Clay are quoted in Appendix 2. These include Eggleston No. 3* (1), No. 4* (2), Arnghyll No. 3* (7), Burfoot Leazes* (12), Woodland Colliery No. 29* (6), Hilton No. 1* (103), Hilton No. 2* (86), Hilton No. 3* (90), Hilton No. 5* (98) and Lutterington Estate* (115). All these boreholes are old, and while in most cases they provide sufficient information to establish correlation, they lack detailed lithological data.

D.A.C.M.

Ganister Clay Coal to base of Brockwell

(shown graphically in Plate VII)

The Ganister Clay Coal is apparently thickest on the southern and western margins of the coalfield. In the Woodland area it ranges between 11 and 28 in, the thicker sections generally being banded. The coal was extensively worked at Gibbsneese Opencast

Site [088 241] north and north-east of High Wood Farm where it ranged between 14 and 19 in thick. Between Gibbsneese and Woodland the section proved in Arnghyll No. 3 Bore* (7) about 300 yd ESE of Hill House, Copley, was coal 12 in, band 2 in on foul coal 6 in.

J.H.H.

Farther east round Butterknowle, Windmill and Oaks the seam is usually between 8 and 10 in thick, while around Cockfield the seam, usually banded, has a composite thickness of coal 14 to 19 in, on band 1 to 9 in, on a thin basal coal 2 to 3 in. Atypically in Gordon House No. 98 Bore* (31) on Cockfield Fell the seam is only ½ in thick. South of Evenwood and around Hilton, Keverstone and Wackerfield and thence north-eastwards towards Bildershaw thicknesses of between 6 and 25 in have been proved. The coal is typically banded especially at the base of the section. The little which is known about the seam in the West Auckland and Shildon areas suggests it is thin and banded. At Norton Fine 'A' Bore (92) at Spring Gardens, West Auckland, the seam section recorded was: coal, inferior 6 in, shale 5 in, coal 2 in, siltstone 1 in, coal 6 in, seatearth 4 in, on coal, pyritic 2 in. The seam is similarly thin and banded at Charles Pit* (159) Middridge.

D.A.C.M.

The measures between the Ganister Clay and Marshall Green (or Bottom Marshall Green) seams are between 30 and 65 ft thick and are recorded from many boreholes in the west and south of the coalfield area. A typical section for the Woodland area is provided by the Burfoot Leazes Bore* (12) where the beds are 47 ft thick with argillaceous measures at the base and top with a median 24 ft 8 in of sandstone. Broadly similar sections were proved in Arnghyll No. 3 Bore* (7), Woodland Colliery No. 29 Bore* (6); and in a borehole (14) west of Kays Lea. In the Woodland Bore* (16) these measures, some 45 ft thick, include a thick medium- to coarse-grained sandstone. The upper part of the sequence is also seen in the beck behind Bogle House [0839 2475] as follows:

	ft	in
Seatearth-mudstone; 6-in ganister rib near top	5	5
Sandstone, grey massive current-bedded feldspathic medium-grained	10	0
Mudstone, dark grey silty micaceous	0	8
Sandstone, grey and white carbonaceous .. .. from 1 in to	0	6
Mudstone, dark grey silty micaceous; sporadic plant fragments	2	0
Sandstone, grey thin- to massive-bedded, sparingly micaceous ..	14	6

J.H.H.

These beds are also exposed in the banks of the River Gaunless 30 yd downstream from an adit [0920 2487] and consist of shale with siltstone and sandstone ribs 19½ ft, sandstone 20 ft, overlain by seatearth-mudstone, 6 ft.

They are also exposed in the River Gaunless* [1032 2430] near Peathrow West where they consist of 55 ft of interbedded sandstones and shales.

North of the River Gaunless between Butterknowle, Windmill and Oaks the typical sequence consists of a thin sandstone overlying the Ganister Clay followed by predominantly argillaceous beds in turn succeeded by sandstone, up to 20 ft thick, which is usually separated from the floor of the Bottom Marshall Green coal by a seatearth. The argillaceous beds sometimes contain mussels. Such a sequence was proved in High Wham No. 32 Bore (27) where the measures totalled 43 ft 2 in. Towards the base of the argillaceous beds the section proved was:

	ft	in
Siltstone, grey; sandy partings; comminuted plant debris; *Planolites sp.* in 2-in band near base ..	8	3
Mudstone, grey, shaly; indet. mussels; *Cochlichnus* cf. *kochi* ..	1	5
Mudstone, grey shaly; planty at top, *Planolites* and mussels near top (at 153 ft 6 in)	6	10

The sandstone, 5 ft 1 in thick, which overlies the Ganister Clay in this borehole appears to thicken towards the north-east.

Around Cockfield, Burnt Houses and north of Wackerfield these measures are generally thicker and more variable, and sandstone or sandy measures tend to predominate. A sandstone is generally present below the Bottom Marshall Green seam; west of Cockfield it is exposed [1082 2446] along with adjacent measures on the south side of the Gaunless valley near Peathrow. Atypical sections have been proved in a number of boreholes. In Gordon House No.

94 Bore (59), east of Esperley Lane Ends, and Keverstone No. 48 Bore* (70), north of Wackerfield, the measures are almost wholly of mixed bed or argillaceous type. In contrast, in Keverstone No. 128 Bore (52), nearly ½ mile NW of Keverstone Grange, they are virtually all sandstone. In some boreholes mussels have been noted in the middle or upper part of the sequence, while in Gordon House No. 98 Bore* (31) mussels, fish debris and worm burrows including *Planolites* are recorded from mudstone and siltstone 14 to 17 ft and 24½ ft below the Bottom Marshall Green; and *Planolites sp.* 2¾ ft above the Ganister Clay. In Gordon House No. 96 Bore (41) at Cockfield a 3-in band containing minute fish fragments is recorded 5 ft above the Ganister Clay horizon.

Around Evenwood Gate sandstone predominates, though argillaceous measures up to 20 ft thick overlie the Ganister Clay. In an underground borehole [1600 2443] west of Evenwood Gate, drilled from the Bottom Marshall Green seam in Randolph Colliery, Evenwood, a 1-in coal is recorded 8 ft 9½ in above the Ganister Clay. Some other bores in adjacent areas prove a seatearth on thin coal at about this position.

Around Hilton, sandstone may be present at two and even three horizons, as exemplified by Hilton No. 51 Bore* (76). The upper two sandstones appear to reach their maximum farther to the north-east as proved by boreholes in the vicinity of Bolton Garths, notably Hilton No. 53 Bore* (97). In this borehole plants, including *Alethopteris* and *Mariopteris* were obtained from mudstone in the roof of the Ganister Clay, while plants together with several small examples of *Curvirimula* were recorded in the bed immediately above.

The little which is known about these measures in the West Auckland and Shildon areas suggests that they are between about 60 and 80 ft thick and consist mainly of a mixed bed sequence, although a sandstone up to 36 ft thick immediately underlies the Bottom Marshall Green locally.

D.A.C.M.

The Marshall Green seam is typically split into top and bottom components and is present as a united seam only around Moorhill Drift Mine [0975 2879] 2½ miles NE of Woodland where the section is: coal 18 in, band 1 in, on coal 4 in.

The Bottom Marshall Green Coal around Woodland and Copley generally ranges between 17 and 21 in, sometimes with a thin band near the top. Locally it is absent. Atypically thick sections of 49 and 67 in respectively were recorded in Woodland Colliery No. 29 Bore* (6) and Cowley Bore* (5). In the latter the coal was originally correlated with the Brockwell because of its thickness.

J.H.H.

Around Butterknowle, Windmill and Oaks the seam generally ranges between 10 and 20 in, with a recorded maximum of 25 in. There is commonly a thin band near the base. The seam thins progressively north-east of Windmill towards Witton Park (Wolsingham (26) district) where it rarely exceeds 10 to 15 in. Around Cockfield and Hilton the seam ranges between 9 and 24 in and tends to be rather thicker in the east than the west. A band between ½ and 3-in thick is sometimes present in the lower part of the section. Between Evenwood and Shildon records suggest that the seam thickness lies within the same general range. In underground boreholes [1644 2789] (78) and [1646 2800*] (77) near Hunter's Hill House, Toft Hill, the section proved was: coal 12 to 14 in, band (sandy) 7 in, on coal 7 to 11 in. In St Helen Auckland Engine Pit (122) south-west of Tindale Crescent it was a single seam of 17 in. In a borehole from the base of Charles Pit* (159) only 135 yd N of Middridge, the seam was apparently absent.

D.A.C.M.

The measures between the Bottom and Top Marshall Green seams generally range between 15 and 50 ft, the extremes being 7 and 80 ft; the thicker sequences including much sandstone. In the Woodland and Copley areas a sandstone generally ranging between 8 and 10 ft rests directly on the Bottom Marshall Green seam; in places argillaceous measures underlie the sandstone. Atypical sections were proved in Arnghyll No. 1 Bore* (9) and Arnghyll No. 3 Bore* (7) where little sandstone is present, while in Cowley Bore* (5) near Cowley the measures are 69 ft 2 in thick and consist of alternating beds of mudstone and sandstone, the latter predominating.

J.H.H.

Around Butterknowle, Windmill and Oaks these beds range between 8 and 33 ft, and are predominantly argillaceous in the west, especially where the Top Marshall Green is absent or very thin, and arenaceous in the east; here the roof of the Bottom Marshall Green seam is generally a grey siltstone with sandy bands, sandstone or a grey fine-grained sandstone as in High Wham No. 31 Bore* (23) and High Wham No. 33 Bore* (28). In boreholes where the measures are thinner, e.g. High Wham No. 32 Bore (27) nearly ½ mile N of High Wham they are more argillaceous although still often sandy at the base.

Around Cockfield and Hilton these measures range from 5 to 55 ft, being greatest between Cockfield and Keverstone Grange. At many localities a thick sandstone forms the roof of the Bottom Marshall Green seam. It crops out in the Gaunless Valley* [1082 2444] north of Peathrow, while in a quarry [1351 2280] in North Wood about ¼ mile WNW of Keverstone Grange it consists of over 20 ft of yellow and purplish massive to thick-, locally false-bedded medium- to coarse-grained sandstone. In Keverstone No. 133 Bore* (63) it is nearly 40 ft thick and apparently cuts out the Top Marshall Green; by contrast, in Keverstone No. 48 Bore (70) west of the Sun Inn, Wackerfield, the separation between the two seams is only 13 ft, mainly of seatearth. Similarly in Hilton No. 51 Bore* (76) these measures, again argillaceous, are only 5¼ ft thick. Little is known about these beds in the West Auckland and Shildon areas; they appear to range between 15 and 21 ft and are predominantly of a mixed bed or sandy character. Flaggy sandstone overlying the Bottom Marshall Green has been worked [2069 2420] north of Royal Oak.

D.A.C.M.

The Top Marshall Green is everywhere thin and occasionally it is absent. The seam is thickest in the Woodland and Copley areas where it ranges up to 16 in and averages about 11 in. At Dent Gate Opencast Site [079 244] it was worked together with the Bottom Marshall Green, the separation between the two seams being 12 to 15 ft.

J.H.H.

South and east of Copley and between Butterknowle, Windmill and Oaks the seam is in general between 4 and 7 in thick. At High Wham No. 32 Bore (27) it is 16 in, including a 5-in band 2 in from the base; by contrast it is absent in High Wham No. 41 Bore (51) nearly ½ mile S of Windmill. The seam rarely exceeds 6 in in the Cockfield, Low Lands, Burnt House and Keverstone Grange areas and is locally absent, its assumed horizon being marked by the top of a seatearth. In Gordon House No. 131 Bore* (57) there are two 1-in coals at this horizon, separated by 11 in of black silty micaceous shale containing ?fish scales while in Keverstone No. 133 Bore* (63) a single coal 10 in thick is present. Around Hilton and towards Bildershaw the seam falls generally within the range of 6 to 14 in, and at most places it is less than 10 in. The few records available in the West Auckland, Shildon and Middridge areas suggest that the thickness ranges between 4 and 11 in. The seam is absent at Charles Pit* (159), Middridge.

D.A.C.M.

The measures between the Top Marshall Green and the Stobswood coal or its assumed position, range generally between 12 and 30 ft, extremes being 2 and 40 ft. In the Woodland and Copley areas a typical sequence was proved by Moorhill No. 121 Bore (20), the details being as follows:

	ft	in
STOBSWOOD (at 88 ft 9 in)		
Shale	0	1
Seatearth-mudstone	3	8
Sandstone	22	6
MARSHALL GREEN (at 116 ft 11 in)		

J.H.H.

Between Butterknowle, Windmill and Oaks these measures range between 2 and 22 ft, the thickest sequences being broadly similar to that proved in Moorhill No. 121 Bore (20) above, or where thinner being predominantly argillaceous in character. In High Wham No. 33 Bore* (28) the two seams are separated by a mere 3 ft 5 in of seatearth-sandstone. South of the River Gaunless and around Cockfield, Wackerfield and Hilton the interval ranges generally between 5 to 10 ft, and consists mainly of argillaceous or silty beds. Exceptionally, at Gordon House No. 131 Bore* (57) the separation is over 11 ft, with median sandy beds.

Near West Auckland, 18¼ ft of measures, mainly sandstone, were recorded, and some distance farther north in Toft Hill Underground Bore 'B'* (77) these measures, 12¼ ft thick, are again mainly of sandstone though an adjacent borehole records only 6¾ ft of fireclay.

The impersistent **Stobswood Coal** is rarely more than 6 in thick. In the Woodland and

Copley areas it reaches up to 5 in, although it is absent in the Woodland Bore* (16). North of the River Gaunless between Butterknowle and West Auckland it is generally absent, its assumed horizon overlying a fireclay. Similarly, south of the River Gaunless it is thin and impersistent, although in Hilton No. 51 Bore* (76) it attains 9 in.

D.A.C.M.

Stobswood to Victoria group of coals. Between the Stobswood Coal, or its assumed position, and the Victoria group of coals are very variable measures containing up to two coals, one of which, locally split, has been termed 'Bottom Victoria'.

In the Woodland and Copley areas these beds are proved in Arnghyll No. 1 Bore* (9), Pioneer Shaft and Bore* (13) and Moorhill No. 118 Bore* (18). The beds thicken northeastwards from 44 to 72 ft, although exceptionally in Pioneer Shaft and Bore they are only about 28 ft thick. Generally the sequence consists of alternating sandstone and shale although in the Woodland Bore* (16) it is predominantly of sandstone. A dark shale containing fish remains overlies the assumed horizon of the Stobswood in this borehole, occupying a similar horizon in Moorhill No. 121 Bore (20).

J.H.H.

North of the River Gaunless between Butterknowle, Windmill and Oaks the thickness of the measures is generally in the range 45 to 55 ft. A typical sequence is recorded in High Wham No. 32 Bore (27). There is a persistent seatearth and/or coal, sometimes in two leaves—the 'Bottom Victoria'—between 14 and 21 ft below the Victoria. Despite its name, the Bottom Victoria is not a split off the Victoria in this district, and always appears as an entirely separate seam. Below this horizon the strata are wholly or predominantly arenaceous to within a few feet of the Stobswood. In places they are underlain by a few feet of argillaceous strata containing fish scales, *Planolites ophthalmoides* and mussels, possibly at the horizon of the Stobswood Marine Band. In Moorhill No. 12 Bore (36) and High Wham No. 40 Bore (39) the 'Bottom Victoria' is present, the section in the latter case being: coal 3 in, band 1 in, coal 3½ in, band 1 in, on coal 4 in. Two miles farther south-east, in Gordon House Nos. 130 (43) and 135 bores (64) [1337 2506, 1469 2546], in the Gaunless valley east-north-east of Low Lands this coal is split into two leaves, an upper leaf 1 to 12 in thick separated by 1½ to 5 ft of strata from a lower leaf 9 to 10 in thick. Two thin coals 5 and 8 in thick respectively separated by 4¼ ft of seggar are present in West Tees No. 1 Bore* (65) near Ramshaw. Other details which demonstrate the wide range of variation in this group of strata include the following:

In Moorhill No. 2 (22), High Wham No. 34 (35) and No. 42 Bores (53), near High Wham and Morley, thin conglomerate bands consisting of sandstone containing pebbles of clay ironstone, shale and siltstone overlie the Stobswood or its position. In High Wham No. 36 (32) and Moorhill No. 12 (36) additional seatearths are present between 14 and 20 ft above the Stobswood horizon. In High Wham No. 33 Bore* (28) a 4-in coal is present at the horizon of the lower seatearth.

Strata between the seatearth or Bottom Victoria and the base of the Victoria group of coals are of a mixed bed character, sandstone being impersistent. The sequence in the area around Cockfield, Burnt Houses and Keverstone Grange shows similar variability. The thickness ranges between 37 and 74 ft, the widest variation being in the north, the separation tending to fall to the east. Mussels have been recorded above the Stobswood only from Gordon House No. 93* (47) and No. 131* (57) bores. The Bottom Victoria is impersistent but where absent its horizon is usually marked by its seatearth. Around Cockfield it is usually thickest and occasionally occurs as two thin coals separated by a few feet of argillaceous measures. For example, in Gordon House No. 97 Bore (33) about ½ mile WNW of Cockfield Church, two coals 1 and 13 in respectively, the latter banded, are separated by 3 ft 4 in of seatearth-sandstone. Around Buck Head and Keverstone Grange this coal occurs as a single seam and attains its maximum thickness of 18 in in Gordon House No. 94 Bore (59) west-south-west of Buck Head. The measures between the Bottom Victoria and the Victoria group of coals are mainly argillaceous with impersistent sandstones. They vary in thickness from 2 to 36 ft, the widest variations being found north of Cockfield. The most typical thickness is around 24 ft. In Gordon House No. 131 Bore* (57) poorly preserved mussels are recorded in the roof of a seatearth some 15 ft below the Victoria, and mussels are also present 17 ft below the Victoria at Gordon House No. 93 Bore* (47). Exceptionally, in Gordon House No. 98 Bore* (31) the separation between the Bottom Victoria and the Victoria is only 2 ft of seatearth-mudstone.

Around Wackerfield, Hilton and Bildershaw the interval between the Stobswood and Victoria averages about 60 ft, a thin coal, the Bottom Victoria, usually being present in the middle or upper part of the sequence. Sandstones are again impersistent but occur more commonly in the upper part of the sequence than elsewhere. In Keverstone No. 48 Bore (70) there is a sandstone over 30 ft thick in the upper part of the sequence underlain by argillaceous measures containing mussel fragments. A bed with fish scales and spines and ?mussel fragments is recorded in this position in a number of boreholes including Hilton No. 55 Bore* (88), and Hilton No. 52 Bore (89) where 2 ft 10 in of shale containing mussel fragments lies above the Stobswood. The Bottom Victoria ranges up to 12 in; in Keverstone No. 48 Bore (70) it is split in two leaves, 8 and 10 in thick, separated by about 4½ ft of seatearth-mudstone and shale, the latter containing large mussels. Mussels and fish scales overlying this coal are recorded in a number of boreholes, notably Hilton No. 50 Bore (73) and Hilton Moor No. 2 Bore* (80).

Around West Auckland scattered boreholes suggest that the sequence is broadly similar to that in the Cockfield and Hilton area, the recorded interval between the Stobswood and the Victoria ranging between about 35 and 47 ft. The lower part of the sequence is mainly argillaceous and the upper part mainly arenaceous or of mixed bed type. A thin coal about the middle of the sequence is sometimes in two leaves separated by up to 2 ft 9 in of fireclay. Toft Hill Underground Bore 'B'* (77) and Norton Fine 'A' Bore (92) exemplify the sequence in this area. There is little reliable information on these beds in the Eldon and Middridge areas.

D.A.C.M.

The Victoria group of coals, that is the Victoria Coal or, where split, the Victoria and its upper member the Victoria Rider together with intervening measures, are seen at the surface only near Copley Lodge north of Butterknowle. In the Woodland and Copley areas the group is highly variable. The Victoria Coal, only locally absent, varies from 1 to 24 in and is separated by as little as 1 in of seatearth to a recorded maximum of 35 ft of variable measures from the Victoria Rider, which has a proved maximum of 8 in. In the Woodland Bore* (16) the Victoria Rider is thought to be represented by a single 3 in seam at 54 ft, the Victoria being absent. Sections through the group are also provided by the Pioneer Shaft and Bore* (13), Moorhill No. 118 Bore* (18) and Cowley Shaft (11).

J.H.H.

Between Butterknowle and Oaks this group is generally either a single united seam or is split into Victoria and Victoria Rider. Occasionally the positions of one or both seams may be absent marked only by seatearths. The combined Victoria and Victoria Rider seams are exposed [1031 2643] in a stream 250 yd SW of Copley Lodge, Butterknowle where they form a thin coal known locally as the Six-Inch. Some 400 yd NW of this locality only the Victoria Rider (7 in) is exposed, though the Victoria is presumed to be present close below. Other records showing the presence of both coals include Moorhill No. 1 Bore* (21): coal 2 in, seatearth 1 ft 7 in, on coal 12 in; and Gordon House No. 130 Bore (43): coal ¾ in, seatearth-mudstone 7¼ ft on coal 13 in. In High Wham No. 32 Bore (27) where only the 9 in Victoria Rider is present, the seam is underlain by 2¼ ft of argillaceous beds, the lower 7 in being black micaceous shale containing fish scales and overlying the assumed horizon of the Victoria.

Around Cockfield, Burnt Houses and Keverstone Grange there are typically two coals. The Victoria ranges generally between 7 and 12 in, the maximum recorded thickness of 19 in (banded) being in Gordon House No. 98 Bore* (31). North and north-east of Keverstone Grange and at Keverstone No. 43 Bore (38) the coal is absent. The measures between the two seams range from 3 ft in Gordon House No. 98 Bore* (31) to an exceptional 35 ft in Keverstone No. 148 Bore (49), a 26-in seatearth being present 18 ft 2 in below the Victoria Rider horizon. Normally the separation ranges from about 9 ft in the west to 25 ft in the east of this area. The measures are generally of shale and seatearth the latter tending to predominate where the separation is least. In Keverstone No. 136 Bore (61) north-east of Keverstone Grange, the two seams are 25 ft apart, 7¾ ft of fireclay overlying the Victoria Coal. The thickness of the Victoria Rider in this area ranges up to 7 in, but in many places it is absent, its assumed horizon overlying a seatearth.

North of Hilton and towards Bildershaw this group occurs both as a single and a split seam. It is a single 6-in seam in Hilton Moor No. 2 Bore* (80) and in Evenwood Gate No. 1 Bore (75) it is 10 in thick. In Hilton No. 55 Bore* (88) it is split. Around West Auckland the sparse records suggest that the coal is typically split, the section in Norton Fine 'A' Bore (92) at Spring Gardens being: coal 3 in (Victoria Rider), seatearth 7¾ ft, coal 3 in (Victoria) and at Toft Hill Pit (67): coal 8 in (Victoria Rider), seatearth 6 ft, on coal with two bands 27 in (Victoria).

D.A.C.M.

Little is known about these coals in the Eldon, Shildon and Middridge area. In Jane Pit* (142) ¾ mile NNW of St John's Church, Shildon, a 12-in coarse cannel 57 ft below the Brockwell may represent both Victoria and Victoria Rider, while at Charles Pit* (159), Middridge the coal is absent, its horizon overlying 4 ft 2 in of seggar clay.

J.H.H.

The measures between the Victoria group and the Brockwell Coal are, except for minor exposures near Butterknowle, recorded only in boreholes. In the Woodland and Copley area there are complete sections of these beds from several boreholes and shafts including Pioneer Shaft and Bore* (13), Woodland Bore* (16), Moorhill No. 118 Bore* (18) and Cowley Shaft (11), the thickness ranging between 44 and 70 ft. A typical section is that from Moorhill No. 118 Bore* (18) where these measures are 48 ft 11 in thick, including a median 24 ft sandstone. This is underlain by the Victoria Shell-Bed—5½ ft of mudstone containing mussels. In this borehole, the roof of the Victoria Rider comprises 2 ft 11 in of shale containing fish scales—the Victoria Fish-Bed. In the Woodland Bore* (16) the Victoria Shell-Bed contained *Planolites ophthalmoides*, *Carbonicola declivis*, *C. pseudorobusta* and *Naiadites sp.* in 12 ft 5 in of mudstone, separated by 8 in of black shale from the Victoria Rider.

J.H.H.

In boreholes between Butterknowle, Windmill and Oaks these measures have an average thickness of about 50 ft. They are of mainly 'mixed beds' lithology, with, in places, a sandstone up to 30 ft thick in the upper part of the sequence. In most places the Victoria Rider, or its horizon, is overlain by the Victoria Fish-Bed, a shale containing fish debris including scales and spines. In High Wham No. 40 Bore (39) 15 in of shale containing *Planolites* cf. *ophthalmoides* intervene between the coal and 3 ft 5 in of black shale containing fish scales and spines, regarded as the Victoria Fish-Bed, which is in turn overlain by the Victoria Shell-Bed. The Victoria Shell-Bed is proved in several boreholes to extend up to 15½ ft above the Victoria Rider Coal. It mainly contains the small mussel *Curvirimula*, but *Carbonicola sp.* and *Planolites ophthalmoides* are also recorded locally. Near the upper part of the sequence, a seatearth, in places carrying an inferior coal up to 4 in thick is present between 8 and 12 ft below the Brockwell coal. This coal may be equated with the ?Bottom Brockwell of the Keverstone area.

South of the River Gaunless around Cockfield, Burnt Houses and north of Wackerfield the group varies between 40 and 82 ft the latter thickness referring to an exceptional section in Gordon House No. 98 Bore (31) where the interval is mostly occupied by sandstone. The Victoria Fish-Bed and the Victoria Shell-Bed are recorded from most boreholes in this area. In Gordon House No. 97 Bore (33) and Gordon House No. 98 Bore* (31) fish remains are recorded above the Victoria Rider Coal. The base of the Victoria Shell-Bed ranges between 9 and 19½ ft above the Victoria group; in all cases however, its contained fossils are recorded only as "mussels" or "small mussels". In Gordon House No. 97 Bore (33) a 2 ft 8 in band, 8 ft 10 in above the Victoria Rider and at the top of the shell bed, contains *Planolites ophthalmoides*. In places, as for example Gordon House No. 131 Bore* (57), a thick seatearth is present up to 5 ft 10 in below the Brockwell and separated from it by argillaceous measures. In some boreholes between Keverstone Grange and Burnt Houses a coal overlies the seatearth and although highly banded and local, it attains a composite thickness of up to 35½ in. The section in Keverstone No. 133 Bore* (63) is an example. This coal is thought to be the equivalent of the Bottom Brockwell Coal of the Durham and West Hartlepool district (Smith and Francis 1967, p. 38).

Little is known of the beds between the Victoria group and Brockwell in the area north-east of Wackerfield towards Bildershaw. Here the only complete sequence is recorded in Evenwood Gate No. 1 Bore (75) south-south-east of Evenwood Gate, where the thickness is 70 ft 1 in including a median 22 ft sandstone. The Victoria Shell-Bed is represented by 4½ ft of silty mudstone containing indeterminate mussels, and separated by 9½ ft of apparently barren mudstone from the Victoria Rider. Partial sections in these measures are recorded in Hilton Moor No. 2 Bore* (80) where neither shell nor fish bed is recorded, and Hilton Moor No. 55 Bore* (88) where fragmentary mussels are present (Victoria Shell-Bed) 17 ft 10 in above the Victoria Rider, the latter being overlain by 2 ft of shale containing fish scales (Victoria Fish-Bed). At Randolph Colliery No. 1 Shaft (72a) the section immediately under the Brockwell coal is: ironstone 10 in, seggar 3 ft, on measures, mainly sandstone, 30 ft.

In the Toft Hill, West Auckland, Shildon and Middridge areas these measures have been recorded in nine boreholes and shafts most of which are old. The group expands eastwards from 48 to 65 ft, and is generally of a mixed bed lithology although sandstones occur impersistently as in Toft Hill Underground 'B' Bore* (77) where both Shell-Bed and Fish-Bed were proved. The Bottom Brockwell has not been recorded in this area, the Brockwell coal being variously floored by thick seatearth or sandstone. In Norton Fine 'A' Bore (92) neither the fish bed nor the shell bed were proved.

Further sections through all or part of these measures include those from Burfoot Leazes Bore* (12), Woodland Colliery No. 29 Bore* (6), Hilton Nos. 1 (103), 2* (86) and 3* (90) bores, Lutterington Estate Bore* (115), Engine Pit, St Helen Auckland (122), Jane Pit, Adelaide Colliery* (142) and Charles Pit, Middridge* (159).

D.A.C.M

Brockwell to Harvey Marine Band

(shown graphically in Plates IX and X)

The Brockwell Coal, known in places as the Main Coal, is the lowest major seam to have been worked in the district. It has been wrought since the earliest times and reserves are all but exhausted. Extraction has taken place from all the major collieries, in addition to the innumerable smaller pits and day holes scattered along the outcrop throughout the northern part of the district. Pillars left by mining from 19th and early 20th century shallow drifts and even earlier 'bell pit' and 'day hole' extraction methods have been removed by opencast working particularly along the northern outcrop, between Woodland and Toft Hill. The coal is now seen only in temporary opencast sections although its outcrop is often well defined by 'sink holes' and other marks of old workings.

D.A.C.M.

In the Woodland and Copley areas the coal ranges between 38 and 84 in. The westernmost recording was in a drift [0356 2467] on Langleydale Common, some 2 miles SW of Woodland, where it was 38 in thick. About a mile farther east Woodland Colliery No. 47 Bore (3; B & S 3096) near Dale Terrace proved: coal 57 in, band 2 in, on coal 20 in; a mile to the south-east in Arnghyll (or Arn Gill) the general seam section is: coal 67 in, band 4 in, on coal 15 in. Comparable seam sections were measured in Cowley Shaft (11; B & S 3086), Woodland Colliery Shaft (10; B & S 2338), and in colliery workings ½ mile S of High Wood Farm [0874 2396]. In Pioneer Shaft and Bore* (13) a single seam section of 72 in was recorded. East of Copley, as for example at Diamond Pit* (24), Butterknowle, the seam is 70 in thick.

J.H.H., D.A.C.M.

North of the river Gaunless between Butterknowle and the area west of Toft Hill, including that around Gordon Beck, the Brockwell is generally between 72 and 75 in thick, the recorded range being 60 and 82 in. The seam was worked at outcrop at a number of sites in the area, the largest being at Cold Hurst [1066 2788], north and east of Morley Farm [1257 2829] and a ¼ mile W [133 283] of Windmill. In an opencast site [1500 2743] over ½ mile NNE of Bowes Close the seam ranged between 68 and 76 in. At Lands Engine Pit [1344 2505] (B & S 1271), Low Lands, the seam section was cannel 11 in, on coal 71 in; while in the nearby Gordon House No. 130 Bore (43) it was coal 45 in, band 11 in, on coal 17 in.

country around Barnard Castle (*M*

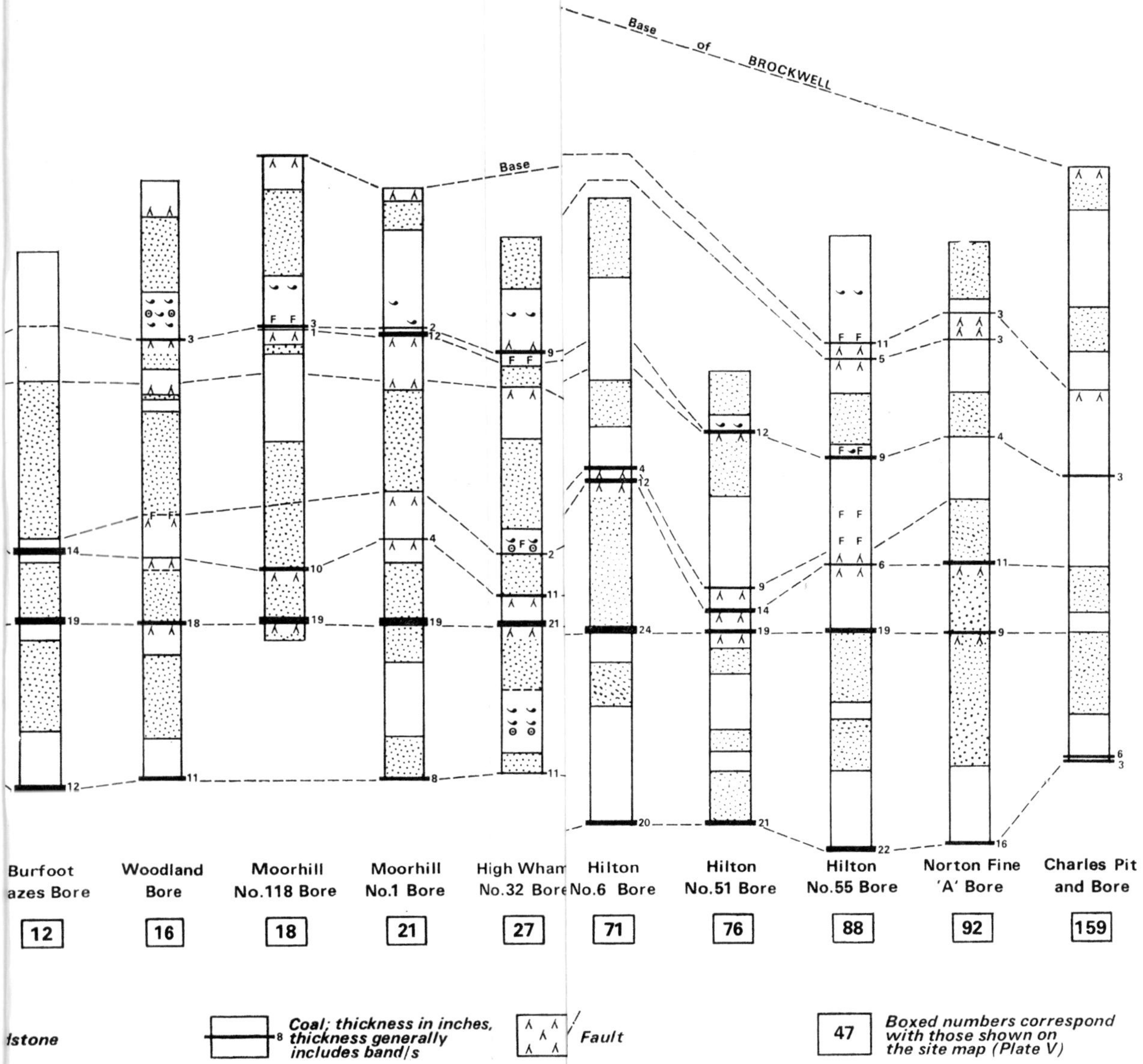

PLATE VII

dstone

Coal; thickness in inches, thickness generally includes band/s

Fault

47 *Boxed numbers correspond with those shown on the site map (Plate V)*

C

Around Cockfield, on Cockfield Fell, and between Shotton Moor [104 237] in the west and Keverstone Grange [1382 2265] in the east, the Brockwell is usually between 70 and 90 in thick, frequently banded, and cannelly at the top. On Shotton Moor [104 237] the section is coal 51 in, band 9 in, on coal 35 in, and at John Pit* (29) cannel 7 in, coal 49¼ in, band ¼ in, on coal 14½ in. On the western side of Cockfield Fell the seam averages up to 81 in including up to 12 in of cannel on top, while farther east on the Fell and north of Hall Pit (45) a typical section is cannel 14 in, coal 60 in, band 4 in, on coal 15 in. East of Keverstone Grange and north-west of Wackerfield [162 227] the seam is thinner, a typical section being cannel 15 in, coal 9 in, band 1 in, coal 30 in, on splint 6 in.

Around Evenwood, and between West Auckland and Wackerfield, the coal ranges generally between 48 and 75 in, a band 2 to 14 in usually being present up to 24 in from the base of the coal. Very thick sections are, however, recorded locally. For instance, at Randolph Colliery No. 2 Shaft* (72b) the seam section is cannel 48 in, coal 66 in, band 3½ in, on coal 6 in, while at the base of Staindrop Field House Drift* over ½ mile ENE of Evenwood Gate the coal section was cannel (including fish fragments) 12 in, coal 48 in, seatearth-mudstone 13 in, coal 12 in, seatearth-mudstone 4 in, on coal 8 in.

Two small areas of Brockwell Coal occur south of the main outcrop. The first is about ½ mile ENE of Wackerfield [162 227] where the coal incrops against the Hilton Fault and Wackerfield Dyke. The coal, up to 60 in thick, is overlain by shale and underlain by fireclay. The strata are highly disturbed by faulting and the coal is baked and cindered along the dyke margins. A second small area of Brockwell occurs north of Todwell House [1667 2129]; here the coal, 60 in thick, was found to be so highly disturbed due to faulting as to be virtually unworkable. The coal is cut off abruptly to the north by one of the faults of the Hilton Fault complex. Considerations of structure suggest that a small area of Brockwell may also exist a short distance south of Wackerfield [154 221].

South and east of Toft Hill, around West Auckland, Tindale Crescent, Hummerbeck and Brusselton, the Brockwell coal generally ranges between 56 and 80 in. At Toft Hill Pit (67) it is 72 in, at Windlestone Pit, West Auckland (106; B & S 2174) 68 in, while at St Helen Auckland Engine Pit* (122) the section is coal 11 in, soft grey metal band 1½ in, coal 16 in, coal splint 2 in, on coal 37 in.

D.A.C.M.

Over the whole of the district farther east and including the South Church, Coundon Grange, Eldon, Old Eldon, Windlestone Park, Fieldon Bridge, Shildon and Middridge areas the coal generally ranges between 55 and 72 in thick. Somewhat thicker sections were proved along the southern outcrop of the seam east of Brusselton Hill [2037 2498], where measurements in shallow workings show a general thickness of the order of 72 in. The seam was worked opencast at New House site south of Hill Top Farm [2123 2487] where a typical sequence was coal 12 in, shale up to 24 in, coal 31 in, shale up to 15 in, on coal 27 in.

J.H.H., D.A.C.M.

The measures between the Brockwell and Busty coals (or where split the Bottom Busty) have been proved in many boreholes, more especially in the central and eastern part of the coalfield area. They are rarely seen at outcrop. The sequence, which overall averages 100 ft thick, may contain up to four sandstones and three thin coals. All these coals tend to be impersistent, although the highest, which is equated with the Three-Quarter seam of the Durham and West Hartlepool district, is proved in most boreholes. Up to four faunal bands are jointly recorded from boreholes through this sequence, notably in the roof measures of the Brockwell, up to 65 ft above the Brockwell, in the roof of the Three-Quarter, and some 16 ft above the Three-Quarter.

D.A.C.M.

In the Woodland and Copley areas these measures, rather thinner than over much of the remainder of the coalfield area, range between 68 ft in the west and 58 ft in the east. The following sequence was proved in Woodland Colliery No. 47 Bore (3) near Dale Terrace:

	ft	in
BOTTOM BUSTY (at 51 ft 8 in)		
Fireclay	14	0
Sandstone	9	6
Grey bed	14	10
Sandstone, grey	23	9
Shale, blue fine	6	0
BROCKWELL (at 126 ft 4 in)		

The above sequence is similar in other records farther east though the proportion of sandstone varies considerably from 80 per cent in Woodland Colliery Shaft (10) where it is in two main "posts", to as little as 27 per cent in Diamond Pit* (24), Butterknowle where it forms three thin bands. Neither the Three-Quarter nor the other two thin coals of the sequence are recorded in this area. A 24-in coal which lies 4 ft above the Brockwell in Woodland Colliery Shaft may be regarded as a split off the Brockwell and as an equivalent of one of the thin roof coals to the seam known in the Durham and West Hartlepool district (Smith and Francis 1967, p. 42). J.H.H.

North of the Butterknowle Fault, between Butterknowle and Windmill, only the lower and occasionally the middle part of the sequence crop out, generally under thick drift, and records of exposures are few. In some areas a sandstone up to 30 ft thick is thought to occur in the lower part of the sequence. Sections immediately above the Brockwell Coal were formerly exposed at a number of opencast sites, and one such section at the south-west corner of an opencast site near Rowntree Farm [1170 2819] was as follows:

	ft	in
Mudstone, dark grey flaggy and shaly highly micaceous and carbonaceous ripple-marked; sandstone ribs becoming abundant to base; worm borings	12	0
Sandstone, dark grey flaggy micaceous and carbonaceous.. ..	8	0
Mudstone, dark grey shaly highly micaceous and carbonaceous ..	4	0
Mudstone, dark grey shaly micaceous and carbonaceous; flaggy bands; poorly preserved plant fragments; disseminated carbonaceous material	6	0
BROCKWELL		

Broadly similar sections were obtained at opencast sites north and east of Cold Hurst [1066 2788] and north and east of Morley Farm [1257 2829]. On the southern edge [1315 2807] of an opencast site 600 yd WSW of Windmill the following section was recorded:

	ft	in
Sandstone, buff grey and yellow fine- to medium-grained locally false-bedded; sporadic micaceous carbonaceous partings	10	0
Mudstone, dark grey flaggy and shaly micaceous; sporadic ribs of muddy carbonaceous micaceous sandstone up to 6 in thick; occasional poorly preserved plant fragments	18	0
BROCKWELL		

At a former opencast site [119 266] south-east of High Wham a 20-in coal about 22 ft above the Brockwell was worked, the intervening beds comprising 1 ft of shale underlain by 21 ft of sandstone.

South of the Butterknowle Fault in the Gordon Beck area the only complete sequence in these measures was recorded in Providence Pit (62; B & S 957) near Bowes Close where they are nearly 67 ft thick, predominantly argillaceous in character and contain four thin ribs of sandstone. A 6-in coal, possibly the Three-Quarter, is present some 28 ft below the Bottom Busty. Farther south-west, at Lands Engine Pit [1344 2505] (B & S 1271), 2 and 9 in coals are present at 34 and 38 ft respectively above the Brockwell.

Around Cockfield, on Cockfield Fell, in the Gaunless valley between Butterknowle and Oaks, and between Shotton Moor [104 237] and Keverstone Grange [1382 2265] these measures range between 38 and 95 ft, the greatest thicknesses being recorded in the west. Typically, a sandstone up to 20 ft thick and sometimes split into two leaves is present between the Brockwell and Three-Quarter, while a further sandstone between the Three-Quarter and Busty locally forms the floor of the latter. The Three-Quarter seam, recorded in most boreholes, is generally between 6 and 15 in thick, but locally round Cockfield thicknesses of up to 30 in have been recorded. An atypical Three-Quarter section is recorded in Gordon House Nos. 95 (42) and 144 (37) bores on Cockfield Fell. In the former, the coal section is coal 5 in, shale 22 in, on coal 13 in; while in the latter the coal is 26 in thick with a 4-in band 8 in from the base. At Gordon House No. 135 Bore (64) in the Gaunless valley north-east of Low Lands, the following sequence was recorded:

	ft	in
BUSTY (at 43 ft)		
Seggar	1	3

	ft	in
Shale, sandy	1	6
Shale, dark; coal traces	0	6
Shale, light grey	17	5
THREE-QUARTER		
Coal	1	2
Shale	2	8
Sandstone	9	0
Shale, sandy	1	6
Sandstone	3	6
Shale, sandy	5	6
Shale, grey	2	6
Shale; dark partings	12	6
Shale, dark	1	3
COAL	0	11
Shale, light grey	3	10
Sandstone, grey	1	3
Shale, sandy	2	9
Sandstone	6	6
Shale, sandy; broken .. about	10	0
BROCKWELL (at 134 ft)		

The 11-in coal in the above section is one of the lower impersistent coals in the Brockwell–Busty sequence and is equated with the 20-in coal at the opencast site [119 266] near High Wham (see p. 96) and the 2- and 9-in coal at Lands Engine Pit [1344 2505] (see p. 96). In Gordon House No. 94 Bore (59) near Buck Head the sequence is as follows:

	ft	in
BUSTY (at 238 ft 8 in)		
Shale, sandy; sandstone bands	13	4
Mudstone, grey; fragmentary mussels at base	1	1
Siltstone, grey slightly sandy; rare mussels and plant remains	3	11
Horizon of ?THREE-QUARTER (at 257 ft)		
Shale, grey sandy; fine argillaceous sandstone partings	3	0
Siltstone, grey shaly; sporadic 'mussel' fragments near top	5	6
Mudstone, grey silty; irony bands; occasional 'mussel' debris; coaly films; micaceous and carbonaceous in basal 6 in	3	0
Sandstone, dark argillaceous carbonaceous; 'mussel' debris, plant remains	2	2
Shale, dark grey splintery micaceous pyritic	0	6
COAL, bright pyritic	0	8
Seatearth, brownish grey	0	6
Siltstone, grey; rooty at top; sporadic ironstone nodules	5	8
Shale, grey sandy; argillaceous sandstone bands	5	0
Sandstone, grey flaggy wispy bedded fine-grained	1	0
Shale, grey sandy; flaggy sandstone bands	4	0
Shale, dark grey sandy; plant fragments; 'mussel' fragments at base	14	0
Cavity	0	6
Shale, predominantly black cannel-oid; some pyrite; irony and hard at base	4	6
BROCKWELL—old workings (at 315 ft)		

The above sequence is notable for a very low sandstone content and for the presence of mussels at three horizons. In Keverstone No. 133 Bore* (63) over half a mile north-east of Keverstone Grange the lower two faunal bands are present and the lowest is 25 ft above the Brockwell seam. This contrasts with Gordon House No. 94 Bore (59) where the lowest mussel band is only about 6 ft above the Brockwell. Except for small exposures of flaggy micaceous mudstone with thin sandstone bands overlying the Brockwell coal on the south side of the Gaunless valley a ¼ mile SSW of the Slack these measures are not seen at the surface.

In the Evenwood area, and between West Auckland and Wackerfield, the Brockwell–Busty interval generally ranges between 95 and 108 ft. Two thin coals are usually present in the sequence. The lower coal, 3 to 8 in, ranges from 37 to 54 ft above the Brockwell; the upper, 8 to 12 in, ranges from 69 to 84 ft above the Brockwell and is thought to be the equivalent of the Three-Quarter. North-east of the Sun Inn [1542 2317] a sandstone up to 30 ft thick is present 10 to 20 ft above the Brockwell. Towards the north-east, this sandstone is absent or sporadic in its distribution. At Randolph Colliery No. 2 Shaft* (72b) the sequence, 88 ft 9 in thick, is mainly arenaceous, but includes the Three-Quarter, 12 in thick, and a lower thin coal. At Staindrop Field House Drift* over ½ mile ENE of Evenwood Gate, these measures are 104 ft thick, the Three-Quarter, 1 ft, being separated from the Busty by 20 ft of mainly argillaceous strata. A faunal assemblage 52 ft above the Brockwell is reminiscent of the *C. declivis* fauna from the Brockwell Ostracod Band of the Durham and West Hartlepool district (Smith and Francis 1967, p. 43) but its position above the coal is stratigraphically very high for this district; moreover ostracods have been proved from near the roof of the Brockwell seam at West Auckland (see below).

Measures overlying the Brockwell occur in three small areas to the south of the main crop. The first is a ¼ mile S and the second ½ mile NE of Wackerfield and the third north of Todwell House, Hilton [153 222, 162 227 and 166 215 respectively]. Here the Brockwell abuts against faults and whilst the measures are not exposed at surface, at the second and third localities they may extend up to and include the Busty, close to the faults.

South and east of Toft Hill these measures range between 80 and 96 ft and contain two or three sandstones separated by argillaceous measures. The Three-Quarter ranges from 1 to 18 in. Etherley No. 4 Bore* (79) provides a typical sequence.

Around West Auckland and Hummerbeck these measures range between about 80 and 100 ft thick and usually include three sandstones and two thin coals. The lower coal is locally split into two leaves and is occasionally absent; the upper coal, the Three-Quarter, usually ranges between 6 and 9 in. Typical of the several borehole sections of the Brockwell–Busty interval is the following proved in West Auckland No. 150 Bore (99) ¾ mile NW of St Helen Auckland:

	ft	in
BOTTOM BUSTY (at 127 ft 6 in)		
Seatearth-mudstone, grey slightly silty	1	6
Siltstone, grey massive	1	6
Sandstone, light grey fine-grained thin and wispy-bedded; dark micaceous partings	19	6
Siltstone, grey; sandy partings; sand-filled worm burrows at base	2	6
THREE-QUARTER		
Coal	0	8
Seatearth-mudstone, dark grey; silty at base	1	1
Siltstone, grey; scattered rootlets and ironstone nodules	3	9
Sandstone, light grey medium-grained; massive at top; fine-grained wispy-bedded at base	9	6
Mudstone, grey; silty at top; mussels in 2-ft band, 3 ft from base	10	6
Mudstone, grey shaly; thin clay ironstone bands; *Curvirimula* 1 ft 8 in from base	2	6
Shale, black; abundant fish scales and spines	0	6
COAL	0	5
Seatearth-mudstone, dark grey; rootlets	2	7
Mudstone, grey silty; scattered rootlets and ironstone nodules	5	0
Sandstone, grey fine-grained false-bedded; sporadically silty	13	6
Siltstone, grey poorly fissile; ironstone layers	10	2
Mudstone, grey shaly; clay-ironstone bands; sporadic mussels including *Naiadites* and *Carbonicola*; ostracods at base	6	8
BROCKWELL (goaf) at 223 ft		

The mudstone band containing ostracods may be regarded as the equivalent of the Brockwell Ostracod Band of the Durham and West Hartlepool district. Other complete sections through this part of the sequence are recorded in West Auckland No. 168 Bore* (104) and Windlestone Pit (106).

Around Tindale Crescent and southwards towards Brusselton these measures are proved in Catherine Pit (116; B & S 1653) and Engine Pit* (122) at St Helen Auckland, Ladysmith Shaft* (118) near Brusselton, Bildershaw 'A' Bore (111) and Bildershaw 'B' Bore (112) near Hummerbeck. The sequence ranges between 88 and 110 ft and most sections show three sandstones, with interbedded argillaceous measures. Two thin coals are also recorded, a lower coal between 3 and 7 in up to 50 ft above the Brockwell and an upper coal, the Three-Quarter, generally ranging between 5 and 6 in. Boreholes in the vicinity of Hummerbeck show that the Three-Quarter is somewhat thicker; in Bildershaw 'A' Bore (111) it comprises: coal 12 in, band 8 in, on coal 5 in.

D.A.C.M.

In the Eldon, Shildon and Middridge areas the Brockwell–Busty interval, up to 150 ft thick contains up to three thin coals, the highest of which may be equated with the Three-Quarter, and a high proportion of sandstone overlying both the Brockwell and Three-Quarter coals. The Three-Quarter is recorded in most boreholes and shafts and has a maximum thickness of 24 in though more usually it is in the range of 7 to 15 in.

In South Pit (129; B & S 2323) Woodhouse Close the sequence recorded was:

	ft	in
BOTTOM BUSTY (at 317 ft 11 in)		
Thill	1	8
Post, white rough strong	46	0
THREE-QUARTER		
Coal	0	7
Band	0	3
Coal	0	2
Thill	1	6
Metal, blue	7	0
Metal, grey strong; ironstone girdles	4	6
Metal, blue	1	0
COAL	0	6
Thill	3	0
Metal, strong grey and blue	16	0
COAL	1	0
Thill, dark	2	6
Post, white strong	21	6
Metal, grey strong	10	0
BROCKWELL (at 439 ft 2 in)		

Generally similar sequences were proved in the adjacent North or Engine Pit (128; B & S 2322) and a nearby borehole (130; B & S 2328). About 1½ miles farther east a thicker sequence with more sandstone was proved in the Machine Pit* (145) north-north-west of Coundon Grange. Here the section is:

	ft	in
BOTTOM BUSTY (at 760 ft 11 in)		
Metal, grey	4	2
Post, white	40	0
Whin	8	0
Post, white	14	8
Ironstone	1	5
?THREE-QUARTER		
Coal	2	2
Blackstone	0	8
Fireclay	7	5
COAL	0	9
Fireclay	5	0
Post and metal, grey	7	0
Post, white	4	6
Post, white flimby	32	0
Metal, grey strong	8	4
BROCKWELL (at 902 ft)		

Broadly similar sequences were proved nearly ¾ mile to ESE and ¾ mile to SSW in North Pit (151; B & S 208); and Jane Pit* (142) respectively, while a more detailed description of the measures between the Busty and Three-Quarter was proved in New Shildon No. 117 Bore* (134). Atypically, at New Shildon No. 146 Bore* (139) the sequence between the Bottom Busty and Three-Quarter was virtually all sandstone. Between Fieldon Bridge and Shildon the only complete record of the measures was proved in New Shildon No. 147 Bore* (141) where a 5-ft mudstone band containing mussels, and lying 35½ ft above the Brockwell may perhaps be equated with the Brockwell Ostracod Band of the Durham and West Hartlepool district. In the Middridge area the Brockwell–Busty interval is rather thicker than to the west as proved in Charles Pit* (159). South of Shildon, the only record of these measures was in West Pit (148; B & S 1966) where the sequence is broadly similar to that proved in Charles Pit.

J.H.H.

East of Brusselton Hill [2037 2498] towards the area north-west of Middridge Grange [2446 2463] where these measures crop out under drift, available evidence suggests that they range between 110 and 120 ft. They include two 6-in coals 60 and 90 ft respectively above the Brockwell, the higher being the presumed Three-Quarter.

The Busty seam, known locally as the Five-Quarter or Crow, has been extensively worked throughout the district. Over much of the area the Busty is split into Top and Bottom Busty seams and it is only west of an irregular line drawn from between Toft Hill and Woodhouses to Spring Gardens and Evenwood Gate that it may be regarded as a single composite seam. Even there, however, it is locally split into Top and Bottom components, and it is also often strongly banded where split, but the two coals are closely associated, and where of suitable thickness, they have been worked together. As the separation increases, the Top Busty deteriorates rapidly and is unsuitable for working. Where the separation is greatest the Top Busty is commonly absent as such, its position being marked in most boreholes by a thin seatearth. Apart from being frequently split the coal group is prone to washout and zones of impoverishment (Fig. 13).

D.A.C.M.

Over most of the Woodland and Copley areas it may be regarded as a single composite seam, though often split into Top and Bottom components. It is a single seam, 24 and 44 in respectively, at two localities in Arn Ghyll. In Woodland Colliery No. 47 Bore (3; B & S 3096) the upper coal (45 in) is

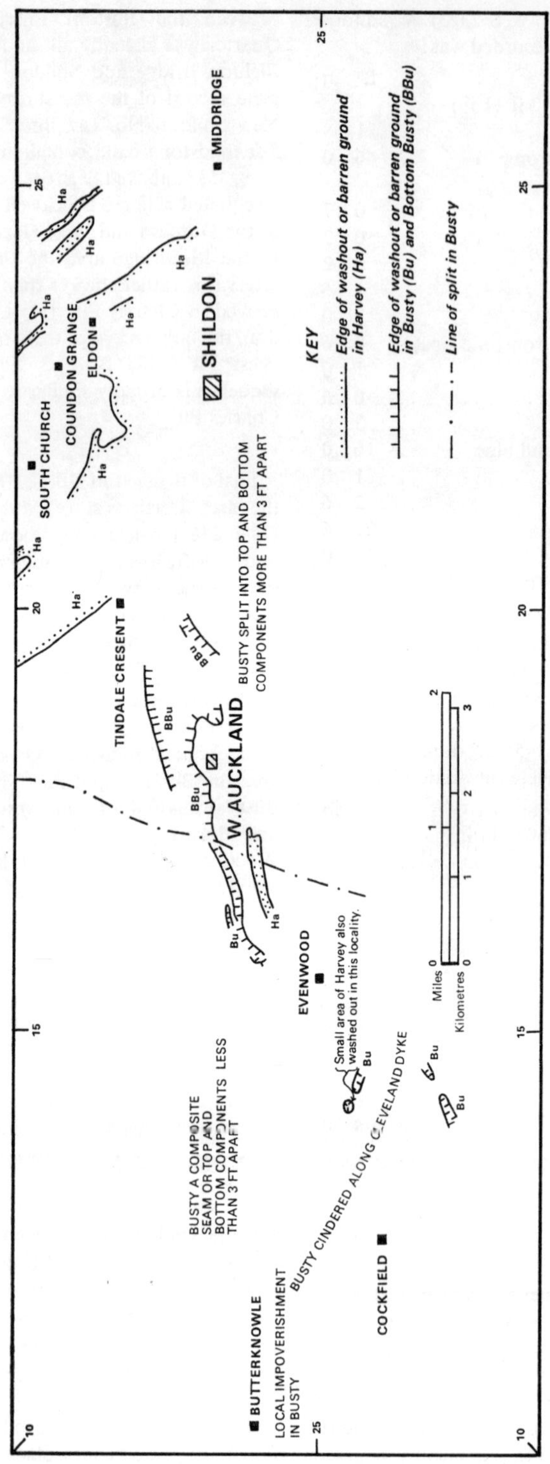

FIG. 13. *Washouts and barren ground in the Busty and Harvey seams*

separated from the lower coal (14 in) by 5 ft 3 in of fireclay. In Woodland Colliery Shaft (10; B & S 2333) the upper coal (coal 4 in, band 5 in, coal 46 in) is separated by 2 ft 8 in of fireclay from the lower coal 15 in thick, the upper 4 in of which is termed 'parrot' coal. In a borehole (15; B & S 370) at Copley the coal section was as follows: coal 10 in, blackstone 6 in, coal 46 in, blackstone 22 in, coal 6 in. At a borehole (17; B & S 2533) near High Wood Farm the Busty is split into distinct Top and Bottom seams, the sequence being as follows:

	ft	in
TOP BUSTY (at 54 ft 1 in)		
Coal	1	2
Shale, dark blue	0	3
Coal	0	3
Shale, dark blue	1	0
Shale, grey sandy	6	6
Sandstone, grey hard	1	3
Shale, grey sandy	1	6
BOTTOM BUSTY (at 68 ft 3 in)		
Coal	3	11

J.H.H.

In the Butterknowle area, the seam is a single united coal ranging between 16 and 65 in. In the vicinity of Diamond Pit* (24) and Gordon Pit (26; B & S 2487) the seam is locally impoverished, while along the margins of the Cleveland Dyke it is cindered.

Around Cockfield, on Cockfield Fell, and between Shotton Moor [104 237] in the west and Keverstone Grange [1382 2265] in the east, the Busty is a composite seam, commonly highly banded. In the west a typical section is: coal 52 in, band 6 in, coal 2 in, band 5 in, on coal 19 in, as proved in John Pit* (29). On the western parts of Cockfield Fell a typical section is: coal 43 to 52 in, band up to 7 in, on coal 9 to 22 in; rarely, a single seam up to 69 in is recorded. On the eastern parts of Cockfield Fell in the vicinity of Gordon House Colliery a typical section is: coal 48 to 49 in, band 7 to 16 in, on coal 16 to 22 in, the band reaching its maximum thickness in the neighbourhood of Esperley Lane Ends [136 243]. Local impoverishment or absence of coal due to washouts occurs in the Cockfield area notably north of Keverstone Grange [1382 2265] and near Buck Head [1456 2444] (Fig. 13). In the latter locality the Busty coal is also cindered along the margins of the Cleveland Dyke.

In the Hilton and Wackerfield areas, two narrow belts of Busty coal are said to incrop [162 228, 167 215] against southward-throwing faults. The areas are so small and the structural setting so complex that the coal, if present, would be unworkable. At the first locality it would probably be cindered by the Wackerfield Dyke. Around the Sun Inn [1542 2317] east-north-east of Keverstone Grange, a typical section is coal 48 in, band 11 in, on coal 15 in.

North of the Gaunless valley, between Butterknowle in the west and Toft Hill in the east, the few records suggest that the Busty is a single seam as in Gordon House No. 135 Bore (64) in the Gaunless valley north-east of Low Lands, or a composite seam as in Providence Pit (62; B & S 957) near Bowes Close where the section recorded was coal 47 in, thill 1 ft 2 in, coal 14 in. In Gordon Gill 'K' Bore* (55) at Snape Foot the seam section was coal 51 in, sandstone 11 in, on coal 12 in. West of Toft Hill the seam section recorded in Toft Hill Pit (67) was coal 30 in, thill 18 in, on coal 18 in. In boreholes south and east of Toft Hill a top coal 9 to 10 in is separated from a bottom coal 48 to 54 in by up to 1 ft of fireclay.

At Evenwood the Busty is a single seam; in Randolph Colliery No. 2 shaft* (72b) it is 38 in thick.

East of an irregular line running from between Toft Hill and Woodhouses to Spring Gardens and Evenwood Gate the Busty seam is generally split into Top and Bottom seams separated by up to a maximum (exceptional) of 54 ft of strata; more generally, however, the separation ranges between 5 and 20 ft.

Between Evenwood and Tindale Crescent the Busty coals are affected by washouts and zones of impoverishment (Fig. 13) which extend in an east-north-east line from near Thrushwood [1494 2590] to north of West Auckland and towards Tindale Crescent. North of this zone of impoverishment the Bottom Busty generally ranges between 36 and 48 in, extremes of 18 and 60 in having been recorded. South of the zone of impoverishment between Evenwood Gate, Hummerbeck and Brusselton the Bottom Busty ranges between 25 and 60 in thick, the thicker sections being recorded around Bildershaw and Brusselton.

D.A.C.M.

Around South Church, Coundon Grange and Eldon, the Bottom Busty, only locally impoverished, generally ranges between 22 and 42 in. In a borehole (130; B & S 2328) near Woodhouse Close the section recorded was: coal, coarse 3 in, splint 6 in, coal 14 in, band ½ in, on coal 3½ in. Atypically, a thickness of only 17 in was recorded in New Shildon No. 125 Bore (135) west of South Church. In Machine Pit* (145) Coundon Grange, the Top and Bottom Busty seams are close together over a small area.

Farther east, in the Old Eldon and Windlestone Park areas, the seam generally ranges between 20 and 42 in. Only very locally is it absent, as in Old Eldon 'B' Bore (156), where its horizon is marked by a seatearth. Farther south between Fieldon Bridge, Shildon and Middridge it has a fairly constant thickness of between 33 and 39 in. A recorded maximum of 48 in was proved in New Shildon No. 145 Bore (132) over a ¼ mile ESE of Fieldon Bridge. Along the outcrop between Brusselton Hill [2037 2498] and New House [2310 2478] the Bottom Busty typically ranges between 36 and 56 in thick; worked opencast east-south-east of Hill Top Farm [2123 2486], it had a typical section of 48 in. Brusselton No. 2 Bore (126; B & S 313) between Ladysmith Shaft* (118) and Brusselton Hill proved an atypical sequence, however, in that Top and Bottom Busty are here virtually united to form a composite banded seam 73 in thick.

J.H.H., D.A.C.M.

The measures between the Top and Bottom Busty seams vary widely in thickness and lithology. Although generally ranging between 5 and 20 ft, they are as much as 54 ft in a limited area north of West Auckland and extending in an east-north-easterly direction towards Tindale Crescent where they are typically sandy in character. In West Auckland No. 150 Bore (99) the measures consist of 1 ft 3 in of seatearth-mudstone, 2 ft 4 in of mudstone and siltstone, and 18 ft of sandstone. Similar general sequences are proved in West Auckland No. 168* (104) and West Auckland No. 177* (109) bores. In the zone of impoverishment the sequence recorded in Windlestone Pit (106; B & S 2174) West Auckland was:

	ft	in
?Horizon of TOP BUSTY		
White post; whin and metal partings	15	0
Grey metal; ironstone balls	6	0
Blue metal	3	0
Grey metal and post	9	0
COAL	0	4
Grey thill stone	7	0
COAL	0	6
Blue metal stone	8	0
COAL (at 244 ft 3 in) ?Bottom Busty	0	6

In West Auckland Colliery Back Drift* the separation was 44 ft 1 in. South of the zone of impoverishment between Evenwood Gate, Hummerbeck and Brusselton the measures between the Bottom and Top Busty are usually argillaceous and range up to a recorded maximum of 19 ft in Ladysmith Shaft* (118) and 15 ft in Bildershaw No. 2 Bore* (114).

D.A.C.M.

In the South Church, Coundon Grange and Eldon areas the interval between the two seams ranges between 5 and 36 ft. In the thicker sequences sandstone predominates while in the thinner sequences sandstone is usually absent, as for instance in New Shildon No. 146 Bore* (139). Atypically in Machine Pit *(145), Coundon Grange, the two seams are so close as to be almost united.

In the Old Eldon and Windlestone Park areas these measures range between 4 and 20 ft and except in the thicker sequences, as for example Old Eldon 'C' Bore (164) and Eldon Moor No. 186 Bore (167), they contain little or no sandstone. In Windlestone 'C' Bore (168) at Park Farm the measures are represented by 7 ft of seatearth-mudstone on 4 ft 3 in of sandstone.

In the Fieldon Bridge and Shildon areas the separation between the two seams rarely exceeds 10 to 12 ft and the beds are usually argillaceous in character. In New Shildon No. 145 Bore (132) over a ¼ mile ESE of Fieldon Bridge these measures were recorded as:

	ft	in
TOP BUSTY (at 816 ft 4 in)		
Seatearth-mudstone	2	3
COAL	0	4
Seatearth-mudstone	1	10
BOTTOM BUSTY (at 824 ft 9 in)		

J.H.H.

Around Middridge from 2 to 5 ft of seatearth-mudstone and mudstone separate Top and Bottom Busty seams, as proved in Charles Pit* (159). Farther north-east in Eldon Moor No. 185 Bore (163) the separation is increased to 13 ft and is mostly sandstone. Along the outcrop between Brusselton Hill [2037 2498] and near New House [2310 2478] the Top and Bottom Busty are separated by 4 to 8 ft of predominantly arenaceous strata.

D.A.C.M., J.H.H.

North of the zone of impoverishment, north of West Auckland and towards Tindale Crescent, the Top Busty is either a thin inferior coal or its horizon is marked by the top of a seatearth. South of the zone of impoverishment between Evenwood Gate, Hummerbeck and Brusselton the seam is variable in both quality and thickness, and ranges up to about 36 in.

D.A.C.M.

In the Eldon, Shildon and Middridge areas, the Top Busty is very variable in thickness. Around South Church, Coundon Grange and Old Eldon it is a thin inferior seam rarely in excess of 9 in, and locally absent. It is similar in the vicinity of Old Eldon, although farther east, towards Windlestone Park, sections up to 20 in thick are recorded. Farther south, in the neighbourhood of Fieldon Bridge, Shildon and Middridge, the seam usually ranges between 24 and 38 in. Immediately south of Shildon, the Top Busty varies from 13 in in Shildon Works Bore (144; B & S 1790) near Shildon Wagon Works, to 34 in farther east in West Pit (148; B & S 1966) about a ¼ mile NE of All Saints Church, Shildon. About 600 yd E of All Saints Church, the seam was proved in the Saints Zone opencast site to range in thickness from 24 to 51 in. Between Brusselton Hill [2037 2498] and north of New House the seam has been worked at several opencast sites along the outcrop. The general thickness was of the order of 45 in, the range being from 24 to 51 in.

J.H.H., D.A.C.M.

The measures between the Busty/Top Busty and Tilley group of coals, generally range between 40 and 50 ft and are characterized in the west by a thick sandstone overlying the Busty seam; farther east this sandstone dies out and a prominent mussel band frequently overlies the Top Busty or its assumed horizon.

D.A.C.M.

In the Woodland and Copley areas these measures are proved in only a few boreholes, the interval ranging up to 24 ft. A sandstone usually underlies the Tilley/Bottom Tilley while in High Copley Bore (15; B & S 370) near Copley virtually the whole sequence is sandstone. Other provings of this sequence have been at Woodland Colliery No. 47 Bore (3; B & S 3096) near Dale Terrace, Woodland Colliery No. 10 Bore (4; B & S 3089) SW of Cowley Farm, and Woodland Colliery shaft (10; B & S 2333).

J.H.H.

Between Butterknowle in the west and the Gordon Beck area in the east these measures abut against the southern side of the Butterknowle Fault and the few records suggest they range between 48 and 71 ft up to the bottom Tilley seam, the thicker sequences being to the east. South-west of the Slack, sandstone which overlies the Busty seam in this area is exposed [1131 2524] above an old adit where it is 15 ft thick and is grey and yellow ferruginous speckled massive to flaggy locally false-bedded, highly micaceous and carbonaceous. In the Gaunless valley below this locality this sandstone is divided by between 10 and 35 ft of shale with only thin sandstone bands, these measures being poorly exposed in the bed of the river below Millfield Grange [1175 2552]. The sequence is also proved in the Diamond Pit* (24), Butterknowle, in Gordon Gill 'K' Bore* (55), where it is faulted, and in Gordon Gill 'L' Bore (56) near Snape Foot Farm. At Providence Pit (62; B & S 957) near Bowes Close the sequence is reduced to 33 ft and sandstone is subordinate.

Around Cockfield, on Cockfield Fell and between Shotton Moor [104 237] in the west, and north of Keverstone Grange [1382 2265] in the east, these measures generally range between 63 and 73 ft, including a thick sandstone above the Busty seam and extending to within a few feet of the Tilley. The lower part of this sandstone, which forms the greater part of Cockfield Fell, is exposed in an old quarry [1190 2433] (Plate VIIIA) 1100 yd west of St Mary's Church, Cockfield. Here over 20 ft of light grey and yellow massive thin- to medium-bedded, occasionally flaggy false-bedded siliceous sandstone are seen; sporadic shaly, micaceous and carbonaceous bands and laminae occur, especially at the top; poorly preserved plant fragments are also present. The sandstone also forms the bulk of the wall rock of the Cleveland Dyke west of Fell Houses [1295 2489], in most cases being hardened by thermal metamorphism. The bed includes coarse-grained, sometimes pebbly, bands, one of which, 3 ft thick, may be seen near the west side of the Cleveland Dyke at a locality [1151 2536] ESE of the Slack [113 254]. Pebbly layers are typical where this sandstone forms the steep bluff on the south side of the River Gaunless at Cragg Wood [143 252]. Most of the shafts and boreholes in this area prove the lower part of the sequence and show that sandstone rests directly on the Busty Coal. Complete sequences include those proved at John Pit* (29) and Gordon House No. 94 Bore (59). In the latter the sequence was:

	ft	in
BOTTOM TILLEY (at 157 ft 6 in)		
Shale, sandy; sandstone bands ..	19	6
Sandstone	57	0
BUSTY (at 238 ft 8 in)		

Little is known of these measures in the country around Evenwood, Staindrop Field House [1691 2471] and north of Wackerfield in the vicinity of the Sun Inn [1542 2317], but mapping suggests that they are slightly thinner and that the sandstone overlying the Busty in the Cockfield area thins towards the north-east and locally becomes separated by shale from the coal itself. The sequence recorded at Randolph Colliery No. 2 Shaft* (72b) is typical.

South and east of Toft Hill these measures range between 40 and 80 ft thick and are mainly argillaceous in character, no sandstone being present in the roof of the Busty. A typical sequence was proved in Etherley No. 4 Bore* (79).

North of West Auckland these measures have been recorded in several boreholes, and range up to a recorded maximum of 55 ft. A typical sequence is recorded in West Auckland Colliery Underground No. 3 Bore (107) north of St Helen's Church as follows:

	ft	in
BOTTOM TILLEY		
Seatearth-siltstone; ironstone nodules; abundant rootlets	3	0
Siltstone, grey thin-bedded; rootlets and stigmaria	7	6
Shale, black silty micaceous; sporadic shaly mudstone, siltstone and ironstone bands; *Carbonicola sp.* (*cristagalli*?), *Geisina arcuata* near base; scattered mussels 5 in to 2½ ft above base	21	5
Horizon of TOP BUSTY (48 ft 7 in above Bottom Busty)		

Broadly similar sequences are proved in adjacent boreholes; the thickness of the mussel bed varies however from as much as 35 ft in West Auckland Underground No. 1 Bore (108), to as little as 2 ft in West Auckland No. 150 Bore (99). In this borehole the measures overlying the assumed horizon of the Top Busty are as follows:

	ft	in
COAL (at 100 ft 9 in)	0	3
Sandstone, black fine-grained carbonaceous	0	9
Shale, dark grey; abundant ostracods at top; poorly preserved mussels 8 in to 1 ft from top; *Gyrochorte* cf. *carbonaria* below; plant debris to base	1	9
Horizon of TOP BUSTY (at 103 ft 3 in)		

Further sequences in these measures are recorded in Spring Gardens No. 2 Bore* (105) West Auckland No. 168* (104) and No. 177* (109) bores. A typical basal *A. modiolaris* Zone fauna was found in mudstones overlying the horizon of the Top Busty in NCB West Auckland Back Drift*. Records from boreholes near Hummerbeck show a broadly similar sequence. The measures, mainly argillaceous in character, range between about 40 and 45 ft thick, but the mussel bed overlying the Top Busty was not found. A typical sequence was recorded in Bildershaw No. 1 Bore (110) near Hummerbeck, the details being as follows:

	ft	in
BOTTOM TILLEY (at 117 ft)		
Fireclay	2	6
Siltstone, grey; sandy laminae and sporadic ironstone bands at top	12	6
Sandstone, grey flaggy fine-grained	1	6
Shale, grey silty; ironstone bands ..	24	2
TOP BUSTY (at 159 ft 10 in)		

A similar sequence is recorded from Bildershaw No. 2 Bore* (114). In the Tindale Crescent and Brusselton areas generally similar sequences are recorded, as in Engine Pit* (122) and Ladysmith Shaft* (118).

D.A.C.M.

In the South Church, Coundon Grange and Eldon areas these measures range between 35 and 44 ft and consist mainly of seatearth on shale and mudstone with only sporadic sand-

(L 351)

A. Massive sandstone overlying the Busty Coal; old quarry on Cockfield Fell

PLATE VIII

B. General view over Coal Measures country, looking north-east from Brusselton Hill, near West Auckland

(L 405)

stone and siltstone ribs. In New Shildon No. 125 Bore (135) over ½ mile E of Woodhouse Close Colliery they were 35 ft 3 in thick and included 1 ft of dark grey mudstone containing mussels and fish debris overlying the horizon of the Top Busty. In Machine Pit* (145) and Jane Pit* (142) similar, though slightly thicker, sections were recorded. In New Shildon No. 146 Bore* (139) the Top Busty, together with the overlying fish bed, are overlain by 5 ft of mussel-bearing mudstone.

In the Old Eldon and Windlestone Park areas these measures range between 30 and 55 ft and are again mostly argillaceous. Fish debris at or near the roof of the Top Busty, as well as the overlying pronounced mussel bed are recorded from a number of boreholes. In Eldon Moor No. 186 Bore (167) the mussel bed is up to 12 ft above the Top Busty.

In the Fieldon Bridge, Shildon and Middridge areas these measures range between 25 and 40 ft; locally a thin sandstone underlies the Bottom Tilley as for example in New Shildon No. 145 Bore (132) over a ¼ mile ESE of Fieldon Bridge where the sequence recorded was:

	ft	in
BOTTOM TILLEY (at 770 ft 5 in)		
Seatearth-mudstone, black; carbonaceous at top, rooty below ..	1	0½
Siltstone, grey massive; thin sandy bands and sporadic rootlets and ironstone nodules	2	3
Sandstone, pale grey very fine-grained thin- and wispy-bedded; silty partings, sporadic rootlets and disseminated plant debris throughout	7	3
Siltstone, grey shaly fissile	5	6
Mudstone, grey; silty at top; shaly to base	12	6
Siltstone, pale brown to grey hard irony	0	10
Mudstone, grey shaly; sporadic ironstone bands and bivalves between 8 and 10 ft from base ..	14	6
TOP BUSTY (at 816 ft 4 in)		

J.H.H.

Along the outcrop between Brusselton Hill [2037 2498] in the west and the area north-east of New House [2310 2478] the Bottom Tilley is between 40 and 45 ft above the Top Busty, the interval being mostly of shale. In Brusselton Drift Mine [2075 2507] a typical basal *A. modiolaris* fauna was obtained from near the roof of the Top Busty, and included *Spirorbis sp.*, *Carbonicola cristagalli*, *C. rhomboidalis*, *Naiadites sp.*, and *Geisina arcuata*.

D.A.C.M.

The Tilley group of coals comprises a complex series of thin impersistent seams ranging over some 48 ft of strata. Over most of the district the Tilley seam is typically split into Top and Bottom coals which are themselves often split into top and bottom leaves. Only in most of the Woodland and Copley areas, and locally elsewhere, is the Tilley a single united seam.

D.A.C.M.

In the Woodland and Copley areas the Tilley, locally known as the Constantine or Beaumont, is thought to be a single seam varying from 30 in at Woodland Colliery No. 47 Bore (3; B & S 3096) near Dale Terrace to 24 inches in the neighbourhood of Woodland. South of the Butterknowle Fault it is 26 in thick in Woodland Colliery No. 10 Bore (4; B & S 3089) near Cowley Farm, and only 15 in thick in the adjacent Cowley Shaft (11; B & S 3086). Near Copley it is generally thin and worthless, but nearly a mile to the east-north-east in Quarry West Pit* (19) west of Butterknowle it has a section: coal 8 in, band 3 in, on coal 7 in.

J.H.H.

The other localities in the district where the Tilley seam is either united or the various leaves are close together include (1) a small area north of the Wigglesworth Fault south of Low Lands [135 250], see Gordon House No. 93 Bore* (47); (2) the immediate neighbourhood of Evenwood, see Randolph Colliery No. 2 Shaft* (72b) (in the adjacent No. 1 Shaft (72a) the Tilley is split into top and bottom coals by 2 ft 9 in of seggar); and (3) the neighbourhood of Brusselton Hill [2037 2498] where Top and Bottom Tilley coals are very close, although not everywhere forming a united seam, see Brusselton No. 2 Bore (126; B & S 313) where the coal section recorded was coal 16 in, grey thill post and metal 4 ft 6 in, coal 3 in, grey metal 9 in, on coal 31 in.

Elsewhere in the district the Tilley seam occurs in Top and Bottom sections, each frequently subdivided into leaves, and the group as a whole between the lowest leaf of the Bottom Tilley and the upper leaf of the

Top Tilley shows much variability. North of the River Gaunless between Butterknowle and the Gordon Beck area the Bottom Tilley occurs as a single thin seam or in two leaves. It varies from a mere 2 in at Diamond Pit* (24) to 24 inches in Olivers Mill Bore (25; B & S 352) about 120 yd to the east. Some distance farther east in Maddisons No. 13 Bore (30; B & S 367) east of Mourning Close, the Bottom Tilley is in two leaves 10 and 12 in thick, separated by 4 ft 3 in of mudstone with sandstone bands. In Gordon Gill 'K' Bore* (55) near Snape Foot Farm the coal is only 3 in thick, although in 'L' Bore (56) only 60 yd to the south it was proved in two leaves of 10 and 8 in, separated by 2 ft 4 in of seatearth-mudstone.

South of Cockfield and west of Evenwood the seam is of no account and little data is available. At John Pit* (29) it forms 12 in of splint coal; at Gordon House No. 131 Bore* (57) over ½ mile NW of Buck Head it is 4 in thick; and in Gordon House No. 94 Bore (59) west-south-west of Buck Head it is a shaly coal 12 in thick. Similarly, little is known about this seam at and to the south of Evenwood. Although at Randolph No. 2 Shaft* (72b) the Tilley is represented by a single united seam, in the adjacent No. 1 Shaft (72a) the Bottom Tilley is 6 in thick and separated from the Top Tilley by 2 ft 9 in of seggar.

South and east of Toft Hill the Bottom Tilley occurs as two or three thin coals separated by up to 16 ft 6 in of strata. The sequence proved in Etherley No. 4 Bore* (79) is typical.

In boreholes north and north-east of West Auckland the Bottom Tilley is typically split into two leaves; the upper leaf, only occasionally absent, ranging from 4 to 13 in, is separated by up to 12 ft of strata, but generally 4 to 5 ft of mostly seatearth-mudstone, from a lower leaf ranging between 2 and 21 in. Very locally it is absent as in Spring Gardens No. 4 Bore (101) over ½ mile W of St Helen's Church, where its assumed horizon is overlying a fireclay some 30 ft above the Top Busty. Other sections through this coal are provided by West Auckland No. 177 Bore* (109), West Auckland Colliery Underground No. 2 Bore* (113) and West Auckland Colliery Back Drift*. West Auckland Colliery Underground No. 1 Bore (108) proved the Bottom Tilley sequence as:

	ft	in
COAL	1	1
Mudstone, dark grey silty; root fragments; '*Guilielmites*' in top 3 in; poorly preserved mussel fragments at top	2	7
COAL (65 ft 11 in above floor of Bottom Busty coal)	0	10

Half an inch of black carbonaceous shale packed with mussels between leaves of the Bottom Tilley was recorded in West Auckland No. 177 Bore* (109). Around Hummerbeck and Brusselton the Bottom Tilley is usually in two leaves separated by 3 ft of shale and fireclay. In Bildershaw No. 2 Bore* (114) the Bottom Tilley is in three thin leaves.

Around Tindale Crescent and towards Brusselton the Bottom Tilley is again in two or three thin leaves, as for example in Catherine Pit (116; B & S 1653), Engine Pit* (122) and Ladysmith Shaft* (118).

Around South Church, Coundon Grange and Eldon, the Bottom Tilley is usually split into two or three thin leaves separated by argillaceous measures. The sequence recorded in New Shildon No. 125 Bore (135) west of South Church was as follows:

	ft	in
COAL, inferior (at 318 ft 5 in) ..	0	4
Sandstone, grey fine-grained wispy-bedded; siltstone partings ..	3	1
COAL, bright sandy	0	5
Seatearth-mudstone, grey; sandy at base	2	1
COAL, bright (at 324 ft 9 in) ..	0	9

Three leaves were also proved in the Machine Pit* (145). Atypically, in New Shildon No. 146 Bore* (139) the Bottom Tilley is a single coal, 20 in thick.

Around Old Eldon and Windlestone Park, the Bottom Tilley is usually a single seam between 20 and 26 in thick. Atypically, in North Pit (151; B & S 208) it is split into two leaves of 8 and 22 in while in Eldon Moor No. 186 Bore (167) it is absent, its assumed horizon being on top of a seatearth.

In the Fieldon Bridge, Shildon and Middridge areas the Bottom Tilley is usually a single seam up to a recorded maximum of 30 in; the following sequence proved in New Shildon No. 145 Bore (132) about ¼ mile ESE of Fieldon Bridge, is atypical:

	ft	in
COAL, inferior (at 763 ft 4 in) ..	0	1
Shale, canneloid; abundant mussels	0	3

	ft	in
Shale, black; vitrain partings ..	0	2
Mudstone, grey shaly micaceous; sporadic silty and sandy bands and ribs	3	1½
COAL, bright	0	2½
Seatearth-mudstone..	1	6
COAL, inferior; ¼-in pyrite band ..	0	1½
COAL, bright (at 770 ft 5 in) ..	1	9

J.H.H.

Along the outcrop east-south-east of Brusselton Hill [2037 2498] the Bottom Tilley, known hereabouts as the Tilley, has a typical section of coal 7 to 9 in, band 4 in, on coal 4 in.

D.A.C.M.

The measures between the Top and Bottom Tilley seams range between 9 and 52 ft in the Butterknowle and Gordon Beck areas and are predominantly composed of alternating flaggy mudstones and shales with subordinate sandstone. A sandstone up to 9 ft thick is occasionally recorded under the Top Tilley. Representative sequences are proved in the Diamond Pit* (24) Butterknowle; Gordon Pit (26; B & S 2487) Butterknowle; Maddison's No. 13 Bore (30; B & S 367) near Mourning Close; Gordon Gill 'K'* (55) and 'L' (56) bores near Snape Foot Farm. In the last two boreholes the separation is 8 ft 10 in and 14 ft 3 in respectively.

Little is known about these measures south of Cockfield and west of Evenwood. In Gordon House No. 94 Bore (59) west-south-west of Buck Head they consist of over 41 ft of shale and sandy shale. In John Pit* (29) the sequence was fireclay 4 ft 10 in, blue shale 29 ft 4 in, on sandstone 5½ ft. In the Evenwood area these measures are recorded in only three places. At Randolph Colliery No. 1 Shaft (72a) Top and Bottom Tilley are separated by 2 ft 9 in of seggar. The sequence is more expanded in Evenwood No. 15 Bore (68):

	ft	in
TOP TILLEY (at 203 ft 4 in)		
Fireclay, greyish brown; sandy at base	2	0
Shale, sandy; irregular sandstone partings	6	6
Shale, silty; ironstone bands ..	25	10
Shale, grey leafy; sporadic mussel fragments	3	0
BOTTOM TILLEY (at 241 ft)		

South and east of Toft Hill these measures range between 16 and 24 ft thick and are wholly argillaceous.

In boreholes north and north-east of West Auckland the measures range between 15 and 26 ft thick and are again predominantly argillaceous, though a thin sandstone is sometimes present. The sequence proved in West Auckland Underground No. 3 Bore (107) north of St Helen's Church was as follows:

	ft	in
TOP TILLEY (bottom leaf)		
Seatearth-mudstone; silty at top ..	1	2
Sandstone, light grey fine-grained micaceous carbonaceous; wavy bedding	4	6
Seatearth-mudstone, grey silty; ironstone nodules at base ..	2	9
Mudstone, grey shaly; sandy and silty at top; poorly preserved plant fragments; *Carbonicola* [frags.] at base	7	9
Siltstone, grey; sandstone at base ..	0	8
BOTTOM TILLEY (lower leaf) (at 83 ft 6 in above Bottom Busty)		

The 2 ft 9 in seatearth band in the above section occurs in at least three other neighbouring boreholes and its top can be regarded as the horizon of a high leaf of the Bottom Tilley. Further records of these measures in this vicinity are: in a borehole (93; B & S 986) over ½ mile ESE of Greenfield House; Spring Gardens No. 6 Bore (94); and in various underground boreholes including West Auckland Underground No. 2 Bore* (113) north of West Auckland. In several boreholes mussel fragments occur in the immediate roof of, or up to two feet above, the Bottom Tilley. Fossils include *Carbonicola oslancis*, *Carbonicola sp.* (*cristagalli*/*pseudorobusta* group), and *Geisina arcuata*. They are consistent with a position near the base of the *A. modiolaris* Zone. Further sections through these measures are also recorded in West Auckland No. 177 Bore* (109) and West Auckland Back Drift*.

Around Tindale Crescent and towards Brusselton the interval ranges between 14 and 20 ft. In Engine Pit* (122) it is 15 ft thick and includes a 6-in "cockle and mussel" shell bed 2 ft 7 in above the Bottom Tilley. At Ladysmith Shaft* (118) the interval is 19 ft 4 in, while about a quarter-mile to the east it has thinned to 4½ ft in Brusselton No. 2 Bore (126; B & S 313).

D.A.C.M.

In the Woodhouse Close area these measures range between 8 and 25 ft and are predominantly argillaceous, though sandstone sporadically occurs as in New Shildon No. 125 Bore (135) where the sequence, 15 ft 8 in thick, includes 6 ft 6 in of sandstone. A 13-in shale band overlying the Bottom Tilley contains mussel fragments and ostracods. Around Coundon Grange these measures range up to 20 ft and may include both fireclay and mudstone; sandstone occurs in places near the base of the sequence. In the area between St Helen Auckland and Coundon Grange a fairly typical sequence was proved in New Shildon No. 117 Bore* (134), east of St Helen's Brickworks. Towards Jane Pit* (142) Adelaide Colliery a thin sandstone is present near the base of the sequence. Farther east in the vicinity of Eldon, Old Eldon 'A' Bore* (147) shows these measures to have thinned to 13 ft, including 5¾ ft of sandstone above the Bottom Tilley. Still farther east, in the neighbourhood of Old Eldon and Windlestone Park, the beds range generally between 6 and 30 ft, and include both sandstone and shale. An exceptional thickness was recorded in Eldon Moor No. 185 Bore (163) north-west of Eldon Moor House, where sandstone at least 37 ft 4 in thick rests directly on the Bottom Tilley, and the Top Tilley may be washed out. The overall trend appears to be one of thickening of these measures to the east and east-north-east.

In the Fieldon Bridge, Shildon and Middridge areas the sequence, argillaceous in the west, becomes progressively more sandy and thicker towards the east. West of Shildon the following sequence was proved in New Shildon No. 145 Bore (132) ¼ mile ESE of Fieldon Bridge:

	ft	in
TOP TILLEY (at 750 ft 8 in)		
Seatearth-mudstone, black silty; rootlets	1	4
Mudstone, grey silty; sporadic sideritic beds and rootlets	4	0
Siltstone, grey thin-bedded, sandy partings; sporadic plant debris	3	0
Mudstone, grey shaly; mussel and 2-in band with fish scales and ostracods near base	4	3
BOTTOM TILLEY (at 763 ft 4 in)		

Other sections in this area record only slight variations in thickness and occasional thin sandstone bands. In Charles Pit* (159) Middridge the interval is all sandstone, the Top Tilley being washed out. In West Pit (148; B & S 1966) Shildon Colliery north-east of All Saints Church the section, by contrast, is: 3½ ft seggar, 32½ ft of blue and grey metal, on 8½ ft of white post. Along the outcrop east and east-south-east of Brusselton Hill [2037 2498] these beds are thought to be predominantly shaly in character and around 20 ft thick.

J.H.H., D.A.C.M.

The Top Tilley is locally a single seam up to a maximum of 31 in, but more usually it is substantially thinner or split into two distinct leaves.

North of the River Gaunless between Butterknowle and the Gordon Beck area it occurs as a single seam or split into two distinct leaves. In Diamond Pit* (24) it is a single seam 10 in thick, while in the neighbouring Gordon Pit (26; B & S 2487) it is split into two 6-in leaves separated by 3 ft 6 in of white post on 7 ft 11 in of mainly argillaceous strata. In Gordon Gill 'K' Bore* (55) and 'L' Bore (56) 1¾ miles to the east-north-east near Snape Foot Farm the Top Tilley is a single seam 8 and 14 in thick respectively.

South of Cockfield and west of Evenwood the seam is of no account and little known. In John Pit* (29) the Top Tilley is split into three leaves. At Evenwood, the coal is 10 in thick in Randolph Colliery No. 1 Shaft (72a), whereas in the adjacent No. 2 Shaft* (72b) both Top and Bottom Tilley seams are united. A banded Top Tilley was recorded in Evenwood No. 15 Bore (68) to the west-north-west, where the sequence was coal 9 in, band 2 in, coal 2 in, band 1½ ft, on coal 4 in.

South of Toft Hill in Etherley No. 4 Bore* (79) the Top Tilley is in two leaves 10 and 16 in thick separated by nearly 4 ft of metal.

In most boreholes north of West Auckland and near Hummerbeck the Top Tilley shows local thickening and typically occurs as a banded seam with top coal 6 to 16 in, band 4 to 12 in, on coal 10 to 21 in. Very locally, as for example in West Auckland Colliery No. 168 Bore* (104) and Windlestone Pit (106; B & S 2174), the Top Tilley is recorded as a single seam. In isolated cases it is absent.

the country around Barnard Castle (*M*

PLATE IX

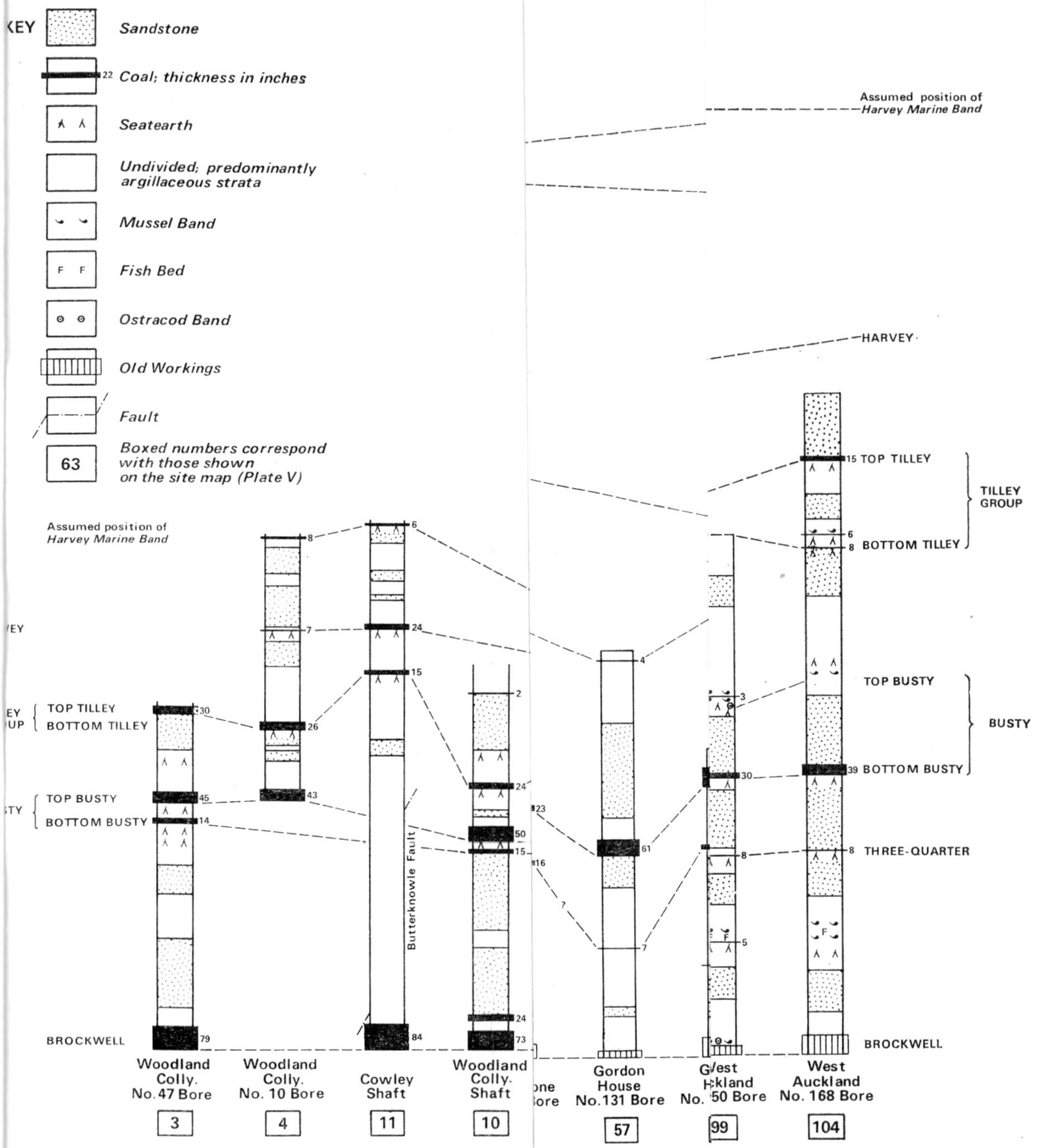

KEY

Sandstone

22 *Coal; thickness in inches*

Seatearth

Undivided; predominantly argillaceous strata

Mussel Band

Fish Bed

Ostracod Band

Old Workings

Fault

63 *Boxed numbers correspond with those shown on the site map (Plate V)*

Assumed position of *Harvey Marine Band*

HARVEY

TOP TILLEY

BOTTOM TILLEY

TILLEY GROUP

TOP BUSTY

BOTTOM BUSTY

BUSTY

THREE-QUARTER

BROCKWELL

Butterknowle Fault

Woodland Colly. No. 47 Bore — 3

Woodland Colly. No. 10 Bore — 4

Cowley Shaft — 11

Woodland Colly. Shaft — 10

Gordon House No. 131 Bore — 57

West Auckland No. 50 Bore — 99

West Auckland No. 168 Bore — 104

MARINE BAND

Around Tindale Crescent and towards Brusselton the coal, usually a united seam, ranges between 8 and 29 in, the thicker sections being generally towards the north.

D.A.C.M.

In the Eldon, Shildon and Middridge areas the seam is relatively thick in some places, but is thin, impoverished or washed out in others. Around Woodhouse Close Colliery it varies from a recorded section of coal 10 in, band 5 in, on coal 20 in, to a mere 6-in single coal, as in New Shildon No. 125 Bore (135). North and north-east of Coundon Grange the seam is usually about 10 in thick, but in some places it is absent or impoverished. Farther east in the vicinity of Eldon and Windlestone Park it usually forms a single, although banded, seam between 10 and 30 in thick. In Windlestone 'C' Bore (168) Park Farm the section recorded was coal 16 in, band 3 in, coal 2 in, band 5 in, on coal 3 in. In Eldon Colliery Shaft* (152) and in Eldon Moor No. 186 Bore (167) the seam was absent.

In the Fieldon Bridge, Shildon and Middridge areas the seam generally ranges between 5 and 11 in. At Eldon Moor No. 185 Bore (163), Charles Pit* (159) and locally elsewhere, the Top Tilley is cut out by sandstone.

J.H.H.

The measures between the Tilley group of coals and the Harvey Coal average about 60 ft, and comprise a variable sequence in which either sandstone predominates or thinner sandstones are separated by argillaceous measures. Up to three thin coals may be present, one up to 16 ft below the Harvey Coal, and two immediately under the Harvey itself. The sequence is relatively unfossiliferous.

In the Woodland and Copley areas, the interval between the two seams ranges between 12 and 78 ft, and shows a thickening towards the east; a sandstone with a maximum of 8 to 10 ft sometimes underlies the Harvey Coal.

J.H.H.

Between Butterknowle and the Gordon Beck area these measures range between about 20 and 60 ft. In the west they are essentially argillaceous with occasional sandstone bands, as recorded in the Diamond Pit* (24).

A thin coal is occasionally recorded a few feet below the Harvey, as in Maddison's No. 13 Bore (30; B & S 367). On the north bank of the River Gaunless south-south-east of Mourning Close the following sequence is exposed [1144 2558] a few feet below the assumed position of the Harvey seam:

	ft	in
Shale, grey	1	0
Sandstone, shaly; shale partings ..	2	3
Mudstone, shaly; sandy at base ..	4	6
Shale, grey micaceous and carbonaceous	4	6
Ironstone, sandy	0	9
Shale, grey micaceous	2	3
Sandstone, grey	0 to 1	6
Shale, grey	1	0

Farther east in the vicinity of High Lands [130 259] and Gordon Beck the sequence contains more sandstone.

South of the River Gaunless thick sandstone overlying the Top Tilley crops out in Cragg Wood [143 252] east of Low Lands. At a locality [1449 2521] nearly ¾ mile E of Low Lands over 30 ft of fine- to medium-grained sandstone with sporadic micaceous carbonaceous shaly bands are seen. This sandstone was formerly exposed in quarries [1527 2561] at Oaks north of Evenwood.

South of Cockfield and west-south-west of Evenwood the sandstone overlying the Top Tilley appears to reach a recorded maximum of 68½ ft thick in Gordon House No. 94 Bore (59) near Buck Head, where its top is within 25 ft of the Harvey seam. West of Cockfield and south of the Wigglesworth Fault the only complete sequence is recorded in John Pit* (29).

The coal underrider to the Harvey seam is also present in Gordon House No. 1* (48a) and No. 2 (48b) shafts where it is 4 in thick.

At Evenwood, Randolph Colliery No. 2 Shaft* (72b) proved these measures to be 48 ft thick, sandy and without coal. In the West Auckland, Hummerbeck, Tindale Crescent and Brusselton areas these measures are again typically sandy except immediately below the Harvey and close above the Top Tilley. A thin coal is locally recorded a few feet below the Harvey and close above the Top Tilley. A thin coal is locally recorded a few feet below the Harvey as seen in Etherley No. 4 Bore* (79).

In Engine Pit* (122) and West Auckland No. 177 Bore* (109) sandstone predominates though in boreholes north of Tindale Crescent, the sequence up to 38 ft thick, includes a greater proportion of mudstone and shale. The section proved in Bildershaw No. 2 Bore* (114) is an example. In Ladysmith Shaft* (118) the separation is 110 ft.

D.A.C.M.

In the Eldon, Shildon and Middridge areas these measures vary considerably both in thickness and lithology, and locally contain thin coals and seatearths. Around South Church, Coundon Grange and Eldon they are up to 58 ft thick and locally contain a thin coal 10 to 12 ft below the Harvey. In New Shildon No. 125 Bore (135) east of Woodhouse Close Colliery the sequence recorded was:

	ft	in
HARVEY (at 258 ft 7 in)		
Seatearth-mudstone, brownish grey; rootlets	3	5
Mudstone, grey; rootlets, ironstone nodules and bands of sphaerosiderite	15	0
Sandstone, white fine-grained micaceous false-bedded	7	0
Mudstone, grey shaly; plant debris, small ironstone nodules and sporadic small-scale slickensides ..	1	0
Sandstone, brown micaceous ..	1	0
Shale, grey micaceous; plant debris and sporadic sandy bands ..	3	9
Shale, black carbonaceous; vitrain partings and slickensides at base	0	3
COAL, fragmentary	0	2
Seatearth-mudstone; slickensides and roots	1	4
Mudstone, grey; rootlets, ironstone nodules and sporadic slickensides	5	6
Sandstone, white fine-grained; argillaceous partings and plant debris	1	0
Mudstone, grey sandy rooty; ironstone bands and plant debris ..	3	11
TOP TILLEY (at 302 ft 5 in)		

A broadly similar sequence was proved at Machine Pit* (145). In Jane Pit* (142) the sequence is virtually all sandstone. The sequence is atypical at New Shildon No. 117 Bore* (134) south-west of Coundon Grange; the Harvey appears to be cut out by a very thick sandstone occupying most of the succession. By contrast, in New Shildon No. 146 Bore* (139) the beds between the Harvey and Top Tilley consist of 20½ ft of shale, and the same beds at Old Eldon 'A' Bore* (147) were proved to comprise 14 ft 8 in of seatearth and siltstone on 54 ft 11 in of sandstone. East of this borehole in the Old Eldon and the Windlestone Park area these measures become progressively thinner due to the wedging-out of the sandstone which is absent altogether in Windlestone 'C' Bore (168) at Park Farm. In this borehole the measures are only 35 ft 4 in thick and contain coal and band 37 in thick, 8 ft 5 in below the Harvey, the remainder of the sequence being shaly and silty. Around Fieldon Bridge, Shildon and Middridge these measures range up to 84 ft thick and are mainly sandy. New Shildon No. 147 Bore* (141) on the outskirts of Shildon, proved a mussel band in the roof of the Top Tilley seam. Between Shildon and Middridge the Tilley group is overlain by a sandstone at least 50 ft thick as proved in a borehole (157; B & S 1964) over a mile east-south-east of St John's Church, Shildon, and in Charles Pit* (159) Middridge.

J.H.H.

The Harvey Coal is known locally as the Yard and has been widely worked in the district. Like the Busty, the seam is prone to washouts especially towards the north-east (Fig. 13). Locally in the eastern part of the district it is split into Top and Bottom coals separated by up to 8 ft of measures.

D.A.C.M.

In the Woodland and Copley areas it is, by comparison with areas farther east, impoverished. It has been proved definitely in only three boreholes and two shafts. At Woodland Colliery No. 10 Bore (4; B & S 3089) near Cowley Farm it has a section, coal 3 in, band 2 in, on coal 4 in; whereas in Cowley Shaft (11; B & S 3086) 600 yd to the east it is 24 in thick. At High Copley Bore (15; B & S 370) it is a mere 7 in, while farther east in Quarry West Pit* (19) west of Butterknowle it measures 26 in.

J.H.H.

Between Butterknowle and the Gordon Beck area the general thickness is between 28 and 45 in usually with little or no band. Some cindering is recorded on the south side of the Cleveland Dyke near the Slack [113 254].

In the Cockfield area a representative section is that recorded in Gordon House No. 1 Shaft* (48a), the section being coal

70 in, splint 18 in, coal 4 in. Atypically for this area, the Harvey seam is split in John Pit* (29) the section being coal 48 in, band 4 in, fireclay 1 ft 4 in, coal 9 in. Farther east around Buck Head [1456 2444] a typical section is coal 4 in, band 2 in, on coal 36 in. In Cragg Wood [143 252] to the north and north-east thickness varies from 26 to 42 in, and two small subcircular washouts affect the seam up to 400 yd WNW of Buck Head.

Around Evenwood the coal section is not banded and is generally around 36 in. Locally, as for example near Copeland House [1667 2606] the seam is affected by washouts (Fig. 13).

South and east of Toft Hill, around West Auckland, Hummerbeck, Tindale Crescent and Brusselton the seam is fairly consistent in thickness, ranging between 35 and 46 in. In the Woodhouses [189 281] area a thin top coal 2 to 7 in is separated by a 6- to 18-in band from a bottom coal 15 to 40 in. Atypically, however, in Rush Pit* (100) to the west of Woodhouses the section is coal 40 in, band 12 in, on coarse coal 6 in. Nearly a mile and a quarter to east-south-east, in Broken Backed House No. 1 Bore (121; B & S 1643) at Tindale Crescent the section measured was coal 4 in, grey metal 2 in, coal 5 in, dark grey metal 2 in, on coal 4 in; whereas in Engine Pit* (122) the coal is 45 in thick. Between Hummerbeck and Brusselton Hill [2037 2498] the seam is usually thicker, although banded, possessing a typical section coal 16 to 24 in, band 2 to 3 in, coal 24 to 49 in. Farther east this band increases to such an extent that the upper coal is termed the Top Harvey which is known locally in areas farther east.

D.A.C.M.

In the Eldon, Shildon and Middridge areas the seam is prone to impoverishment and washouts in some localities (Fig. 13); elsewhere it is fairly consistent and generally ranges between 3 and 4½ ft including bands. North of Coundon Grange and Eldon respectively the seam is generally 40 to 42 in thick. Slightly thicker sections are recorded still farther east at Old Eldon and in the neighbourhood of Windlestone where a split seam locally occurs. In Windlestone 'C' Bore (168) at Park Farm the Harvey seam is split which is more characteristic of the Durham (27) district (Smith and Francis 1967, pp. 53–6) than of the rest of this area. The section recorded was:

	ft	in
TOP HARVEY (at 318 ft 6 in)		
Coal (cannel)	0	10
Shale, black carbonaceous; mussel fragments at top, fish remains and worm tubes below	0	6
Coal (cannel)	1	4
Mudstone, dark grey shaly; plant debris	1	8
Siltstone, pale grey argillaceous; abundant plant debris	3	0
Mudstone, dark grey; sporadic slickensides and abundant plant debris; some pyrite	2	0
BOTTOM HARVEY (at 333 ft 4 in)		
Coal, bright	4	0
Seatearth-mudstone, dark grey slickensided	0	9
Coal, inferior; shaly partings ..	1	7

In the Fieldon Bridge, Shildon and Middridge areas records show variations from a united seam up to 53 in thick to a banded seam with typical section coal 20 in, band 21 in, on coal 39 in. The Shildon Works Bore (144; B & S 1790) said to be near the Shildon Railway Works Engine Shed proved the Harvey split into Top and Bottom seams, the section recorded being coal 18 in, seggar clay 4 ft 9 in, on coal 48 in. Average thicknesses of 20 in for the top leaf and 37 in for the lower leaf were recorded in the adjacent Thickley Opencast site [223 251].

J.H.H.

East of Brusselton Hill [2037 2498] along the outcrop towards the area north of Southfield House [2229 2460] the Harvey is split into two, a lower coal averaging 36 in separated by up to 4 ft of often sandy strata from a top coal (Top Harvey) up to 18 in thick.

The measures between the Harvey and the base of the Harvey Marine Band are usually between 40 and 60 ft thick and consist of a variable sequence of strata in which sandstone and shaly beds are often equally proportioned. In the Woodland and Copley areas these measures range from about 42 ft in the west to 31 ft near Cowley Farm [0640 2530], the only exception to this being in High Copley Bore (15; B & S 370) where they are only 17 ft, and almost wholly of sandstone. A typical section through these beds is recorded in Quarry West Pit* (19). The thin coal underlies the assumed position of the Harvey Marine Band throughout the area.

South of the Butterknowle Fault between Butterknowle in the west and the Gordon Beck area in the east the measures range between 18 and 40 ft, the sequence in most cases being broadly similar to that proved in Quarry West Pit. The coal at the top of the sequence ranges between about 8 and 14 in; only locally is it absent.

South of the Wigglesworth Fault around Cockfield these measures were proved in John Pit* (29) and Gordon House No. 1* (48a) and No. 2 (48b) shafts. In Evenwood No. 16 Bore (66) the presumed Harvey is overlain by an 8-in band of shale containing well-preserved ostracods. This horizon is the probable equivalent of the Hopkins Band (Armstrong and Price 1954, p. 977; Smith and Francis 1967, pp. 15, 56).

Evenwood No. 15 Bore (68) proved a sequence typical of these measures in the Evenwood area as follows:

	ft	in
?HARVEY MARINE BAND (at 52 ft 2 in)	—	—
Fireclay, grey; ankeritic nodules ..	4	10
Shale, sandy and silty	10	0
Shale, sandy; passes down into argillaceous sandstone	8	6
Sandstone, current-bedded; shale pellets at base	2	6
Shale, sandy; ankeritic bands; sporadic plant remains	9	4
Sandstone, fine-grained flaggy at base; sporadic shale lenses ..	14	2
Shale, grey micaceous; ostracods and *Spirorbis sp.* (Hopkins Band)	0	5
Fireclay, greyish brown sandy ..	2	1
HARVEY (at 104 ft 2 in)	—	—

In Randolph Colliery No. 2 Shaft* (72b), Evenwood, these measures were 39 ft thick, mainly sandy in character, the upper coal being 3 in thick. In Gordon Bank Fore Drift [1502 2695] about 1 mile NNE of Ramshaw Village, *Spirorbis sp.*, *Anthraconaia ?* [frag.], *Naiadites sp.* [frag.] and *Geisina arcuata* were obtained from the roof of the Harvey Coal.

South and west of Toft Hill these measures, ranging between 25 and 30 ft thick, are mainly argillaceous but contain thin sandstone ribs. In the vicinity of Hunter's Hill House Etherley No. 4 Bore* (79) proved two thin closely associated seams 5 and 10 in thick in the middle part of the sequence.

In boreholes and shafts around Woodhouses [189 281] the sequence generally averages around 35 to 40 ft and consists of shales and mudstones with a median sandstone, as in Rush Pit* (100).

North-east of West Auckland the interval ranges between 28 and 44 ft, a typical section being recorded in West Auckland No. 177 Bore* (109) although here the thin coal usually present at the top is absent. In Spring Gardens No. 2 Bore (105) the Harvey seam is overlain by 6 in of dark grey shale containing *Spirorbis sp.*, *Naiadites sp.*, *Carbonita humilis* and *Geisina arcuata*, correlated with the Hopkins Band. A typical development of the Hopkins Band was recorded in West Auckland Back Drift* at West Auckland. Old borehole and shaft records from between West Auckland and Tindale Crescent give a broadly similar though slightly thinner lithological section. The Hopkins Band was also recorded in more recent boreholes farther south in the vicinity of Hummerbeck notably in Bildershaw No. 2* (114) and Bildershaw No. 3 (102) bores, where a foot of shale overlying the Harvey seam yielded *Spirorbis sp.*, *Naiadites sp.* and abundant *Geisina arcuata*. Boreholes on the north-western slopes of Brusselton Hill [2037 2498] suggest that these measures range between 30 and 40 ft. At Ladysmith Shaft* (118), Brusselton, the sequence is similar.

D.A.C.M.

In the South Church, Coundon Grange and Eldon areas a fairly typical sequence was proved in New Shildon No. 125 Bore (135) ½ mile E of Woodhouse Close Colliery, as follows:

	ft	in
Horizon of HARVEY MARINE BAND (at 200 ft 6 in)	—	—
Seatearth mudstone, grey micaceous; slickensides	0	10
Sandstone, dark grey fine-grained micaceous; rooty at top; ironstone nodules	5	6
Seatearth-mudstone, pale brown highly slickensided	0	6
Mudstone, dark grey silty	1	6
Sandstone, pale grey fine-grained wispy-bedded micaceous; becoming argillaceous towards base ..	2	0
Mudstone, black shaly	2	0
Shale, carbonaceous; vitrain partings	1	2

PLATE X

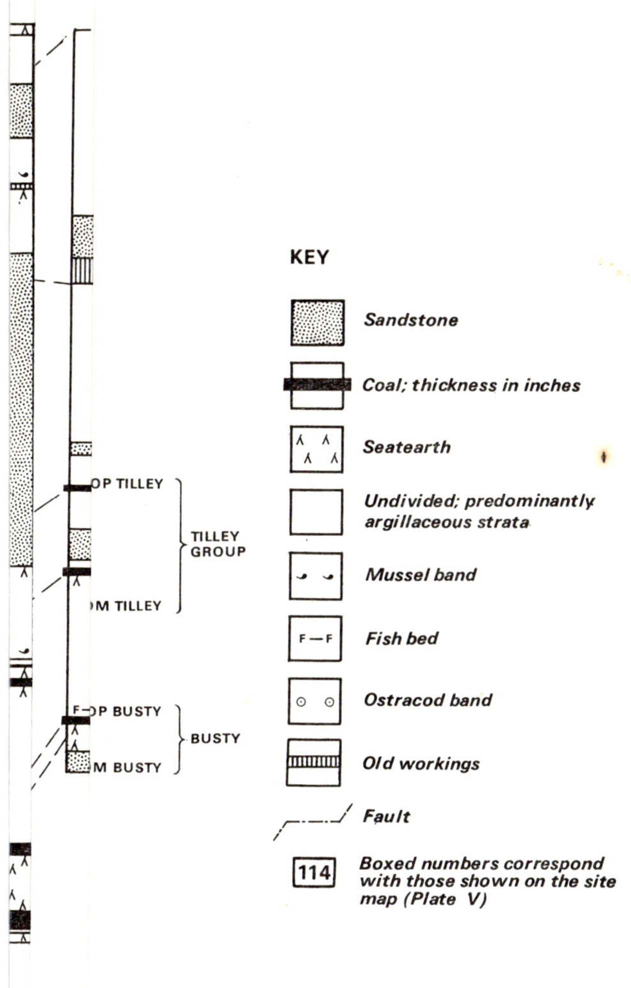

OP TILLEY
TILLEY GROUP
M TILLEY
F–
OP BUSTY
BUSTY
M BUSTY
QUARTER
OCKWELL
rsh
B
Old El
'B' B
14
15
BA
KEY
Sandstone
Coal; thickness in inches
Seatearth
Undivided; predominantly argillaceous strata
Mussel band
F—F
Fish bed
Ostracod band
Old workings
Fault
114
Boxed numbers correspond with those shown on the site map (Plate V)

	ft	in
Seatearth-mudstone, pale brown slickensided; ironstone nodules . .	4	10
Mudstone, silty grey massive; shattered 6 ft from base (?fault) . .	8	0
Siltstone, dark grey argillaceous; plant debris and sporadic flaggy bedding	11	0
Sandstone, pale grey to white fine-grained micaceous; wispy-bedded and silty partings in top 15 ft . .	19	0
HARVEY (at 258 ft 7 in)	—	—

Broadly similar sequences were proved around Woodhouse Close Colliery, although the sandstone overlying the Harvey Coal varies considerably in thickness, but atypically in Woodhouse Close North Pit (128; B & S 2322) sandstone occupies the whole sequence. Again, farther east and north of Coundon Grange and Eldon, sequences basically similar to that proved in New Shildon No. 125 Bore are recorded. The sandstone overlying the Harvey coals appears to thin progressively eastward and is absent altogether in Eldon Colliery Shaft* (152). New Shildon No. 117 Bore* (134) about $\frac{1}{2}$ mile E of St Helen's Brickworks showed the Harvey seam to be washed out by a thick sandstone, overlain by $48\frac{1}{2}$ ft mainly of siltstone and mudstone up to the horizon of the Harvey Marine Band.

In the Old Eldon and Windlestone Park areas these measures are slightly thinner than farther west and are mainly argillaceous. In Windlestone 'C' Bore (168) at Park Farm the sequence recorded was:

	ft	in
HARVEY MARINE BAND (at 284 ft 4 in)	—	—
COAL, bright	0	6
Siltstone, grey; laminae of sand; very rooty with ironstone nodules in top 2 in	3	8
Mudstone, grey shaly; sporadic ironstone nodules; 6-in mussel band 1 ft 6 in from top	5	0
Seatearth-mudstone, grey slickensided	7	0
Sandstone, white fine-grained wispy-bedded; abundant silty micaceous partings	6	9
Mudstone, grey to dark grey; black and carbonaceous in basal 4 in . .	3	1
COAL, bright	1	0
Seatearth-mudstone, dark grey slickensided; ironstone nodules . .	5	10
Shale, dark grey carbonaceous; sporadic coalified plant fragments and pyritic partings	1	0
TOP HARVEY (at 318 ft 6 in)		

Between St Helen Auckland and Shildon these measures range between 30 and 50 ft and consist essentially of coal, often banded, 2 to 18 in, on 30 to 50 ft of shale and siltstone with sandstone ribs in places including thin seatearths. In the western outskirts of Shildon, New Shildon No. 147 Bore* (141) proved these beds to consist of 8 ft of sandy shale with sandstone ribs on 31 ft of sandstone.

J.H.H.

Comparative sections of the Lower Coal Measures between the Brockwell and Harvey Marine Band are shown in Plates IX and X, while further sections and synopses of sections which show all or part of these measures are quoted in Appendices 1 and 2.

MIDDLE COAL MEASURES

The main outcrop of the Middle Coal Measures is in the north-east part of the district to the east and north-east of West Auckland and around Eldon and Shildon. Farther west these beds are restricted to two narrow belts, one extending along the south side of the Butterknowle Fault as far as Cowley near Woodland and the other along the south side of the Wigglesworth Fault to a mile beyond Cockfield. They are rarely exposed at the surface, but have been observed in a number of former opencast workings.

Harvey Marine Band to Main Coal
(shown graphically in Plate XI)

The Harvey Marine Band at the base of the Middle Coal Measures is generally represented by a few inches of black or very dark grey shale or shaly mudstone sometimes overlying a thin coal. Although the horizon has been penetrated in many boreholes and shafts, few faunas have been recorded. In Gordon Gill 'F' Bore* (54) it consists of 4 in of black shale containing *Lingula*; in Evenwood No. 16 Bore (66) 4¾ ft of dark grey finely micaceous shale overlying the Harvey Marine Band coal is recorded in this position. Broadly similar sequences are proved in the Eldon and Shildon areas where fish debris is occasionally recorded as in New Shildon No. 117 Bore* (134) and Old Eldon Bore 'A'* (147). In Windlestone 'C' Bore (168) the Harvey Marine Band is represented by 8 in of slightly silty mudstone containing abundant *Lingula*, and this is the only good record of the Harvey Marine Band in this district. Elsewhere the horizon can only be inferred, by correlating either with the coal or seatearth which commonly underlie the marine band or with the shell bed which often overlies the marine band.

The measures between the Harvey Marine Band and the Hutton (Bottom Hutton) generally range between 100 and 130 ft, and form a variable sequence of strata including up to six sandstones and a number of thin, generally impersistent coals. The highest coal often quite thick, and locally split in the western part of the coalfield, is known locally as the Jubilee seam, and is the probable equivalent of the Ruler seam of north-west Durham and the Plessey seam of Northumberland. A mussel band is recorded locally in the roof of the Harvey Marine Band, as elsewhere in the Northumberland and Durham Coalfield. Mussels are also recorded from a number of other horizons between the Harvey Marine Band and the Ruler Coal and from the roof of the Ruler or its equivalent.

D.A.C.M.

In the Woodland and Copley areas the westernmost record of these measures is from Cowley Shaft (11; B & S 3086) where they are over 107 ft thick, and mainly argillaceous, with a 3-in coal 14 ft 5 in below the Ruler. The Ruler is a banded coal with section: coal 9 in, band 2 in, coal 12 in, band 4 in, splint coal 4 in. At High Copley Bore (15; B & S 370), only the lower part of the sequence is recorded. Here the measures between the Ruler and the assumed horizon of the Harvey Marine Band are chiefly sandstone, but a 16-in coal, the presumed equivalent of the 3-in coal in the Cowley Shaft, lies 23 ft below the Ruler, the latter being a single coal 18 in thick.

J.H.H., D.A.C.M.

Broadly similar sequences to that proved in the Cowley Shaft are recorded in boreholes and shafts in the neighbourhood of Butterknowle, although the proportion of sandstone differs slightly in each case. One significant difference however is that the Ruler seam is frequently split, one of the wider separations being proved in the Engine Pit (44; B & S 350), the sequence being as follows:

	ft	in
TOP RULER (at 36 ft 6 in)		
Coal	1	1
Hard sill	9	0
Blue metal and grey beds	9	10
BOTTOM RULER (at 57 ft 6 in)		
Coal	1	8

In the Diamond Pit* (24) the Top Ruler, 18 in, is separated from the Bottom Ruler, 20 in with a 4-in stone band, by 5 ft of 'thill'. Between Butterknowle and the area south of Toft Hill there are few records of these measures along the south side of the Butterknowle Fault. The recorded range in thickness is 120 to 132 ft, and argillaceous strata appear to predominate. In Gordon Gill 'F' Bore* (54), near Snape Foot Farm, a band containing small mussels was recorded immediately above the Harvey Marine Band. In Gordon Gill 'A' Bore (58) the Ruler seam is split into Top and Bottom coals, 4 and 6 in respectively, separated by 1 ft 10 in of fireclay and 1 ft 3 in of shale, the latter containing abundant mussels with large forms in the basal 3 in. Mussels are also present in the roof of the Top Ruler, immediately below the Bottom Ruler and at scattered intervals in the 35 ft of argillaceous strata underlying this seam; in most cases they are associated with ironstone bands.

South of the Wigglesworth Fault in the neighbourhood of Cockfield and Evenwood the sequence, proved in only a few boreholes and shafts, ranges between 100 and 150 ft. A thin sandstone with shaly bands sometimes lies beneath the Ruler seam, while another sandstone is present between the Ruler and Hutton. The sequence recorded in John Pit* (29) is typical.

In a similar sequence at Gordon House No. 1 Shaft* (48a) the Ruler was 38 in thick, with a 3-in band 20 in above its base. In the adjacent No. 2 Shaft (48b), however, it is split into two leaves 18 in thick separated by 12 ft 2 in of shale. An atypical sequence was recorded in Evenwood No. 16 Bore (66). Here the measures are 88 ft thick and except for a thin band of sandstone immediately underlying the Hutton (?Bottom Hutton) they are wholly argillaceous. A mussel band overlies the presumed position of the Harvey Marine Band. The Ruler is 26 in thick with poorly preserved mussels in its roof; a 1-in coal lying 10 ft 9 in below the Ruler may be a thin Bottom Ruler. Midway between the Ruler and the Hutton is a band containing poorly preserved mussels. In Randolph Colliery No. 2 Shaft* (72b) a 4-in coal is separated by 3 ft 7 in of seggar and 4 ft 10 in of hard grey post from the Hutton seam. This coal is not present in the adjacent No. 1 Shaft (72a).

Between Toft Hill and the area around Woodhouses [189 280] the Harvey Marine Band to Hutton interval is usually about 130 ft, and contains up to four thin impersistent coals. The measures are usually of a mixed bed character but thin impersistent sandstones up to 12 ft thick occur throughout the sequence. The Ruler is thin or locally absent. A typical sequence is recorded in Rush Pit* (100). Sub-Hutton measures have been proved in Greenfields Nos. 1 and 2 bores (84 and 85 respectively), south-west of Greenfield. In Greenfields No. 1 Bore *Anthracosphaerium* aff. *exiguum* was present, some 16 ft below the Hutton. At Greenfields No. 2 Bore three bands below the Hutton yielded fossils: *Anthracosphaerium exiguum* at 24 ft 6 in; *Anthraconaia sp.* [*modiolaris* group] and *Anthracosphaerium?* at 50 ft; *Spirorbis sp.*, *Anthracosia spp.*, *Naiadites* cf. *triangularis*, and *Carbonita sp.* at 72 ft. A further band contained mussels 83 ft 9 in below the Hutton.

Between West Auckland and Tindale Crescent these measures are around 120 ft thick and are mainly argillaceous. Up to three thin coals are recorded, the highest of which, the presumed Ruler, is represented either by 'coal scars' or a coal up to 6 in thick. The lowest coal, between 40 and 60 ft above the presumed Harvey Marine Band, thickens locally; in Tindale Pit (120; B & S 2320) it is represented by 24 in of splint or jet, while in Catherine Pit (116; B & S 1653) it is 11 in thick inclusive of band. West of Brusselton Hill [2037 2498] the greater thickness of these measures is largely accounted for by a greater proportion of sandstone, as in Brusselton No. 4 Bore (124; B & S 315). The Ruler Coal hereabouts is thought to be absent or thin as in the Ladysmith Shaft* (118). Atypically, the sequence some distance to the west in Bildershaw No. 3 Bore (102) is more argillaceous than at Brusselton and contains fossils at several horizons. The details are as follows:

	ft	in
Shale, silty finely micaceous ..	1	3
Sandstone, light grey fine-grained massive argillaceous	3	6
Shale, silty	4	6
Sandstone, fine-grained massive jointed	18	0
Shale, sandy micaceous; plant debris	6	0
Sandstone, fine-grained massive ..	2	1
Shale, grey; sporadic plant debris ..	5	4
Sandstone, light grey fine-grained; dark micaceous partings ..	2	9
Shale, grey silty micaceous; poorly preserved mussels	7	7
Shale, grey sandy micaceous; plant debris	1	4
Shale, black splintery finely micaceous; fragmented mussels ..	1	4
Shale, silty grey	0	7
Shale, grey; sporadic mussel fragments including *Anthracosia sp.*, *Anthracosphaerium* cf. *affine* ..	5	0
Shale, grey silty micaceous; sandstone partings	2	6
Sandstone, grey massive fine-grained ankeritic	2	0
Shale, grey silty; sporadic mussel fragments	13	6
Shale, grey barren	1	0
Shale, grey silty	19	0
Shale, dark grey barren finely micaceous	9	6
HARVEY MARINE BAND (presumed horizon at about 171 ft 6 in) ..	—	—

In Bildershaw No. 2 Bore* (114) north-west of Hummerbeck the mussel band overlying the assumed horizon of the Harvey Marine Band is represented by 4 ft 8 in of grey shale containing mussels and *Spirorbis*.

D.A.C.M.

In the South Church, Coundon Grange and Eldon areas these beds form a variable sequence up to about 120 ft thick in which sandstone and shale are in roughly equal proportions; in some sinkings, as in New Shildon No. 146 Bore* (139) nearly 50 ft of sandstone underlie the Bottom Hutton, while in Machine Pit* (145) sandstone occupies much of the middle part of the sequence. There are up to two thin coals, neither of which may be equated with the Ruler coal of areas farther south-west. In New Shildon No. 117 Bore* (134) mussels occur in the roofs of both coals, though they are not recorded at these levels in Old Eldon 'A' Bore* (147). In both boreholes however mussels were obtained above the Harvey Marine Band, fish scales also being present on the latter. A broadly similar sequence was proved in New Shildon No. 125 Bore (135).

Farther east in the Old Eldon and Windlestone Park areas the sequence is similar to that around South Church, Coundon Grange and Eldon, except that the proportion of sandstone is substantially less. In North Pit (151; B & S 208) a coal with section coal 6 in, band 6 in, coal 12 in, some 65 ft below the Hutton seam, is thought to be the Ruler. Still farther east these measures are mainly argillaceous and the Ruler, where present, is very thin. In Old Eldon 'B' Bore (156) the measures, all shale or mudstone, are 90 ft thick, whereas in Old Eldon 'C' Bore (164) they are about 111 ft, with a thick sandstone at and below the horizon of the Bottom Hutton. Nearly ¾ mile to the east-north-east Windlestone 'C' Bore (168) at Park Farm proved most of these beds beneath the Permian unconformity, and is notable for its record above the Harvey Marine Band of over 28 ft of mudstone with mussels, and with *Spirorbis* in the basal 2 ft.

Between Fieldon Bridge and Shildon, these measures range up to 145 ft thick. At New Shildon No. 145 Bore (132) they are mainly of a sandy mixed bed type, and include three thin coals, 8, 7 and 15 in thick in upward order, the highest one of which may be equated with the Ruler. The 7-in coal is equated with the 16-in seam midway between the Harvey Marine Band and the Bottom Hutton in New Shildon No. 147 Bore* (141) in the western outskirts of Shildon.

J.H.H., D.A.C.M.

The Hutton seam, sometimes known locally as the Four Foot, or in the north-east part of the district the Bottom Hutton, generally occurs as a single composite seam up to 51 in thick. In the Woodland and Copley areas the single seam ranges from 40 in at Cowley Shaft (11; B & S 3086) to 46 in at Quarry West Pit* (19) while farther east between Butterknowle and Toft Hill records range from 42 to 50 in. In the Cockfield area and to the west of Evenwood similar thicknesses are recorded. Atypically, in Evenwood No. 16 Bore (66) the presumed Bottom Hutton is impoverished and represented by two coals, 2 in and 4 in respectively, separated by 7¾ ft of fireclay. Around and to the east of Evenwood the seam ranges between 36 and 45 in, a thin splint band being present in places near the base of the coal. In small opencast sites [1731 2559 and 175 256] south-east of Copeland House the seam showed an average section of 31 in; in nearby Finlays Back Drift it ranged between 39 and 42 in.

Between Toft Hill and Woodhouses [189 281] the Hutton varies from 30 to 36 in, and is banded only locally as for example at Rush Pit* (100) where it occurs in two 18-in leaves separated by 2 in of stone band. Local impoverishment and small washouts affect the seam north-east of Woodhouses. Further impoverishment appears to affect the seam near Woodhouse Close, since in Woodhouse Close No. 1 Bore (127; B & S 2325) the presumed Bottom Hutton is only 5 in thick. A large area has been worked by opencast up to nearly ¾ mile NW of Tindale Crescent [200 274], the seam showing an average section of coal 34 in, often splinty near the base. Farther south between Tindale Crescent [200 274] and Hummerbeck [1883 2540] and Brusselton Hill [2037 2498] the seam usually ranges from 34 to 42 in.

D.A.C.M., J.H.H.

Over most of the north-east part of the district the seam is only 24 to 36 in thick and is locally impoverished and affected by washouts. In the area between a ¼ and ½ mile SW of South Church, and in New Shildon No. 125 Bore (135) it is only 12 in thick. In the extreme north-east, in the neighbourhood of Old Eldon and Windlestone Park, however, sections up to 42 in are recorded.

Atypically, in a limited area to the east of Old Eldon 'B' Bore (156) the Bottom Hutton is absent, and in Old Eldon 'C' Bore (164) west of Park House, its horizon can be recognized by the occurrence of a seatearth overlain by sandstone, the coal having been washed out. West of Shildon the seam is thicker, attaining a recorded maximum of 42 in. Farther south the Bottom Hutton ranged in thickness from 30 to 37 inches in the New House opencast site west and south of High West Thickley [2150 2516].

J.H.H., D.A.C.M.

The interval between the Bottom and Top Hutton ranges from 24 to 70 ft, the more usual range being 30 to 40 ft. The beds are mainly argillaceous, although sandstones occur in the thicker sequences. In the extreme west the only complete record of these measures is in Quarry West Pit* (19).

J.H.H.

At Cockfield the log of Gordon House No. 1 Shaft* (48a) shows over 70 ft of mainly argillaceous strata with two sandstones. The 45-ft sequence proved in Evenwood No. 16 Bore (66) was mainly arenaceous, though a basal 11 ft of argillaceous strata contained poorly preserved mussels at top and base. Around Evenwood this interval ranges between 40 and 54 ft; the strata proved in Randolph 'A' Bore (83) is regarded as typical of the thinner sequences, the details being as follows:

	ft	in
TOP HUTTON (at 168 ft 2 in) ..	—	—
Seatearth, grey	1	0
Sandstone, fine-grained irony ..	1	1
Siltstone, grey sandy	1	0
Shale, grey silty rooty	13	9
Mudstone, grey shaly; irony bands; sporadic mussels	4	9
Sandstone, light grey wispy-bedded	7	0
Shale, grey	1	3
Sandstone, light grey; shale partings	1	9
Mudstone, grey shaly; sporadic mussel impressions	7	3
HUTTON (Bottom Hutton) (at 210 ft)	—	—

The slightly thicker sequences elsewhere in this vicinity may be accounted for by an additional thin sandstone in the upper part, as for example in Randolph No. 2 Shaft* (72b). At Finlays Back Drift a mile to the east of Evenwood fossils collected from the roof of the Hutton coal included *Spirorbis sp.*, *Anthracosia sp.* (*phrygiana ?*), *Naiadites quadratus* and *N. triangularis*. A richer fauna was collected from a band 35 ft above the Hutton coal and included *Spirorbis sp.*, *Anthraconaia ?*, *Anthracosia* cf. *aquilina*, *A. ovum*, *A. sp.* intermediate between *ovum* and *phrygiana*, *Anthracosphaerium exiguum*, *Naiadites quadratus* and *N.* cf. *triangularis*.

Between Toft Hill and Woodhouses and in the area north-east of Tindale Crescent the sequence is argillaceous, apart from a thin sandstone locally under the Top Hutton seam as in Rush Pit* (100).

Between Tindale Crescent and Brusselton a typical sequence is recorded in Brusselton No. 181 Bore (117) on the north-western slopes of Brusselton Hill. The details are:

	ft	in
Horizon of ?TOP HUTTON (at 138 ft)	—	—
Seatearth-siltstone, grey; clay ironstone nodules	6	0
Sandstone, grey fine-grained massive	5	0
Siltstone, grey thin-bedded fissile ..	7	0
Sandstone, grey fine- to medium-grained massive	11	0
Mudstone, grey shaly; sporadic plant debris; abundant 'mussels' 7 ft from top	8	3
HUTTON (Bottom Hutton) (at 188 ft 3 in)	—	—

A broadly similar sequence is proved in Brusselton No. 179 Bore* (123).Mussels are commonly recorded in more recent boreholes from the roof of the Hutton. *Naiadites* and *Anthracosia* occur 9 and 11 ft respectively above the roof of the Hutton in Brusselton No. 184 Bore (125), *Naiadites sp.*, *Anthracosia* and *Anthracosphaerium* near the roof of the Hutton in Brusselton No. 180 Bore (119).

D.A.C.M.

In the South Church, Coundon Grange and Eldon areas these measures generally range between 24 and 36 ft thick. A thin sandstone locally underlies the Top Hutton or its assumed horizon, and another thin sandstone also occurs in places in the middle part of the sequence. Otherwise the beds are chiefly argillaceous. A representative section is recorded in Old Eldon 'A' Bore* (147). Mussels are recorded at a number of bores in the area, and in New Shildon No. 117 Bore* (134) they occur up to 11 ft 5 in above the Bottom Hutton. Similar sequences are proved farther east towards Old Eldon, and Windlestone 'B' Bore (166) recorded similar strata to those proved in Old Eldon 'A' Bore (see above). In the area between Fieldon Bridge and Shildon these beds range from 30 to 50 ft thick and are predominantly argillaceous, though in the thicker sections sandstone occurs beneath the Top Hutton or its assumed horizon. In New Shildon No. 147 Bore* (141) the sequence consists of 12 ft of grey shale on 24 ft of sandy shale. The measures proved in Old Shildon No. 99 Bore (149) is notable in that virtually the whole sequence contains mussels, and ostracods are also present between 5 and 12 ft below the Top Hutton. The detailed record is as follows:

	ft	in
TOP HUTTON (at 122 ft 4 in) ..	—	—
Seatearth-mudstone, grey slickensided rooty; ironstone nodules, joints	4	8
Shale, dark grey; abundant mussels and ostracods	7	0
Mudstone, grey; sporadic irony bands	8	3
Shale, dark grey; mussels	0	6
Mudstone, dark grey shaly; irony bands	9	3
BOTTOM HUTTON (at 155 ft)	—	—

J.H.H., D.A.C.M.

The Top Hutton is a thin impersistent coal throughout the district and of no economic value. Occasionally it is absent altogether or split into two or three thin coals. Its most westerly recorded occurrence is in Quarry West Pit* (19) where coals 3 and 6 in thick are separated by 5 in of seggar clay. The seam is absent in Gordon House No. 1 Shaft* (48a), its assumed horizon being at the roof of a fireclay. In Evenwood No. 16 Bore (66) it is 11 in thick, while in Randolph Colliery No. 2 Shaft* (72b) it forms coals 6 and 3 in thick separated by a 3-in band. In Randolph 'A' Bore (83) Evenwood, coals 2 and 4 in thick are separated by 6½ ft of seatearth on 1½ ft of grey mudstone. In Finlays Back Drift east of Evenwood the Top Hutton is absent. Between Toft Hill and Woodhouses the coal ranges between 6 and 7 in thick. In a number of boreholes between Tindale Crescent and Brusselton Hill, two thin coals have been proved in this position. The lower coal is absent in places, the assumed horizon being underlain by a seatearth as in Brusselton No. 179 Bore* (123) and in nearby Brusselton No. 181 Bore (117) the details of which are as follows:

	ft	in
TOP HUTTON—upper leaf (at 126 ft 7 in)		
Coal	0	7
Seatearth-mudstone, silty at base ..	3	0
Siltstone, grey; plant debris at base	8	5
Horizon of TOP HUTTON—lower leaf (at 138 ft)	—	—

D.A.C.M.

In the South Church, Coundon Grange, Eldon and Shildon areas the Top Hutton is recorded in most boreholes and shafts as a single seam split into two, or occasionally three, thin leaves; in places it is absent. In New Shildon No. 117 Bore* (134), New Shildon No. 146 Bore* (139) and Jane Pit* (142) for instance, it is absent; in Machine Pit* (145) the sequence recorded was coal 4 in, dark thill 5½ ft, on coal 4 in. Nearly ¾ mile ESE in Gurney Pit (150; B & S 2437) the split is wider, the section being coal 3 in, seggar clay 7 ft 8 in, on coal 7 in. In Old Eldon 'A' Bore* (147) the separation is small, the section being coal ½ in, seatearth-mudstone 1 ft 11½ in, on inferior coal 6 in. In Eldon Colliery Shaft* (152) the Top Hutton is split into three leaves. In the area to the east of Eldon Colliery Shaft, the leaves of the coal tend to come together to form a composite seam, with a thickness, usually including a thin band, of up to 18 in. In Old Eldon 'C' Bore (164) the section is coal 5 in, band 6 in, on coal 7 in. In the area between Fieldon Bridge and Shildon the Top Hutton is thin and impersistent. It is absent in New Shildon No. 147 Bore* (141) while in Old Shildon No. 99 Bore (149) the seam section is coal 2 in, carbonaceous shale 2 ft 2 in, on coal 4 in.

J.H.H., D.A.C.M.

The measures between the Top Hutton and Brass Thill (or Bottom Brass Thill) seams range between 20 and 60 ft. They form a variable sequence in which argillaceous strata tend to predominate. Sandstones occur locally, especially in the western part of the district. The most westerly proved record of these beds is in Quarry West Pit* (19) west of Butterknowle where they are 28 ft thick and include two sandstones in the upper part of the sequence. At Cockfield and on the south side of the Wigglesworth Fault, the measures are sandy, the sequence proved in Gordon

House Colliery No. 2 Shaft (48b) being as follows:

	ft	in
BRASS THILL (at 284 ft 6 in) ..	—	—
Fireclay	2	6
Grey post	5	6
White post, with whin	8	0
Black stone, with cannel coal ..	5	0
Fireclay	2	0
Grey post	5	0
Blue shale	5	0
TOP HUTTON—upper leaf (at 319 ft 7 in)	—	—

In the adjacent No. 1 Shaft* (48a) the sequence contains less sandstone and the Top Hutton is absent, its horizon overlying a seatearth. Thicker sequences were proved in Evenwood No. 16 (66) and Evenwood No. 14 (69) bores, the greater part of the measures being sandstone, or sandy in character. At the latter locality poorly preserved mussels were present some 12 ft above the Top Hutton. In Randolph Colliery No. 2 Shaft* (72b) the measures are mainly argillaceous and include a 1½-in coal 6 ft below the Brass Thill. A typical section near Evenwood is recorded in Randolph 'A' Bore (83) about ½ mile S of Copeland House, the details being:

	ft	in
BRASS THILL (at 128 ft 11 in) ..	—	—
Shale, black carbonaceous rooty ..	1	0
Seatearth, grey	0	5
Sandstone, light grey wispy-bedded; rooty at top, shale bands at base	10	8
Shale, grey silty; sandy at top; plant remains	4	6
Shale, dark grey; fauna of mussels, especially 7 ft from base ..	11	6
Sandstone, argillaceous	0	6
Shale, dark sooty; pyritic inclusions; mussel impressions	2	2
TOP HUTTON—upper leaf (at 160 ft)	—	—

Broadly similar sequences are recorded farther east. Mussels are recorded from mudstone overlying the assumed horizon of the Top Hutton in Finlays Back Drift.

Little is known about this sequence on the south side of the Butterknowle Fault from south of Toft Hill to north of Wigdan Walls [1871 2842], but the few records suggest that they range between 24 and 32 ft thick and are typically argillaceous. In Greenfields No. 1 Bore (84) nearly ¼ mile SW of Greenfield the interval is an exceptional 56 ft, the Brass Thill being underlain by 32 ft of sandstone with some shaly bands. South of Tindale Crescent and north and north-west of Brusselton Hill a typical sequence was recorded from Brusselton No. 181 Bore (117), the details being as follows:

	ft	in
BRASS THILL (at 95 ft 5 in) ..	—	—
Shale, black carbonaceous; coaly partings	1	0
Seatearth-mudstone, grey broken	1	7
Sandstone, light grey fine-grained; grey siltstone interlaminae; mainly siltstone at base	6	0
Siltstone, grey; argillaceous partings; comminuted plant debris ..	8	6
Shale, black silty finely micaceous; abundant mussels	0	6
Seatearth-sandstone	2	0
Siltstone, dark grey micaceous fissile	3	0
Mudstone, dark grey shaly finely micaceous; abundant large mussels (*Naiadites*) with *Spirorbis* especially in upper 3 ft; plant debris in basal 4 ft	7	0
Siltstone, grey thin-bedded; thin sandstone bands and partings; plant debris at base	7	0
TOP HUTTON—upper leaf (at 126 ft 7 in)	—	—

A broadly similar sequence is recorded in Brusselton No. 179 Bore* (123). In Brusselton No. 180 Bore (119) *Naiadites* is present in an 8-in band in the roof of the Top Hutton, and again in a 3-ft band, 8½ ft below the Brass Thill.

D.A.C.M.

In the South Church, Coundon Grange and Eldon areas these measures are, except at one locality, wholly argillaceous. One of the thinner though more detailed sequences was recorded in Old Eldon 'A' Bore* (147). Mussels are recorded a short distance below the Brass Thill here and in New Shildon No. 117 Bore* (134).

In the Old Eldon and Windlestone Park areas the sequence is again mainly argillaceous and usually about 30 ft thick. In Old Eldon 'B' Bore (156) mussels including *Naiadites* and *Anthracosia* occur over most of the 29 ft 11 in interval between the Brass Thill and Top Hutton. Their distribution is broadly similar in Eldon Underground No. 1 Bore (161), although here as in Windlestone 'B' Bore (166), the sequence is more arenaceous especially in its middle part.

In the area from Fieldon Bridge towards Shildon the sequence, which ranges between 24 and 58 ft, is in all but one locality wholly argillaceous. In some boreholes mussels are recorded in the middle or upper part of the sequence. A sandstone some 4 ft thick is recorded from Old Shildon No. 99 Bore (149).

J.H.H., D.A.C.M.

The Brass Thill may be regarded as a single seam throughout most of the district, although locally it is split into Top and Bottom coals separated by up to 11 ft of strata.

A small area of Brass Thill is present on the south side of the Butterknowle Fault south and west of Butterknowle where Quarry West Pit* (19) shows the seam to be split.

J.H.H.

South of Little Moor Farm, Toft Hill and on the south side of Butterknowle Fault opencast workings [1465 2715] in the seam exposed a section up to 42 in thick, often banded.

South of the Wigglesworth Fault, at Cockfield, and in the area to the west of Evenwood the seam varies in both quality and thickness. At Gordon House Colliery No. 1 Shaft* (48a) it is 36 in thick, and in Evenwood No. 16 Bore (66) a quarter mile to the east-north-east it has a section coal 21 in, band 12 in, on coal and shaly coal 22 in. Locally, however, it shows considerable impoverishment, as for example south of Garden House [1208 2415], Cockfield, where the section recorded was coal 9 in, band 3 in, on coal 6 in. In Evenwood No. 14 Bore (69) the ?Brass Thill was only 6 in thick. Sections at and to the east of Evenwood show a thickness in the general range 30 to 39 in, sometimes with a thin band near the base.

On the south side of the Butterknowle Fault east of Toft Hill, the coal, usually banded, generally ranges between 14 and 28 in, but locally is impoverished and split as at Rush Pit* (100).

In the Gaunless valley east of West Auckland and north and north-west of Brusselton, the seam is generally thicker, the range being 25 to 51 in. Again a thin band is present in the lower part of the section.

In the South Church, Coundon Grange, and Eldon areas the seam generally ranges between 36 and 48 in, including a band up to 10 in thick in the lower part. Farther east in the Old Eldon and Windlestone Park areas it is either a single seam, sometimes banded, up to 56 in thick, or split into Top and Bottom coals. North of Eldon Drift the section proved in Eldon No. 2 Underground Bore (160) is as follows:

	ft	in
TOP BRASS THILL	—	—
Coal	2	7
Mudstone	4	3
Sandstone	2	1
Shale	1	9
BOTTOM BRASS THILL	—	—
Coal	1	1

Three-quarters of a mile to the east Windlestone Supplementary 'B' Bore* (165) proved a similar sequence.

In the area between Fieldon Bridge and the eastern outskirts of Shildon the seam is generally undivided and between 36 and 57 in thick. Thicknesses between 47 and 57 in were proved in opencast sites south and south-east of Low West Thickley [2102 2556].

J.H.H., D.A.C.M.

The measures between the Brass Thill and the Durham Low Main seams range between 20 and 60 ft and are generally argillaceous. Sandstones are sporadic and usually absent; only rarely do they occupy virtually the whole sequence.

Nothing is known of these measures on the south side of the Butterknowle Fault south of Butterknowle and south and south-west of Toft Hill, although in the latter locality they are thought to be about 50 ft, including a median coal up to 8 in thick.

South of the Wigglesworth Fault between Cockfield and Evenwood these measures, ranging between 42 and 48 ft, are predominantly argillaceous, but a thin sandstone locally underlies, or is separated by, a few feet from, the base of the Durham Low Main. In Evenwood No. 14 Bore (69) the Brass Thill Shell-Bed is represented by 14¼ ft of shale and silty shale containing mussels overlying the Brass Thill Coal. A typical sequence through these measures in and to south and east of Evenwood is recorded in Randolph 'A' Bore (83). The details are as follows:

	ft	in
DURHAM LOW MAIN (at 73 ft 3 in) ..	—	—
Sandstone, argillaceous rooty wispy-bedded	2	9
COAL	0	3

	ft	in
Mudstone, grey rooty; irony bands	4	9
Sandstone, light grey fine-grained . .	4	0
Sandstone, wispy-bedded; shale partings	5	0
Shale, grey silty; sandy partings and sporadic irony bands	22	0
Mudstone, grey shaly; sporadic mussel debris at base (Brass Thill Shell-Bed)	14	3
BRASS THILL (at 128 ft 11 in) . .	—	—

The thin coal underlying the Durham Low Main, a possible split off the base of the coal itself, is very impersistent. In nearby Randolph 'B' Bore (82) 32 ft of mainly argillaceous beds overlying the Brass Thill Coal contain mussels. In Finlays Back Drift a fauna including *Anthraconaia sp.* (*?salteri* juv.), *Anthracosia disjuncta?* and *Naiadites quadratus* was collected from 4 ft of silty mudstone overlying the roof of the Brass Thill coal.

Between Toft Hill and the area north-west of Woodhouses [189 280] these measures are thought to be between 50 and 55 ft thick and mainly argillaceous. Atypically, in Greenfield No. 1 Bore (84), 37½ ft of predominantly sandy measures, in which a 1 ft 6 in band of sandy shale with sporadic *Naiadites* is present, overlie the Brass Thill.

North-west of Brusselton Hill the sequence proved in Brusselton No. 181 Bore (117) may be regarded as typical:

	ft	in
DURHAM LOW MAIN (at 51 ft)	—	—
Seatearth-mudstone, yellowish grey	7	0
Mudstone, grey shaly	4	0
Seatearth-sandstone, brownish grey medium-grained	2	0
Sandstone, light grey fine-grained; siltstone interlaminae; comminuted plant debris	4	0
Siltstone, grey; massive at top, bedded below	10	0
Mudstone, grey shaly; mussels and mussel fragments (Brass Thill Shell-Bed)	15	0
BRASS THILL (at 95 ft 5 in)		

In adjacent Brusselton No. 180 Bore (119) 260 yd to SSE, the Brass Thill Shell-Bed, 7 ft 8 in thick, contained mussels including '*Anthracosia*' and '*Naiadites*'. A broadly similar section to that from Brusselton No. 181 Bore was recorded in Brusselton No. 179 Bore* (123). D.A.C.M.

In the South Church, Coundon Grange and Eldon areas the measures are virtually all argillaceous. In some places the Durham Low Main is absent, probably due to washout. Mussels occur at two distinct horizons in New Shildon No. 117 Bore (134) and Old Eldon 'A' Bore* (147). Atypically in New Shildon No. 146 Bore* (139) the whole sequence is sandy.

In the Old Eldon and Windlestone Park areas, argillaceous strata predominate and mussels are abundant in places. In Old Eldon 'B' Bore (156) 37 ft of strata overlying the Brass Thill coal contained mussels including *Naiadites* and *Anthracosphaerium*. In a few places, as for example Windlestone 'B' Bore (166), the Durham Low Main is underlain by some 20 ft of sandstone. In Eldon Underground No. 1 Bore (161) the sequence proved was:

	ft	in
DURHAM LOW MAIN (at 244 ft 1 in)	—	—
Seatearth-mudstone	1	7
Siltstone	1	4
Shale, sandy; roots	1	0
Mudstone; roots and ironstone bands	7	0
Mudstone; abundant mussels and *Spirorbis*	14	0
Mudstone; plant debris	5	6
Shale, black carbonaceous . .	0	4
Seatearth-siltstone	0	6
Shale, silty	6	8
Mudstone; ironstone nodules and mussel fragments at the top . .	4	4
TOP BRASS THILL (at 288 ft 2 in)	—	—

In the area between Fieldon Bridge and Shildon the strata are argillaceous.

J.H.H., D.A.C.M.

The Durham Low Main (locally known as the Chatham) generally ranges between 20 and 36 in, but it is locally prone to washout or impoverishment.

South of the Butterknowle Fault between Coleburn [144 273], Toft Hill and north of Woodhouses [189 281], the seam is between 20 and 48 in and is locally banded. It was between 21 and 41 in at small opencast sites [1791 2814 and 1815 2847] near Woodhouses.

South of the Wigglesworth Fault, at Cockfield and towards Evenwood the seam is up to 63 in thick, a typical section being coal and splint 26 in, band 3 in, on coal 33 in. Zones of impoverishment exist, as for example south of Garden House [1208 2415], Cockfield, where the section is coal 9 in, band 3 in, on coal 6 in.

At and to the east of Evenwood the seam is generally thinner, ranging between 24 and 45 in. Typically, a top coal 6 to 9 in is separated by a band 4 to 6 in from a bottom coal 24 to 27 in. South of Tindale Crescent [200 274] and north-west of Brusselton Hill [2037 2498] it is usually a single coal between 25 and 40 in thick.

D.A.C.M.

In the South Church, Coundon Grange and Eldon areas the seam thickness is up to about 30 in, but around and to the east of Auckland Park as at Jane Pit* (142) and Machine Pit* (145) it is absent. Farther east in the Old Eldon and Windlestone Park areas the general thickness is between 26 and 33 in, though the seam is in places banded and locally absent. In Windlestone 'B' Bore (166) it is 51 in thick.

In the area between Fieldon Bridge and Shildon, the seam, again locally absent, ranges generally between 18 and 30 in. East of Shildon, Old Shildon No. 99 Bore (149) proved a section: coal 18 in, band 2 in, coal 10 in.

J.H.H., D.A.C.M.

The measures between the Durham Low Main and Main coals attain a thickness of up to 220 ft, and in most places are distinguished by a thick sandstone, the Durham Low Main Post, overlying the Durham Low Main.The Maudlin, a composite seam up to a maximum thickness of 77 in, and up to 70 ft below the Main, is only locally present.

South of the Butterknowle Fault between Greenfield [1700 2804], Toft Hill, to the area north of Woodhouses [189 281], these measures are inferred to be up to 220 ft thick. The Durham Low Main Post is thought to be relatively thin, and north-east of Etherley Dene [193 286] is probably less than 20 ft. The Maudlin seam has not been proved but may be locally represented by a 7-in coal up to 60 ft below the Main.

South of the Wigglesworth Fault and east-south-east of Cockfield these measures are about 170 ft thick, the greater part being sandstone (the Durham Low Main Post); the Maudlin seam has a maximum composite banded thickness of 77 in, lying up to 30 ft below the Main. The measures between the Maudlin and the Main are argillaceous, commonly containing bands rich in ironstone nodules. The section at Gordon House No. 1 Shaft (48a) is typical. Farther east, workings in the Maudlin Coal from Esperley Lane Drift showed that the coal was typically split into three leaves, the lowest usually being the thickest. The range of variation proved was coal 6 to 22 in, fireclay up to 2 ft, coal 17 to 18 in, on coal 10 to 46 in. In Evenwood No. 14 Bore (69) farther east, the Durham Low Main Post is 65 ft thick and separated from the base of the Maudlin by some 30 ft of predominantly argillaceous strata. The Maudlin section is coal 22 in, fireclay and coal shaly 3 ft 5 in, coal 17 in, band 20 in, on coal 10 in. East of here the coal dies out, and the apparent interval between the Durham Low Main and Main dwindles to less than 100 ft.

East of Evenwood, only the lower part of this sequence is proved, but in each record the Durham Low Main Post overlies the Durham Low Main seam and reaches a recorded maximum of 95 ft 6 in at Randolph 'B' Bore (82) south of Copeland House. It is a predominantly grey and brown, fine-grained wispy-bedded sandstone, with sporadic shaly partings. It is overlain by 30 ft of grey shale and mudstone with coal traces 19 ft above the base which probably represent the Maudlin. Thick sandstone with shaly partings overlies the Durham Low Main coal north and north-west of Brusselton Hill [2037 2498].

D.A.C.M.

In the South Church, Coundon Grange and Eldon areas these beds are between 90 and 150 ft thick. Most of the sequence is of sandstone, the Durham Low Main Post, being locally split into two or three leaves. In some places the Durham Low Main is separated from the Durham Low Main Post by a few feet of argillaceous measures. The Maudlin Coal is absent.

Farther east in the Old Eldon and Shildon areas these beds are between 112 and 156 ft thick. The Durham Low Main Post occurs either as a single thick bed or split by shale partings into two or three leaves with a composite thickness of up to 60 ft. In some localities, a thin band of argillaceous measures, sometimes mussel-bearing, occurs in the roof of the Durham Low Main, as in Old Eldon 'B' Bore (156) where the mussels include *Anthracosia*, and in Eldon 'A' Bore (158) where large *Anthracosia* were also

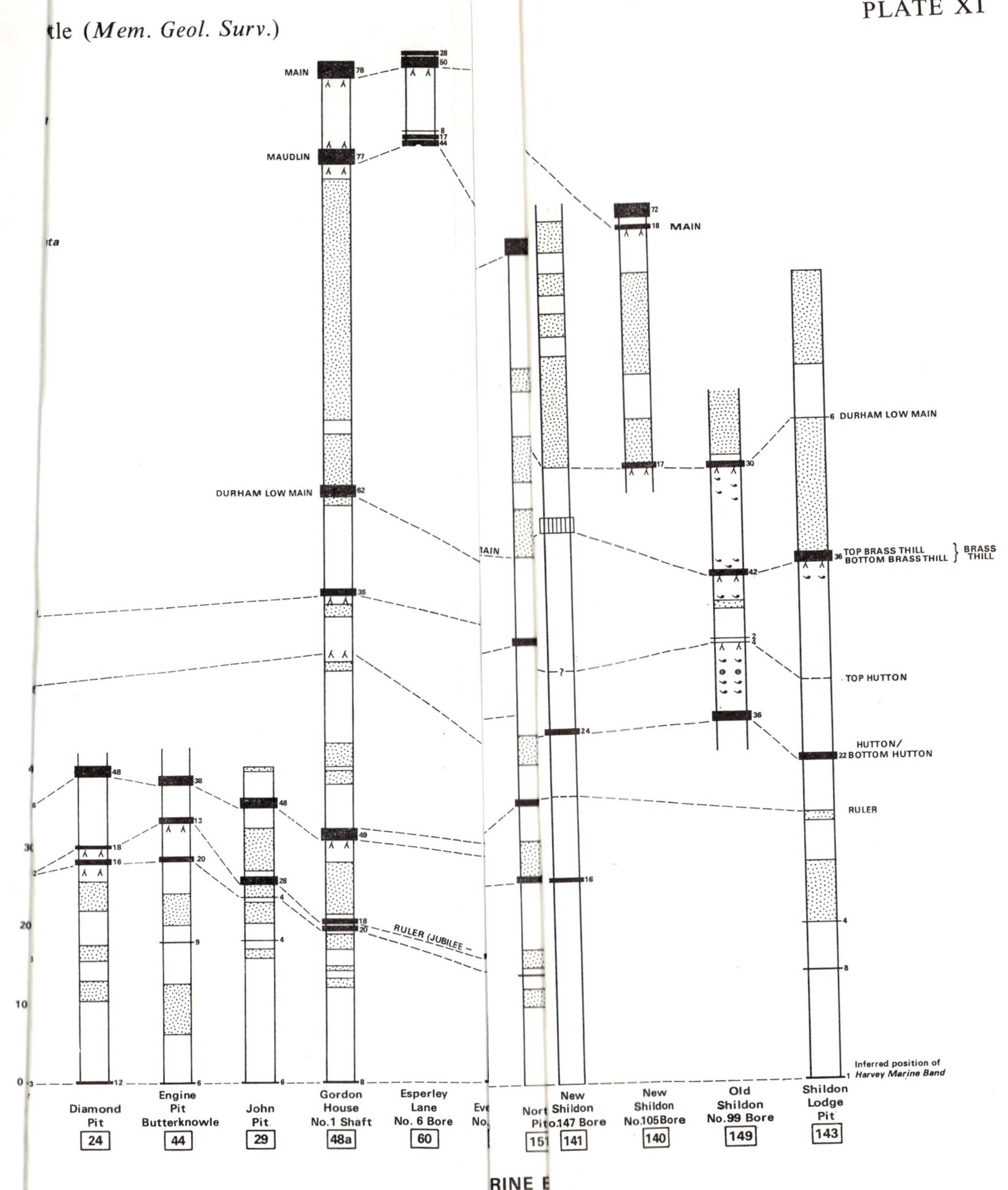
MAIN
MAUDLIN
DURHAM LOW MAIN
RULER (JUBILEE
MAIN
DURHAM LOW MAIN
TOP BRASS THILL
BOTTOM BRASS THILL
BRASS THILL
TOP HUTTON
HUTTON/
BOTTOM HUTTON
RULER
Inferred position of
Harvey Marine Band
Diamond Pit
24
Engine Pit Butterknowle
44
John Pit
29
Gordon House No.1 Shaft
48a
Esperley Lane No. 6 Bore
60
New Shildon No.147 Bore
141
New Shildon No.105 Bore
140
Old Shildon No.99 Bore
149
Shildon Lodge Pit
143
RINE

reported from this position. Throughout most of the area the Maudlin is absent, but locally a thin coal is proved between the Durham Low Main Post and an overlying sandstone. Still farther east, these measures include more coals at or near the assumed Maudlin horizon, and are also notable for the low content or absence of sandstones. A typical sequence, and one in which the Maudlin is presumed to be split, was recorded in Windlestone Supplementary 'B' Bore* (165).

In the area between Fieldon Bridge and Shildon these measures range in thickness between 100 and 145 ft and are predominantly sandy, with argillaceous strata occurring near the top of the sequence. New Shildon No. 145 Bore (132) proved a 9-in coal some 20 ft below the base of the Main, and here the Durham Low Main Post is 114 ft thick.

J.H.H., D.A.C.M.

The Main occurs as a single coal of between 36 and 132 in, or as a composite seam of between 80 and 171 in, including bands.

South of the Butterknowle Fault and east of Greenfield [1700 2804], Toft Hill, the Main, only locally banded, has been proved in small opencast sites to be between 54 and 72 in thick.

South of the Wigglesworth Fault around Gordon House Colliery [1334 2403] a typical section through the Main seam is coal 28 in, band 4 in, on coal 50 in; farther east-south-east in Esperley Lane Drift a typical sequence is coal 3 in, band 13 in, coal 5 in, band 6 in, on coal 57 in. About 2½ miles to the east-north-east in a small opencast site near Low Staindrop Field House the Main had a typical section coal 24 in, band 1 in, coal 16 in, band 1 in, coal 45 in, on splint coal 11 in.

D.A.C.M.

In the South Church, Coundon Grange and Eldon areas, the seam is generally between 84 and 134 in exclusive of bands. The minimum thickness recorded was 36 in at New Shildon No. 146 Bore* (139). The section recorded in Jane Pit* (142) was coal 17 in, slaty band 1 in, coal, good 42½ in, coal, coarse 4½ in, splint 17½ in, blackstone and thill 1 ft 6 in, coal, coarse, with bands every 3 or 4 in 42 in, blackstone 1 ft 2 in, on coal 10 in.

In the Old Eldon and Windlestone Park areas broadly similar thicknesses are recorded. In Eldon Colliery Shaft* (152) the section was coal 80 in, band 10 in, coal 17 in, band 6 in, coal 24 in, band 14 in, on coal 11 in. In Old Eldon 'B' Bore (156) the section was coal, durainous at base 85 in, mudstone 8 in, coal 17 in, mudstone with vitrain partings 7 in, coal 18 in, mudstone with coaly partings at top 19 in, inferior coal and bands 16 in, mudstone with thin coaly partings 2 in, coal 4 in, mudstone 2 in, coal 10 in. Between Old Eldon and Shildon opencast site-proving boreholes in the vicinity of Eldon Hill [240 272] showed the Main to be a banded seam in the general range 150 to 171 in. Between St Helen Auckland and Adelaide Colliery the seam is generally either united and between 36 and 63 in, or banded and between 102 and 167 in.

Between Fieldon Bridge and Shildon the Main has been extensively worked and most borehole records indicate old workings at this horizon. In New Shildon No. 105 Bore (140) the section recorded was coal 72 in, band 30 in, on coal 18 in. Boreholes north and north-east of Brusselton Farm [2027 2551] consistently proved a united Main of between 84 and 90 in.

J.H.H., D.A.C.M.

Comparative sections of the Middle Coal Measures between the Harvey Marine Band or its assumed horizon and the top of the Main Coal are shown in Plate XI, and sections and synopses of sections of boreholes and shafts which show all or part of these measures are quoted in Appendix 2.

Middle Coal Measure above the Main

(shown graphically in Plate XII)

The measures between the Main and Five-Quarter coals, mainly argillaceous, range between 47 and 120 ft, and are thickest where locally they include sandstones. Three thin and impersistent coals occur in the sequence.

South of the Butterknowle Fault east-north-east of Greenfield [1700 2804], Toft Hill, these beds are between 60 and 70 ft thick and predominantly argillaceous. Two thin coals, the lower up to 15 in, lie about midway between the two seams. The following sequence was formerly exposed at an opencast site [1867 2860] 180 yd N of Wigdan Walls, Woodhouses:

	ft	in
Ironstone band	0	3
Mudstone, brownish grey blocky	5	6
Shale, grey fissile carbonaceous	2	2½
COAL, dirty	0	7
Fireclay	0	3
Shale, grey carbonaceous	0	6
Shaly mudstone; becoming blocky to base	3	8
Shale, grey carbonaceous	3	0
Shale, clayey carbonaceous	4	0
COAL, dirty	1	0
Shale, dark grey carbonaceous fissile	1	6
Mudstone, grey; plant remains	0	6
Shale, grey sparingly micaceous; plant remains	4	4
Shale, grey	1	6
Mudstone, light grey	0	6
Shale, grey micaceous	3	0
Mudstone, shaly; ironstone nodules	1	8
Mudstone, grey; ferruginous at base	1	10
Mudstone, shaly grey	5	4
not exposed (estimated)	10 to 15	0
MAIN	—	—

D.A.C.M.

In the area between Auckland Park and Eldon Colliery a typical section was proved in Machine Pit* (145). In Eldon Colliery Shaft* (152) a 10-in coal is present in the middle part of the sequence overlying a sandstone. North Pit (151; B & S 208), proved two sandstones in the lower part of the sequence. In the extreme north-east of the district these measures were penetrated in Windlestone Supplementary 'B' Bore* (165) east of Windlestone Grange where fish debris and mussels were obtained above an intermediate coal and fish fragments in the roof of the Main. Fish debris, and mussels including *Anthracosia*, but no coal, are also noted at an intermediate horizon in Eldon 'A' Bore (158), and *Spirorbis sp.*, *Anthracosia* aff. *caledonica*, *A.* cf. *planitumida* and *Naiadites sp.* were collected from the roof measures of an intermediate but higher coal at Middridge No. 195 Bore (153).

Between St Helen Auckland and Old Eldon these measures are more argillaceous with an average thickness of about 105 ft. Thin intermediate coals were proved in New Shildon No. 117 Bore* (134) and in Jane Pit* (142). A 34-ft sandstone lying 25 ft above the Main in Old Eldon 'B' Bore (156) suggests these measures become more arenaceous to the east.

In the Fieldon Bridge and Shildon area a typical section was proved in New Shildon No. 123 Bore (131) where this sequence was recorded:

	ft	in
FIVE-QUARTER (at 45 ft 6 in)	—	—
Shale; sandy in top 3½ ft and basal 14 ft with sandstone lenses	50	6
Sandstone	2	0
Shale	25	6
MAIN (at 134 ft)	—	—

Farther east these measures, though lithologically similar, become progressively thinner, only 55¼ ft of strata being recorded in New Shildon No. 145 Bore (132) and 47½ ft in New Shildon No. 105 Bore (140).

In the area north-east of Brusselton Farm [2027 2551] only the lower part of these measures are preserved, and the section from New Shildon No. 139 Bore (133) west-north-west of Low West Thickley indicates a higher proportion of sandstone in this part of the sequence. The details are:

	ft	in
Shale	4	2
COAL	0	4
Shale, canneloid	0	2
Mudstone, silty; some ironstone bands	7	4
COAL; shale partings	0	3
Shale, carbonaceous	0	9
Siltstone, rooty at top; plant debris	14	0
Sandstone, fine-grained thin-bedded; silty partings	18	0
Siltstone; sandy partings	7	6
Mudstone; sparse ironstone nodules and bands	14	6
Shale; sporadic rootlets	4	7
MAIN (at 111 ft)	—	—

J.H.H., D.A.C.M.

The Five-Quarter is a single united seam in this district and generally ranges from 60 to 72 in. Locally it is united also with the overlying Metal seam.

Small areas of Five-Quarter coal are thought to occur on the south side of the Butterknowle Fault ¾ mile ENE of Greenfield [1700 2804], Toft Hill, and 250 yd N of Wigdan Walls [1875 2845], but there are no records.

D.A.C.M.

In the area between Auckland Park and Eldon Blue House [2442 2805], east of Coundon Grange, the seam ranges between 60 and 72 in. Farther east records suggest a similar thickness though in Windlestone Supplementary 'B' Bore* (165), where the coal is near the base of the Permian unconformity, the thickness is only 10 in.

In the area south of South Church and towards Eldon the thickness is between 64 and 72 in; atypically, however, at New Shildon No. 146 Bore* (139) the coal is absent, probably washed out.

Between Fieldon Bridge and Shildon it has been extensively worked at a general thickness of 60 in. Locally, however, it is thinner; for example, in New Shildon No. 105 Bore (140) it is a mere 24 in.

J.H.H., D.A.C.M.

The measures between the Five-Quarter and Metal seams show wide variation in thickness and lithology. In the area between Auckland Park and Eldon they are about 50 ft thick, a typical sequence being proved in Machine Pit* (145). In Gurney Pit (150; B & S 209) nearly ¾ mile to east-south-east, these measures, 40 ft thick, are wholly argillaceous. East of this area they thin progressively and in Eldon Colliery Shaft* (152) they comprise only 3 ft of seggar clay. Still farther east in Windlestone Supplementary 'B' Bore* (165), they are only 15 ft 9 in thick. In Eldon Drift Mine the following fauna was collected from between 19 ft 8 in and 23 ft 2 in above the Five-Quarter coal, *Naiadites sp.*, *Carbonita humilis*, fish remains including *Rhabdoderma sp.* and *Rhizodopsis sp.* Below this, *Spirorbis sp.*, *Anthracosia?* and *Naiadites sp.* were recorded.

In the vicinity of Shawbrow Hill [217 273], to the south-west, there is evidence to suggest that the Five-Quarter and Metal are virtually united. Farther east in Jane Pit* (142) the seams are only 6½ ft apart and still farther east, in Eldon 'A' Bore (158), the separation is a mere 11 in of seatearth-mudstone.

In the area between Fieldon Bridge and Shildon these measures range from 14 ft of argillaceous strata as proved in New Shildon No. 145 Bore (132) to 47½ ft of shale, sandstone and shaly sandstone, as proved in New Shildon No. 105 Bore (140).

J.H.H., D.A.C.M.

The Metal Coal, locally called the Jet, Dicky Dant, or Top Five-Quarter, is, in a few localities, united with the underlying Five-Quarter seam. In the area between Auckland Park and Eldon it ranges between 29 and 36 in, being banded at the former locality. Farther east at Windlestone Supplementary 'B' Bore* (165) it is only 6 in thick.

In Jane Pit* (142) the Metal has a section coal 48 in, ironstone band 2 in, on coal 16 in; at Eldon 'A' Bore (158) the seam is 43 in thick.

West of Shildon, the Metal and Five-Quarter are sufficiently close together to have been worked extensively as one coal and boreholes generally prove old workings. Opencast site-proving boreholes to the north of Blue House [2142 2615] have proved the Metal to range between about 20 and 25 in.

J.H.H., D.A.C.M.

The measures between the Metal and High Main (or Bottom High Main) are between 40 and 81 ft thick, are predominantly argillaceous with sporadic thin sandstones and locally include a coal up to 26 in. Between Auckland Park and Eldon Colliery, the following typical section was proved in Gurney Pit (150; B & S 209):

	ft	in
BOTTOM HIGH MAIN (at 158 ft 6 in)	—	—
Thill stone	3	0
Stone, grey; ironstone girdles	21	0
Stone, black	1	0
Jet, black	3	0
Stone and metal	8	0
Post	2	0
Stone, grey; ironstone girdles	21	0
METAL (at 217 ft 6 in)	—	—

A similar, essentially argillaceous, sequence has been proved both east and west of this locality, although farther west it is somewhat thinner. To the south-south-east Middridge No. 194 Bore (155) proved an 8-ft bed containing fish scales and poorly preserved mussels including *Anthraconaia*, *Anthracosia* and *Naiadites* 34 ft above the goaf of the united Metal and Five-Quarter seams.

The 63 ft section recorded in Jane Pit* (142) was mainly argillaceous and included a 26-inch coal 25 ft 2 in below the Bottom High Main. In Eldon 'A' Bore (158) the sequence, 45½ ft thick, is again argillaceous and contains fossils at two horizons; a few fragmentary mussels are present in the immediate roof of the Metal, while fish debris and *Anthracosia* are present in 5 ft of shale some 23 ft higher, and 25 ft below the High Main.

In the area west of Shildon a typical sequence through these measures was recorded in New Shildon No. 105 (140), the details being as follows:

	ft	in
BOTTOM HIGH MAIN (at 143 ft 3 in) ..	—	—
Shale; sporadic sandstone lenses ..	27	3
Shale, black	2	6
Shale, grey	5	0
Sandstone	8	0
Shale, sandy	9	0
Sandstone	2	0
Shale, sandy	27	0
METAL (at 230 ft 6 in)	—	—

This is the maximum Metal-High Main separation proved in this district.

J.H.H., D.A.C.M.

The High Main is generally a split seam but because the Top High Main is thin, or perhaps locally absent, it has been mapped as a single seam. For descriptive purposes the terms Top and Bottom High Main are used. The High Main is thought to form a single seam in several localities and to range between 18 and 28 in thick. In North Pit (151; B & S 208) it is absent.

The Bottom High Main (locally known as the Willy Winter) is generally a single seam, ranging in thickness from 18 to 38 in. West of Shildon it is thicker; in New Shildon No. 145 Bore (132) the section recorded was coal 18 in, shale 25 in, coal 2 in, shale 2 in, on coal 38 in. In opencast site-proving boreholes in the vicinity of Coppy Crook [2095 2650] thicknesses from 36 to 66 in including bands were recorded; in this area the coal appears to consist of three leaves, and in places even four.

The measures between the Bottom and Top High Main seams range up to a maximum of 24 ft; more usually, however, the separation is in the range 10 to 15 ft. The strata are essentially argillaceous, but one or two impersistent sandstones occur in the thicker sections. In places, also, the Top High Main and some of the beds below are cut out by the overlying sandstone, the transgressive High Main Post. In the area between Auckland Park and Eldon Colliery the measures between Bottom and Top High Main range from 9 ft 6 in of grey shale, as proved in a borehole on the Pasture Houses Opencast site, east of St Philip's Church, Auckland Park, to 13 ft of blue metal and stone in Gurney Pit (150; B & S 209). In Middridge No. 194 Bore (155) fish fragments were proved in the roof of the Bottom High Main, with scattered mussel fragments some 2 ft higher.

Between St Helen Auckland and Shildon the only definite record of these measures is from Jane Pit* (142) where they consist of 14 ft of thill with post girdles, on 4 ft of grey metal with ironstone. Elsewhere in this area the Top High Main is absent but comparison with other sections which include the High Main Post suggests that they range between about 10 and 22 ft thick and are argillaceous. West of Shildon, New Shildon No. 105 Bore (140) proved the following sequence:

	ft	in
TOP HIGH MAIN (at 117 ft 6 in) ..	—	—
Shale, grey	2	6
Sandstone	8	0
Shale, grey	7	0
Sandstone	4	0
Shale, grey	2	6
BOTTOM HIGH MAIN (143 ft 3 in) ..	—	—

This is the maximum separation between Bottom and Top High Main recorded in this district. In the southern part of this area opencast site-proving boreholes south-east of Coppy Crook [2095 2650] proved the interval to be between 4 and 19 ft.

The Top High Main has a maximum recorded thickness of 33 in; more commonly it is 6 to 21 in thick and locally it is cut out by the transgressive High Main Post. In the area between Auckland Park and Eldon Colliery the only record of this seam is from Gurney Pit (150; B & S 209) where it is only 3 in thick. To the south between Shawbrow Hill [217 273] and Old Eldon the maximum recorded thickness is 28 inches in Jane Pit* (142). Boreholes between Fieldon Bridge and Shildon show the following seam thicknesses: New Shildon No. 145 (132) 21 in; Shawbrow Hill No. 5 Bore* (137) 7 in; New Shildon No. 105 Bore (140) 6 in. Boreholes near Coppy Crook [2095 2650] show a range between 13 and 33 in, the latter the maximum recorded for this district. In other localities the seam is frequently very thin or absent.

J.H.H., D.A.C.M.

A maximum of 160 ft of strata have been proved in this district overlying the Top High Main. They are confined to the area north of Coundon Grange where they are mainly arenaceous, and around Shawbrow Hill [217 273] on the synclinal axis of the Coal Measures where they are more argillaceous.

Over 140 ft of measures overlying the Top High Main were recorded in Gurney Pit (150; B & S 209), the details being as follows:

	ft	in
Soil	2	0
Metal, sandy	11	0
Freestone; clay, whin and grey partings and some coal pipes ..	50	0
Metal, grey	10	0
Metal, blue; ironstone girdles ..	4	0
HIGH MAIN POST		
Freestone; red in basal 7 ft; coal fragments	67	0
TOP HIGH MAIN (at 144 ft)		

Boreholes to the north-west in the adjacent Wolsingham (26) district suggest that there should be a 6 to 24 in coal overlying the High Main Post. In the area north and east of Shawbrow Hill [217 273] typical sections were recorded in New Shildon No. 146 Bore* (139) and Jane Pit* (142). In Eldon New Winning (154; B & S 770) the measures, 130 ft thick, are mainly argillaceous, but over 35 ft of sandstone is present at the top of the shaft. No coal is proved above the High Main but 2 ft of what is described as 'jet' underlain by 1 ft of 'thill' is present over 45 ft above the High Main. The High Main Post is absent. Eldon 'A' Bore (158) proved no coals in a 98-ft section. Fish debris were noted in the roof of the High Main; the High Main Post here is 21 ft 10 in thick and is overlain by some 7 ft of shale and shaly mudstone containing *Anthracosia* and *Spirorbis* at the top and abundant fish debris at the base. Shawbrow Hill No. 5 Bore* (137) provides the record of the highest Middle Coal Measures in this district above the Top High Main. None of the thin coals proved in this borehole is marked on the one-inch map of the district. Broadly similar sequences were proved in Shawbrow Hill No. 1 Bore (138) and in New Shildon No. 105 Bore (140), though in the latter no coals were proved.

Opencast site-proving boreholes in the vicinity of Coppy Crook [2095 2650] show a 17- to 30-in coal 90 to 120 ft above the Top High Main and it would seem possible that this coal, probably the Ryhope Little, may also be equated with the 12-in coal known to occur locally in the Auckland Park area. A borehole [2119 2654] east of Coppy Crook recorded the following:

	ft	in
COAL	1	10
Shale	13	2
Sandstone	2	1
COAL	0	7
Shale	4	2
Sandstone	2	7
Shale	6	5
Sandstone	7	8
Shale	19	3
COAL	0	3
Shale	37	3
TOP HIGH MAIN	—	—

Comparative sections of the Middle Coal Measures above the Main coal are shown in Plate XII, and a selection of relevant sections are included in Appendix 2.

D.A.C.M., J.H.H.

References

ANDERSON, W. and DUNHAM, K. C. 1953. Reddened beds in the Coal Measures beneath the Permian of Durham and South Northumberland. *Proc. Yorks. geol. Soc.*, **29**, 21–32.

ARMSTRONG, G. and PRICE, R. H. 1954. The Coal Measures of north-east Durham. *Trans. Instn Min. Engrs*, **113**, 974–97.

BORINGS AND SINKINGS. 1878–1910. Council of the North of England Institute of Mining and Mechanical Engineers. An account of the strata of Northumberland and Durham as proved by borings and sinkings. 7 vols. in 4 books. Newcastle-upon-Tyne.

CALVERT, R. 1884. *Notes on the geology and natural history of the County of Durham.* Bishop Auckland.

HOPKINS, W. 1929. The distribution and sequence of the non-marine lamellibranchs in the Coal Measures of Northumberland and Durham. *Trans. Instn Min. Engrs*, **78**, 126–44.

KIRKBY, J. W. and DUFF, J. 1872. Notes on the geology of part of South Durham. *Nat. Hist. Trans. Northumb.*, **4**, 151–98.

LAND, D. H. 1974. Geology of the Tynemouth district. *Mem. geol. Surv. Gt Br.*

MAGRAW, D., CLARKE, A. M. and SMITH, D. B. 1963. The stratigraphy and structure of part of the south-east Durham coalfield. *Proc. Yorks. geol. Soc.*, **34**, 153–208.

MILLS, D. A. C. and HULL, J. H. 1968. The Geological Survey Borehole at Woodland, Co. Durham (1962). *Bull. geol. Surv. Gt Br.*, No. 28, 1–37.

RICHARDSON, G. 1965. In *Summ. Prog. geol. Surv. Gt Br. for 1964*, 51.

SMITH, D. B. and FRANCIS, E. A. 1967. Geology of the country between Durham and West Hartlepool. *Mem. geol. Surv. Gt Br.*

WOOLACOTT, D. 1923. A boring at Roddymoor Colliery, near Crook, Co. Durham. *Geol. Mag.*, **60**, 50–62.

PLATE XII

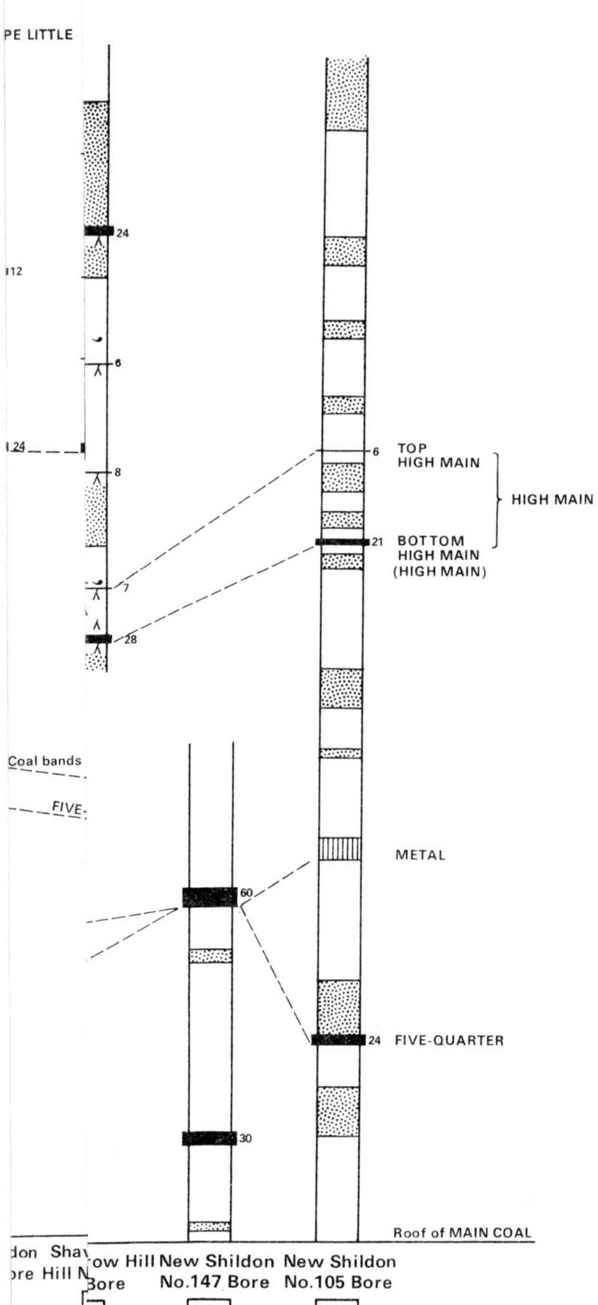

COMP

Chapter 5

PERMIAN

INTRODUCTION

PERMIAN rocks, mainly Magnesian Limestone, unconformably overlie Carboniferous strata in the eastern part of the district and form a south-western part of the main Permian outcrop of Durham county. The western boundary is irregular, faulted and reaches its maximum extension in the Tees valley north and south of Gainford (Fig. 14). Only in the north-east part of the district, notably at Eldon Hill (500 ft) and Shackleton Beacon (600 ft), is there a well-marked escarpment like that of other parts of the county; elsewhere the formation exercises only a broad topographic control on the landscape. In the south-east and along the eastern margin of the district the Permian rocks are almost wholly obscured by thick drift.

North of the River Tees most exposures are to be found near the western limit of the formation; they include Eldon Hill Quarry, Thickley Quarry (East Thickley), Middridge Quarry, Old Towns Quarry near Middridge Grange, Shackleton Beacon and High Side Bank Quarries near Heighington; Denton Quarry and Summerhouse Quarry. An almost continuous section through the lower part of the sequence is exposed in the River Tees between Piercebridge and High Coniscliffe, but south of the river, Permian strata can be seen in only three quarries north of Eppleby and in a cutting on the Durham motorway south-west of Cleasby. In drift-covered ground Permian rocks have been proved in several boreholes.

The greatest recorded thickness of strata is 196 ft in a borehole on the western outskirts of Darlington. This figure, however, is less than the maximum, for the uppermost members of the sequence are absent. An inlier of Carboniferous rocks is exposed south-west of Cleasby, while larger inliers occur along faults in the Redworth, Summerhouse and Denton areas. It is possible that due to faulting and variation in sub-drift topography other inliers may occur.

Previous Research. Winch (1817) sketched the western limit of the Magnesian Limestone, and referred to a working at Denton, but Sedgwick (1829) was the first to examine the formation in detail. Later works most pertinent to the district are by Kirkby and Duff (1872) and Trechmann (1921), while brief references to the area have also been made by Browell and Kirkby (1865–67), Wilson (1881), Trechmann (1914, 1925, 1931) and Fowler (1956). Fossils previously recorded from the Permian of this district include plants, brachiopods, bivalves, fish and reptiles from the Marl Slate in the railway cutting and adjacent quarries at Middridge and at Thickley Quarry, together with brachiopods and bryozoa from the Lower Magnesian Limestone at Thickley (Sedgwick 1829; King 1850; Hancock and Howse 1870a, b; Howse 1890; Trechmann 1921; Stoneley 1958). Brady (1876) mentioned the occurrence of foraminifera in the Lower Magnesian Limestone at several localities in the central part of the Permian outcrop of this district.

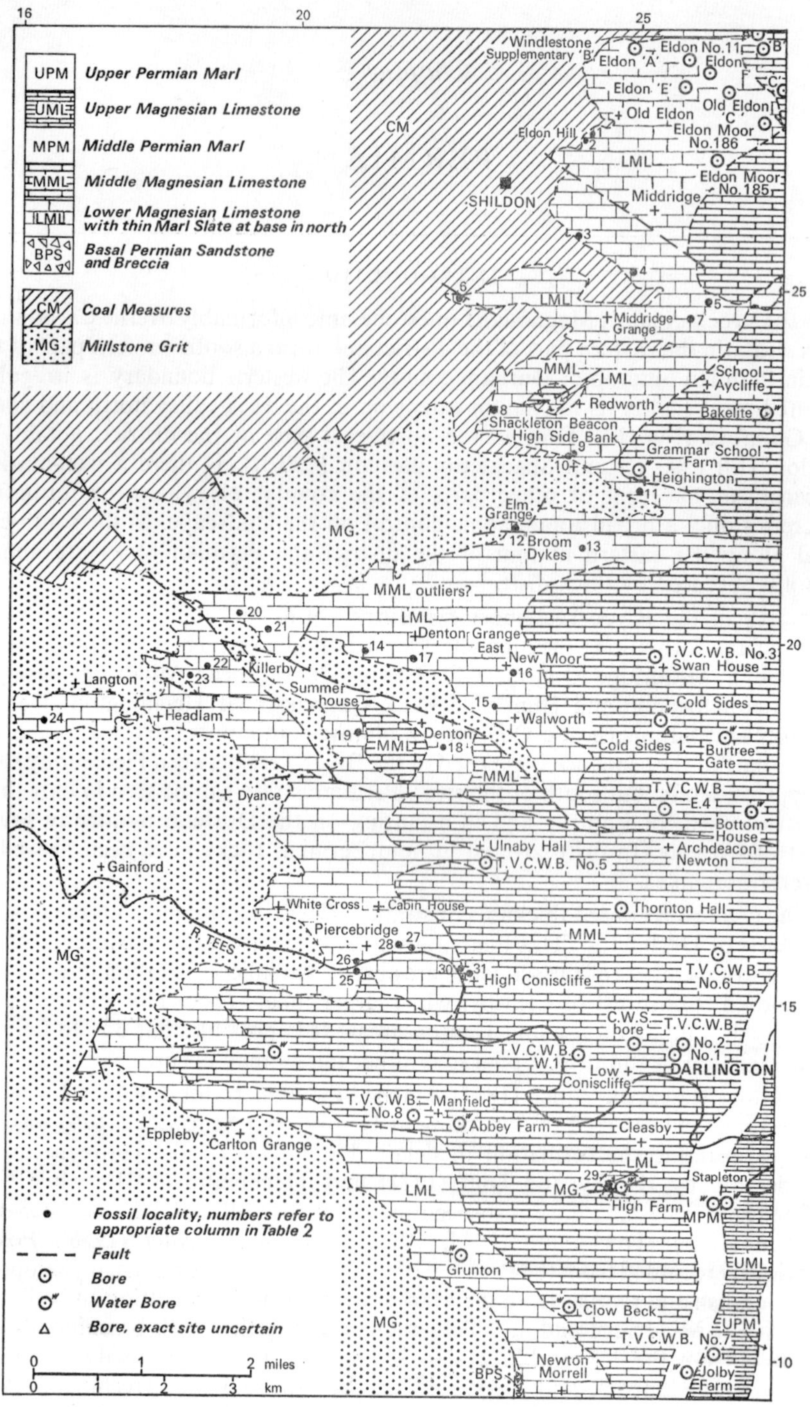

FIG. 14. *Distribution of the Permian rocks in the Barnard Castle district showing position of the more important fossiliferous localities and boreholes*

Classification. The formational classification adopted here is a modified form of that used in the Durham and West Hartlepool District (Smith and Francis 1967, fig. 17, p. 91), the unqualified figures in the table below being maximum thicknesses:

		ft
Upper Permian Marl	at least	40
Upper Magnesian Limestone		60
Middle Permian Marl		30
Middle Magnesian Limestone		
Shelf/Lagoon facies } Transitional Beds }		160
Lower Magnesian Limestone		130
Marl Slate		0–13
Basal Permian Sandstone and Breccia		0–44

Recent work has further refined classification and nomenclature, and this has been summarized by Smith (1970, p. 67). His classification is based on that of County Durham, and incorporates data from adjoining parts of Yorkshire.

The Upper Permian Marl has not been proved within the district and its presence is only inferred. By comparison with the Durham and West Hartlepool sequence the divisions are thinner.

Conditions and History of Deposition

The Permian palaeogeography of the district accords with that of the Durham and West Hartlepool district described by Smith and Francis (1967, pp. 91–4). By late Lower Permian times the Carboniferous rocks had been uplifted and eroded to form a wide shallow embayment the mid-line of which extended east-south-east from Piercebridge. This formed a westward extension of the desert of east and south Durham and was probably bounded by ranges of low hills to the north, west and south. In general there was a westward rise in the floor of the embayment, the plane of unconformity having a low relief. Variations in the thickness of the Lower Magnesian Limestone suggest that the bed of the embayment was not that of a simple saucer-like depression, but one in which distinct, but subdued, topographical relief was present. An exposure of the unconformity (Plate XIV; Fig. 15) in a cutting on the Durham motorway at Cleasby suggests that the desert surface was marked by irregularities possibly formed by temporary watercourses.

The lowest Permian deposits are breccias and sandstones containing fragments chiefly of limestone, sandstone and shale. These deposits were probably formed in an arid environment and redistributed by ephemeral sheet floods. Wind action may have helped in their formation, but no dune structures like those in the Yellow Sands between Durham and West Hartlepool have been found in this district. The basal deposits are discontinuous and are commonly about two feet thick; where they are thicker they may fill old watercourses or form residual talus accumulations.

Continental conditions were ended by the transgression of the Zechstein Sea in which the Marl Slate and Magnesian Limestone were deposited. Because deposition took place for the most part in an embayment, rather than in the more open shelf-sea conditions to the north-east, the sequence is thinner than in the Durham and West Hartlepool district, though transitional thicknesses are recorded in the extreme north-eastern part of the district.

The three evaporite cycles recognized to the north-east (Smith 1970, p. 67) can be found in this district although in modified form. The lowest cycle comprises the Marl Slate and Lower and Middle Magnesian Limestone, and possibly the lower part of the Middle Permian Marl, representing the sulphate phase. As in the Durham and West Hartlepool district, there is no halite phase.

From fossil content and sedimentary structures it is inferred that the Marl Slate, generally a silty dolomitic shale, shale or shaly dolomite was deposited in a neritic inshore environment at depths of between three hundred and six hundred feet (Schuhert *in* Love 1962, p. 354). Locally, beds of Lower Magnesian Limestone lithology are interbedded with Marl Slate. The succeeding Lower Magnesian Limestone consists mainly of dolomite or dolomitic limestone, but near the base at some localities it includes limestone which seems to have been deposited in shallow calm water receiving little or no terrigenous sediment. Disturbed beds and turbidites also occur locally—probably the result of slumping and sliding which was initiated by an earthquake shock. At the beginning of Middle Magnesian Limestone times the embayment sea was shallow and its floor may even have been emergent locally, giving rise to disconformities. The lowest or transitional beds consist of variably bedded open-textured dolomites that are in turn overlain by similar beds containing stromatolites, indicative of a shelf/lagoon environment. Towards the end of Middle Magnesian Limestone times, the Zechstein Sea again became filled with sediment and the succeeding gypsiferous Middle Permian Marls are deposits of the sulphate phase of the first and possibly also the second evaporite cycles; gypsum was proved in a borehole in the extreme south-east. The Upper Magnesian Limestone constitutes part of a third evaporite cycle, and evidently represents a recurrence of the conditions indicated by the carbonates of the first cycle.

Basal Permian Sandstone and Breccia

The Basal Permian Sandstone and Breccia are best known from boreholes in the north-eastern part of the district. They consist generally of a thin layer of breccia or sandstone. The sandstone, when unweathered, is typically bluish grey, hard, and well cemented, with small pebbles of quartz or, more commonly, fragments of sandstone, siltstone or shale. They generally carry pyrite, probably introduced by downward-percolating waters of the Marl Slate sea (Smith and Francis 1967, p. 96). At or near outcrop the pyrite is often oxidized to limonite, which imparts a yellowish, orange or orange-brown colour to the rock.

The breccias are highly variable and range from rocks composed of small angular fragments of Carboniferous mudstone, sandstone or limestone set in a matrix of sandstone, to breccio-conglomerates consisting of subangular to subrounded fragments of chert and limestone in a hard ankeritic sandstone matrix. The matrix is normally composed of fine- to medium-grained sand, but there are also sporadic coarse ‘millet-seed’ grains, presumably wind-rounded.

In the north-eastern part of the district, the basal Permian deposits are commonly about 2 ft thick, but they are locally absent. The local maximum is recorded in a borehole near Eldon Moor House, where the Marl Slate is underlain by 11 ft of sandstone with included fragments resting on whitish grey pyritic mudstone presumed to be weathered Coal Measures. South of Heighington evidence from boreholes suggests that the Basal Permian Breccias are rarely more than five feet thick; in places they are thinner, but are only rarely absent.

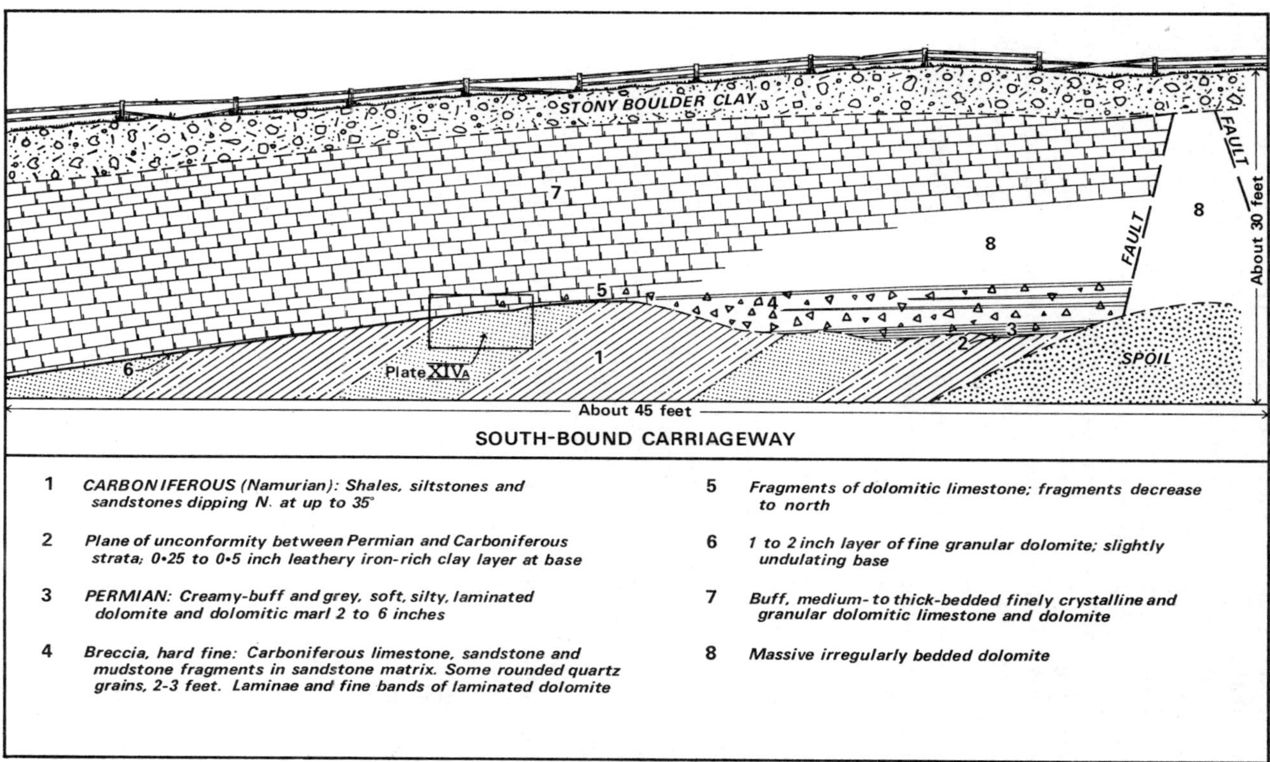

FIG. 15. *Diagrammatic representation of relationship between Permian and Carboniferous strata in cutting on east side of Durham motorway near Cleasby*

K

In the extreme south-east, however, notably in a borehole at Grunton, the thickness may be as much as 44 ft.

The unconformity between Permian and Carboniferous rocks is exposed at only three localities, notably Eldon Hill Quarry (Plate XIIIA), Thickley Quarry (Plate XIIIB) at East Thickley, and in a cutting on the east side of the Durham motorway south-west of Cleasby (Plate XIVA; Fig. 15). At the latter locality the basal breccia is up to 3½ ft thick, and fills a hollow, possibly a watercourse, on the eroded Carboniferous surface. Irregular bands of dolomitic marl in the breccia probably represent clays formed in temporary ponds. At Eldon Hill Quarry, a former opencast coal working, much of the face is overgrown and the unconformity is largely obscured. A borehole rather less than half a mile south-east of Windlestone Grange in the north-east part of the district penetrated a 3-in "limestone" presumed to be of a similar origin, and interbedded with sandstone.

Former exposures of the unconformity at Middridge Quarry and in the adjacent railway cutting are now obscured. D.A.C.M.

DETAILS

Of 17 cored boreholes sunk through the base of the Permian in the north-eastern part of the district north of Heighington (Fig. 14), five proved no Basal Breccias. Details of the deposits proved in five boreholes are given below to exemplify the range of variation.

National Coal Board (NCB) Windlestone 'B' Bore [2667 2834] 950 yd E of Windlestone Grange and Windlestone 'C' Bore [2715 2785] 750 yd SE of the above: Bluish grey fine to medium-grained hard pyritic sandstone with small rock fragments, 6 to 7 in.

NCB Old Eldon 'C' Bore [2623 2776] 800 yd SSE of Windlestone Grange: Grey hard medium-grained sandstone with millet seed grains and abundant pebbles of shale and siltstone, 1½ in; on grey semi-porcellanous limestone, 3 in; and bluish grey hard dolomitic sandstone with abundant siltstone pebbles and some millet seed grains, 2½ in.

NCB Eldon 'E' Bore [2559 2787] 550 yd SSW of Windlestone Grange: Pale grey hard well-cemented sandstone with sporadic rounded grains, 1 ft 10 in.

NCB Eldon Moor No. 185 Bore [2611 2685] near Eldon Moor House: Orange-brown medium- to coarse-grained soft sandstone with small rounded quartz pebbles, 2 in; on pale yellowish grey medium-grained micaceous sandstone with orange-yellow bands, 7 ft 10 in; and white and brown speckled sandstone with reddened shale partings, passing into breccia-conglomerate below in 3- to 6-in bands of alternately hard bluish grey pyritic rock and of softer mottled purple rock, 3 ft.

At the north end of Eldon Hill Quarry [2427 2725] 6 in of dolomitic conglomerate overlie pale grey Carboniferous mudstone with rootlets.

At the north end of the east face of Thickley Quarry [2407 2565] (Plate XIIIB) at East Thickley, 4 in of breccia rests on the eroded surface of cross-bedded Coal Measures sandstone. The bed is very hard and the breccia fragments stand out on the weathered surface to give the rock a rubbly appearance. Trechmann (1921, p. 538) obtained well-preserved colour-banded specimens of *Lingula credneri* from these beds E.A.F., D.A.C.M.

In shallow boreholes [2444 2556 and 2451 2543] between East Thickley and Middridge Quarries 2 and 3 ft respectively of "yellow sands" was penetrated underlying the Marl Slate; in the latter, the deposit was not bottomed.

On the west side of the abandoned Eldon Drift Mine [2526 2821], Coal Measures mudstone is overlain by 9 in of pale grey fine- to medium-grained sandstone containing pyrite blebs and films, coarse well-rounded grains of quartz, and isolated sandstone fragments up to half an inch across. The sandstone has a minutely irregular upper surface.

(L 380)

A. Eldon Hill Quarry. Lower Magnesian Limestone (with disturbed bedding at top), Marl Slate and Basal Permian Sandstone and Breccia unconformably overlying Middle Coal Measures

PLATE XIII

B. Thickley Quarry at East Thickley near Shildon. Lower Magnesian Limestone, Marl Slate and Basal Permian Sandstone and Breccia unconformably overlying false-bedded Lower Coal Measures sandstone

(L 375)

A. Bedded dolomite and dolomitic limestone in the Lower Magnesian Limestone unconformably overlying Namurian sandstone and shale in cutting on east side of Durham motorway near Cleasby, Darlington (1964) (L 36

PLATE XIV

B. Brecciated calcitic dolomite in Lower Magnesian Limestone, Shackleton Beacon near Heighington (L 38

In a recent borehole [2559 2453] drilled below the floor of Old Towns Quarry 1 ft of Basal Permian Deposits was proved at the base of the hole consisting of 6 in of "conglomerate" on 6 in of Permian Yellow Sand; in a further borehole [2523 2453], nearly ¼ mile to the west, 4 in of "conglomerate" on 8 in of Permian Yellow Sands was recorded at the base of the hole. Further examination of specimens from these boreholes has shown the "conglomerate" to be a grey to dark grey breccio-conglomerate containing small angular and subangular fragments of sandstone, dark grey elongate pieces of siltstone and mudstone set in a siliceous matrix which also contains angular to rounded fragments of quartz.

In a disused water borehole [2497 2246] at Grammar School Farm, Heighington, 2 ft of 'grey grit' are assumed to represent Basal Permian breccias.

Basal Permian breccias have been proved in four boreholes west and north-west of Darlington as follows:

Tees Valley and Cleveland Water Board (TVCWB) Archdeacon Newton E.4 Bore [2532 1771] north-north-west of Archdeacon Newton: Breccia with subangular to subrounded fragments up to half an inch long, set in a matrix of pale grey fine-grained sand, 7 ft 11 in (only 11½ in recovered).

Thornton Hall Bore* [2480 1634] ¼ mile SE of Thornton Hall, High Coniscliffe: 'white post', 3 ft 5 in.

TVCWB No. 5 Bore [2283 1693] near Ulnaby Hall: Breccia, pale grey with buff patches, composed of angular and subangular fragments of Carboniferous rocks up to 1 in in a fine sand matrix containing 20 to 30 per cent coarse frosted sand grains, 2 ft 11 in.

TVCWB No. 1 Bore* [2552 1420] at Broken Scar Pumping Station on the western outskirts of Darlington: Grey medium-grained hard sandstone containing small angular fragments, locally pyritized, of Carboniferous limestones and sandstone, 9 in; on breccia consisting of Carboniferous sandstone, limestone and siltstone in a fine- to medium-grained sandstone matrix with sporadic coarse frosted grains, 5 ft. The lowest 1 in of the breccia is composed of flaky fragments of pale grey limestone in a matrix of grey silty micaceous sandstone.

South of the River Tees in a cutting [2460 1223] on the Durham motorway south-south-west of Cleasby (Plate XIVA; Fig. 15) yellow, brown and buff locally hard breccia infills a hollow in minature dip and scarp topography on the eroded surface of Carboniferous strata. Where it abuts against a fault the breccia is 3½ ft thick, but it thins northwards over a distance of some 30 yd, to become represented by only a few small fragments. A few thin irregular bands of buff laminated dolomite and dolomitic marl, 2 to 6 in thick are interbedded with, and locally underlie, the breccia. These bands are softer than the breccia and have sharp irregular bases and eroded tops. The breccia contains abundant elongate fragments of dull white sandstone, rarer fragments of Carboniferous limestone and mudstone, and derived crinoid stems. Rounded and subrounded grains of quartz are scattered through the fine- to medium-grained ankeritic sandstone matrix. Between the breccia and the underlying Carboniferous rocks (p. 222) is an irregular layer, ¼ to ½ in thick, of leathery iron-rich clay.

Elsewhere south of the River Tees, Basal Permian breccias are recorded from only three water boreholes as follows:

Abbey Farm Water Bore [2239 1331], Mansfield: 'Gritstone' 7 ft 9 in.

Grunton Water Bore* [2239 1148], Mansfield:

Conglomerate, containing subrounded pebbles of chert and grit up to one inch long in a matrix of sandy dolomitized limestone with secondary calcification, 29 ft. Fifteen feet of 'gravel' overlying this limestone may be interpreted either as a basal Permian deposit or as Drift.

Clow Beck Farm Water Bore [2404 1064] Newton Morrell: yellow sandy rock, described as 'not true yellow sand', 4 in.

D.A.C.M.

MARL SLATE

The Marl Slate, too thin to show on Fig. 14, is thought to be present in only the eastern and north-eastern parts of the district where it is up to 13 ft thick; west and north-west of Darlington it has been recognized in five boreholes and in these its thickness measured between 1 ft 7 in and 8 ft 8 in. South of the Tees it has not been proved.

In the north-east the deposit is continuous with and similar to the Marl Slate of the Durham and West Hartlepool district (Smith and Francis 1967, pp. 102–6). It is a finely laminated commonly bituminous silty dolomitic shale which, when freshly fractured, smells of oil. At outcrop, it is a dark yellowish orange or yellowish brown commonly fissile rock, but where unweathered it is hard and compact, with alternating grey and black laminae. As in the south-west of the Durham and West Hartlepool district it is locally interbedded with thin beds of dolomite and dolomitic limestone. More rarely, as at Eldon Hill, and in a number of boreholes it is interbedded with limestone resembling that in the overlying Lower Magnesian Limestone. The top of the Marl Slate is taken at the top of the highest laminated layer.

The Marl Slate in places includes rounded sand grains and minerals such as sphalerite, galena, chalcopyrite and malachite. These and other features, together with a bibliography, have been summarized by Smith and Francis (1967, pp. 102–4). Westoll (1941) records that some of the fish remains from the South Durham outcrop are associated with syngenetic mineralization.

D.A.C.M.

Palaeontology. The Marl Slate contains a characteristic fauna of palaeoniscoid fishes with the brachiopod *Lingula credneri* Geinitz and a flora which mainly consists of pteridosperms and early conifers. Stoneley (1958) describes the quarries south of Middridge as the source of the best-preserved Permian plants yet found in England. From the same locality several species of fish and the shells *Liebea* and *Peripetoceras* have been collected. The beds typical of Marl Slate lithology do not usually yield invertebrates other than *Lingula*, whereas beds which lithologically resemble the Lower Magnesian Limestone contain abundant invertebrate remains, including bryozoa, brachiopods and bivalves.

In general, the organic remains are similar to those found in the Kupferschiefer at the base of the Upper Permian (Thuringian) of Central Europe with which the Marl Slate is usually equated. Principal works dealing with the flora and fauna are by Sedgwick (1829), King (1850), Hancock and Howse (1870a, b), Howse (1890), Trechmann (1921), Westoll (1941), and Stoneley (1958). J.P.

DETAILS

In Eldon Hill Quarry [242 272] (Plate XIIIA) the base of the Permian sequence was formerly exposed along a 350-yd face but is now largely obscured. The Marl Slate ranges in thickness between 8 in and 3 ft 10 in. A section at the extreme northern end [2427 2725] of the east face comprised 8 in of mudstone, 6 in of brown mudstone with thin impersistent bands of dolomitic limestone, 4 in of hard fossiliferous dolomitic limestone and 9 in of earthy dolomite, on Basal Permian Breccia. The brown mudstone and dolomitic limestone yielded the following fauna: *Agathammina pusilla; Acanthocladia anceps, Batostomella crassa, ?Fenestella retiformis, Protoretepora ehrenbergi, ?Thamniscus dubius; Dielasma elongatum, Horridonia horrida, Streptorhynchus pelargonatus*, indet. spiriferoid; and *Bakevellia ceratophaga*. The 4-in band of fossiliferous dolomitic limestone underlying this bed is rich in bryozoa in the topmost 1 in: *Batostomella crassa, Fenestella retiformis, Protoretepora ehrenbergi, ?Thamniscus dubius; Bakevellia?* also occurs. As in the overlying bed, the bryozoan doubtfully assigned to *Thamniscus dubius* is abundant here, but entirely fragmentary. In the lower part of this band, the fauna is dominated by bivalves and the bryozoa form a separate assemblage, being found mainly on bedding planes other than those on which most of the bivalves lie; the fauna consists of *Acanthocladia anceps*, *Batostomella crassa*, *Fenestella retiformis; Bakevellia binneyi*, *B. ceratophaga*, *Permophorus costatus*, and *Schizodus?*. E.A.F., J.P.

About 90 yd farther south-west along the face of Eldon Hill quarry, the Marl Slate consists of shale 1 ft 8 in, on a dark brown earthy layer 1 in, on hard grey dolomitic limestone with crystal-filled cavities at the top. A broadly similar sequence was visible some 120 yd still farther south-west where 3 ft 10 in of shale overlie 1 ft 8 in of dolomitic limestone. The Marl Slate can generally be picked out by virtue of the underlying harder bands which stand out on the face a little above the unconformity. E.A.F.

In Eldon Drift Mine [2526 2821], the Marl Slate comprises 4½ ft of dark grey and black laminated shale, slightly sandy near the base, and contains *Lingula credneri*. It has a minutely irregular lower contact with the Basal Permian Breccia and a sharp junction with the overlying dolomitic limestone. The Marl Slate here is seen in thin section (E 33058) to consist of an even-grained (0.007 mm) interlocking mosaic of xenomorphic dolomite containing thin, subparallel opaque streaks and filaments, scattered granules of angular quartz and sparse polygranular quartz, finely disseminated specks of sulphide, mainly pyrite and including pyritospheres, and wisps of illite. The dolomite, (refractive index $\omega = 1.684 \pm 0{\cdot}002$) contains minute vermicular inclusions. The laminated character is due principally to streaks of bitumen alternating with carbonate-rich layers, and in this the rock resembles other specimens of the Marl Slate (Hirst and Dunham 1963, pp. 917–21). D.A.C.M., R.K.H.

In Thickley Quarry [2407 2565] (Plate XIIIB) at East Thickley, the Marl Slate comprises 6 ft of yellowish brown fine dolomitic mudstone and silty mudstone with dolomite laminae. A 1-in soft brown layer is present at the top. Some bands are more resistant to weathering than others and stand out. Deans (1950, p. 340) noted that galena is dispersed through the dolomite, often replacing it. Detrital quartz is also present. Galena has replaced some of the detrital quartz and is in places accompanied by blende. E.A.F.

Fish scales are abundant and the fish species recorded from this locality include *Acentrophorus glaphyrus*, *Acrolepis sedgwickii*, *Coelacanthus granulatus*, *Palaeoniscus comtus*, *P. elegans*, *P. longissimus*, *P. macrophthalmus*, *Platysomus macrurus* and *Pygopterus mandibularis* all by King 1850, *Wodnika striatula* (Howse 1890) and *Janassa bituminosa* (Trechmann 1921). King also noted the presence of the brachiopod *Lingula credneri* and the plants listed by Stoneley (1958) from the Marl Slate of this quarry include *Ullmannia frumentaria*, *Pseudovoltzia liebeana* and *Hiltonia rivuli* [= *Ullmannia bronni*].

Plants and fish were formerly found in abundance at Middridge Quarry [248 253] (now filled in) from which most of the museum specimens of Marl Slate fossils from this district were obtained. The upper 2 ft of Marl Slate can be seen only at an exposure [2471 2527] alongside the former railway sidings. Plants recorded from this locality, mostly by Howse (1890) have been redetermined by Stoneley (1958) as follows: *Algites virgatus*, *Taeniopteris eckhardti*, *?Psygmophyllum cuneifolium*, *Sphenobaiera digitata*, *Pseudoctenis middrigensis*, *Ullmannia bronni*, *Ullmannia frumentaria*, *?Pseudovoltzia liebeana*, *Hiltonia rivuli* [= *Ullmannia bronni*]. King (1850) recorded the pteridosperm *Neuropteris huttoniana* from Middridge as well as the fish species *Acentrophorus glaphyrus*, *Coelacanthus granulatus*, *Palaeoniscus comtus* [= *P. freieslebeni*], *P. elegans*, *P. longissimus* and *P. macrophthalmus*. Further fish from this locality were listed by Howse (1890) as *Janassa bituminosa*, *Pygopterus mandibularis*, *Acrolepis exculptus*, *A. sedgwickii*, *Platysomus striatus*, *Globulodus* [*Platysomus*] *macrurus* and *Dorypterus hoffmanni*. In addition he also recorded the reptiles *Protorosaurus speneri* and *P. huxleyi*, and the invertebrates "*Lingula mytiloides* Sowerby" [sic], *Liebea squamosa*, and *Peripetoceras* [*Nautilus*] *freieslebeni*. The flora and fauna of this quarry are the most prolific for the Marl Slate in Durham, and Howse concluded that many of the remains were swept out by rivers and deposited with marine shells in a littoral environment. E.A.F., J.P.

A new quarry [249 252], some 150 yd E of the old Middridge Quarry (now filled in) has recently been opened. The lowest 8 ft of beds on the quarry face contain thin bands and laminae of silty dolomitic shale interbedded with dolomitic limestone. Most of the floor of the quarry is in limestone and the development of Marl Slate lithology is nowhere so near apparent as in Old Towns Quarry farther to the south-east.

In Old Towns Quarry [257 246] east of Middridge Grange some 6 ft of Marl Slate are exposed along the base of the face on the western and southern sides of the quarry, and in addition most of the floor of the quarry is on Marl Slate. It consists of thin bands of carbonaceous micaceous shale interbedded with grey and brown limestone and dolomitic limestone and silty dolomite shale. The proportion of shale decreases upwards and the highest band, some 6 ft above the base of the quarry floor, consists of a shale film between beds of limestone. D.A.C.M.

In the north-eastern part of the district, Marl Slate is recorded in boreholes (see Fig. 14), as follows:

NCB Windlestone Supplementary 'B' Bore* [2664 2866]: Dark grey finely laminated shale with brown-speckled bedding planes, 7 ft 8 in (faulted at base) on dark grey finely laminated shale with dolomitic bands, 3 ft 4 in.

NCB Windlestone 'B' Bore [2667 2834]: Dark grey finely laminated shale with sporadic fish scales, 4 ft.

NCB Eldon Moor No. 186 Bore [2676 2734]: Marl Slate with pyrite and galena on bedding planes and in veins, 8 ft.

NCB Eldon Moor No. 185 Bore [2611 2685]: Black and brown fissile shale with soft brown bands, 6 ft, including a massive 6-in dolomite band 3 ft above base.

NCB Eldon 'F' Bore [2600 2803]: Pale brown dolomitic shale, 2½ ft; grey argillaceous dolomitic limestone, 3 ft; brown dolomitic shale, 1 ft; on grey dolomitic shale with hard dolomitic bands, 4½ ft.

NCB Eldon No. 11 Bore [2563 2821]: Grey laminated shale with rounded sand grains in basal 6 in, 9 ft 10 in.

NCB Eldon 'A' Bore [2484 2843]: Beds of Marl Slate lithology including fish scales and *Lingula sp.*, with thin intercalations of round-grained sand (cf. Basal Permian Sand) near top, 5 ft 10 in; on dolomitic limestone, 2 ft.

In five shallow boreholes between Thickley Quarry [2407 2565] and Middridge Quarry [248 253] the Marl Slate generally appears to range between 4 and 6 ft although in one borehole [2456 2565] it was at least 7 ft, its base not having been proved.

A series of twelve shallow boreholes at and to the west of Old Towns Quarry [257 246] penetrated, but in most cases did not bottom, the Marl Slate which in this area is thought to be about 7½ to 8 ft thick. Nearly ¼ mile to the west of the quarry one such borehole [2523 2453] which penetrated the whole sequence proved 10½ ft of beds of Marl Slate lithology described as weathered brown in parts.

Bakelite Water Bore* [2680 2329], Aycliffe Trading Estate: Black shale, 13 ft.

Grammar School Farm Bore [2497 2246], Heighington: hard black shale, 5 ft.
E.A.F., D.A.C.M.

Marl Slate has been proved in five boreholes west and north-west of Darlington, including three drilled for the Tees Valley and Cleveland Water Board:

TVCWB No. 3 Bore* [2520 1982] at Swan House: Grey argillaceous dolomitic mudstone, alternating with thin beds of limestone lithology, 2 ft 1 in; on argillaceous dolomite with fine laminae of dolomitic limestone, 6 ft 7 in (base not proved).

TVCWB Archdeacon Newton E.4 Bore [2532 1771]: Grey finely laminated siltstone with fine dolomitic interlaminae and 6-in limestone band at base, 5 ft 4 in.

Thornton Hall Bore* [2480 1634]: 'Brown shale' 4 ft.

TVCWB No. 5 Bore [2283 1693] near Ulnaby Hall: Grey finely crystalline dolomitic limestone with a few carbonaceous patches, 1 ft 7 in.

Cold Sides Water Bore [2536 1889]: Black shale, 2 ft (Fowler 1956, p. 255).

At Quarry Hole [2289 1910] north-north-west of Walworth, and in Hobgate Quarry [1646 1889] near Langton, thin shaly bituminous interlaminae of Marl Slate lithology occur at the base of the sequence. D.A.C.M.

Lower Magnesian Limestone

Although this formation crops out (Fig. 14) in a wide belt along the whole of the western margin of the Permian in this district, exposure is limited. North of the River Tees sections are mainly confined to quarries. An almost continuous section of Lower Magnesian Limestone is seen along the River Tees between Piercebridge and High Coniscliffe while south of the river it is exposed only in three quarries north of Eppleby and in a cutting near Cleasby on the Durham motorway (Plate XIVA; Fig. 15). The formation is proved in boreholes scattered over the Permian outcrop with concentrations in the north-east part of the district, and west and north-west of Darlington.

Lack of a well-defined horizon to mark the top of the Lower Magnesian Limestone makes assessment of thickness difficult. In the north-east, boreholes near Windlestone Park proved about 130 ft, the thickness decreasing southwards, while over the remainder of its area of occurrence in the district thicknesses generally range between 40 and 90 ft, being least in the marginal parts of the embayment and greatest in the deeper and more central parts. Thicknesses of about 93 and 94½ ft in boreholes near Swan House and Cold Sides are probably exceptional, as is also a thickness of considerably less than 40 ft in the extreme south-east of the district where deposition is thought to have taken place on a fluctuating sea bed adjacent to the Middleton Tyas Anticline. Limited overlap of the main part of the Lower Magnesian Limestone over the basal parts of the sequence is present in the south-east and probably to a lesser extent elsewhere.

The rocks consist largely of dolomitic limestones, calcitic dolomites and dolomites and vary in colour from white to grey, buff, brown and yellow. Laminated argillaceous layers, commonly of brown leathery clay, are present, especially near the base of the sequence. Pure limestones although present are rare. For instance, at Thickley Quarry (Plate XIIIB) the lowest beds of the formation, thick-bedded dolomitic limestones, are overlain by fossiliferous limestone texturally and faunally similar to that at Raisby Hill Quarry in the Durham and West Hartlepool District; but while at Raisby Hill Quarry the limestone is up to 90 ft thick (Smith and Francis 1967, p. 107), at Thickley it is only 3 ft. At Langton Quarry near Gainford highly fossiliferous limestones occur at what is thought to be a similar horizon.

Disturbed beds or turbidites are present at one or two localities north of the Tees and are thought to be the result of penecontemporaneous earthquake shocks. They exist as 'balled up' beds of dolomitic limestone with a maximum thickness of 7 ft at one, or locally two, horizons in the Lower Magnesian Limestone. The beds show some evidence of small-scale slumping and sliding, though in comparison with equivalent beds in north-east Durham and south-east Northumberland (Smith 1970) the extent and degree of movement is small.

The bedding of the formation is variable, although in general it ranges between thin-bedded (2 to 3 in) and thick-bedded (6 in to 2 ft) and false-bedding is occasionally present. Flaggy or laminated beds are common both near the base and at the top of the formation, while more thickly bedded units tend to occur in the middle of the sequence. The bedding as a whole is even and regular in general aspect, but shows small-scale irregularities and undulations. Many beds include stylolites, some of them sub-horizontal, accompanied by a thin coating of a carbonaceous or bituminous residue. Atypically, the lowest bed seen at outcrop in the Tees valley at Piercebridge is of massive rubbly nodular irregularly bedded and apparently unfossiliferous calcitic dolomite. The weathered surfaces have a brecciated appearance, an effect probably produced by penecontemporaneous solution. Irregularly bedded dolomite also occurs locally near the top of the sequence.

Texturally the rock ranges between porcellanous and very finely crystalline and granular. Coarse saccharoidal bands occur near the top of the sequence, and except for these the formation is in general dense or very dense, hard and lacking in porosity. Recrystallization is common throughout.

Cavities and vugs, some empty, others lined or infilled with carbonate crystals, are common throughout. Some of the vugs and associated veinlets are filled with baryte, calcite and pyrite, less commonly sphalerite, and possibly chalcopyrite. Where the rock is an autobreccia interstices between the fragments sometimes harbour these minerals. The beds as a whole are often netted by fine ramifying calcite and dolomite veins. D.A.C.M.

Palaeontology. The Lower Magnesian Limestone outcrop of the district has been systematically examined for fossils during the course of the survey. The collections have revealed a surprisingly varied fauna, the distribution of which is shown in Table 2 and Fig. 14, localities 1 to 29.

The Zechstein Sea, in which these remains were deposited, and in which all the animals and at least some of the plants lived, occupied a totally or partly enclosed basin in what is now north-central and western Europe. It was subject to conditions of abnormal and variable salinity which prevented the entry of certain groups of characteristic Permian marine organisms such as fusulinid foraminifera and ammonoids, and inhibited the growth of most others. The only invertebrate species in the Lower Magnesian Limestone of this district which commonly have a maximum dimension exceeding 10 mm are the brachiopod *Horridonia horrida* and the bivalve *Janeia biarmica*.

North-eastern England was on the western margin of the Zechstein basin and most of the county of Durham including the north-eastern part of this district (localities 1 to 13 in Table 2) was occupied at the time of the deposition of the Lower Magnesian Limestone by open shelf sea. In most of the rocks deposited in these shelf areas fossil remains are rare and consist largely of calcareous, imperforate foraminifera and species of the bivalve genus *Bakevellia*. Logan (1967, p. 10) suggested that *Bakevellia* was a shallow water attached benthonic organism, fairly tolerant of variable salinity, and it has been inferred (Crespin 1958, pp. 29–30) that foraminifera similar to those found in this area lived on the floor of a warm to temperate moderately shallow sea.

At some localities, especially in the more calcitic rocks, the Lower Magnesian Limestone of the shelf sea area contains a richer fauna including brachiopods, bryozoa and nodosariid foraminifera. Among these localities are Eldon Hill and Thickley Quarries (localities 1, 2, and 3 in this district); Raisby Hill and the Fishburn area of the Durham and West Hartlepool district to the north-east (Smith and Francis 1967, pp. 111, 173–6). The commonest brachiopods at these localities are the productoid forms *Horridonia horrida* and *Strophalosia morrisiana*. Trechmann (1921) commented on the excellent preservation of the brachiopods at Thickley and on how they demonstrated the manner in which they had lived on the bed of the sea. He figured specimens of *H. horrida* apparently preserved as they lived, partly buried in fine-grained sediment and anchored by spines to which smaller brachiopods had attached themselves. The spines are

TABLE 2

Distribution of fossils in the Lower and Middle Magnesian Limestone

Numbers refer to localities listed at the end of this table. Their positions are shown in Fig. 14. Marl Slate fossils are listed in the stratigraphical account.

Genera and species	LOWER MAGNESIAN LIMESTONE																													MML	
	1	2	3	4	5	6	7	8	9	10	11	12	13	14	15	16	17	18	19	20	21	22	23	24	25	26	27	28	29	30	31
Plantae																															
Algites sternbergianus (King)	.	.	3	.	.	.	.	.	.	.	.	.	.	.	.	.	.	.	.	.	.	.	.	.	.	.	.	.	.	.	.
A. virgatus (Munster)	.	.	3	.	.	.	.	.	.	.	.	.	.	.	.	.	.	.	.	.	.	.	.	.	.	.	.	.	.	.	.
Calcinema [*Tubulites*] *permiana* (King)	.	.	.	.	.	.	.	?	.	.	.	.	.	?	.	.	.	.	.	?	.	.	.	.	.	.	.	.	?	.	.
Plant remains [undetermined]	.		.	.	5	6	7	8	9	10	.	12	13	.	15	16	17	18	19	20	21	.	23	24	.	.	.	.	29	.	.
Foraminifera																															
Agathammina milioloides (Jones, Parker & Kirkby)	1	.	3	.	.	.	?	.	.		.	?	?	14	.	.	.	.	.	?	.	.	23	24	.	26	.	.	.	.	.
A. pusilla (Geinitz)	1	.	3	.	5	?	7	.	9	.	?	?	13	14	?	.	17	18	19	20	21	22	23	24	25	.	27	28	29	.	.
'*Ammodiscus*' *sp.*	1	.	3	.	?	.	.	.	.	.	.	.	?	14	15	.	?	18	19	20	.	22	23	24	?	.	.	.	?	.	.
Calcitornella? minutissima (Howse)	1	.	3	.	5	6	7	.	.	.	.	.	13	14	15	.	17	18	19	20	.	22	23	24	.	.	27	.	.	.	.
Uniserial, chambered foraminifera	1	.	3	.	.	6	.	.	.	.	.	.	.	14	15	.	17	.	19	20	.	.	23	24	.	.	.	.	.	.	.
including *?Geinitzina jonesi* (Brady)	.	.	.	.	.	.	.	.	.	.	.	.	.	14	15	.	.	18	19	20	.	.	.	24	.	.	27	.	.	.	.
Bryozoa																															
Acanthocladia anceps (Schlotheim)	?	.	3	.	?	.	.	.	.	.	.	.	.	?	.	.	.	.	19	20	.	?	23	?	?	.	?	.	.	.	.
Batostomella columnaris (Schlotheim)	.	.	.	.	.	.	.	.	.	.	.	.	.	14	15	?	.	18	?	.	.	.	23	24	.	.	.	.	.	.	.
B. crassa (Lonsdale)	1	2	.	.	.	.	.	.	9	.	.	.	.	14	.	.	.	18	19	20	.	.	.	.	.	.	.	.	?	.	.
Fenestella retiformis (Schlotheim)	1	.	3	.	.	6	.	.	9	.	.	.	.	14	15	16	.	18	19	.	.	?	23	24	.	.	.	.	.	.	.
Hippothoa? voigtiana (King)	.	.	?	.	.	.	.	.	.	.	.	.	.	.	.	.	.	.	.	.	.	.	.	.	.	.	.	.	.	.	.
Penniretepora waltheri (Korn)	.	.	.	.	5	.	.	.	?	.	?	.	.	.	.	?	.	?	.	.	.	.	.	.	.	.	.	.	.	.	.
Protoretepora ehrenbergi (Geinitz)	1	.	3	.	.	.	.	.	.	.	.	.	.	.	.	?	.	.	?	.	.	.	.	.	.	.	.	.	.	.	.
P.? cf. *solida* (Korn)	.	.	.	.	5	.	.	.	.	.	.	.	.	.	.	.	.	.	.	.	.	.	.	.	.	.	.	.	.	.	.
Synocladia sp.	.	.	.	.	5	.	.	.	.	.	.	.	.	.	.	.	.	.	.	.	.	.	.	.	.	.	.	.	.	.	.
Thamniscus dubius (Schlotheim)	.	.	.	.	5	.	.	.	.	.	.	.	.	.	.	.	.	.	.	?	.	.	.	.	.	.	.	.	.	.	.
Undetermined bryozoan species (see Smith and Francis 1967)	1	.	3	.	.	.	.	.	.	.	.	.	.	.	.	.	.	.	.	.	.	.	.	.	.	.	.	.	.	.	.
Cryptostome bryozoa, indet.	.	2	.	.	.	.	7	.	.	.	.	.	13	14	.	.	.	.	19	.	21	.	.	.	.	.	27	.	29	30	.
Trepostome bryozoa, indet.	.	.	.	.	.	.	.	.	.	.	.	.	.	.	.	.	17	.	.	.	.	22	.	.	.	.	.	.	.	.	.
Annelida																															
Spirorbis sp.	.	.	.	.	.	.	.	.	.	.	.	.	.	?	.	.	.	.	.	.	.	.	.	.	.	.	.	.	.	.	.
Brachiopoda																															
Crurithyris clannyana (King)	.	.	.	.	.	.	.	.	.	.	.	.	.	?	.	.	.	.	.	.	.	.	.	24	.	.	.	.	.	.	.
Dielasma elongatum (Schlotheim)	1	.	3	.	?	.	.	.	.	?	.	.	.	?	.	.	.	?	19	20	.	?	?	24	.	.	.	.	.	.	.
Discina konincki (Geinitz)	1	.	3	.	.	6	.	.	.	.	.	.	.	.	.	.	.	.	.	.	.	.	.	.	.	.	.	.	.	.	.
Horridonia horrida (J. Sowerby)	1	.	3	.	.	6	.	.	.	.	.	.	.	.	.	.	.	.	19	20	.	.	.	24	.	.	.	.	.	.	.
Howseia latirostrata (Howse)	.	.	3	.	.	.	.	.	9	.	.	.	.	.	.	.	.	.	.	.	.	.	.	.	.	.	.	.	.	.	.
Lingula credneri Geinitz	1	.	3	.	.	6	.	.	.	.	.	.	.	.	15	.	.	.	.	.	.	.	.	.	.	.	.	.	.	.	.
Pterospirifer alatus (Schlotheim)	1	.	3	.	.	.	.	.	.	.	.	.	.	.	.	.	.	.	.	.	.	.	.	.	.	.	.	.	.	.	.
Stenoscisma schlotheimi (von Buch)	.	.	3	.	.	.	.	.	.	.	.	.	.	.	.	.	.	.	.	.	.	.	.	.	.	.	.	.	.	.	.

TABLE 2 (*Continued*)

Genera and species	LOWER MAGNESIAN LIMESTONE																													MML	
	1	2	3	4	5	6	7	8	9	10	11	12	13	14	15	16	17	18	19	20	21	22	23	24	25	26	27	28	29	30	31
Streptorhynchus pelargonatus (Schlotheim)	.	.	3	.	.	.	.	.	.	.	.	.	.	.	.	.	.	.	.	.	.	.	.	24	.	.	.	.	.	.	.
Strophalosia morrisiana King	.	.	3	.	.	.	.	.	.	10	.	.	.	14	.	.	17	18	.	.	.	22	23	24	.	.	.	.	.	.	.
Productoids, indet.	.	.	.	.	.	6	.	.	.	.	.	.	.	14	.	16	.	.	19	.	.	.	.	.	.	.	.	.	.	.	.
Spiriferoids, indet.	.	.	.	.	.	.	.	.	9	.	.	.	.	.	.	.	.	.	.	.	.	.	.	.	.	.	.	.	.	.	.
Gastropoda																															
Glyptospira? tunstallensis (Howse)	.	.	.	.	.	.	.	.	.	.	.	.	.	.	.	.	.	.	.	.	.	22	23	.	.	.	.	?	.	.	.
Glyptospira? sp.	.	2	.	.	.	.	.	.	.	.	.	.	.	.	.	.	.	.	.	.	.	.	.	?	.	.	.	.	.	.	.
Mourlonia? sp.	.	.	.	.	.	.	.	.	.	.	.	.	.	.	.	.	.	.	.	.	.	.	.	?	.	.	.	.	.	.	.
Naticopsis sp.	.	.	.	.	.	.	.	.	.	.	.	.	.	.	.	.	.	.	.	.	.	.	.	.	.	.	.	28	.	.	.
Straparollus sp.	.	.	.	.	.	.	.	.	.	.	.	.	.	.	.	.	.	.	.	.	.	.	?	?	.	.	.	.	.	.	.
Strobeus geinitzianus (King)	.	.	.	.	.	.	.	.	.	.	.	.	.	14	.	.	.	.	19	.	.	22	.	.	.	.	.	?	.	.	.
Turreted gastropods, indet.	1	.	.	.	.	.	.	.	.	.	.	.	.	.	15	.	.	.	.	.	.	.	.	24	.	.	.	.	.	.	.
Gastropods, indet.	.	.	.	.	.	.	.	.	.	.	.	.	.	14	15	.	.	18	19	20	.	.	.	.	.	.	27	.	.	.	.
Bivalvia																															
Astartella vallisneriana (King)	.	.	3	.	.	.	.	.	.	.	.	.	.	.	.	.	.	.	.	.	.	.	.	.	.	.	.	.	.	.	.
Astartella sp.	.	.	.	.	.	.	.	.	.	.	.	.	.	?	.	.	.	.	19	.	?	.	23	?	.	.	.	.	.	.	.
Aviculopinna? pinnaeformis (Geinitz)	.	.	.	.	.	.	.	.	.	.	.	.	?	.	.	.	.	.	.	.	.	.	.	.	.	.	.	.	.	.	.
Bakevellia (Bakevellia) binneyi (Brown)	.	.	.	.	5	.	.	.	9	10	11	.	13	14	15	16	17	18	19	20	21	22	23	24	25	.	27	.	29	.	.
B. (B.) ceratophaga (Schlotheim)	1	2	.	.	.	.	.	.	?	?	.	.	.	14	15	16	17	18	19	20	21	22	23	24	.	.	.	28	.	.	.
Bakevellia (Bakevellia) sp.	.	.	.	.	.	.	7	.	.	.	.	.	.	.	.	.	.	.	19	.	.	.	.	.	.	.	.	.	.	30	31
Elimata permiana (King)	.	.	.	.	.	?	.	.	.	.	.	.	.	14	?	16	17	18	19	20	.	22	23	24	.	.	.	.	.	.	.
Janeia biarmica (de Verneuil)	.	.	.	.	.	.	.	.	.	.	.	.	.	.	.	16	17	.	?	.	21	.	.	24	.	.	.	.	.	.	.
J. normalis (Howse)	.	.	3	.	.	.	.	.	.	.	.	.	.	.	.	.	.	.	.	.	?	.	.	.	.	.	27	.	.	.	.
Janeia sp.	.	.	.	.	.	.	.	.	.	.	.	.	.	?	?	.	.	?	.	.	.	.	.	.	.	.	.	.	.	.	.
Liebea squamosa (J. de C. Sowerby)	.	.	.	.	.	.	.	.	.	.	.	.	.	.	.	.	.	.	19	.	.	.	.	.	.	.	.	.	.	.	?
Parallelodon striatus (Schlotheim)	.	.	.	.	.	.	.	.	.	.	.	.	.	.	.	.	.	18	.	.	.	.	.	24	.	.	.	.	.	.	.
Permophorus costatus (Brown)	.	.	.	.	.	.	.	.	.	.	.	.	.	.	15	.	.	?	19	20	.	.	23	24	.	.	.	28	.	.	.
Pseudomonotis speluncaria (Schlotheim)	.	.	3	.	.	.	.	.	.	.	.	.	.	.	.	.	.	18	19	20	.	.	.	24	.	.	.	.	.	.	.
Schizodus obscurus (J. Sowerby)	.	.	.	.	.	.	.	.	.	.	.	.	?	14	.	.	.	.	19	20	21	.	23	24	.	.	27	.	.	.	.
Schizodus sp.	.	.	.	.	.	.	.	.	.	.	.	.	.	14	.	.	?	?	.	20	.	.	.	.	25	.	.	.	.	.	.
Streblochondria? pusilla (Schlotheim)	1	.	.	.	.	.	.	.	.	.	.	.	.	.	.	.	.	.	?	.	.	.	.	.	.	.	.	.	.	.	.
Bivalves, indet.	.	.	.	.	.	.	.	.	.	.	.	12	.	.	.	.	.	.	.	.	.	.	.	.	.	.	.	.	.	.	.
Nautiloidea																															
Peripetoceras freieslebeni (Geinitz)	.	.	3	.	.	.	.	.	.	.	.	.	.	.	.	.	.	.	.	.	.	.	.	.	.	.	.	.	.	.	.
Ostracoda																															
Ostracods	1	.	3	.	.	6	.	.	9	10	11	.	13	14	15	16	17	18	19	20	21	22	23	24	25	.	27	28	29	30	31

TABLE 2 (*Continued*)

Genera and species	LOWER MAGNESIAN LIMESTONE																													MML	
	1	2	3	4	5	6	7	8	9	10	11	12	13	14	15	16	17	18	19	20	21	22	23	24	25	26	27	28	29	30	31
Crinoidea																															
Crinoid columnals	1	.	.	.	.	6	.	.	9	.	.	.	.	.	.	.	.	.	.	.	.	22	.	.	.	.	.	.	.	.	.
Pisces																															
Fish scales	.	.	.	.	.	6	.	.	.	.	.	.	.	.	15	.	.	.	.	.	.	.	.	24	.	.	.	.	.	.	.
Amphibia																															
Lepidotosaurus duffii Hancock & Howse	.	.	.	4	.	.	.	.	.	.	.	.	.	.	.	.	.	.	.	.	.	.	.	.	.	.	.	.	.	.	.
Problematica																															
Palaeophycus insignis (Geinitz)	.	.	3	.	.	.	.	.	.	.	.	.	.	.	.	.	.	.	.	.	.	.	.	.	.	.	.	.	.	.	.

1. Eldon Hill Quarry [2426 2725]
2. Eldon Hill Quarry [2408 2712]
3. Thickley Quarry [2408 2564]
4. Middridge Quarry [248 253]
5. Old Quarry [2589 2481] near Greenfield House
6. Old Quarry [2240 2479] near Southfield House
7. Old Towns Quarry [257 246]
8. Old Quarry [2293 2330] near Shackleton Beacon
9. High Bank Quarry [2400 2273]
10. Exposure [2386 2272] near High Bank Side
11. Old Quarry [2503 2210] near Heighington
12. Old Quarry [2325 2161] near Elm Grange
13. Old Quarry [2416 2138] near Broom Dykes
14. Denton Quarry [2100 1996]
15. Quarry Hole, Walworth [2289 1910]
16. Old Quarry [2314 1955] near New Moor
17. Old Quarry [2170 1979] near Denton Grange East
18. Old Quarry [2213 1853] near St Mary's Church, Denton
19. Summerhouse Limekilns [2093 1875]
20. Morton Limekilns [190 205]
21. Old Quarry [1964 2013] north-east of Killerby
22. Killerby Garths Quarry [1873 1961]
23. Old Quarry [1860 1951] south-west of Killerby
24. Hobgate Quarry [1646 1888] near Langton
25. River Tees, south bank [2093 1544] west of Piercebridge
26. River Tees, north bank [2100 1557] west of Piercebridge
27. River Tees, north bank [2162 1572] east of Piercebridge
28. River Tees, north bank [2146 1583] east of Piercebridge
29. Motorway cutting [2460 1232 to 2467 1249] near Cleasby
30. River Tees, north bank [2243 1529] near High Coniscliffe
31. Old Quarries [2254 1527] at High Coniscliffe

intact though almost three inches long, suggesting a calm water environment and preservation *in situ*. The bryozoa at Thickley are also largely unbroken. The foraminifera, brachiopods and, to a lesser extent, the bryozoa in these more calcitic rocks are forms, which, it may be inferred from their association with stenohaline organisms elsewhere in the Magnesian Limestone, were intolerant of brackish or hypersaline water.

The bryozoan faunas from Thickley and near Greenfield House (locality 5) include several forms either unrecorded or recognized at few localities elsewhere in the Magnesian Limestone but this is probably due more to the application of fairly recently erected names than the discovery of new forms.

In the remainder of this district, south and south-west of Heighington, the Lower Magnesian Limestone was deposited in a shallow embayment of the Zechstein Sea. Here the molluscan content of the fauna is dominant and more varied than in the areas to the north and south. As elsewhere, *Bakevellia* is the most abundant bivalve but it is commonly accompanied by *Elimata permiana*, a form which Logan (op. cit.) inferred was free-swimming and stenohaline. Other common bivalves (with inferred palaeoecology, after Logan) are: *Janeia biarmica*, a stenohaline burrower; *Permophorus costatus* and *Pseudomonotis speluncaria*, euryhaline shallow-water forms, the former possibly and the latter certainly living attached to the substrate by means of a byssus; and *Schizodus obscurus* a euryhaline shallow-water burrower.

The other common elements in the fauna are calcareous imperforate foraminifera similar to, but more numerous than, those to the north, and ostracods. The latter are mostly forms with rounded peripheries and smooth carapaces such as the Bairdiids but they also include the reticulate species *Kirkbya permiana*. Brachiopods, bryozoa and nodosariid foraminifera are fairly common and are locally abundant, especially near the base of the formation at the western margin of the outcrop. Near Langton (locality 24) the productoids *Horridonia horrida* and *Strophalosia morrisiana* are particularly common. This assemblage probably represents the remains of a shell bank near the western shore of the embayment. The richness of the fauna in the whole of the embayment in comparison with that in other marginal parts of the Lower Magnesian Limestone sea was probably due to the shallowness of the water. It is analogous in some respects to that in the Middle Magnesian Limestone lagoonal beds of the Durham district (Smith and Francis 1967, p. 120).

Throughout this district and especially in the embayment, the Lower Magnesian Limestone contains scattered carbonaceous plant debris consisting largely of flattened ovoid bodies about 1·0 mm in diameter. The origin of these is obscure. J.P.

Details

For the purpose of this description the area occupied by the Lower Magnesian Limestone is divided into four parts: north of Heighington, between Heighington and the River Tees, the River Tees, and south of the River Tees. Each locality from which a collection has been made and a fauna identified has been numbered and is shown in Fig. 14, and the same numbers are used in Table 2. To avoid repetition the occurrence of wavy and irregular bedding, and the presence of cavities are not mentioned unless there is some special distinguishing feature.

North of Heighington. In Eldon Hill Quarry [242 272] (localities 1, 2) some 30 ft of Lower Magnesian Limestone were formerly exposed (Plate XIIIA). At the top the beds are locally affected by folds of 10 to 12 ft amplitude which Smith (1970) has suggested could be due to slumping and sliding although an alternative hypothesis by Mr E. A. Francis has suggested that they may be due to the effects of Pleistocene drag folding. The lower portion comprises 10 to 12 ft of medium-bedded fine-grained dolomitic limestone in which individual beds are generally 4 to 5 in thick and these are overlain by up to about 20 ft of thin-bedded dolomitic limestones. This part of the sequence is thought to correspond broadly in horizon to the fossiliferous limestone of Thickley (p. 139) and Raisby Hill Quarries (Smith and Francis 1967, pp. 107, 110), and it contains a comparable faunal assemblage. Fossils collected both from the north-east corner [2426 2725] of the quarry, and [2408 2712] some 250 yd to the south-west included foraminifera, crinoid columnals, bryozoa, brachiopods, a gastropod, bivalves and abundant ostracods (see Table 2). The faunal assemblage collected from the second locality is unusual in that it is dominated by the trepostomatous bryozoa *Batostomella crassa*, and fragments of cryptostomatous bryozoa probably referable to *Acanthocladia* or *Thamniscus*.

At Thickley Quarry [2408 2564] (locality 3) at East Thickley the section on the east side of the quarry (Plate XIIIB) is:

	ft
Dolomite and dolomitic limestone, thin-bedded flaggy fine-grained ..	10
Limestone, white and light grey lens-like semi-porcellanous; fossiliferous .. .. up to	3
Dolomite and dolomitic limestone, yellowish brown and buff thick-bedded fine-grained	8

D.A.C.M., E.A.F.

Smith (1970) has suggested that the limestone may be a shelly turbidite like those described in north-east Durham.

Fossils collected from the limestone include foraminifera, bryozoa, brachiopods and ostracods. Howse (1890) recorded many of these, as well as two species of plants and Trechmann (1921, p. 540) remarked that he had "nowhere seen the calcareous Lower Magnesian Limestone so fossiliferous as it is at East Thickley". Stoneley (1958, p. 299) noted that the type locality of *Algites sternbergianus* (King) is at Thickley Quarry and that the holotype probably came from the Lower Magnesian Limestone, although King stated that it was found in the Marl Slate.

Five shallow boreholes between Thickley Quarry [2407 2565] and Middridge Quarry [248 253] have proved up to 61 ft of interbedded dolomites and dolomitic limestones, including thin calcitic limestones near the base. The rock is grey and dark grey, buff and yellow and the degree of dolomitization is very variable throughout. At the time of the resurvey 36 ft of thin-bedded dolomitic limestone were visible in Middridge Quarry [248 253] (loc. 4). This quarry has since been completely filled, but in a new quarry [249 252], opened about 150 yd to the east some 45 ft of grey, yellow and buff thin-to medium-bedded interbedded dolomitic limestone and dolomite are exposed at the time of writing. The upper 6 ft of beds are more rubbly and irregularly bedded. Thin discontinuous bands of limestone occur lower down in the sequence. Near the base of the face laminae and bands of Marl Slate lithology occur interbedded with the limestone. At a nearby railside exposure [2471 2527] Marl Slate is overlain by 3½ ft of thick-bedded dolomitic limestone, comparable with that in Thickley Quarry at the same stratigraphical position, followed by 24 ft of thin-bedded and flaggy dolomitic limestone. Of the fossiliferous limestone of Thickley Quarry there is no trace. The dolomitic limestone in and around the old Middridge Quarry yielded no fossils during the survey, but Howse (1890) records the amphibian *Lepidotosaurus duffi* from the "lowest limestone" at Middridge, presumably this quarry.

Lower Magnesian Limestone, formerly exposed in a quarry [245 275] north-west of Old Eldon, is no longer visible there, but a few feet of yellowish brown thin-bedded dolomitic limestone exposed nearby suggest a position near the base of the thin-bedded sequence. Small quarries, now obscured, were formerly worked in the basal beds of the division at positions [2375 2685; 2440 2823 to 2448 2835] north-east of Red House.

E.A.F., D.A.C.M.

Exposed in an old quarry [2589 2481] (loc. 5) near Greenfield House are 18 ft of grey and buff thin-bedded finely crystalline dolomitic limestone with a few bands of softer granular calcitic dolomite. The fauna, restricted in the main to the upper 10 ft of the section, includes foraminifera, bryozoa, brachiopods and bivalves; plants are also present.

In an old quarry [2240 2479] (loc. 6) near Southfield House, the following sequence is exposed:

	ft	in
Dolomitic limestone and dolomite, grey and yellow thinly bedded or flaggy; predominantly finely crystalline; abundant stylolites, locally bituminous or carbonaceous; plant remains and foraminifera .. .. .. up to	8	0
Dolomitic limestone and dolomite, buff grey and yellow; thin-bedded; predominantly dense fine grained and hard; thin carbonaceous bituminous and clayey interlaminae especially near base; plant remains, foraminifera, bryozoa, brachiopods, bivalves and ostracods	10	0
Dolomitic limestone, dull yellow soft finely laminated; carbonaceous plant fragments and rare fish scales .. .. at least	1	3

D.A.C.M.

A specimen (E 33054) collected 6 ft from the base of the exposure is yellowish buff, fine-grained, with drusy cavities (1 cm diameter) infilled with secondary calcite. While the dolomite forming the bulk of the rock fluoresces a yellowish cream colour under long-wave (3665 Å) ultraviolet light, a secondary calcite momentarily phosphoresces yellowish green, a phenomenon noticed in all specimens of Lower Magnesian Limestone which contain secondary calcite. The rock consists of a mosaic of interlocking even-grained (0.03 to 0.05 mm) xenomorphic and euhedral dolomite (refractive index ω = 1.684), with minor secondary calcite in the matrix and stringers as well as vugs. The presence of both euhedral and anhedral dolomite may perhaps indicate that replacement of the primary calcite has taken place in two phases (Carozzi 1960, p. 283). R.K.H.

The composite sequence visible in Old Towns Quarry [257 246] (loc. 7) east of Middridge Grange is as follows:

	ft	in
Dolomite, calcitic flaggy; irregular bedding; more massive at top ..	12	0
Dolomite, calcitic, mottled buff and greyish buff, unevenly bedded; carbonate-lined cavities ..	10	0
Dolomitic limestone, mottled buff and grey, unevenly bedded, locally stylolitic; partially auto-brecciated near top (?turbidite)	10	0
Dolomitic limestone and limestone, variably bedded mainly thin-bedded to flaggy but locally irregularly and rubbly bedded dark grey carbonaceous; dolomitic clay interlaminae near base; some baryte and sphalerite; fossiliferous in lower part	35	0

The limestone in this quarry is often irregular and patchy in its development, but most of it is contained in a 20-ft band, some 15 ft above the quarry floor.

The fossils in the basal 14½ ft of strata include foraminifera, bryozoa, bivalves and plant remains.

Twelve shallow boreholes at and to west of Old Towns Quarry [257 246] proved thicknesses of up to 78 ft of dolomitic limestones, dolomites, and limestone overlying the Marl Slate. There is in places a thin porcellanous limestone near the base of the sequence. The irregularity of the bedding, the degree of dolomitization and cohesion tend to decrease upwards. The highest beds proved may be at or near the base of the Middle Magnesian Limestone.

In an old working [2508 2449] east-south-east of Middridge Grange, 6 ft of thin-bedded shaly dolomitic limestone, resting on 6 ft of brownish yellow massive finely crystalline dolomite, are probably equivalent in part to the upper 30 ft of beds in Old Towns Quarry.

Lower Magnesian Limestone crops out in a number of small quarries in the neighbourhood of Redworth, Heighington and School Aycliffe. Near Shackleton Beacon [2293 2330] (locality 8) the sequence includes up to 30 ft of buff yellow and greyish brown thinly bedded predominantly finely crystalline calcitic dolomite and dolomite, probably near the top of the division. The uppermost 2 to 3 ft of beds are locally irregularly brecciated and consist of medium or coarse angular blocks of loosely consolidated calcitic dolomite (Plate XIVB). Plant fragments and tubular hollows, up to 4 mm long and 0.4 mm wide (cf. *Calcinema* [*Tubulites*]) are the only fossils noted.

At High Bank Quarry [2400 2273] (loc. 9), north-west of Heighington, 15 ft of compact, finely granular laminated and flaggy calcitic dolomite and dolomitic limestone containing sporadic clayey or bituminous laminae lie near the base of the Lower Magnesian Limestone. Fossils include foraminifera, bryozoa, brachiopods, bivalves, ostracods, crinoid columnals, and plants.

Beds of similar lithology but slightly lower in the sequence are exposed [2386 2272] (loc. 10) on the south side of the road adjacent to this quarry and 150 yd NW of High Side Bank. Similar beds are also exposed [2376 2274] on the north side of the road 240 yd WNW of High Side Bank. This exposure is of interest in that it contains, in common with some other quarries south of Heighington, a 2-ft irregular slumped wavy lenticular band 5 to 6 ft above the base.

About 16 ft of grey and yellow finely granular calcitic dolomite and dolomitic limestone are exposed in an old quarry [2555 2378], 410 yd NW of School Aycliffe. The rock is locally microporous and has undergone patchy recrystallization. The beds are predominantly thin- to medium-bedded but are locally irregular, and the lithology suggests a position near the top of the Lower Magnesian Limestone. Similar strata are exposed in another old quarry [2527 2370] 320 yd to the west-south-west. D.A.C.M

North of Heighington the Lower Magnesian Limestone is proved in many boreholes where the formation is underlain by productive Coal Measures. The following two boreholes exemplify the main features of the sequence:

	ft
NCB Windlestone 'B' Bore [2667 2834]:	
Dolomitic limestone, yellow to buff fine-grained; abundant stylolites; sporadic small vugs and veinlets . .	70
Dolomitic limestone, grey; abundant small vugs with dolomite, pink baryte and ?chalcopyrite	41
Dolomitic limestone; highly brecciated in dark grey argillaceous slickensided matrix; interstices in top 5 ft contain pink baryte and pyrite	19
Dolomitic limestone, thin-bedded; interbedded dark dolomitic shale bands at base	2
NCB Windlestone 'C' Bore [2715 2785]:	
Dolomitic limestone, greyish brown crystalline; stylolites; sporadic crystal-lined vugs chiefly near top and base with associated reddish brown bands	45
Dolomitic limestone, creamy yellow with brownish bands fine-grained, flaggy; vugs and veinlets . .	51
Dolomitic limestone, light grey semi-porcellanous; thin-bedded; dark grey slickensided partings; scattered large vugs	20
Dolomitic limestone, light grey fine-grained autobrecciated; some pyrite in interstices . .	10
Dolomitic limestone, grey semi-porcellanous; thin-bedded and fissile; dark partings; a strong smell of oil	2

Similar sequences of all or part of the Lower Magnesian Limestone were proved in other boreholes, as for example: NCB Eldon Moor No. 186 Bore [2676 2734], 119 ft; NCB Eldon 'E' Bore [2559 2787], 69 ft; NCB Eldon No. 11 Bore [2563 2821], 42½ ft. The formation was also penetrated in NCB Windlestone Supplementary 'B' Bore* [2663 2866], and Charles Pit* [2503 2634] Middridge.

In the Heighington and School Aycliffe areas the Lower Magnesian Limestone has been proved only in the following poorly documented boreholes: Grammar School Farm Bore [2497 2246], Heighington, about 50 ft of 'limestone', the lower 12 ft of which is described as soft; Bakelite Water Bore* [2680 2329], Aycliffe Trading Estate, about 53 ft of 'hard limestone'.

Recent evidence from the adjacent Stockton (33) district to the east suggests that the Lower Magnesian Limestone may extend in a narrow tongue to the eastern margin of the district along Woodham Burn [271 253] north of Newton Aycliffe.

Between Heighington and the River Tees. Although there are several exposures mainly in overgrown quarries along the western edge of the outcrop between Heighington in the north and the River Tees in the south, most show the same Lower Magnesian Limestone sequence. Description, therefore, can be limited to examples to illustrate their general charactersitics.

A quarry [2325 2161] (loc. 12) 300 yd ESE of Elm Grange shows strata apparently near the top of the Lower Magnesian Limestone:

	ft
Dolomite, calcitic massive finely crystalline and recrystallized; sporadic irregular bands of breccia-conglomerate (?turbidite); traces of sphalerite, magnetite and baryte..	11
Dolomite, dull yellow and buff compact medium- to thin-bedded granular porous; plant debris, foraminifera and bivalves	20

The sequence, which is steeply dipping, was referred to by Kirkby and Duff (1871–72, p. 187). D.A.C.M.

A specimen [E 33055] of the breccia-conglomerate contains poorly sorted, rounded to angular, pinkish grey dolomite phenoclasts set in a reddish brown matrix, the whole rock being vuggy and porous. The phenoclasts consist of mosaics of interlocking xenomorphic dolomite (0.03 to 0.08 mm) with conspicuous secondary calcite impregnation and replacement. Some dolomite mosaics appear to have been partially calcitized prior to cementation. The matrix is composed of finely comminuted particles of granular dolomite with much clear coarser grained calcite cement. Hematite forms sparse discrete particles and ferric oxide generally is prominent in the matrix. The rock thus represents brecciated and eroded dolomite, perhaps representing a local reworking at shallow depth, cemented and veined by secondary calcite accompanied by ferric oxide. R.K.H.

In a quarry [2416 2138] (loc. 13) south-east of Broom Dykes, greyish brown and yellow compact dense crystalline medium-bedded dolomite, equivalent to the lower part of the above quarry, was formerly exposed. In addition to plants it contained a fauna of foraminifera, bryozoa, bivalves and ostracods. D.A.C.M.

A specimen (E 33056) from this quarry, is hard, dense, vuggy and porous, with a purplish grey colour, and is composed of a fine (0·01 to 0·02 mm), interlocking mosaic of xenomorphic dolomite grains. Minor secondary calcite occurs mainly in fine veinlets and drusy cavities. R.K.H.

At Denton Quarry [2100 1966] (loc. 14), nearly ¾ mile N of Denton crossroads, the following sequence is exposed:

	ft
Dolomite, grey and buff mottled medium- to thick-bedded porcellanous to finely granular microporous; algae and sparse fauna of foraminifera, a ?gastropod, bivalves and ostracods	6
Dolomite, grey and brown, massive with small-scale turbulent bedding or slumping; scattered large cavities	2
Dolomitic limestone and dolomite, thin-bedded grey and yellow finely crystalline	½
Dolomitic limestone, grey and yellow hard finely crystalline to porcellanous; turbulently and irregularly bedded with "balled-up" structures at top	4
Dolomitic limestone, grey, yellow and brown mainly thin- to medium-bedded finely crystalline to porcellanous;	4
Dolomitic limestone, grey thick-bedded variably textured porous; plant debris, foraminifera, bryozoa, brachiopods, bivalves and ostracods. The fossils which range over the above 12 ft of strata become progressively more sparse upwards	1
Dolomitic limestone, yellow and brown soft thin-bedded and flaggy; incoherent finely granular, sporadic sandy clay interlaminae; calcite on joint faces; plant remains including ovoid bodies; foraminifera and ostracods abundant, a worm tube, bryozoa, brachiopods, a gastropod and bivalves	7

Fine bituminous films occur between bedding planes. The uppermost part of this section is probably very close to the base of the Middle Magnesian Limestone. Similar though less well exposed sections are visible in this part of the sequence at Quarry Hole [2289 1910] (loc. 15) north-north-west of Walworth where the Marl Slate is represented by shaly interlaminae at the base of the section, and in New Moor Quarry [2314 1955] (loc. 16) 780 yd N of Walworth Castle. At both these quarries turbulently bedded disturbed bands are present. In the former, a 2½ ft band occurs 6 ft from the base, while in the latter the middle and upper part of the section consists of granular, porous thick-bedded or massive turbulently bedded dolomite. At Quarry Hole the basal four feet contains *Lingula credneri*, indeterminate bivalve fragments and rare palaeoniscoid scales. In the remainder of the section the fauna, though poorly preserved, is more varied and includes foraminifera, bryozoa, gastropods, bivalves and ostracods. Plant remains are also present. At New Moor Quarry, the fauna is similar, except that gastropods and foraminifera have not been found.

In a quarry [2170 1979] (loc. 17) ¼ mile WSW of Denton Grange East, 18 ft of light yellow, off-white and dull grey compact mainly medium- to thick-bedded finely granular microporous, dolomitic limestones crop out. For the most part they underlie beds seen at the base of the Denton Quarry (p. 148). The fauna (in which *Bakevellia* is the most common species) becomes more abundant towards the top and includes foraminifera, bryozoa, brachiopods, bivalves and ostracods. Plants are also present. A similar section is also visible in an old quarry [2213 1853] (loc. 18), ¼ mile SSE of St Mary's Church, Denton.

In a quarry [2093 1875] (loc. 19) ½ mile ESE of Summerhouse the following sequence is exposed on the east face (Plate XVA):

	ft	in
Dolomite, coarsely granular and porous; thick-bedded and massive with fretted and rubbly appearance locally disturbed; rare bivalves ..	7	
Dolomite, massive; sparse bivalves ..	5	
Dolomite, grey and brown, thick-bedded; ostracods, sparse bryozoa and bivalves	6	
Dolomite, massive disturbed; irregular lenses of crystalline dolomitic limestone; abundant foraminifera and plant remains; also ostracods ..	2½	
Dolomitic limestone, with discontinuous irregular brown crystalline bands; thin to medium bedding; plant remains, foraminifera [abundant], brachiopods, bivalves and ostracods		2
Dolomitic limestone, locally concretionary; irregular base; plant remains, foraminifera, gastropods		1¼
Dolomitic limestone, predominantly finely crystalline, flaggy; sporadic films of bituminous clay; plant remains and calcareous imperforate foraminifera throughout and especially near base, *Bakevellia ceratophaga* and *Fenestella retiformis* also common; fauna also includes nodosariid foraminifera, bryozoa, brachiopods, gastropods, bivalves and ostracods	10	

Including a section [2071 1875] visible in the south-west corner of the quarry the basal member of the sequence is thought to be about 18 ft thick. D.A.C.M.

Two specimens (E 33049, 33050) collected 1 ft and 8 ft respectively, above the base of the east face of the quarry are buff- to pink-grey, rather porous, fine-grained dolomites. They consist of very fine (2 to 3 microns) dusty dolomite aggregates, with local areas, partly related to veinlets, of recrystallized, coarser and clearer dolomite. The veinlets and some of the pores are occupied by secondary calcite, a common feature of friable dolomitized rocks (Trechmann 1914, p. 254). R.K.H.

Summerhouse Quarry is one of the few localities in this area to have been mentioned in previous literature. Sedgwick (1829, p. 77) states . . . "at Summerhouse Quarry, near Denton, a thin-bedded limestone . . . alternates with thin bands of black micaceous shale [?Marl Slate] which is sufficiently bituminous to be regarded as an impure coal". The occurrence of fossils here was noted by Browell and Kirkby (1865–67, p. 213). They gave an analysis and stated that the quarry provided a "moderate road metal" as well as burnt lime.

Sections broadly similar to that of Summerhouse Quarry are also seen in the following quarries around Killerby:

Morton Limekilns [190 205] (loc. 20) north-west of Killerby: about 35 ft of dolomite and dolomitic limestones with thin laminae of dolomitic lime sand especially near the base. At the west end of the quarry deep irregularly shaped cavities have been eroded into the rock and some of these contain a friable sand. Kirkby and Duff (1872, p. 197) described the upper surface of the quarry, now overgrown, and noted that it had a curious eroded appearance which they attributed to chemical action. The rock is in places extensively fretted by calcite veins. Although the bulk of the rock is not fossiliferous a collection made from the east face [1920 2041] near the base of the sequence included plant remains, foraminifera, bryozoa, brachiopods, gastropods, bivalves and ostracods.

Quarry [1964 2013] (loc. 21) north-east of Killerby: dolomite and dolomitic limestones 30 ft, including a turbulently bedded disturbed band, 2 ft thick, 12 ft from the base. Fossils include plants, foraminifera, bryozoa, bivalves, ostracods and a worm burrow. Foraminifera and ostracods are rare while two species of *Bakevellia* are the most common forms present. A similar sequence including a disturbed bed is visible in the quarry [1955 2025] on the opposite side of North Lane.

Killerby Garths Quarry [1873 1961] (loc. 22) 620 yd SW of Killerby: dolomite and dolomitic limestones 29 ft. The fauna includes foraminifera, crinoids, bryozoa, brachiopods, gastropods, bivalves and ostracods; the bivalves *Bakevellia binneyi* and *B. ceratophaga* are the most common forms.

Quarry [1860 1851] (loc. 23) about ½ mile SW of Killerby: dolomite and dolomitic limestone 20 ft. The rock is extensively fretted with calcite veinlets. The fauna is dominated by calcareous imperforate foraminifera, bivalves and ostracods, while bryozoa, brachiopods and gastropods are rare; plant remains are also present.

Small exposures of sparsely fossiliferous, finely granular, predominantly thinly and irregularly bedded dolomites and dolomitic limestones crop out at Dovecote Holes [1789 1915] near Headlam. At an old quarry [1986 1865] about ¾ mile NE of Dyance 5 ft of massive dull grey and yellow crudely bedded dolomite crops out. Its lithology suggests a position near the top of the Lower Magnesian Limestone.

Hobgate Quarry [1646 1888] (loc. 24) near Langton, shows 12½ ft of pale brown to buff yellow and grey dolomitic limestone with bands of impure limestone. The occurrence of very fine laminae of shale, bituminous in places, near the base of the sequence implies the existence of the Marl Slate at or below quarry level. The rock varies from a finely granular or powdery locally microporous dolomite, to a finely crystalline hard dense dolomitic limestone. Typically it is medium- to thick-bedded, with scattered crystal-lined cavities mainly in the more granular bands. The sequence is notable for its rich fauna including a relatively large number of brachiopods in the middle beds of the section. Foraminifera and ostracods are abundant, while *Bakevellia* and *Elimata* are the most common bivalves. Similar occurrences have also been noted at Thickley Quarry (p. 145) and Raisby Hill Quarry in the Durham and West Hartlepool district (Smith and Francis 1967, pp. 111–2). The fauna also includes bryozoa, gastropods, other bivalves and fish fragments; plant remains are also present.

North of Cabin House, Piercebridge, 12 ft of dense fine-grained medium- to thick-bedded dolomitic limestone is exposed in the north-east corner of a quarry [2131 1656]. The section passes upwards into a more rubbly, less dense, and vuggy dolomite with irregular bedding, recalling beds in the upper part of the Lower Magnesian Limestone.

Hard compact grey thin-bedded dolomitic limestone was formerly exposed in an old working [1985 1618] at White Cross.

Several boreholes have proved Lower Magnesian Limestone:

TVCWB No. 3 Bore* [2520 1982]: Lower Magnesian Limestone, about 93 ft 1 in. The uppermost beds indicate a position near the base of the Middle Magnesian Limestone.

Cold Sides Bore* [about 254 188]: Lower Magnesian Limestone 94 ft 5 in.

Cold Sides Water Bore [2536 1889]: Fowler (1956, p. 255) records 3 ft of hard grey and brown limestone with patches and inclusions of whitish to pinkish massive baryte.

(L 362)

A. Bedded dolomite and dolomitic limestone in the Lower Magnesian Limestone. Summerhouse Quarry

PLATE XV

B. Algal macrostructures in Middle Magnesian Limestone. north bank of River Tees, High Coniscliffe

(L 398)

	ft	in
TVCWB No. 5 Bore [2283 1693]:		
"Limestone"	9	6
Dolomite, calcitic, grey and buff hard crystalline predominantly dense; black carbonaceous stylolitic clay films on some bedding planes; abundant fauna of foraminifera, bryozoa, small brachiopods and bivalves; plant fragments	13	6
Dolomite, calcitic, pale buff with grey patches finely crystalline; black carbonaceous films on bedding planes; fauna as above	2	3
Dolomite, buff with grey patches finely granular to finely crystalline; black speckling; scattered foraminifera and plant filaments; bivalves common 1 ft 3 in from top; a productoid near base ..	8	11
Dolomite, as above, but mainly finely crystalline; scattered poorly preserved foraminifera; plant fragments and debris	10	10
Dolomite, calcitic, buff finely crystalline; foraminifera, brachiopods and bivalves; fauna decreases to base	5	3
Dolomite, grey fine-grained; clay films on most bedding planes; abundant fine plant debris and sporadic foraminifera	2	3
Marl Slate	—	—

Thornton Hall Bore* [2480 1634]: 'broken ground' (probably soft granular dolomitic limestone) 36 ft 3 in, on brown limestone 51 ft 6 in.

	ft	in
TVCWB Archdeacon Newton E.4 bore [2532 1771]:		
Dolomite, buff-brown saccharoidal (?oolitic): locally recrystallized..	1	0
Dolomite, buff-brown saccharoidal predominantly dense; sporadic small-amplitude stylolites; sporadic fauna includes foraminifera, bryozoa, brachiopods and bivalves; plant debris, fronds and filaments at base	29	8
Dolomite, buff-grey to yellow porous; sporadic stylolites and small irregular cavities often crystal-lined; bryozoa and brachiopods at base; plant debris and carbonaceous material throughout	12	0
Dolomite, cream-grey dense; comminuted plant debris ..	1	4
Dolomite, buff laminated fine-grained soft carbonaceous; ?bivalves; plant debris ..	1	5

TVCWB No. 1 Bore* [2552 1420]: dolomite and limestone, over 41 ft 5 in. Due to very poor recovery between a depth of 245 ft and 290 ft the position of the boundary between the Middle and Lower Magnesian Limestone is doubtful.

TVCWB W.1 Bore [2419 1422], near Low Coniscliffe: dolomite, calcitic buff-grey finely crystalline; locally abundant dolomite-lined cavities up to 1 in; sporadic flecks of carbonaceous matter. Thickness unknown due to poor recovery.

The fault east of Broom Dykes shown on Fig. 14 is not now thought to exist, and the structure is as shown on the published map.

River Tees. Most of the Lower Magnesian Limestone sequence is exposed along the River Tees between Piercebridge and High Coniscliffe. The lowest exposed beds crop out in a bluff [2093 1544] (loc. 25) on the south bank of the river between 100 and 250 yd above the Pierce Bridge where the section recorded is as follows:

	ft
Dolomitic limestone and dolomite, brownish yellow and buff predominantly massive, granular to saccharoidal, locally rubbly; abundant irregular cavities; hollows and streaks of dolomitic marl; black speckling; grades into bed below at least	10
Dolomitic limestone, light yellow and buff mainly thick-bedded finely granular to saccharoidal locally soft and powdery; sporadic discontinuous brown laminae of leathery clay; black speckling	8
Dolomitic limestone, thick-bedded finely granular hard; otherwise as above	4
Dolomitic limestone and dolomite, thick-bedded fine-grained; residual leathery clay laminae; black speckling	3½

	ft
Dolomitic limestone and dolomite, yellow and buff medium- to thick-bedded; finely granular locally soft and powdery with some hard bands; recrystallized in places; fossils not common but include plant fragments, foraminifera, bryozoa, bivalves and ostracods	3¼
Dolomitic limestone, brownish yellow or grey finely granular microporous; rubbly bedding .. at least	3½

D.A.C.M.

Two specimens (E 33051, 33052) were examined from 1 ft and 20 ft respectively above the base of the exposure. The first is a powdery, pale creamy yellow dolomite riddled with vugs and traversed by veinlets bearing secondary calcite. In section, the rock consists of a very fine-grained aggregate of calcite and particles of shells. Porphyroblastic clusters of bladed calcite crystals are probably pseudomorphous after anhydrite. The second specimen differs from the first mainly in the coarser (0.01 mm) overall granularity of its constituent dolomite mosaic, which contains patches of clear, coarser, secondary calcite. The dolomite grains become a little coarser and porphyroblastic at their contacts with the calcite. Granules of hematite are scattered throughout the rock. R.K.H.

The lowest member of the sequence described above is better exposed on the opposite side of the river [2100 1557] (loc. 26), 80 yd W of the Pierce Bridge where it consists of 8 ft of grey and brown rubbly finely granular microporous, in part banded, dolomite and dolomitic limestone. The weathered rock is much broken and shattered, giving a brecciated appearance to bedding surfaces. Cavities, up to 4 in, and commonly elongate, occur throughout. Calcareous imperforate foraminifera are occasionally found.

The sequence, in part equivalent to that exposed at locality 25, is continued on the north side of the river between Piercebridge and High Coniscliffe, [2130 1589 to 2224 1555]. The composite sequence is as follows:

	ft
Middle Magnesian Limestone ..	—
Dolomite, dull white soft powdery and granular, with thin flaggy or irregular bedding; indeterminate productoid	6
Dolomite, cream buff and yellow, medium- to thick-bedded; coarser saccharoidal bands near top, otherwise predominantly finely granular and microporous; black speckling; foraminifera, bryozoa, bivalves, a gastropod and ostracods (loc. 27) [2162 1572]	25
Dolomite, buff to dull white finely granular microporous, locally soft and powdery; sporadic recrystallized patches and bands; sporadic carbonate-lined cavities; jointed vertically and horizontally; foraminifera, gastropods, bivalves, ostracods (locality 28) [2146 1583]	10
Dolomitic limestone, buff to dull yellow; thin- to medium-bedded, finely granular; ostracods ..	2
Dolomitic limestone, cream-brown and light yellow, medium-bedded finely granular microporous; locally soft and powdery; interbedded ½- to 1-in fine-grained recrystallized bands; ostracods	6½
Dolomitic limestone, cream to light yellow, bedded in 3- to 9-in units; interlaminated with low-angled cross-bedded finer-grained ½- to 1-in bands becoming less common at top; finely crystalline; ostracods ..	12
Dolomitic limestone, cream-buff thin- to medium-bedded; fine-grained 1- to 2-in pale grey buff bands occur throughout and show convergent bedding; finely granular and microporous; scattered stylolites of small amplitude; sporadic vugs; black speckling; ostracods [2130 1589] ..	5

D.A.C.M.

A specimen (E 33057) of soft, pale cream, unevenly bedded dolomite from the 25 ft bed near the top of the sequence consists largely of an aggregate of dusty, microcrystalline dolomite, with vuggy infillings and irregular patches of secondary calcite. The dolomite is, in places, a little coarser and evidently shows incipient recrystallization. R.K.H.

South of the River Tees. The main outcrop in the Caldwell and Eppleby area lies east of a fault which strikes N 25° E 40 yd E of Caldwell Mill [1625 1301], and north of Caldwell and Eppleby villages. The relationship between the basal Permian strata and the underlying Carboniferous is obscured by thick glacial deposits, but it is thought that about 70 ft of Lower Magnesian Limestone rest directly on Namurian rocks.

Pale grey and dull yellow compactly bedded dolomitic limestone was formerly exposed at the base of the old railway cutting [1894 1538] 170 yd NW of Low Field Farm. Some 8½ ft of fawn thinly bedded fine-grained dolomite with sporadic casts and vugs crop out in a quarry [1836 1481] about ¼ mile E of Low Chapel House. About ¼ mile W of Holme House 14 ft of dull yellow and buff, predominantly finely granular, porous dolomite is exposed in a quarry [2164 1496]. Massive at the base, it passes up into more regularly bedded units up to 5 in thick. There is a band of porous saccharoidal dolomite near the top of the sequence. This section almost certainly includes the top beds of the Lower Magnesian Limestone and probably also basal beds of the Middle Magnesian Limestone J.H.H.

Sections in Lower Magnesian Limestone are visible in a cutting on the Durham motorway south-west of Cleasby. They are exposed [2460 1232 to 2467 1249] over a distance of 220 yd along the east side of the motorway (Plate XIVA; Fig. 15). The section (loc. 29) consists of about 55 ft of medium to thick-bedded predominantly finely crystalline and granular microporous dolomitic limestone and dolomite; thin bands and lenticles of hard finely crystalline limestone and recrystallized rock occur throughout the sequence. A few thin leathery clay laminae occur, especially near the base of the section. Coarsely granular saccharoidal bands are present at the top of the sequence and in a position near to or at the base of the Middle Magnesian Limestone. At the south end of the exposure the rock becomes more massive and poorly bedded adjacent to a fault. Small irregular subangular to subrounded crystal-lined cavities occur throughout while black speckling is locally abundant. The sequence as a whole is not very fossiliferous but the following have been collected: carbonaceous plant remains, calcareous ?algal debris (3 to 4 ft from base), foraminifera, worm burrows, bryozoa, bivalves and ostracods.

At the south end [2460 1232] of the section, the cutting is traversed by three faults separating the above sequence from beds of predominantly Middle Magnesian Limestone age (p. 156). Between the main fault [2440 1232] and the southern end of the rock cutting [2458 1229], there are exposures of thinly bedded dolomitic limestone and dolomite, which become more massive near faults. A former exposure [2453 1234] 180 yd W 10° N of High Farm, showed 13 ft of dull yellow, massive, irregularly bedded, granular and microporous, locally recrystallized, dolomite and dolomitic limestone resting on 8 ft of thin-bedded dolomitic limestone. This sequence is overlain by 3 ft of Middle Magnesian Limestone. Other sections previously seen at the base of quarries on the west side of the cutting were in similar beds overlain by basal Middle Magnesian Limestone.

Old quarries immediately south-west of High Farm are almost wholly filled in, but small sections are visible along the northern margin of the workings. For example, [2466 1219] 100 yd S by W of the farm, over 10 ft of grey, brown, dull yellow and buff, thin-bedded flaggy, finely granular microporous dolomitic limestone are exposed, and are separated from beds of Middle Magnesian Limestone age by a fault running close to the south face.

Lower Magnesian Limestone is proved in five boreholes south of the River Tees. Most of them are poorly documented and the thickness of the formation is estimated, as follows:

Water bore [1972 1422], ¾ mile NNE of Carlton Grange, Eppleby: yellow limestone 51 ft, harder blue limestone 6 ft, on brownish yellow dolomitic limestone 12 ft.

TVCWB No. 8 Bore [2176 1346], Manfield: limestone 3 ft, very soft limestone 10 ft, on dolomite 17 ft 4 in. The lowest bed forms a yellowish grey or yellow thick-bedded to massive finely crystalline microporous dolomite; sporadic more rubbly broken thin bands are also present, especially towards base.

Abbey Farm Water Bore [2239 1331], Manfield: bastard limestone and sand seams 12 ft, 'white running marl' 10 ft, on hard shaly marl 28 ft.

High Farm Water Bore [2472 1231], Cleasby: yellow limestone 17 ft.

Clow Beck Farm Water Bore [2404 1064], Newton Morrell: about 40 ft of dolomitic limestone recovered 'as a lime-sand except for 6 ft of soft yellow rudely bedded dolomite at the base'. D.A.C.M.

MIDDLE MAGNESIAN LIMESTONE

Although the Middle Magnesian Limestone crops out over a wide area in the eastern part of the district (Fig. 14), good sections are seen only in the Tees valley between Piercebridge and High Coniscliffe, in quarries near Eppleby, and in a cutting and adjacent quarries on the Durham motorway south-west of Cleasby. As already indicated quarries in the Lower Magnesian Limestone may well be capped by a few feet of Middle Magnesian Limestone. Middle Magnesian Limestone is also recorded in scattered boreholes in the eastern part of the district.

The western limit of the Middle Magnesian Limestone may be rather less than shown on the one-inch map of the district and its boundaries must therefore be regarded as tentative. A borehole [2283 1693] near Ulnaby Hall has shown the presence of Lower Magnesian Limestone (Fig. 14) in an area previously thought to be underlain by a thin cover of Middle Magnesian Limestone. It is also possible that the Lower Magnesian Limestone may well be present elsewhere either as small inliers or as a result of unsuspected faults.

The formation is represented throughout by dolomites, calcitic dolomites and dolomitic limestones, the greatest recorded thickness being 148½ ft in a borehole on the western outskirts of Darlington. The full thickness probably does not exceed 160 ft.

The earliest deposits, the transitional beds, form a thin group of variably textured dolomites and dolomitic limestones possessing features associated with deposition in shallow moving water, for example, current bedding and oolites. The junction of these beds with the Lower Magnesian Limestone is not in this district marked by the distinctive group of mudstones and dolomites recognized at this horizon farther south. A slight discontinuity is present at High Coniscliffe and a group of flaggy dolomites with thin clay bands is present in some boreholes and exposures in the adjacent Stockton (33) district.

The transitional beds are succeeded by the shelf/lagoon facies, a series of variably textured dolomites, in places saccharoidal or oolitic. At High Coniscliffe they include a lenticular reef-like limestone, the only known such development in the district. D.A.C.M.

Palaeontology. The only organic remains recorded from beds of definite Middle Magnesian Limestone age in this district are algal macrostructures in stromatolitic growths, a few bivalves, cryptostomatous bryozoa fragments and ostracods. As an indicator of environment the occurrence of colleniform stromatolites is important since these organisms can exist only within the intertidal zone. J.P.

DETAILS

North of the River Tees. The Middle Magnesian Limestone is not thought to crop out at the surface although the beds at the top of some quarry faces described above are thought to be virtually at or near the base of the formation. The formation is, however, recorded from fourteen boreholes, as follows:

NCB Windlestone 'B' Bore [2667 2834]: Buff to cream-yellow thin-bedded finely crystalline dolomitic limestone, with sporadic cavities, 62 ft.

NCB Windlestone 'C' Bore [2715 2785]: Light brownish grey flaggy coarsely crystalline dolomitic limestone, with sporadic 1- to 10-in porous bands, 15 ft, on cream-yellow, less flaggy dolomitic limestone with scattered crystal-lined cavities 13 ft.

Bakelite Water Bore* [2680 2329]: Limestone about 38 ft.

Grammar School Farm Bore [2497 2246], Heighington: 'Marl' 10 ft, on hard limestone 38 ft.

Cold Sides Bore* [about 254 188]: Middle Magnesian Limestone, 83 ft.

Cold Sides Water Bore [2536 1889]: Hard limestone about 10 ft.

Thornton Hall Bore* [2480 1634]: Middle Magnesian Limestone 22 ft 10 in.

Burtree Gate Water Bore [2618 1868]: Yellow 'marl' with limestone 8 ft; on hard limestone 2½ ft.

Bottom House Farm Water Bore [2662 1768]: Limestone and marl 22 ft 7 in.

TVCWB Archdeacon Newton E.4 Bore [2532 1771]: Predominantly buff saccharoidal dolomites about 30 ft.

TVCWB No. 6 Bore [2903 1563], Mowden Bridge: Dolomites and dolomitic limestones about 60 ft.

TVCWB No. 1 Bore* [2552 1420], Darlington: Dolomites of shelf/lagoon facies 114 ft, on transitional beds, 34½ ft.

TVCWB No. 2 Bore [2568 1442], Darlington: Mostly granular dolomites of shelf/lagoon facies, 70 ft.

Co-operative Nurseries Water Bore [2497 1445], Merrybent: Limestone about 120 ft (recovered mostly as lime-sand).

Tees valley. Transitional beds at the base of the Middle Magnesian Limestone crop out on the north bank of the river from [2224 1551] 500 yd WNW of High Coniscliffe Church to [2243 1529] 240 yd further downstream; the sequence (loc. 30) is as follows:

	ft
Dolomite, dull buff to cream, predominantly massive to thick-bedded, finely granular and saccharoidal, microporous; very rubbly and cavernous towards base with large, often irregular crystal-lined cavities and hair veins of calcite; at top becomes more massive compact and recrystallized with prominent 4- to 12-in cryptozoon stromatolites (Plate XVB); other fossils include bivalves and ostracods in upper 4 ft	25
Dolomite, yellow saccharoidal granular porous; laminated or bedded in units of 2 to 6 in	6
Lower Magnesian Limestone ..	—

D.A.C.M., E.A.F.

A vuggy and porous, cream-buff specimen from the lower bed (E 33053) contains much powdery carbonate and crystalline secondary calcite. Dolomite occurs mainly as aggregates of dusty, microcrystalline (0.005 to 0.015 mm) granules charged with inclusions and showing in places oolitic development, in a cement of secondary calcite. In places coarser, clearer dolomite grains exhibit subhedral forms and indicate recrystallization. R.K.H.

Strata overlying this sequence crop out [2254 1527] (loc. 31) on the east and north faces of old quarries north-west of High Coniscliffe Church, as follows:

	ft
Limestone, greyish white predominantly massive reef-type finely crystalline; irregular porcellanous bands and lenticles; ?algal structures; irregular, often large, crystal-lined cavities throughout; irregular base .. .. .. up to	15
Dolomite, grey yellow and buff; finely oolitic, porous; predominantly thin- to medium-bedded in 3- to 5-in units locally cross-bedded (foresets to south); isolated crystal-lined cavities; fauna includes bivalves and ostracods .. .. .. up to	15

D.A.C.M.

A specimen (E 33047) from 6 ft above the base of the exposure is a cream-coloured, porous, dolomitized oolitic limestone, with much secondary calcite filling cavities. The ooliths range considerably in mean diameter (0.1 to 1.0 mm) and shape (spheroidal to ellipsoidal and irregular). They are outlined by rims of dark, microcrystalline dolomite, which coalesce and form an interconnecting network, with nuclei of clear, coarser dolomite and irregular interstitial patches of calcite. The original concentric structure of the ooliths is preserved in places. Fragments of organic origin are sparsely scattered between the ooliths. The rock evidently represents shallow-water deposition, subject to current- and wave-action.

A second specimen (E 33048) collected from the upper reef-like limestone about 4 ft above the base of the bed differs in lithology from the first, being a grey, crystalline limestone consisting of an interlocking, equigranular (0.2 mm) mosaic of anhedral calcite with fine interstitial calcitic granules. There is also dusty, cryptocrystalline semi-opaque material, perhaps of organic origin, scattered between the calcite granules. R.K.H.

In the old quarry [2261 1526] north-east of the church, 3 to 4 ft of reef-like limestone overlie about 20 ft of cream or buff flaggy dolomites probably incorporating algal beds at the base, but the section is so overgrown that it is difficult to examine in detail. D.A.C.M.

South of the River Tees. The Middle Magnesian Limestone is recorded from five main localities. In the area north and north-east of Eppleby village it is thought to be about 60 ft thick. The transitional beds are visible in a quarry [1920 1474] east of Low Chapel House, as follows:

	ft	in
Dolomitic limestone, yellow to fawn thin-bedded nodular fine-grained recrystallized; sporadic shell casts	2	6
Dolomitic limestone, yellow to fawn, massive jointed; fine-grained recrystallized; shell casts	4	8

A slightly higher sequence in the transitional beds is seen in a quarry [1856 1404] south-west of Rennison:

	ft	in
Dolomite, yellow thin-bedded fine-grained saccharoidal; rubbly in places	1	0
Dolomite, calcitic yellow fine-grained; rubbly and nodular ..	2	0
Dolomite, decalcified soft fine-grained saccharoidal rubbly; hard calcitic dolomite lenses; shells	6	0
Dolomitic limestone, white and yellow soft fine-grained ..	1	10

J.H.H.

In an old quarry [2164 1496] about ¼ mile W of Holme House a porous saccharoidal dolomite at the top of the sequence is almost certainly near the base of the transitional beds of the Middle Magnesian Limestone.

Middle Magnesian Limestone is exposed in the quarries and in the adjacent cutting on the Durham motorway near High Farm, south-west of Cleasby (Fig. 15). Granular, microporous, irregular medium- to thick-bedded dolomite, 3 ft thick, is exposed overlying Lower Magnesian Limestone 180 yd W 10° N [2453 1234] of High Farm. Similar sequences crop out near the top of partly filled quarries on the west side of the motorway at the south end of the cutting. At a point [2450 1227] 180 yd SW of High House, over 15 ft of hard dull yellow massive to thick-bedded finely granular dolomitic limestone overlie 9 ft of soft yellow granular and oolitic limestone. Transitional beds were formerly exposed on the south side of the old quarry [2463 1216] 140 yd SSW of High Farm. The section exposed on the south side of a fault showed 16 ft of calcitic predominantly finely crystalline irregularly bedded massive dolomite. Irregular bands and lenses of pseudobreccia, finely granular and in places oolitic microporous dolomite and bedded dolomite occur throughout.

In the old quarry [2488 1223] 150 to 270 yd ESE of High Farm, grey yellow-brown and buff dolomites are visible. The beds are faulted and locally discoloured. They are also very variable in lithology, being composed of finely crystalline massive and irregularly thick-bedded calcitic dolomites with streaks, lenses and irregular masses of highly porous open textured saccharoidal granular and oolitic dolomite. Carbonate-lined cavities are abundant throughout.

South of the River Tees, Middle Magnesian Limestone is recorded in three poorly documented boreholes as follows:

Abbey Farm Water Bore [2239 1331], Manfield: bastard limestone and sand seams, about 28 ft.

Clow Beck Farm Water Bore [2404 1064], Newton Morrell: limestone, about 19 ft, recovered mostly as lime sand.

Jolby Farm Water Bore [2570 0970], Newton Morrell: soft limestone 14 ft, on broken harder limestone 18 ft.

D.A.C.M.

Higher Permian Measures

Although nowhere visible at the surface the presence of Middle Permian Marl, Upper Magnesian Limestone and Upper Permian Marl is inferred from meagre borehole information and from evidence in the adjacent Richmond (41) and Stockton (33) districts to south and east respectively. The mapped positions of the outcrops in the south-east corner of the district must therefore be regarded as tentative (Fig. 14). The Middle Permian Marl is unlikely to be over 30 ft thick and may thin gradually to the north-east, while the Upper Magnesian Limestone is probably about 60 ft. The Upper Permian Marl (nowhere proved) is probably at least 40 ft thick.

Details

Beds above the Middle Magnesian Limestone are recorded only in three boreholes:

Stapleton Water Bore [2641 1214]: Rather soft limestone (Upper Magnesian Limestone) 24 ft, on red marl (Middle Permian Marl) 21 ft.

TVCWB No. 7 Bore* [2610 0993] near Jolby Manor: Upper Magnesian Limestone, 41 ft, on Middle Permian Marl, 19 ft.

Jolby Farm Water Bore [2570 0970]: red marl (Middle Permian Marl) 19 ft.

D.A.C.M.

References

BRADY, H. B. 1876. A Monograph of Carboniferous and Permian Foraminifera. *Palaeontgr. Soc.* [*Monogr.*].

BROWELL, E. J. and KIRKBY, J. W. 1866. On the chemical composition of various beds of the Magnesian Limestone and associated Permian rocks of Durham. *Nat. Hist. Trans. Northumb.*, **1**, 204–30.

CAROZZI, A. V. 1960. *Microscopic Sedimentary Petrography.* New York; Wiley & Sons, viii + 485 pp.

CRESPIN, IRENE. 1958. Permian Foraminifera of Australia. *Bull. Bur. Miner. Resour. Geol. Geophys. Aust.*, No. 48, 1–207.

DEANS, T. 1950. The Kupferschiefer and associated lead-zinc mineralization in the Permian of Silesia, Germany and England. *Rept. XVIII Int. Geol. Congr.* (London), Pt. 7, 340–52.

DUNHAM, K. C. 1960. Syngenetic and diagenetic mineralization in Yorkshire. *Proc. Yorks. geol. Soc.*, **32**, 229–84.

——1961. Black shale, oil and sulphide ore. *Advmt Sci. Lond.*, **18**, No. 73, 284–299.

EDWARDS, W., WRAY, D. A. and MITCHELL, G. H. 1940. Geology of the country around Wakefield. *Mem. geol. Surv. Gt Br.*, viii + 215 pp.

FOWLER, A. *in* Manuscript. Note on a borehole at Merrybent, Darlington.

——1956. Minerals in the Permian and Trias of north-east England. *Proc. Geol. Ass.*, **67**, 251–65.

HANCOCK, A. and HOWSE, R. 1870a. On a new labyrinthodont amphibian from the Magnesian Limestone of Middridge, Durham. *Q. Jnl geol. Soc. Lond.*, **26**, 556–64.

—— ——1870b. On *Proterosaurus speneri*, Van Meyer, and a new species, *Proterosaurus Huxleyi*, from the Marl Slate of Middridge, Durham. *Q. Jnl geol. Soc. Lond.*, **26**, 656–72.

HIRST, D. M. and DUNHAM, K. C. 1963. Chemistry and petrography of the Marl Slate of S.E. Durham, England. *Econ. Geol.*, **58**, 912–40.

HODGE, M. B. 1932. The Permian Yellow Sands of north-east England. *Proc. Univ. Durham Phil. Soc.*, **8**, 410–58.

HOWSE, R. 1890. Catalogue of the local fossils in the Museum of Natural History. *Nat. Hist. Trans. Northumb.*, **11**, 227–88.

KING, W. 1850. A monograph of the Permian fossils of England. *Palaeontgr. Soc.* [*Monogr.*].

KIRKBY, J. W. and DUFF, J. 1872. Notes on the geology of part of South Durham. *Nat. Hist. Trans. Northumb.*, **4**, 151–198.

LOGAN, A. 1967. The Permian Bivalvia of Northern England. *Palaeontogr. Soc.* [*Monogr.*].

LOVE, L. G. 1962. Biogenic primary sulphide of the Permian Kupferschiefer and Marl Slate. *Econ. Geol.*, **57**, 350–66.

SCHOUTEN, C. 1946. The role of sulphur bacteria in the formation of the so-called sedimentary copper ores and pyritic ore bodies. *Econ. Geol.*, **41**, 517–38.

SEDGWICK, A. 1829. On the geological relations and internal structure of the Magnesian Limestone, and the lower portion of the New Red Sandstone Series in their range through Nottinghamshire, Derbyshire, Yorkshire and Durham to the southern extremity of Northumberland. *Trans. geol. Soc. Lond.*, Ser. (2) **3**, 37–124.

SMITH, D. B. 1970. The Permian and Trias *in* Geology of Durham County. G. A. L. Johnson, Ed. *Trans. Nat. Hist. Soc. Northumb.*, **41**, 66–91.

——1970. Submarine slumping and sliding in the Lower Magnesian Limestone of Northumberland and Durham. *Proc. Yorks. geol. Soc.*, **38**, 1–36.

——and FRANCIS, E. A. 1967. The geology of the country between Durham and West Hartlepool. *Mem. geol. Surv. Gt Br.* (contains an extensive list of references to the Permian). xii + 354 pp.

STONELEY, HILDA M. M. 1958. The Upper Permian flora of England. *Bull. Br. Mus. Nat. Hist.*, **3**, No. 9, 295–337.

TRECHMANN, C. T. 1914. On the lithology and composition of Durham Magnesian Limestone. *Q. Jnl geol. Soc. Lond.*, **70**, 232–65.

——1921. Some remarkably preserved brachiopods from the Lower Magnesian Limestone of Durham. *Geol. Mag.*, **58**, 538–43.

——1925. The Permian Formation in Durham. *Proc. Geol. Ass.*, **36**, 135–45.

——1931. The Permian *in* Contributions to the geology of Northumberland and Durham. *Proc. Geol. Ass.*, **42**, 246–52.

WESTOLL, T. S. 1941. The Permian fishes, *Dorypterus* and *Lekanichthys*. *Proc. Zoo. Soc.*, **111**, 39–58.

WILSON, E. 1881. The Permian Formation in the north-east of England. *Mid. Nat.*, **4**, 97–101; 121–124; 187–191; 201–208.

WINCH, N. J. 1817. Observations on the geology of Northumberland and Durham. *Trans. geol. Soc. Lond.*, **4**, 1–101.

WOOLACOTT, D. 1912. The stratigraphy and tectonics of the Permian of Durham (northern area). *Proc. Univ. Durham Phil. Soc.*, **4**, Pt. 5, 241–311.

Chapter 6

STRUCTURE

Introduction

THE CONTRAST between the tectonically stable upland areas east of the Pennine and Dent fault lines and the highly faulted and folded rocks to the west was seen many years ago as evidence that the Carboniferous of the northern Pennines rested upon stable blocks. An initial concept of a single rigid block of Lower Palaeozoic rocks underlying the Carboniferous strata of the northern Pennines suggested by Marr (1921, p. 63) was later modified into a northern Alston Block (Trotter and Hollingworth 1928, p. 433) and a southern Askrigg Block (Hudson 1938, p. 309). The two blocks are separated by the broad and shallow structural depression including Stainmore and Cotherstone Moor referred to by Versey (1927) and later termed the Cotherstone Syncline (Reading 1957, p. 27). Subsequent geophysical work in this and neighbouring districts, fully summarized by Bott (1967), suggested that the stable blocks were underlain by granite batholiths, and separated by areas in which Lower Carboniferous and possibly Devonian rocks might unconformably overlie older Palaeozoic sediments. Two granites of end-Silurian age have since been proved by drilling; they are the Weardale Granite beneath the Alston block (Dunham and others 1965, p. 383) and the Wensleydale Granite beneath the Askrigg Block (Dunham 1974). The Barnard Castle district would appear to fall mainly within the area between the two blocks, though the extreme north-western part of the district lies on the Alston Block.

No single fault in this district can be regarded as delineating the southern margin of the Alston Block. The easterly continuation of the Lunedale Fault, here known as the Lunedale–Staindrop Fault, may have operated to a limited extent as a hinge-line in Namurian (E_2) times, for on its southern (downthrow) side the strata between the Upper Felltop and the Botany limestones are relatively thick compared with the equivalent beds on the Alston Block (Fig. 5). The Butterknowle Fault has been suggested as the southern boundary of the block (Dunham 1948, p. 63; Bott and Masson-Smith 1957, p. 112; Bott 1967, fig. 5) and geophysical evidence for a southward thickening of Carboniferous strata across this line has been presented by Bott (1961, p. 8), though there is no directly supporting geological evidence. The evidence from the Woodland Bore (Mills and Hull 1968, p. 11) indicates that the hinge-line, in E_1 times at least, lay close to the relatively minor Eggleston–Woodland Fault, which joins the Butterknowle Fault north-east of Eggleston. In the Woodland Bore the sequence in the lower Namurian is lithologically more akin to that of the Cotherstone Syncline than that of the Alston Block. A hinge-zone rather than a hinge-line is therefore envisaged as the south-eastern margin of the Alston Block.

The chief elements in the structure of this district are set out in Fig. 16. In the south the Carboniferous strata have a gentle northerly regional dip, for they form the northern limb of the Middleton Tyas–Sleightholme Anticline (Wells 1957, p. 248). The northern half of the district is part of a complex and much

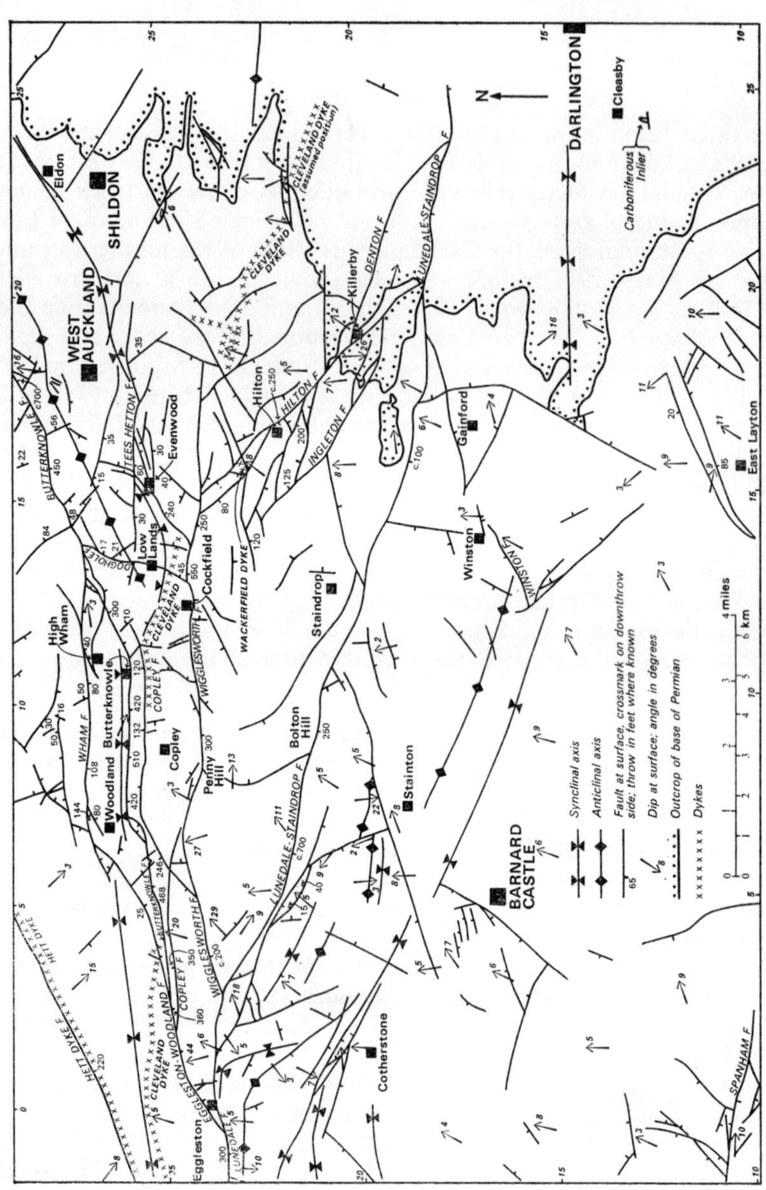

FIG. 16. *Chief elements of the structure of the Barnard Castle district*

faulted broadly synclinal region in which are preserved Coal Measures that form a westward extension of the main Durham coalfield. The Hercynian age of these structures is clearly demonstrated by their relationship with the overlying relatively undisturbed Permian strata. The gentle eastward regional dip of the Permian reflects the regional tilting to which the whole of north-eastern England was subjected in post-Jurassic, presumably Tertiary times.

STRUCTURAL HISTORY

There is no direct geological information within the district which throws light on its Lower Palaeozoic history but much evidence can be adduced from adjoining regions. Evidence from the Cronkley (Teesdale) Inlier (Harry 1950; Johnson and Dunham 1963) to the west and the Roddymoor Bore (Woolacott 1923) to the north suggests that Carboniferous rocks in the Barnard Castle district rest unconformably on Lower Palaeozoic beds partly of Ordovician age.

The earth-movements of the Caledonian orogeny, and subsequent erosion, resulted in the production of a surface on which Carboniferous, and locally possible Devonian, beds were deposited. Over this surface the Lower Carboniferous sea transgressed and deposited sediments, mainly cyclic, from at least C_1 times onwards (Burgess and Harrison 1967, table 1). The major structural elements that had been initiated in the Caledonian orogeny remained active, although subdued, through much of the Carboniferous, and profoundly affected sedimentation. The variations in thickness of individual cyclothems are thought to illustrate the variable rate of subsidence of the northern Pennines and from such evidence it has been adduced that the Alston Block (Dunham 1948, p. 63; Reading 1957, p. 51; Bott 1967, p. 160) and the Askrigg Block (Rowell and Scanlon 1957a, p. 32) were subsiding at a lesser rate than the intervening Cotherstone Syncline. It has been shown that some of the thickness variations, especially on the margins of the blocks, are associated with faulting. Examples of this with respect to the Alston Block are the Stublick Fault (Trotter and Hollingworth 1928, p. 433), the Swindale Beck Fault (Shotton 1935, p. 675; Burgess and Harrison 1967, p. 203) and the Lunedale Fault (Reading 1957, p. 51).

Unconformities and non-sequences are further manifestations of intra-Carboniferous earth-movements. A marked 'sub-Millstone Grit' unconformity has been recognized for some time on the southern margin of the Askrigg Block (Dunham and Stubblefield 1945, p. 233). More recent work (Rowell and Scanlon 1957b, p. 89; Wilson 1960, p. 302) has shown this intra-E_1 unconformity to be ubiquitous on the Askrigg Block, and it oversteps older beds in a generally southerly direction. Rowell and Scanlon (1957b, p. 89) state that the Grassington Grit passes northwards into the Mirk Fell Ganister, the base of the joint formation representing the unconformity. In this district the equivalent of the Mirk Fell Ganister is taken to be the group of sandstones beneath the Lower Felltop Limestone; though erosion is only locally demonstrable at their base these sandstones are nevertheless equivalent to transgressive sandstones overlying the 'Rogerley Transgression' on the Alston Block (Dunham 1948, p. 36).

On the Alston Block north-west of the present district a minor unconformity related to the 'Tanhill Transgression' has been recognized by Carruthers (1938, p. 237). The sandstones immediately overlying this transgression, which were subsequently called the Coalcleugh Transgression Beds (Dunham 1948, p. 40), were recognized in the Cotherstone Syncline by Reading (1957, p. 43), who attributed this unconformity to earth-movements in early E_2 times. No evidence exists in the Barnard Castle district for the absence of any strata below these sandstones; on the contrary, the overall thickness of these beds appears to be greater than elsewhere. A major unconformity has been postulated by some authors (e.g. Reading 1957, p. 50, and fig. 2, p. 31) at the base of the 'Millstone Grit' of the primary Survey. In this district, however, the evidence suggests that all the Namurian stages are present at least in part (Hull 1968, p. 306).

Hercynian movements caused most of the folding and faulting (Figs. 16 to 18) to which the Carboniferous beds have been subjected, and their uplift, erosion and intrusion by igneous rocks (p. 170). These movements ended before the deposition of the Basal Permian Sandstone and Breccia.

Post-Upper Permian movements (presumably Tertiary) resulted in slight modifications of the earlier structures, together with the uplift and gentle eastward tilting of the Permian strata. Fault movements from this time appear for the most part, to be along pre-existing lines. In their later stages these movements were locally accompanied by igneous activity as witnessed by the intrusion of the Cleveland Dyke (p. 173). If it is assumed that the dyke was intruded along a tensional fracture at about the same time as or slightly later than the faulting, some indication of the relative ages of the structures can be inferred. The age of the dyke has been calculated at 40 million years (see p. 180), i.e. close to the Eocene–Oligocene boundary (Funnell 1964, p. 188).

Folds

The folds in this district are primarily the result of late Hercynian earth-movements and are illustrated in Figs. 16 to 18. The southern half of the district is occupied by the northern limb of the Middleton Tyas–Sleightholme Anticline (Wells 1957, p. 248). In this area dips in Carboniferous strata, which include a component of post-Permian deformation of approximately 2° to the east, range from 2° to 11° in directions generally between north and north-east.

Versey (1927) had referred to the broad and shallow east–west trending structural depression lying mainly to the west of the district and including Stainmore and Cotherstone Moor. This fold was called the Cotherstone Syncline by Reading (1957, p. 27). Traced eastward into the district towards Cotherstone village and Barnard Castle the fold becomes less distinct, and it apparently dies out eastwards on the broad northern flank of the Middleton Tyas Anticline. Some reflection of the syncline however remains east of Barnard Castle since a very gentle monocline appears to be present on the northern flank of the anticline (Figs. 16 and 17).

In the northern part of the district a broad east-north-easterly plunging syncline is complicated by faulting, and with it are associated a number of complementary synclines and anticlines sub-parallel to the Lunedale–Staindrop, Wigglesworth and Butterknowle faults. The folds related to the Lunedale Fault are relatively gentle with surface dips which range from 2° to 22°. The folds associated with the Wigglesworth and Butterknowle faults include an anticline and a complementary syncline. These folds continue in association with the Butterknowle Fault into the Durham district (Smith and Francis 1967, p. 193; Magraw and others 1963, p. 181), and generally plunge, with minor culminations, to the north-east. Westwards, however, they die out in the vicinity of Cockfield beyond which the Butterknowle Fault is flanked on its south side by the parallel Copley Fault. Between the Wigglesworth and Butterknowle

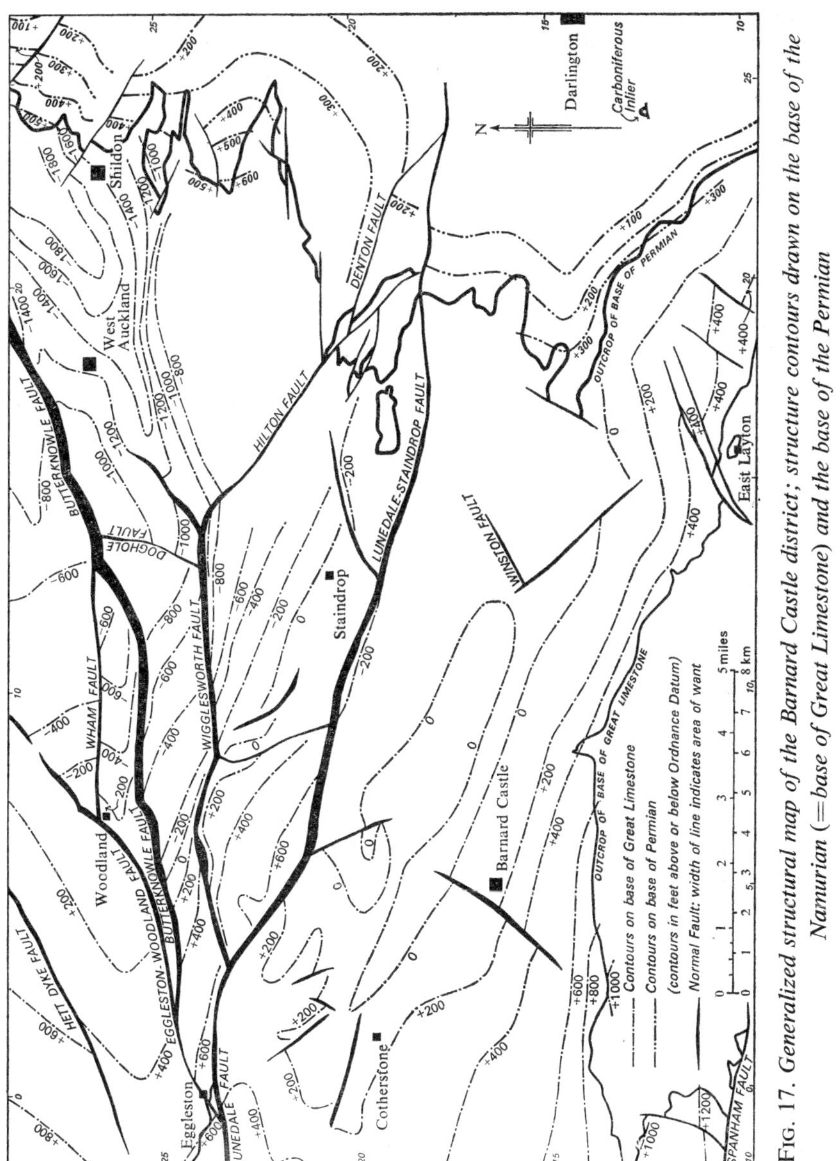

FIG. 17. *Generalized structural map of the Barnard Castle district; structure contours drawn on the base of the Namurian (=base of Great Limestone) and the base of the Permian*

Faults information from underground workings show that the strata dip at angles between 11° and 30°, the steeper dips generally being found on the southern limb of the main syncline. Immediately south of the Wigglesworth Fault the strata generally dip into the fault at angles between 20° and 30° but at John Pit [1156 2397] near Cockfield there is a slight upward flexure producing a shallow syncline against the south side of the fault. The highest recorded tectonic dip in the Carboniferous strata of the district of 44° to the north-north-west was recorded at the surface south of the Eggleston–Woodland Fault west of where the Butterknowle and Copley faults originate.

In the north-west of the district strata of high Namurian and low Westphalian age commonly dip at low angles of 2° to 3° generally in directions between south-east and south-south-east; exceptionally dips up to 15° are recorded locally (Fig. 16). Farther east dips of the same magnitude are observed but directions are more variable especially adjacent to the Wham, Butterknowle and associated faults.

The tilting and gentle folding which resulted from the presumed Tertiary earth-movements are shown by the structural contours on the base of the Permian (Fig. 17). North of the Lunedale–Staindrop Fault the Permian beds form a low easterly plunging anticline, whereas about 2 miles S of the fault there is a minor easterly plunging syncline. Dips in Permian strata are generally low, values of between 1° and 3° being common; locally, however, higher dips are recorded in the vicinity of faults.

FAULTS

The majority of the faults proved in the district are shown in Fig. 16. The evidence, adduced mainly from the adjoining Durham district (Smith and Francis 1967, p. 195), of differential throws between Carboniferous and Permian strata suggests that most of the faults were initiated in pre-Upper Permian times with subsequent post-Permian movements along the pre-existing lines. In the north-western part of the district one fault is associated with the Hett Dyke (p. 170), which has been dated, by comparison with the Whin Sill, at 295 ± 6 million years, suggesting an early Stephanian age (Fitch and Miller 1967, p. 247). Assuming, as seems likely, that the dyke was intruded along an existing fault plane, the faulting and the earlier folding must have occurred in late Westphalian times. Some of the major faults, however, which are thought to have controlled sedimentation in Carboniferous times, must also have moved at earlier dates, probably along pre-existing Caledonoid lines.

Because the important Lunedale–Staindrop Fault does not affect productive Coal Measures it is known in less detail than other faults. Strongly faulted strata were encountered in the lowest 110 ft of a borehole [0015 2342] on the line of the fault north-west of West Barnley Farm, Eggleston. On the western margin of the district in the River Tees [9955 2338] the throw of the fault is about 300 ft to the south, whereas about 5 miles to the east-south-east the throw reaches a possible maximum of the order of 700 ft. Farther east, in the area north-east of Selaby Hall [1530 1830], Gainford, the throw diminishes to about 100 ft; north-west of Darlington where the fault effects Permian strata, the throw may be even less.

The main faults in the northern and central parts of the district are all branches of the Lunedale Fault, and have a general trend between north and east. They are all normal faults with downthrows to the south, except for the Copley Fault

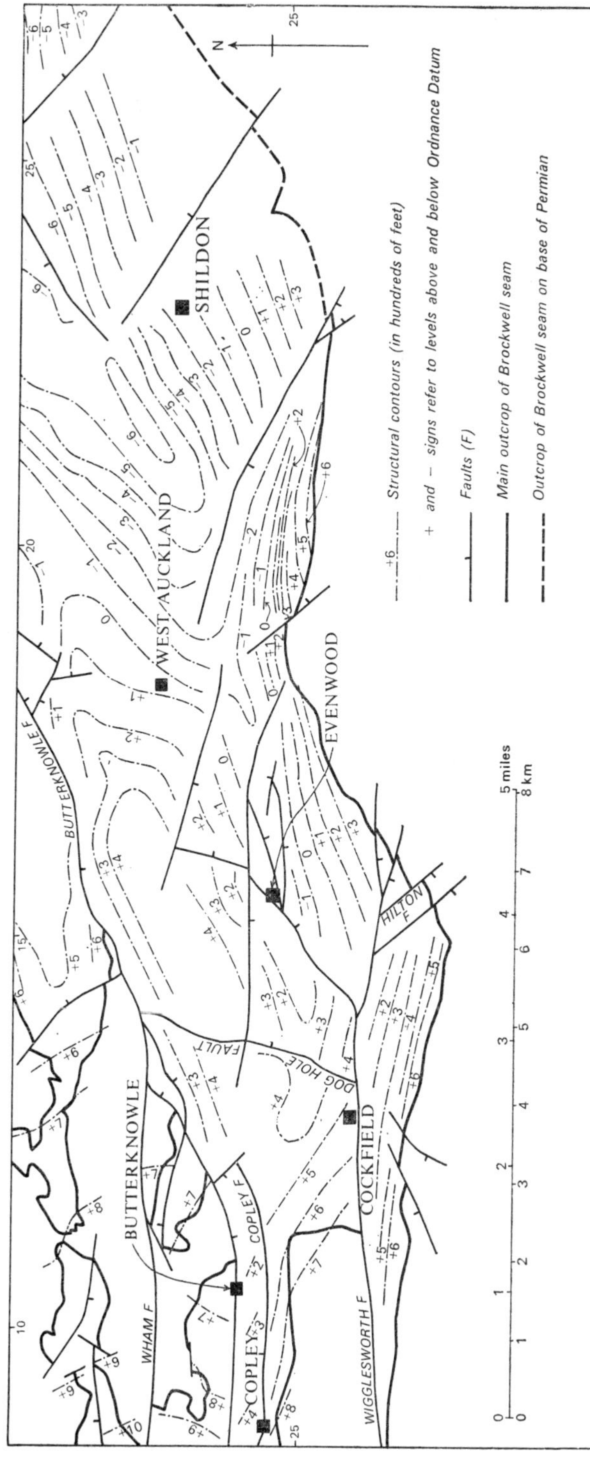

FIG. 18. *Plan of the Brockwell coal, showing chief folds and faults of the main coalfield area, Barnard Castle district*

which throws north towards the Butterknowle Fault. Most of them affect productive Coal Measures and consequently their characteristics are relatively well known.

The Eggleston–Woodland Fault trends between north-east and east-north-east and from it other major faults such as the Copley, Butterknowle and Wham faults diverge. The throw of the fault is generally small, of the order of 25 ft, but locally it may exceed 400 ft. Its near coincidence with the hinge-zone bounding the Alston Block has already been referred to.

The Hett Dyke Fault trends parallel to the Eggleston–Woodland Fault, and lies on its northern side being, therefore, the only significant fault in this district on the Alston Block; its throw, which is south-easterly, ranges from nothing in the area west of Eggleston Burn to about 250 ft where it crosses the northern margin of the district.

The Wham Fault follows an almost due easterly course for 4½ miles from north of Woodland to where it coalesces with the Butterknowle Fault; along this line the throw of the fault decreases eastwards from a maximum of about 144 ft in the west to 80 ft north of High Wham [1121 2666] and to approximately 50 ft further east. Not much is known about the hade of this fault except that it is probably low and at one point near its western extremity vertical.

The Butterknowle Fault has in general an east-north-easterly trend and its throw increases irregularly eastwards from an average of about 430 ft in the west to more than 800 ft near Bishop Auckland, with locally high values of 510 ft near Copley [0858 2583] and over 700 ft near Wigdan Walls Farm [1827 2845]. The hade of the fault generally ranges from 40° to 45°, as it does in districts to the north-east (Clarke 1962, p. 213; Smith and Francis 1967, p. 194). North-east of Copley [092 256] underground information suggests a hade of about 69° at an upper level decreasing with depth to 45°.

At the surface the Copley Fault branches from the Eggleston–Woodland Fault about ½ mile WSW of the Butterknowle Fault, and then trends east-north-east for some 6½ miles before dying out near Millfield Grange [1175 2553], Butterknowle; farther east its displacement is taken up by folding. By contrast with the Butterknowle Fault, the Copley Fault throws north, the amount diminishing eastwards, ranging from between 300 to 350 ft in the west to about 120 ft near The Slack [113 254], Butterknowle. The hade measured in underground workings tends to increase eastwards from about 14° near Cowley Farm [0640 2531] to approximately 27° east of Copley, where the Cleveland Dyke is intruded along it and irregular 'sill-like' intrusions have been recorded in abutting coal seams.

The Wigglesworth Fault, from its junction with the Lunedale–Staindrop Fault, follows a sinuous easterly course and has a southerly throw. The amount of throw varies but in general steadily increases from about 200 ft in the west to 550 ft in the east at Gordon House Colliery [1334 2403]. The hade of the fault is variable generally between 18° and 48°, while locally 59° has been recorded. At Cockfield the fault splits into two components; the northern branch has a throw of 240 ft and is associated with a monocline as far east as Evenwood, where it joins the Tees Hetton Fault. The southern branch, has a throw of 250 ft near Cockfield but this diminishes quickly eastwards, where the fault splits into several components some of which link up with the complex of faults around Hilton.

The Doghole or Lands Fault, throwing east, trends east-north-east for almost 2 miles from the eastern part of Cockfield Fell and links the Butterknowle and Wigglesworth faults. The throw is about 60 ft near Low Lands [135 250], decreasing both north and south to 17 and 45 ft respectively at its extremities. The hade decreases southwards from 32° in the north to 27°. On its eastern side it displaces the Wigglesworth Fault by 60 yd to the north and on Cockfield Fell [1331 2467] the Cleveland Dyke is shifted by its own width to the north.

A similar 'linking' fault joins the Wigglesworth and Lunedale–Staindrop faults; it follows a curved line, concave to the east, from near Penny Hill [0820 2386] southwards to near Bolton Hill. Its throw is also down east but, in contrast to the Doghole Fault, it ranges from 300 to 350 ft at both extremities and as little as 20 ft, locally, in between. These 'linking' faults are thought to result from accommodation of differential movements along the Butterknowle, Wigglesworth and Lunedale–Staindrop faults.

In addition to the main faults and the accommodation faults that link them there are two additional sets of normal faults. One set ranges in trend between north-north-east and east-north-east, with throws of up to 120 ft. An example is the Winston Fault, which, where it crosses the River Tees [1426 1627] east of Winston Bridge, has a throw of approximately 35 to 40 ft. The second set includes faults which trend approximately north-west to south-east and are often connected by accommodation faults which form low angles with them. Over most of the district their throws are less than 100 ft either to the north-east or south-west though, locally, in the north-east part of the district throws of over 100 ft are known. The Hilton Fault which has a throw to the south-west of between 200 and 300 ft is exceptional. This fault (Fig. 16), which is truncated by the southern branch of the Wigglesworth Fault (p. 166), forms the main component of a complex of little known faults in the Hilton, Ingleton and Killerby areas. It cuts the Wackerfield Dyke [1619 2284] north-west of Hilton and displaces it by half its width to the south, while farther south-west the dyke is terminated by a north-westerly trending branch of the Hilton Fault.

The Spanham Fault (Wells 1957, p. 250) has a trend broadly parallel to the Lunedale–Staindrop Fault and is associated with lead mineralization. It is further notable because its throw changes eastwards, in the vicinity of Eller Beck [998 103], from 150 ft to the north to 150 ft to the south.

In addition to the normal faults so far described a reversed fault affects the Busty and Brockwell seams in the workings of Tees Hetton Colliery, about 300 yd E of Field Houses [1774 2548], West Auckland and a wrench fault displaces an anticlinal axis (Fig. 16) associated with the Lunedale–Staindrop Fault about a mile north-west of Stainton [073 186]. This fault additionally exhibits changes in direction of throw.

Two faults with an ENE trend near East Layton in the south-east of the district have throws of 20 ft and 85 ft respectively in opposite directions, thus producing a small horst; the smaller of the two faults carries copper mineralization.

Superficial Structures

Structures due mainly to gravity such as landslips, solifluxion, hill creep, etc. are common in the district. In the main these structures are to be found in Pleistocene or more recent deposits. One notable exception occurs east of Eldon Hill Quarry [242 272] where the Lower Magnesian Limestone has been forced into tight irregular folds of 10 to 12 ft amplitude (Plate XIIIA). Smith (1970) has suggested that this could be due to the effects of slumping and sliding of the Lower Magnesian Limestone as a result of earthquake movements, although an alternative hypothesis by Mr. E. A. Francis has suggested that they may be interpreted as Pleistocene drag folds due to pressure of ice from the west.

D.A.C.M., J.H.H.

REFERENCES

BOTT, M. P. H. 1961. A gravity survey off the coast of north-east England. *Proc. Yorks. geol. Soc.*, **33**, 1–20.

——1967. Geophysical investigations of the northern Pennine basement rocks. *Proc. Yorks. geol. Soc.*, **36**, 139–68.

——and MASSON-SMITH, D. 1957. The geological interpretation of a gravity survey of the Alston Block and the Durham coalfield and interpretation of a vertical field magnetic survey in north-east England. *Q. Jnl geol. Soc. Lond.*, **113**, 93–117.

BURGESS, I. C. and HARRISON, R. K. 1967. Carboniferous basement beds in the Roman Fell district, Westmorland. *Proc. Yorks. geol. Soc.*, **36**, 203–25.

CARRUTHERS, R. G. 1938. Alston Moor to Botany and Tanhill: an adventure in stratigraphy. *Proc. Yorks. geol. Soc.*, **23**, 236–53.

CLARKE, A. M. 1962. Some structural, hydrological and safety aspects of recent developments in south-east Durham. *Trans. Instn Min. Engrs*, **122**, 209–31.

DUBEY, V. S. and HOLMES, A. 1929. Estimates of the ages of the Whin Sill and Cleveland Dyke by the helium method. *Nature, Lond.*, **124**, 477.

DUNHAM, K. C. 1948. The geology of the Northern Pennine Orefield; Vol. 1, Tyne to Stainmore. *Mem. geol. Surv. Gt Br.*

——1974. Granite beneath the Pennines in north Yorkshire. *Proc. Yorks. geol. Soc.*, **40**, 191–4.

——and STUBBLEFIELD, C. J. 1945. The stratigraphy, structure and mineralization of the Greenhow mining area, Yorkshire. *Q. Jnl geol. Soc. Lond.*, **100**, 209–68.

——DUNHAM, A. C., HODGE, B. L. and JOHNSON, G. A. L. 1965. Granite beneath Viséan sediments with mineralization at Rookhope, northern Pennines. *Q. Jnl geol. Soc. Lond.*, **121**, 383–417.

FITCH, F. J. and MILLER, J. A. 1967. The age of the Whin Sill. *Geol. Jnl.*, **5**, 233–50.

FUNNELL, B. M. 1964. The Tertiary period. *Q. Jnl. geol. Soc. Lond.*, **120** **(S)**, 179–91.

HARRY, W. T. 1950. Basement Carboniferous in Upper Teesdale, N. Yorks. *Geol. Mag.*, **87**, 297–9.

HILLS, E. S. 1953. *Outlines of structural geology*. London.

HUDSON, R. G. S. 1938. The geology of the country around Harrogate. *Proc. Yorks. geol. Soc.*, **49**, 306–30.

HULL, J. H. 1968. The Namurian stages of north-eastern England. *Proc. Yorks. geol. Soc.*, **36**, 297–308.

JOHNSON, G. A. L. and DUNHAM, K. C. 1963. *The geology of Moor House*. Monograph of the Nature Conservancy. xviii + 182, London: H.M.S.O.

MAGRAW, D., CLARKE, A. M. and SMITH, D. B. 1963. The stratigraphy and structure of part of the south-east Durham coalfield. *Proc. Yorks. geol. Soc.*, **34**, 153–208.

MARR, J. E. 1921. On the rigidity of north-west Yorkshire. *Naturalist*, 63–72.

MILLS, D. A. C. and HULL, J. H. 1968. The Geological Survey borehole at Woodland, Co. Durham (1962). *Bull. geol. Surv. Gt Br.*, No. 28, 1–37.

READING, H. G. 1957. The stratigraphy and structure of the Cotherstone Syncline. *Q. Jnl geol. Soc. Lond.*, **113**, 27–56.

ROWELL, A. J. and SCANLON, J. E. 1957a. The Namurian of the north-west quarter of the Askrigg Block. *Proc. Yorks. geol. Soc.*, **31**, 1–38.

—— ——1957b. The relation between the Yoredale Series and the Millstone Grit on the Askrigg Block. *Proc. Yorks. geol. Soc.*, **31**, 79–90.

SHOTTON, F. W. 1935. The stratigraphy and tectonics of the Cross Fell Inlier. *Q. Jnl geol. Soc. Lond.*, **91**, 639–704.

SMITH, D. B. 1970. Submarine slumping and sliding in the Lower Magnesian Limestone of Northumberland and Durham. *Proc. Yorks. geol. Soc.*, **38**, 1–36.

——and FRANCIS, E. A. 1967. The geology of the country between Durham and West Hartlepool. *Mem. geol. Surv. Gt Br.*

TROTTER, F. M. and HOLLINGWORTH, S. E. 1928. The Alston Block. *Geol. Mag.*, **65**, 433–48.

VERSEY, H. C. 1927. Post-carboniferous movements in the Northumbrian fault block. *Proc. Yorks. geol. Soc.*, **21**, 1–16.

WELLS, A. J. 1955. The development of chert between the Main and Crow Limestones in north Yorkshire. *Proc. Yorks. geol. Soc.*, **30**, 177–96.

——1957. The stratigraphy and structure of the Middleton Tyas–Sleightholme Anticline, north Yorkshire. *Proc. geol. Assoc.*, **68**, 231–54.

WILSON, A. A. 1960. The Carboniferous rocks of Coverdale and adjacent valleys in the Yorkshire Pennines. *Proc. Yorks. geol. Soc.*, **32**, 285–316.

WOOLACOTT, D. 1923. A boring at Roddymoor Colliery, near Crook, Co. Durham. *Geol. Mag.*, **60**, 50–62.

Chapter 7

INTRUSIVE IGNEOUS ROCKS

THE distribution of the outcrops of the igneous intrusions known in the Barnard Castle district is shown in Fig. 19. At several localities dykes are seen to extend laterally to form sills, while one dyke expands locally to form a small laccolith. The dykes are all of basic rocks and are associated either with the emplacement of the Great Whin Sill during late Carboniferous times or with the emplacement, in early Tertiary times, of the great system of dykes radiating from volcanic centres in western Scotland.

The Great Whin Sill does not crop out but is believed to be present at depth in the north-west part of the district. It has been proved in one borehole (Mills and Hull 1968, p. 1).

A description of individual intrusions follows; their ages are discussed in a separate section at the end of the chapter.

THE HETT DYKE

The Hett Dyke (Fig. 19) is the principal member of a group of dykes well known in north-east England. The group includes the Brandon and Ludworth dykes of the Durham (27) district to the north-east (Smith and Francis 1967, p. 187) and the Connypot Dyke, the Greengates Dyke and two small dykes seen on the Pennine Escarpment near Long Fell Mine in the Brough under Stainmore (31) district to the west.

The Hett Dyke, striking N 70° E with an average width of 40 ft, has been intruded along a southward-throwing fault, and is exposed in Eggleston Burn [9857 2582] and in quarries north of Knotts Plantation [9896 2605 to 9910 2610], 1½ miles NNW of Eggleston Parish Church. Neither dyke nor fault is known at the surface any farther west than a locality 300 yd W of Eggleston Burn. Typically a fine-grained quartz-dolerite, the dyke is crudely layered, exhibits spheroidal structure, and is well-jointed. At Knotts Plantation it contains xenoliths of white-bleached sandstone. The normally dark grey shales in contact with and up to 200 ft from the dyke have become olive-green, baked and porcellanous. Sandstones have been bleached and rendered hard and quartzitic, though the alteration is not always obvious since unmetamorphosed white quartzitic sandstones occur away from the intrusion.

The width of the zone of alteration of sandstone varies from a few inches up to 200 ft, and this presumably reflects differences of porosity in the sandstones. One contact of dyke and sandstone in Knotts Plantation [9896 2605] is shattered, and the shatter-belt is enriched in iron which has since weathered to limonite. J.H.H.

Petrography. The petrography of the dyke has been described by Teall (1884a, pp. 228–30) and by Holmes and Harwood (1928, pp. 502, 526–8). The rock is dark greyish green with a sub-conchoidal fracture. It is generally fine-

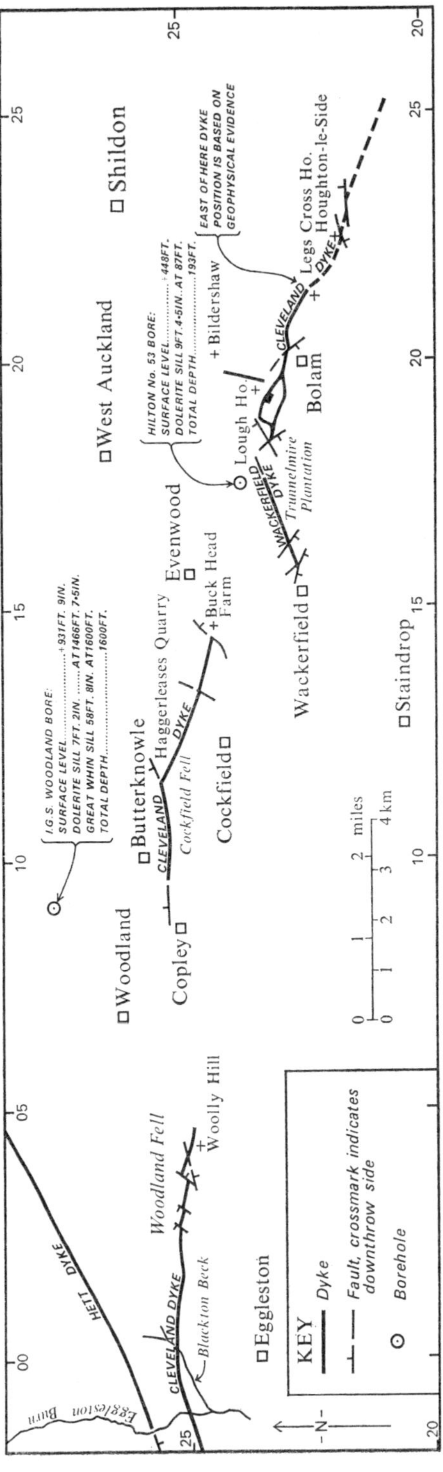

FIG. 19. *Sketch-map showing the distribution of igneous rocks in the Barnard Castle district*

grained, although the centre of the intrusion is commonly coarser and apparently more prone to weathering than the margins. Sporadic amygdales are filled with calcite (Teall 1884a, p. 229). A specimen (E 33151)[1] from Knotts Plantation Quarry [9907 2609], approximately 1½ miles N of Eggleston, consists of an even-grained ophitic felt of labradorite laths (0·4 to 0·6 by 0·1 mm). These have a refractive index (β) = 1·563 ± 0·003 and are near Ab_{42} An_{58} in composition. Pale green clinopyroxene, near augite or hypersthene-augite, is mainly anhedral and granular, with refractive index (β) = 1·696 ± 0·003. Sparser, coarser clinopyroxene shows euhedral outlines. Orthopyroxene forms scattered prisms with frayed terminations, up to 1 mm in length, and is largely altered to clay-minerals. Skeletal crystals and cubes of titanomagnetite are plentiful. Quartz, chlorite, alkali feldspar and cryptocrystalline mesostasis occur in the interstices of the ophitic felt, the mesostasis indicating a tholeiitic trend (Dunham 1948, p. 59). A modal analysis of this specimen gave, in volume per cent: plagioclase 47, clinopyroxene 28, enstenite 1, quartz 1, sulphide 1, mesostasis 8, titano-magnetite 7, chlorite 6, carbonate ½, hornblende and biotite, trace. As noted by Teall (1884a, p. 229) feldspar laths penetrate the clinopyroxene, and clearly crystallized before it. Phenocrysts are generally sparse, though Holmes and Harwood (1928, p. 502) described from this dyke a limpid glomeroporphyritic aggregate of highly calcic plagioclase near Ab_{10} An_{90} zoned out to Ab_{70} An_{30} and cited it as evidence of reaction with the enclosing magma until a state of physico-chemical equilibrium was reached. This also points to an initial more cafemic parent magma which permitted the crystallization of the highly calcic plagioclase and phenocrysts. R.K.H.

The Wackerfield Dyke

This dyke (Fig. 19) is exposed about 2 to 2½ miles NE of Staindrop. It was first referred to by Kirkby and Duff (1872, pp. 195–6) who considered it an off-shoot of the Cleveland Dyke. Later, a detailed account of the field relations and petrography was given by Holmes and Smith (1921, pp. 440–54), but the exposures which they saw are now largely obscured. The quarry [1586 2269] at the western end of the dyke was worked for road metal as early as 1830, but it was apparently worked out by 1872. The dyke can be traced for nearly a mile from an old quarry [1586 2269] north-east of Wackerfield to the western edge of Trunnelmire Plantation [1732 2325] where it forms a well-marked feature, but its continuation beyond these localities in either direction has not been proved. Except for a minor deflection near the western end of the dyke the strike of N70°E is similar to that of the Hett Dyke. The Wackerfield Dyke, like the Hett Dyke, is intruded along a fault which throws to the south; it is 20 to 25 yd wide along most of its outcrop, but at either end the intrusion swells out laterally to form sill-like structures, the full limit of which cannot be accurately established, but which are evidently associated with a thinning of the dyke itself. At the western end the apparent termination of the intrusion at the surface coincides with a series of faults, the largest of which strikes south-east and passes to the north of Hilton village. Mining information suggests that the dyke has a slight hade to the south, and is itself cut by a fault [1621 2284] west-south-west of Hilton Moor. The metamorphic effects on the adjacent sedimentary rocks are

[1]Sliced Rock Collection of the Institute of Geological Sciences.

similar to those noted for the Hett and Cleveland dykes. The Brockwell Coal is cindered for an unknown distance on the south side of the dyke, 500 yd WSW of Hilton Moor. D.A.C.M.

Petrography. The rock is typically dark grey quartz-dolerite, of fairly fine and even grain, becoming slightly finer nearer the margins. Sparse plagioclase phenocrysts, up to 3 mm long, show rounded margins indicating partial absorption, and approach anorthite in composition, some crystals being zoned outwards to oligoclase (Holmes and Smith 1921, p. 447). Groundmass plagioclase laths in subophitic texture are labradorite (near $Ab_{40}An_{60}$), as confirmed optically in a specimen (E 33152) from a quarry [1611 2279], ½ mile NNE of Wackerfield. Pyroxenes comprise granular to subhedral pale greyish green augite or hypersthene-augite scattered between the labradorite laths, and coarser prisms of hypersthene. Holmes and Smith (1921, p. 448) noted that granular enstatite-augite commonly fringes the orthopyroxene. Accessory minerals include skeletal titanomagnetite, intersertal quartz and sparse apatite. Mesostasis is mainly cryptocrystalline and charged with chloritic and other inclusions. Other late-stage or hydrothermal components include carbonates, chlorite and pyrite. Modal and chemical analyses are given by Holmes and Smith (1921, pp. 449–52) who also note that two main late-stage processes have affected the dyke; these are silicification and carbonatization. The alteration is mainly marginal, locally showing deep and irregular penetration into the intrusion. These alteration processes were probably of late-magmatic or hydrothermal origin (to which the infilling of amygdales in the Great Whin Sill and Hett Dyke has been attributed), and they have rendered the dyke more prone to weathering. R.K.H.

The Cleveland Dyke

The *en echelon* suite of dykes collectively known as the Cleveland–Armathwaite Dyke traverses northern England in a discontinuous line from just south of Carlisle to the North Yorkshire moors. East of the Pennines it is usually referred to as the Cleveland Dyke. Within the present district the dyke (Fig. 19) was first referred to by Kirkby and Duff (1872, pp. 193–5), while Teall (1884a, pp. 209–27) gave the first detailed account of its field relations and petrography. The following account traces the course of the dyke from west to east across the district.

Between the western margin of the district and Eggleston Burn the dyke has a strike averaging N75°E, approximately parallel to, and half a mile south of, the Hett Dyke. It is well exposed in Bell Sike [9857 2496] and again in Eggleston Burn [9897 2512] where it is about 40 ft wide. Beyond Blackton Beck, the strike veers round to about S5°E, and the dyke follows a somewhat irregular course for a distance of 2¾ miles to its intersection with the Butterknowle Fault about 600 yd E of Woolly Hill [0442 2494]. On the southern part of Woodland Fell the outcrop is broken in three places, probably by faults, into segments *en echelon*, the displacement in each case being to the north. The dyke is not seen at the surface again for nearly 3½ miles and it reappears in the Grewburn area, ¾ mile E of Copley, where it is intruded along the northward-throwing Copley Fault, which strikes N86°E. In the intervening ground, the dyke may be present at no great depth from the surface, since at Copley and again between Copley and Grewburn it has been proved in workings in the Brockwell Coal. J.H.H.

Between Grewburn and The Slack, ¼ mile ESE of Butterknowle village, the dyke, though nowhere exposed at the surface, has been recorded with a maximum width of 6 ft in colliery workings. Sheets of 'whinstone' and zones of cindered coal are reported in old workings in the Busty seam just south of Grewburn [1004 2540] and near Gaunless Mill [1142 2552]. J.H.H.

Near where the dyke crosses the Gaunless valley [1154 2549] it changes direction abruptly to the east-south-east, and diverges from the Copley Fault. Near such changes of strike lateral apophyses and strings commonly occur. For instance, on the north side of the Gaunless valley [1097 2463], about ¾ mile S of Butterknowle village, a dyke stated to be 9 ft in width, and striking N 58° E, is recorded for a distance of 230 yd in old workings in the Marshall Green Seam, but the relationship between this intrusion and the Cleveland Dyke is not established.

East of The Slack [113 254], the outcrop can be traced continuously for 1¾ miles across the northern part of Cockfield Fell, terminating against the Wigglesworth Fault Belt just west of Buck Head [1439 2443]. Except for a short distance on either side of the former railway line, the dyke has been extensively quarried in this section, chiefly for road metal. The earliest recorded workings were between 1780 and 1790, and all the quarries are now abandoned. At Haggerleases Quarry [1169 2540], at the western end of Cockfield Fell, the dyke has been worked for a distance of about 360 yd E of the River Gaunless. Teall (1884a, p. 211) states that "where the quarry was opened near the river, the dyke was only 15 ft in width; 170 yd to the east it was 29 ft wide, hading south-south-west at 20°"; at the eastern limit of working its width is now seen to be 75 ft. Teall also recorded a 4-ft dyke to the north of the main dyke near the quarry entrance. The jointing displayed on the quarry wall (Plate XVIB) is irregular, and contrasts strongly with that at Bolam Quarry (Plate XVII) where it is columnar.

The dyke has been worked to a depth of up to 60 ft for a distance of over 2460 yd E of the old railway line, and considerable quantities of rock have been removed (Plate XVIA). Near Fell Houses [1295 2489] it is 90 ft wide, and is intruded along a small northward-throwing fault. About ¼ mile E of Fell Houses the dyke is displaced by its own width to the north by the Doghole or Lands Fault.

North and west of Esperley Lane Ends, the dyke is vertical but the crop ends abruptly about 200 yd W of Buck Head [1439 2443]. Teall (1884a, p. 210) states "At the time of our visit (Jan. 1882). . . we saw a phenomenon which must be of frequent occurrence, but which cannot very often be directly observed, viz. the vertical dying-out of the dyke beneath the stratified rocks. The whinstone was seen to terminate upwards very abruptly in the form of a low and somewhat irregular dome, over which the Coal Measures shales passed without any fracture and only with a slight upward arching." This unfortunately is no longer visible. Kirkby and Duff (1872, p. 195) state that "a drift was put through the dyke near to Buckhead Farm, at which point it was not found in one mass, but split up into three or more walls of whinstone with pieces of coal strata between." The disappearance of the dyke is clearly related to the Wigglesworth Fault Belt just to the east of this locality. Of great significance is the marked expansion of the intrusion immediately west of the surface termination; and although direct observation is no longer possible, information from mining sources suggests that it has expanded into a minor apophysis, the expansion being wholly on the

(L 348)

A. Old quarry in Cleveland Dyke, View west-north-west across Cockfield Fell

PLATE XVI

B. Cleveland Dyke. South wall of Haggerleases Quarry near Butterknowle. Note irregularity of jointing

(L 347)

A. Bolam Laccolith (Cleveland Dyke). South face of western pit, Bolam Quarry (1963). Note columnar jointing. Westphalian sediments form quarry floor (L 35?

PLATE XVII

B. Bolam Laccolith (Cleveland Dyke). North face of western pit, Bolam Quarry. Tholeiitic dolerite split into two leaves by Westphalian sediments (L 35

south side of the assumed dyke line. West of Buck Head the dyke appears to have a distinct hade to the north, as it is absent in workings from the south in the Brockwell Coal, which lie directly beneath the surface position of the dyke.

Between Buck Head and Sharpley Plantation, 1 mile NW of Bolam, there is no indication of the dyke. Furthermore, it has not been encountered in colliery workings despite the fact that seams have been extensively worked almost down to sea-level to the north and north-west of Wackerfield. The log of Hilton No. 53 Bore* [1748 2383] near Bolton Garths and slightly north of the projected surface trace of the Cleveland Dyke records 9 ft 4½ in of bluish grey jointed whinstone with its base at 87 ft from the surface. The relationship between this occurrence and the Cleveland Dyke is not known.

To the north-west of Bolam the dyke, with an overall strike of S80°E, crops out for a distance of about 1900 yd between Sharpley and Crag Plantations. Throughout this area it is vertical and does not exceed 30 ft in width, but west of Crag Lane it has expanded laterally to form an intrusion of laccolith type, now largely quarried away, up to 1150 yd long, 375 yd wide, and with a maximum thickness of 50 ft. The upper surface of this intrusion, here termed the Bolam Laccolith (Plate XVII, Fig. 20), appears to have formed a low broadly elongated dome. Although the laccolith was thickest on the southern side of the dyke (or feeder) its lateral extension was greatest to the north. Its base, broadly conformable with the sedimentary floor, showed only local steps and minor irregularities. Parts of the floor are still visible throughout much of the quarry, and the sediments dip north at between 1° and 4°. Along the northern edge [1888 2333] of

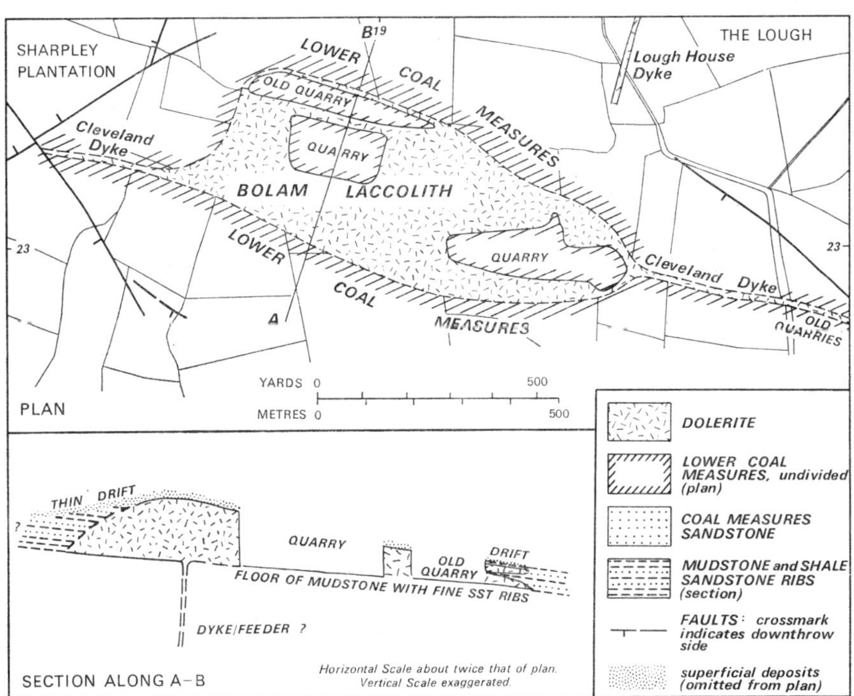

FIG. 20. *Plan and diagrammatic section of the Cleveland Dyke and Bolam Laccolith at time of survey* (1961).

the north-western quarry, the intrusion is split into two sills up to 6 ft thick and 8 ft apart, the upper sill dying out before the lower (Plate XVIIB). This exposure is now largely obscured. Kirkby and Duff (1872, p. 195) state that "fragments of coal are sometimes found in the whin very beautifully coked". In the eastern quarry, a small fracture trending east-south-east apparently throws down the base of the sill 6 to 12 ft to the north, but elsewhere the intrusion does not appear to be affected by faulting. The dyke feeding the intrusion is nowhere exposed. At the western end, the laccolith becomes attenuated and forms a narrow dyke just east of Sharpley Plantation, and this is apparently cut off by a fault [1828 2320] near the edge of the plantation.

Quarry operations have ceased and only limited reserves of stone remain. However, good exposures of both the dolerite and the sedimentary floor are still visible in the quarry which now covers the whole intrusion.

A minor dyke, up to 5 yd wide, here named the Lough House Dyke, strikes north-north-east from near the margin of the laccolith for a distance of 1000 yd, dying out south-east of Bildershaw Grange [1969 2420]. No connexion with the laccolith or the Cleveland Dyke can be proved. Poor exposures of this dyke can be seen [1953 2342] 150 yd SE of Lough House and 470 yd SE of Bildershaw Grange.

East of Crag Plantation, exposure of the Cleveland Dyke is limited to two quarries. In the first, at Brownside [203 228], the dyke strikes east-north-east over some 250 yd and does not exceed 30 ft in width. In the second and more easterly quarry the workings extend over some 500 yd E of the B6275 Piercebridge to West Auckland road on an east south-east line. In both quarries the dyke tapers at each end of the workings, but features and an isolated intervening exposure [2075 2272] suggest that they are connected by a thin stringer. The arcuate course of the dyke system in this area is probably the result of tensional strains set up by neighbouring faulting. There is no known exposure of the dyke between Leg's Cross House and Coatham Stob, 12 miles to the east-south-east on the Stockton (33) Sheet, where it cuts Triassic rocks. An aeromagnetic and magnetometer survey of the area by the Geophysical Department, however, has established the line of the Cleveland Dyke east-south-east of Leg's Cross House, and this has been indicated on the maps. In this sector the dyke probably lies at no great depth from the surface. Although the evidence is slight, the survey indicates a deviation from normal in the course of the dyke south-east of Houghton-le-Side, and this is probably due to faulting.

Metamorphic effects of the dyke are chiefly confined to baking and bleaching. In Haggerleases Quarry [1155 2549] shales near the entrance to the quarry and within a few feet of the assumed dyke wall are baked, and highly altered. At the base of the workings on the south side of the quarry 300 yd from the entrance, an old adit [1175 2537] (now obscured by rubbish) penetrated the dyke wall into alternating mudstones and micaceous sandstones, which show very little alteration, except for a hardening effect up to 2 ft from the dyke wall which is vertical hereabouts with only minor irregularities, and is separated from the country rock by a ferruginous trap up to 1 in thick. Similar minor effects are visible along the course of the Cleveland Dyke across Cockfield Fell. West of Fell Houses [1295 2489] the southern face of the quarry cut is formed of a massive sandstone, hardened locally and bleached.

The coking effect of the dyke on coal seams has been noted at several places. The description by Witham (1856, p. 438) of such alteration is one of the few remaining records relating to coal workings on Cockfield Fell. The dyke may account for a zone up to 100 yd wide of bad coal, and in the particular locality (unknown) to which Witham refers he states "the coal spoilt by the dyke on one side is 25 yd bad coal, 16 yd cinder, 10 yd sooty substance" picturesquely described as 'dawk' or 'swad'. In Hilton No. 53 Bore* [1748 2383] a coal, 1½ in thick overlying a dolerite sill possibly associated with the Cleveland Dyke (p. 175), was altered to a hard, dark grey, prismatic mineralized coke possessing a columnar structure. A proximate analysis made in 1957 by the Coal Survey Laboratory, Durham Division, National Coal Board, showed: moisture 11·5, volatile matter (less moisture) 4·3, fixed carbon 59·4, ash 24·8, CO_2 3·23.

Shale and mudstone beneath the floor of the Bolam Laccolith show strong alteration. The shale is baked, slickensided and locally has a polished appearance; the sandstone is hardened. At the western end of the dyke workings [2087 2264] north of Leg's Cross House, the north wall of the dyke is a bleached, highly decomposed, quartz-sandstone with iron-rich concentrations.

Petrography. The rock is typically a massive, hard, dark grey to bluish grey porphyritic dolerite, slightly finer-grained near the margins of the dyke. It rapidly acquires a darker bloom on exposure to the air. Both its colour and porphyritic nature readily distinguish this rock from that of the Hett Dyke. The dyke is well-jointed commonly with a strongly developed columnar structure (Plate XVIIA), a feature clearly seen in the quarries north-west of Bolam. In places it is spheroidal with pillow-like structures as for example in Eggleston Burn; where its mode is that of a dyke, the jointing is often irregular (Plate XVIB). The porphyritic nature of the rock is very evident towards the margins of the dyke, where the phenocrysts are especially prominent. Near the margins, the rock is commonly carbonate-rich, and this was attributed by Teall (1884a, p. 213) to secondary processes. Teall describes a more advanced state of alteration at the contact, where a brown layer, rich in iron oxides, is developed, commonly sharply divided from a paler grey band rich in kaolinized feldspar several inches thick which shades off into the main mass of the dyke. D.A.C.M.

Specimens of the Cleveland Dyke from the Eggleston and Bolam areas show broadly similar characters, identifying them with the tholeiites described previously (Teall 1884a, pp. 213–27; Holmes and Harwood 1929, pp. 36–41; Dunham 1948, p. 60). The specimen (E 33150) from a quarry [9895 2511] in Eggleston Burn, 1 mile NW of Eggleston, is, however, highly saussuritized. A markedly porphyritic texture, with pellucid feldspar phenocrysts up to 5 mm across, is the main characteristic of the dyke. The groundmass is mainly microcrystalline to cryptocrystalline. The phenocrysts are euhedral or subhedral and show in places marginal corrosion and replacement by the groundmass. Oscillatory zoning is common with inner zones charged with submicroscopic inclusions. Refractive index measurements indicate a main composition in the labradorite range. Holmes and Harwood (1929, p. 36) following Teall (1884a, pp. 217–8) noted more calcic (anorthite) phenocrysts, but these are apparently rarer than labradorite. Orthopyroxene forms rather sparse phenocrystic prisms, up to 3 mm in length, with frayed terminations. These show weak pleochroism where unaltered: X=pale brown; Z=pale green. The groundmass comprises an ophitic framework of labradorite laths, near $Ab_{40}An_{60}$ (refractive index β= 1·565), granular clinopyroxene, skeletal titanomagnetite, and an intersertal devitrified mesostasis with late-stage and secondary products. The granular pyroxene is pale greyish green augite while sparser, coarser (up to 0·5 mm diameter) euhedral or subhedral grains approach pigeonite. Titanomagnetite is the principal accessory mineral, with minute amounts of apatite and biotite. Quartz is commonly associated with carbonates and may in part be secondary. Pyrite occurs with calcite in vesicles and veinlets. The mesostasis occurs in a variety of forms; as cryptocrystalline, devitrified, pale grey aggregates charged with microlites, globulites and feldspar laths, as darker turbid areas traversed by feldspar microlites; and as rather rounded subvesicular aggregates up to 5 mm diameter, of late-magmatic origin and composed of feldspar needles, pyroxene granules and opaque dust. The feldspar microlites commonly show flow-structure and the segregations are fringed with acicular feldspar. Details of mesostasis structures are given by Teall (1884a, pp. 221–3). Modal analyses of six specimens from the Bolam area are given in Table 3, and show close agreement with published data. A chemical analysis of a specimen from Bolam is given in Holmes and Harwood (1929, p. 39).

R.K.H.

TABLE 3

Modal Analyses and Specific Gravity Determinations of the Cleveland–Armathwaite Dyke

	Composition in volume per cent, of six specimens[1]			*Mean values of four modal analyses of the Cleveland Dyke*[2]
	Range	*Mean*	*Standard deviation*	
Plagioclase	24–36	34	4·4	34·1
Clinopyroxene	17–26	22	2·8	24·7
Orthopyroxene	tr–1	—	—	—
Titanomagnetite	4–9	6	1·7	—
Quartz	1–6	2	1·8	—
Carbonates	tr–2	½	—	—
Devitrified mesostasis ..	31–43	35	4	36·3
Augite-feldspar colonies ..	tr–2	½	—	—
Specific gravity[3]	2·76–2·80	2·78	0·02	2·77[4]

[1]All from the Bolam area.
[2]Holmes and Harwood (1929, p. 38).
[3]Determination by Mr M. D. Traynor on five specimens from the Bolam area.
[4]Holmes and Harwood (1929, p. 39); determination on analysed specimen.

THE GREAT WHIN SILL AND ASSOCIATED SILLS

The Great Whin Sill does not crop out within the district, but is exposed about a mile to the west of the district boundary in a quarry south of Middleton-in-Teesdale. Here over 40 ft of quartz-dolerite are exposed representing the upper part of an intrusion known to be at least 140 ft thick (Dunham 1948, p. 52). The sill appears to have been intruded at about the horizon of the Single Post Limestone.

The Great Whin Sill was penetrated immediately beneath the upper leaf of the Three-Yard Limestone at a depth of 1541 ft 4 in by the Woodland Bore* [0910 2770] (Mills and Hull 1968, p. 9) and 58 ft 8 in of dolerite had been proved on completion of the borehole at 1600 ft. In hand specimen, the dolerite is massive, dark bluish grey and mainly of fine- to medium-grain. The rock was coarser and porphyritic towards the bottom of the borehole, suggesting that less than half the total thickness was proved. The metamorphic effects on the strata in the borehole extended to little more than 18 ft above the contact; elsewhere in the north of England, alteration is seen up to 100 ft above the sill (Dunham 1948, p. 58). A higher quartz-dolerite sill, 7 ft 2 in thick, was encountered in the borehole at a depth of 1459 ft 5½ in, its base some 75 ft above the Great Whin Sill and about 38 ft below the Four-Fathom Limestone. This intrusion may be compared with the 5 ft 10½ in of Little Whin Sill proved in the Three-Yard Limestone, some 403 ft above the Great Whin Sill, in the Rookhope Bore in the Alston (25) district (Dunham and Kaye 1965, p. 234). At both localities the two sills are petrologically similar. Full details of petrography and metamorphic effects of both the sills in the Woodland Bore have been published (Harrison 1968, pp. 38–54).

The petrography and chemistry of the Great Whin Sill were described by Teall (1884b, pp. 640–57), Holmes and Harwood (1928, p. 494), Smythe (1930, p. 16), Tomkeieff (1929, p. 100), Dunham (1948, pp. 51–8) and Harrison (1968, pp. 38–54). They are also described in the Geological Survey memoirs on Belford, Holy Island and the Farne Islands (Gunn and others 1927, pp. 110–2) and Brampton (Trotter and Hollingworth 1932, pp. 114–6). D.A.C.M., J.H.H.

AGE OF THE INTRUSIONS

The igneous rocks of this district belong to two episodes of intrusion; an earlier one associated with the emplacement of the Great Whin Sill and the Hett and Wackerfield dykes and probably of Upper Carboniferous age; and a later one of Tertiary age which gave rise to the great system of dykes radiating from volcanic centres in western Scotland and which includes the Cleveland Dyke. Dyke trends in northern England have long been recognized as reliable indicators of age and this district is no exception to the general case; ENE for late Carboniferous, i.e., the Hett and Wackerfield dykes, and WNW for Tertiary, i.e., the Cleveland Dyke.

The youngest strata cut by the Hett Dyke in the present district are of Lower Coal Measures age; elsewhere the dyke transgresses Middle Coal Measures, but does not penetrate Upper Permian or later rocks (Dunham 1948, p. 59). Stratigraphical evidence alone therefore places the age of the dyke between Middle Coal Measures and Upper Permian.

The Wackerfield Dyke resembles the Hett Dyke in both texture and composition, and is a dolerite mineralogically identical with the Whin Sill (Holmes and Smith 1921, p. 446). It is intruded into rocks of Lower Coal Measures age, and although not proved, contemporaneity with the Hett Dyke and the Whin Sill is implied.

Recent palaeomagnetic studies on the Wackerfield Dyke (Tarling and Mitchell 1973, p. 427) confirm an Upper Carboniferous age for this intrusion. There seems little doubt therefore that it is contemporaneous with emplacement of both the Hett Dyke and Great Whin Sill.

The age of the Great Whin Sill is deduced in a similar manner to that of the Hett Dyke. In Lunedale, and in the Woodland and Rookhope bores, the sill is intruded into Viséan rocks, and associated sills are found in beds of Coal Measures age in the Brampton district (Trotter and Hollingworth 1932, p.114). These facts, together with the evidence of derived pebbles of alleged Whin Sill (Holmes and Harwood 1928, pp. 532–3; Dunham 1932, p. 425) in the Upper Brockram of the Vale of Eden, set stratigraphical limits of Upper Carboniferous and Permian for the time of intrusion.

An independent isotopic age determination using the helium method by Dubey and Holmes (1929, pp. 794–5) gave an age of 182 m.y. Holmes (1959, p. 199) made a correction for loss of helium and arrived at an age of 280 m.y., which would place the date of the intrusion near the Carboniferous–Permian boundary. More recent work by Miller and Mussett (1963, pp. 547–53) using potassium-argon methods gave an age of 295$\pm$19 m.y. when taking the average of the best of three ages. This age, accepted by Fitch and Miller (1964, p. 169), suggests intrusion shortly after the maximum of the principal Variscan earth-movements, and early Stephanian in age.

The youngest rocks traversed by the Cleveland Dyke in this district are of early Middle Coal Measures age. Outside the district, however, the dyke cuts Middle Jurassic rocks in the Cleveland area of Yorkshire (Wilson 1948, p. 70) and Triassic rocks in both the Vale of Eden (Trotter and Hollingworth 1932, pp. 120–1; Dunham 1948, p. 60) and the lower part of the Tees Valley at Coatham Stob near Stockton-upon-Tees. An isotopic age determination using the helium method, by Dubey and Holmes (1929, pp. 794–5) gave a minimum age of 26 m.y., but this was later modified by Dunham (1948, p. 60) to 28 m.y. The minimum age must be corrected for helium loss, and assuming (Holmes 1959, p. 199) that 70 per cent of the helium generated has been removed, a more realistic age would be around 40 m.y. which would place the date of the intrusion near the Eocene–Oligocene boundary on Holmes' revised scale. J.H.H., D.A.C.M.

References

Dubey, V. S. and Holmes, A. 1929. Estimates of the ages of the Whin Sill and Cleveland Dyke by the helium method. *Nature, Lond.*, **123,** 794–5.

Dunham, K. C. 1932. Quartz-dolerite pebbles (Whin Sill type) in the Upper Brockram. *Geol. Mag.*, **69,** 425–7.

——1948. The geology of the Northern Pennine Orefield; Vol. 1, Tyne to Stainmore. *Mem. geol. Surv. Gt Br.*, 51–61.

Dunham, A. C. and Kaye, M. J. 1965. The petrology of the Little Whin Sill, Co. Durham. *Proc. Yorks. geol. Soc.*, **35,** 229–76.

Fitch, F. J. and Miller, J. A. 1964. The age of the paroxysmal Variscan orogeny in England. *Q. Jnl geol. Soc. Lond.*, **120(S),** 159–73.

Gunn, W. 2nd ed. by Carruthers, R. G., Dinham, C. H., Burnett, G. A. and Maden, J. 1927. Geology of Belford, Holy Island and the Farne Islands. *Mem. geol. Surv. Gt Br.*

Harrison, R. K. 1968. Petrology of the Little and Great Whin Sills in the Woodland Borehole, Co. Durham. *Bull. geol. Surv. Gt Br.*, No. 28, 38–54.

Holmes, A. 1959. A revised geological time scale. *Trans. Edinb. geol. Soc.*, **17,** 183–216.

——and Harwood, H. F. 1928. The age and composition of the Whin Sill and related dykes of the north of England. *Mineralog. Mag.*, **21,** 493–552.

HOLMES, A. and HARWOOD, H. F. 1929. The tholeiite dykes of the north of England. *Mineralog. Mag.*, **22,** 1–52.

——and SMITH, S. 1921. The Wackerfield Dyke, Co. Durham. *Geol. Mag.*, **58,** 440–54.

KIRKBY, J. W. and DUFF, J. 1872. Notes on the geology of South Durham. *Nat. Hist. Trans. Northumb.*, **4,** 150–98.

MILLER, J. A. and MUSSETT, A. E. 1963. Dating basic rocks by the potassium-argon method. *Geophys. Jnl.*, **7,** 547–53.

MILLS, D. A. C. and HULL, J. H. 1968. The Geological Survey borehole at Woodland, Co. Durham (1962). *Bull. geol. Surv. Gt Br.*, No. 28, 1–37.

SMITH, D. B. and FRANCIS, E. A. 1967. Geology of the country between Durham and West Hartlepool. *Mem. Geol. Surv. Gt Br.*

SMYTHE, J. A. 1930. A chemical study of the Whin Sill. *Trans. nat. Hist. Soc. Northumb.*, **7,** 16–150.

TARLING, D. H. and MITCHELL, J. G. 1973. A palaeomagnetic and isotopic age for the Wackerfield Dyke of northern England. *Earth Planet Sci. Letters*, **18,** 427–32.

TEALL, J. J. H. 1884a. Petrological notes on some north of England dykes. *Q. Jnl geol. Soc. Lond.*, **40,** 209–47.

——1884b. On the chemical and microscopical characters of the Whin Sill. *Q. Jnl geol. Soc. Lond.*, **40,** 640–57.

TOMKEIEFF, S. I. 1929. A contribution to the petrology of the Whin Sill. *Mineralog. Mag.*, **22,** 100.

TROTTER, F. M. and HOLLINGWORTH, S. E. 1928. The Alston Block. *Geol. Mag.*, **65,** 443–8.

—— ——1932. Geology of the Brampton district. *Mem, geol. Surv. Gt Br.*, 113–21.

WILSON, V. 1948. East Yorkshire and Lincolnshire. *Br. reg. Geol.*

WITHAM, H. T. M. 1856. *History, topography and directory of the County Palatine of Durham.* M. Whellan and Co.

Chapter 8

MINERALIZATION

ORE MINERALIZATION

IN THE south-western part of the district ore veins containing lead and associated calcite cut the Great Limestone in the Spanham area. These veins (see p. 216) are probably an extension of the epigenetic mineralization of the Askrigg Block, which is attributed to the addition of deep-seated juvenile waters to actively circulating ground waters during the Hercynian (pre-Upper Permian) orogeny (Dunham 1959, p. 24).

In the south-central part of the district near East Layton epigenetic copper veins and associated deposits are found beneath the base of the Great Limestone which appears to have acted as a cap rock. In a few localities the mineralizing fluids have penetrated along joints and fractures up into the limestone itself; one such joint bears copper minerals at a lower level but the mineralization is terminated upwards by an argillaceous parting in the limestone. These occurrences suggest that the copper mineralizing fluids originated at depth and that they were probably connected with those responsible for the mineralization of the Askrigg Block generally. On the Alston Block it has been shown (Dunham 1948, p. 97) that copper minerals are generally located with the fluorite zone, the innermost zone of mineralization, but that there is a possible outer-zone recrudescence of copper mineralization (Dunham 1934, p. 698). As the copper minerals of this district are remote from the fluorite zone centres of the Askrigg Block, they may well represent a similar reactivation of copper mineralization. An alternative suggested by Deans (1950, p. 350; 1951) is the downward percolation of mineralized fluids from the syngenetic copper deposits in the adjacent overlying Marl Slate and Lower Magnesian Limestone. The present authors agree with Dunham (1959, p. 24), however, that the field relationship of the copper deposits to the host rock militates against the idea of epigenetic mineralization from above.

DOLOMITIZATION

Many of the limestones of Carboniferous age are patchily dolomitized and dolomite is not uncommon on joint faces. Though the relationship of dolomite to calcite suggests that the former is secondary, the apparent restriction of the dolomitization to the limestones suggests that the mineralization is diagenetic rather than epigenetic. Many authors, including Fairbridge (1957, p. 125) and Teodorovitch (1960, p. 74), have discussed mechanisms by which the diagenetic replacement can be effected.

In addition to being patchily dolomitized throughout, the basal few feet of the Great Limestone are often strongly dolomitized. At this horizon is a colony of compound corals which are known to be associated with dolomite elsewhere in northern England (Johnson 1958, p. 149)—evidence which would seem to accord with the widely held belief that dolomite associated with coral reefs is diagenetic (Rankama and Sahama 1955, p. 455). Alternatively, as the dolomitization at this horizon is more widespread than the coral colonies, it is conceivable that the corals preferred to inhabit a dolomite-depositing environment, and this would suggest that the dolomitization of the matrix may be syngenetic. The degree of dolomitization in the Great Limestone is higher in the widespread basal layer than in the patchy distribution at higher levels. It is here suggested, therefore, that the basal dolomitization may fall between the extremes of syngenetic and diagenetic, a conclusion broadly in accord with the views of Fairbridge (1957, p. 170) who recognized different levels of dolomitization.

Silicification

Within this district strata ranging from the Four-Fathom Limestone (P_2) up to at least the horizon of the Grindstone Limestone (E_2) are notable for the local development of chert and for the silicification of parts of the sequence. The silicification appears to be of at least three types. These are:

1. The diagenetic formation of siliceous nodules and concretions by the re-arrangement of silica deposited at the same time as the other constituents of the rock. An example of such concretions can be seen near the base of the Four-Fathom Limestone (see Plate II).
2. The formation of brown to black or grey, bedded, banded chert with geodes and fractures infilled with quartz crystals and chalcedonic silica. The best example of this type of deposit is above the Great Limestone and has been described in this and adjacent districts by Wells (1955, p. 177).
3. The production of siliceous limestones and interbedded siliceous shales. Strata of this type occur in this district above the Four-Fathom Limestone (see p. 16), above the Little Limestone, where they have been called the "Richmond Chert Series" (Wells 1955, p. 180), above the Crag (Crow) Limestone in the Richmond area, where they have been described by Hey (1956, p. 289), and above the horizon of the Knucton Shell-Beds and Rookhope Shell-Beds respectively. Silicification at higher horizons than those mentioned led to the development of chert in the Lower Felltop Limestone and locally the complete replacement of the Grindstone Limestone by silica to form a quartzite. The distribution of these various types of deposit is described in detailed sections in Chapters 2 and 3.

The position of the bedded cherts in relation to the other sediments of the cycles of which they form a part, their restricted lateral development and the absence of silica from the sediments above and below the cherts suggest that they are essentially primary deposits; a conclusion also reached by Wells (1955, p. 191). The source of the silica and the mode of deposition, however, are in doubt. As the cherts are for the most part present only in areas where the sandstones of the Coal Sills are absent, it may be that the sand grains which would normally constitute the sandstones have been converted to a silica gel on entering a peculiar depositional environment. The work of Correns (1950, p. 49) indicates that the solubility of silica increases with increasing pH so that an environment with a pH of 8 or 9 would probably be sufficient to cause any silica to go into solution. Wells (1955, p. 193) reviews the means by which silica might then be precipitated and concludes, having discounted an organic origin, that these deposits resulted from the "simultaneous deposition of $CaCO_3$ and colloidal silica as a gel, in a shallow epicontinental sea, off the mouth of a large river". Apart from the statement of Rankama and Sahama (1955, p. 216) to the effect that the inorganic origin of siliceous deposits in the sea under normal conditions is unlikely, a weakness in Wells's argument is the lack of a source of silica in sufficient quantity and its sudden availability for deposition at restricted localities. One environment in which silica can be precipitated in geographically restricted areas is where juvenile thermal waters issue, in the form of springs or geysers (Rankama and Sahama 1955, pp. 216, 556), directly into the sea. This hypothesis is thought to be the most likely and furthermore it is tentatively suggested that the source of the silica-rich juvenile waters may be related to the overall mineralization of the northern Pennines.

The widespread occurrence of silica in the Namurian limestones and associated shales of the southern part of this district may suggest the primary deposition of silica in an environment which locally forms a normal part of the cyclothemic sequence. To explain the observed cross-cutting field relationships (p. 31) between silicified and less silicified parts of the sequence, however, it would be necessary to envisage the subsequent mobilization of this primary silica and its intrusion into the overlying strata. This concept contrasts with that of Wells (1955, p. 191) who considered both the Main Chert and Richmond Chert Series to have a similar primary origin, and Hey (1956, p. 297) who considered the silicification of the Crow cherts to be secondary, but nevertheless penecontemporaneous. Hey concluded (loc. cit.) that the difference of interpretation between himself and Wells resulted from real differences in lithology between the strata above the Little Limestone and those above the Crag (Crow) Limestone. In this district, however, there appears to be little lithological difference between the strata above the Little Limestone and those above the Crag Limestone, or indeed the strata which occur above the succeeding limestones. Hey (op. cit., p. 299) considers the silicification to have resulted from the direct inorganic precipitation of silica in sea water, but both Wells (1955, p. 191) and Rankama and Sahama (1955, p. 216) have shown this to be geochemically unlikely in normal sea water. Perhaps, once again, the hypothesis of hydrothermal fluids may be invoked to explain either the mobilization of existing silica or the introduction of additional silica, especially since the areas in which silicification occurs lie between the orefields of the Alston and Askrigg blocks.

One additional type of silicification which has a less speculative origin is that caused by the intrusion of the Hett Dyke. Where the quartz-dolerite dyke is in contact with sandstones they become thermally metamorphosed and converted to quartzites. Where the dyke cuts shales or mudstones these are baked but not apparently enriched in silica, consequently there is no reason to suppose that silica has been introduced into the sandstones, but only that the original sand grains have been modified by the temperature of the intrusion. J.H.H.

References

CORRENS, C. W. 1950. Zur Geochemie der Diagenese. 1 Das Verhalten von $CaCO_3$ und SiO_2. *Geochim. cosmochim. Acta*, **1**, 49–54.

DEANS, T. 1950. The Kupferschiefer and associated lead-zinc mineralization in the Permian of Silesia, Germany, and England. *Rept. XVIII Int. Geol. Congr.* (London) Pt. 7, 340–52.

——1951. Notes on the copper deposits of Middleton Tyas and Richmond. Abstract in *Mineralog. Soc. Notice 74* on the meeting of June 7th.

DUNHAM, K. C. 1934. Genesis of the Northern Pennine ore deposits. *Q. Jnl geol. Soc. Lond.*, **90**, 689–720.

DUNHAM, K. C. 1948. Geology of the Northern Pennine Orefield; Vol. 1, Tyne to Stainmore. *Mem. geol. Surv. Gt Br.*

——1959. Epigenetic Mineralization in Yorkshire. *Proc. Yorks. geol. Soc.*, **32,** 1–29.

FAIRBRIDGE, R. W. 1957. The Dolomite Question in Regional aspects of Carbonate Deposition. *Am. Ass. petrol. Geol.* publication.

HEY, R. W. 1956. Cherts and limestones from the Crow Series near Richmond, Yorkshire. *Proc. Yorks. geol. Soc.*, **30,** 289–99.

JOHNSON, G. A. L. 1958. Biostromes in the Namurian Great Limestone of Northern England. *Palaeontology*, **1,** 147–57.

RANKAMA, K. and SAHAMA, TH. G. 1955. *Geochemistry*. Univ. of Chicago Press.

TEODOROVICH, G. I. 1960. O procsklozhdenii osodochinogo dolomita. *Sovetskaya Geologiya*, No. 5, 74–87.

WELLS, A. J. 1955. The development of Chert between the Main and Crow Limestones in North Yorkshire. *Proc. Yorks. geol. Soc.*, **30,** 177–96.

——1957. The Stratigraphy and Structure of the Middleton Tyas–Sleightholme Anticline, North Yorkshire. *Proc. Geol. Ass.*, **68,** 231–54.

WILSON, A. A. 1960. The Carboniferous Rocks of Coverdale and adjacent valleys in the Yorkshire Pennines. *Proc. Yorks. geol. Soc.*, **32,** 285–316.

Chapter 9

PLEISTOCENE AND RECENT

INTRODUCTION

THE superficial deposits of the district, widespread and diverse in character, include boulder clay, sand and gravel, laminated clay, morainic drift, alluvium and peat. A maximum thickness of 194 ft is recorded in a buried valley. The mapped Pleistocene deposits appear to be the product of one main glacial episode in which the main ice stream originated west of the Pennines in the vicinity of the Lake District. In addition, penecontemporaneous ice from more local sources, notably Tan Hill and Swaledale to the south-west and Cross Fell and Upper Teesdale to the north-west impinged on parts of the district. Distribution of erratics, notably Shap Granite, has enabled approximate boundaries to be drawn delimiting the extent of these ice masses (Plate XVIII). At the time of maximum glaciation it is thought that the district was wholly covered by ice.

Both the direction of ice-movement and the erratics indicate a correlation with the "Second or Main Glaciation" of Eastern Edenside and the Alston Block (Trotter 1929b, p. 560), the "Main Dales Glaciation" (Penny 1964, p. 400) and the Lower Boulder Clay of the Durham (27) district (Smith and Francis 1967, p. 239). There is no direct evidence within the district for dating this episode in terms of the continental European Pleistocene sequence. Elsewhere dates have been suggested ranging from Saale (Smith and Francis 1967, p. 242) to Early or Main Würm (Weichselian) (Penny 1964, p. 400). Accepting the correlation (Catt and Penny 1966, p. 404; Smith and Francis 1967, p. 242) of the Lower Boulder Clay of Durham with the Drab Till (Boulder Clay) of Holderness, then more recent work (Penny and others 1969, p. 65; Francis 1970, pp. 137, 139–49) suggests an upper Devensian age for the glaciation of this district.

The drift sequence along part of the eastern margin of the district more closely resembles that in the adjacent Stockton (33) district where the deposits are more typical of those in the lower Tees valley, and a complete understanding of them awaits the completion of survey of the Stockton district.

Apart from research by Dwerryhouse (1902), Fawcett (1916) and brief notes on the drift in the coalfield area by Kirkby and Duff (1872, pp. 196–8) the glacial and recent deposits of this district have not received detailed attention, although the district as a whole has been included in more general accounts of adjacent areas notably by Howse (1864), Goodchild (1875), Woolacott (1905, 1921), Trotter (1929a and b) Raistrick (1931), Carruthers (1938) and Wells (1957). Francis (1970) has reviewed the Quaternary of County Durham, and Taylor and others (1971, pp. 83–90) have summarized the Quaternary of northern England.

The drift is discussed, following an account of the history of Pleistocene and Recent events, under the following main headings: Buried Valleys, Boulder Clay, Drumlins, Morainic Drift, Sand and Gravel, Glacial Lake Deposits, Glacial

Drainage Channels and Recent Deposits. Distribution of the main types of glacial deposits is shown in Plate XVIII.

History of Pleistocene and Recent Deposits

GLACIATION

The district was affected by three penecontemporaneous ice sheets whose limits of influence are shown in Plate XVIII. In the north-west ice containing only local erratics moved down the Upper Tees valley above Eggleston, and at about the same time ice from the Lake District entered from the west *via* Stainmore (Dwerryhouse 1902, p. 572; Trotter 1929b, p. 557). These two ice sheets coalesced near Eggleston, the less active Teesdale ice being diverted north-eastwards through the valley of Blackton Beck into the River Wear drainage system. At about this time, ice, which is thought to have accumulated on the high ground near Mallerstang, also entered the south-western corner of the district *via* Upper Swaledale and impinged against the southern margin of the Stainmore ice. The more active Stainmore ice effectively deflected the Swaledale glacier towards the south-east beyond the southern margin of the district, and south of the Barningham area. Thus, while ice from these three centres formed separate glaciers, at the time of maximum glaciation all three merged to form a single ice sheet of which the Stainmore component covered by far the greater part of the district. The Stainmore ice front extended steadily eastward, fanning out in doing so and probably becoming less active. Throughout the district affected by Stainmore ice, Shap Granite erratics are common.

As the glaciation waned from its maximum the ice withdrew from interfluves in the western part of the district. Retreat of the ice led to the construction of moraines such as those at Romaldkirk and Carr House, near Staindrop, and also to the reworking of glacial material to form outwash deposits. Between Darlington and Houghton-le-Side, an ill-defined NW–SE belt of kame moraine was deposited along an ice margin during dissolution of the Stainmore ice. In general, however, moraines are not widespread in this district. This suggests that the ice, during its dissolution, became for the most part stagnant and was accompanied by few of the features associated with retreat and re-advance of an active ice front. In the east a mass of stagnant melting ice was left, the eastern edge of which probably retreated slowly and irregularly westwards. It was then that sub-glacial streams became specially active, laying down sub-glacial and outwash deposits such as sands and gravels, silts and laminated clay. Although the latter are present along the eastern margin of the district they are much more widespread in the adjacent Stockton (33) district.

Some of the more irregular features on the boulder clay surface in the district may be attributed to the decay of stagnant ice, as for example the occurrence of self-contained hollows, often containing laminated clays, brick clays and peat. Kettleholes also occur, sometimes being inset into the older gravel terraces. These may be attributed to collapse following the melting of buried masses of ground ice.

Development of Drainage

It has long been recognized (Fawcett 1916, pp. 318–19; Trotter 1929a, p. 171) that the district once formed part of an uplifted Tertiary peneplain with a general easterly tilt. The consequent drainage pattern was partly governed by the tilt of the peneplain and partly by the generally east–west 'grain' of the country underlain by Carboniferous rocks. The result was a series of eastward flowing consequents, the most important of which are shown in Fig. 21. The broad outcrop of Permian rocks in the eastern part of the district has a general north–south strike, but the surface exhibits no marked relief. The consequents were therefore able to maintain a generally eastward course across the Permian outcrop.

This broadly parallel system of consequents, developed in early Tertiary times, appears to have been modified before the onset of glaciation. The main modification was by the development of active subsequent streams (Fig. 22) which extended by headward erosion in a general north-westerly direction, capturing portions of consequent streams; the subsequents are, as Fawcett (1916, p. 312) first pointed out, commonly located along faults. The subsequent streams which had a major effect on the drainage are listed below (see also Trotter 1929a, p. 178):

1. A tributary of the Greta–Hutton Beck–Aldbrough Beck consequent which cut back from Barnard Castle to Eggleston in a gorge, beheading a number of consequents and ultimately capturing the Upper Tees–Langley Beck consequent.
2. A tributary of the Eller Beck–Nor Beck consequent which captured the Upper Greta west of Tutta Beck, forming the arc of the River Greta which is now incised and forms Brignall Gorge [056 114].
3. A subsequent, penecontemporaneous with or slightly later than 1 and 2 extended to the north-west by headward erosion in the area east of Newsham, ultimately capturing the River Greta and River Tees near Greta Bridge [0861 1317] and diverting a major part of the drainage of the western part of the district into Swaledale.
4. A major subsequent of the Upper Tees, the Eggleston Burn.
5. A subsequent of the proto-Lower Tees which cut back and captured the Sudburn–Selaby–Walworth consequent, and at a somewhat later stage, following the blocking of the valley by moraine, captured the Langley Beck section of the Langton–Cocker Beck consequent.
6. Another subsequent of the proto-Lower Tees which cut back northwards from near Piercebridge intercepting the Sudburn–Walworth consequent *via* the present Dyance Beck north-west of Headlam.

The above subsequents appear to have become well established in pre-glacial times. Further substantial modification took place during the ensuing glaciation particularly in the southern and south-central part of the district. Former valleys became choked with drift and blockages caused by ice and accumulated debris caused water to be diverted.

Probably the greatest single event in the development of the drainage system of this area took place at a late-glacial or early post-glacial stage when the Upper Tees and Greta drainage systems were captured by a tributary of the proto-Lower Tees (Fig. 22). This stream was probably a minor tributary to a subsequent in immediately pre-glacial times and extended headwards toward Winston, but on the waning of the glaciation it assumed an importance out of proportion to its original state. It is suggested that this was caused as follows. The major river that had drained the Greta and Upper Teesdale catchments and flowed into Swaledale was blocked, possibly by stagnating local ice or, more likely, by a minor isostatic upwarp creating a low watershed. This resulted in the formation

750

IEWSHAM

Whorlton B

500

NOR BE

Elements c

of a lake between Thorpe Farm and Hutton Magna and in a reversal of drainage in the area immediately to the south near Smallways Bridge [111 111]. It is likely that ice was still standing in what is now the Tees valley at and below Greta Bridge and that this lake was for a short time drained by Hutton Beck and its eastward continuation. At some stage, however, the water of this lake broke through an impounding barrier, probably ice, and linked up with the small tributary of the Lower Tees flowing north-eastwards towards Winston. The volume of water cut a deep gorge which not only drained the lake but effectively rendered any previous course of the river obsolete. In this way, the link between the Upper Tees and the district farther south was severed, and the main elements of the present-day River Tees system were established. In immediately post-glacial times the Tees was graded to between 40 and 70 ft above present datum. The river became incised through knickpoints on the older profile, and gorges such as that at Rokeby [085 144] (Fawcett 1916, fig. 2) and above Winston Bridge were incised to their present depth. The cutting down of the river is attributed to isostatic elevation of the land after the disappearance of ice from this and adjacent districts.

BURIED VALLEYS

The above account of the history of drainage is based on the major geomorphological features of the district. There are places, however, where pre-glacial channels are infilled by thick drift deposits, thus forming buried valleys.

Little information is available regarding the buried valleys of the district outside the outcrop of the Coal Measures. Even here information on their exact course is limited and the lines indicated on Fig. 22 must be regarded as tentative. In the higher reaches their sides appear to be steep with slopes of between 45° and 60°, whereas in their lower reaches they are less steep. Their courses generally closely follow present-day drainage patterns. Where for example the course of the modern channel is north–south the pre-glacial buried valley is commonly offset to the west, whereas in the case of present-day streams flowing from west to east the modern and pre-glacial drainage channels are generally co-axial.

In the coalfield area a major system of buried channels is associated with the River Gaunless. Thc main channel appears to originate in the neighbourhood of Low Lands pursuing a course and gradient closely corresponding to the present day river system, except that it has a much steeper profile near its inception. Another system of buried channels forming a tributary of the Gaunless appears to originate near the Sun Inn [1541 2317], Wackerfield and extends to near Hummerbeck. Its main line approximates to that of the present Hummer Beck but is off-set to the north-west, and there are several minor tributary valleys.

At a locality [2235 1540] west-north-west of High Coniscliffe Church, where the River Tees bends to the south, the rockhead surface on the north bank descends to below river level from about 50 ft above it over a zone 60 yd wide. This probably represents a buried channel infilled with grey and brown stony boulder clay.

South of the Tees, the Wycliffe Hall Bore* [1228 1328] proved 76 ft of boulder clay. Since limestone crops out close by, this suggests the presence of a small buried channel which was probably cut by a tributary of the pre-glacial river which flowed south-eastwards towards Ravensworth.

Pleistocene Deposits

BOULDER CLAY

The greater part of the district is covered with Pleistocene deposits and of these by far the most widespread is boulder clay. The largest drift-free area is in the north-western part of the district and includes the higher part of Eggleston Common, Woodland Fell and Langleydale Common. Smaller patches of drift-free ground are scattered throughout the district especially in the north, and in many cases along the crests of escarpments. Elsewhere solid rock occurs at the surface only where rivers and streams have cut through the drift cover. The boulder clay varies considerably in thickness being generally thicker in the valleys and thinner over the interfluves. It ranges in composition from a gravelly, or sandy deposit with a small amount of clay to a true clay without boulders or fragments of any kind. Between these two extremes the deposit consists typically of bluish grey clay enclosing angular to rounded, often striated rock fragments, the bulk of which are derived from local sources, together with a few from west of the Pennines. In the east the boulder clay is locally dull red in colour and rather less stony than in the west.

Boulder clay produced by the Swaledale ice is confined to the extreme south-west, notably along the north side of the fells forming the watershed between Teesdale and Arkengarthdale and the western part of Gilmonby Moor. The northern limit is ill defined but the maximum extent is probably on Gilmonby Moor south of the River Greta. South-east of Bowes the ice was deflected back into Arkengarthdale by the Stainmore ice. The absence of Shap Granite erratics in this area suggests that the boulder clay, which contains local rocks only, was probably deposited from a glacier originating in the Tan Hill area. On the higher ground the boulder clay is often much weathered, and in some parts much of the clay fraction has been removed by leaching. On Sleightholme Moor in the Brough-under-Stainmore (31) district south-west of Bowes there are mounds of drift consisting mainly of unsorted sandstone debris and Wells (1957) has suggested that they were found during a period when the flow of this local ice mass was held up by the Stainmore ice to the north. These hills extend into the extreme south-western part of the district on the southern side of Gilmonby Moor.

The boulder clay from the Teesdale ice is found north of a line which runs from where the River Tees crosses the western margin of the district, thence along the north side of Blackton Beck passing to the south of Woodland and subsequently north-eastwards towards the Linburn valley and the Wolsingham (26) district (Plate XVIII). The contained erratics consist almost entirely of Carboniferous limestones and sandstones, many of which appear to have been derived from the upper reaches of Teesdale by ice emanating from the Cross Fell area. In areas covered by ice originating from either Swaledale or Teesdale no deposits attributable to the Stainmore ice are found.

Boulder clay produced by the Stainmore ice is widespread throughout the district and its nature varies considerably from place to place. In the west and north of the district, the deposit ranges over the interfluves from 5 to 20 ft in thickness whereas in the valleys the range is generally from 20 to 54 ft. Locally, however, much greater thicknesses have been recorded, especially in buried valleys.

. 22. El

East of an irregular line from near Heighington to Walworth, Thornton Hall, Hall Moor and Cleasby to Newton Morrell records show that the general range in thickness is of the order of 45 to 60 ft.

Typically, the boulder clay consists of a stiff blue to grey stony clay in which material of local origin is mixed with many far-travelled erratics. The latter include blocks of red sandstone (probably Triassic), Permian Brockram, rocks of the Borrowdale Volcanic suite and Shap Granite. Large rounded boulders of Shap Granite are scattered widely throughout most of the area.

Along a restricted zone on the eastern margin of the district, south of Newton Aycliffe and extending to the southern margin, the deposits are more diverse in character. Irregular lenses and bands of sand, gravel, and laminated clay too thin to be mapped individually, are interstratified with boulder clays which differ from the type described above, being dull red, silty and often relatively stoneless, what stones there are being mostly of local origin. These clays appear to mantle much of the lower ground and to taper out westwards against a belt of hummocky topography interpreted as kame-moraine and described on p. 196.

Although the Stainmore Boulder Clay is widespread there are comparatively few sections in the deposit, and the greater part of the information relating to its nature and thickness comes from boreholes. The thicknesses proved, together with more generalized notes, are shown on Plate XVIII. J.H.H., D.A.C.M.

Details

Cotherstone, Bowes, Brignall and Barningham. Between the River Balder and Deepdale Beck, extending eastwards to the River Tees, the ground is largely covered by boulder clay ranging in thickness from a few inches to over 50 ft. Boreholes at the site of the Filter Plant [0114 1823] at Lartington proved thicknesses in excess of 38 ft.

South of Deepdale Beck, the Mount Pleasant Bore* [0328 1508] near Barnard Castle proved 79 ft 3 in of boulder clay, but farther south near Boldron village dip and scarp slopes of the Great Limestone are generally drift-free. Boulder clay re-appears again south of the scarp at Kilmond Scar [0279 1343] but it is probably of no great thickness between here and the River Greta.

South of the Greta there are considerable deposits of boulder clay mostly of Stainmore derivation. A thickness of 194 ft of boulder clay proved in a borehole [0152 1246] at White Close Farm probably indicates the presence of a pre-glacial valley, and another borehole [0322 1178] north-west of Scargill proved 120 ft of boulder clay with some gravel at the base. A number of other buried channels exist between the River Tees and Barningham and in these considerably greater thicknesses of drift, including boulder clay may be expected.

Woodland, Hindon Beck, Langleydale Common, Barnard Castle, Westwick and Wycliffe. South-west of Woodland, between Hindon Beck and Arn Gill, boulder clay ranges in thickness from 8 to 15 ft, erratics in this area including blocks of red (Triassic) sandstone, Lake District granites (excluding Shap) and possibly Whin Sill. South-east of Woodland and north of the River Gaunless the boulder clay generally ranges in thickness from 15 to 20 ft, being thicker locally, as for example at Cowley Shaft [0676 2542] where 46 ft 8 in were recorded.

In the area east of the River Tees and north of Langley Beck, the boulder clay cover is thin and impersistent, but it thickens east of Langleydale Common. Immediately south of Arn Gill, for example, it ranges from 3 to 10 ft and eastwards, south of the River Gaunless, the range is from 12 ft to a recorded maximum of 54 ft. Similar thicknesses may be expected in the area between Langley Beck and the approximate line of the Lunedale–Staindrop Fault.

In the northern outskirts of Barnard Castle, the boulder clay is recorded as being 8 ft thick, whereas at a locality near the town centre 32 ft is proved. East of Barnard Castle and north of Westwick the boulder clay cover ranges from 5 to 17 ft, but is locally thicker in buried valleys trending parallel to the Tees. For example a borehole [0747 1678] near Mount Eff proved 5 ft of boulder clay whereas another hole 100 yd to the south proved 22 ft on the edge of a buried channel. To the south-east 76 ft of boulder clay was proved in the Wycliffe Hall Bore* [1228 1328].

J.H.H.

Butterknowle, Morley, Toft Hill, Cockfield and Evenwood. The boulder clay thickness in this area is generally within the range 10 to 20 ft. It is thin or absent on north- and west-facing slopes such as Cragg Wood [145 253] and Cockfield Fell [125 245] while other relatively drift-free areas include the high ridge between Woodland and Toft Hill, an area north of Ramshaw Hall [153 268], and the Gaunless valley with associated tributaries above Low Lands [136 250].

An east–west belt of boulder clay up to a mile in width, extends from the Gibbsneese area south-west of Cockfield to Esperley Lane Ends [138 242] and Evenwood, thicknesses in excess of 25 ft generally being recorded. Along the southern margin and approximately corresponding with the watershed between the valleys of the River Gaunless and Langley Beck a belt of low amplitude hummocky topography, consisting in the main of thick deposits of coarse boulder clay extends from Shotton Moor [105 236] in the west through Burnt Houses to Keverstone Grange, thence swinging to the north-east and finally dying out near the headwaters of Crook Beck [153 240]. Thicknesses in this belt range in general between 30 and 40 ft, increasing somewhat towards the east, exceptionally to 54 ft and 71 ft 6 in in boreholes [1205 2349 and 1521 2318] at Burnt Houses and near the Sun Inn, Wackerfield. Rare natural sections show it to be a stiff bluish grey and yellow clay mostly containing fragments of local rocks but greywacke, agglomerate and ash are also reported. Sporadic lenses of sand and gravel are recorded in the boulder clay in several boreholes. In one [1205 2349] near Burnt Houses the section recorded was: Soil 1 ft, on Clay and Boulders 8 ft, Sand 5 ft, Clay and Boulders 16 ft, Sand and Gravel 10 ft, and Clay with Boulders 14 ft.

Staindrop, Hilton, South of Bolam, Ingleton, Langton, Whorlton and Gainford. The south-facing scarp, which extends along the northern part of this area is covered by thin boulder clay over all but the steeper parts of the feature. A small area of moundy gravelly boulder clay is present east of Trunnelmire Plantation [180 230]. The few records of boulder clay thicknesses available for this area suggest a thickness between 10 and 25 ft, thicker deposits being present locally along the base of the scarp as for example in the Raby Castle Water Bore* [1288 2218] which proved 65 ft. Thin lenses of sand and gravel occur in the drift. Boulder clay is present throughout most of the southern part of Raby Park and, together with other deposits, around Staindrop in the neighbourhood of Grainger Barn, New Raby, Sink House, and in the Alwent Valley.

The drift is relatively thin over the two major interfluves between Staindrop and Winston. At Dunn House Quarry [1136 1938] south-west of Staindrop the boulder clay, up to 12 ft thick, consists of stiff grey and blue clay. Included in it are fragments of local rocks, erratics of Carboniferous Limestone greywacke, red sandstone (? Triassic) quartzite, dolerite, Shap granite and tuff. In the old railway cutting 500 to 800 yd E of the former Winston Station, bouldery clay with layers of gravel included brockram and green slate erratics. A borehole [1415 1782] near Winston Station proved 64 ft of clay, the basal 6 ft being laminated. Thicknesses of boulder clay are variable between Whorlton and Winston, but except on scarp slopes and edges, the general range is between 15 to 20 ft. However, a borehole [1123 1614] north-west of Etherley House penetrated over 100 ft of superficial deposits.

East of the Alwent valley drift on the interfluves is generally thin and locally absent, except along the lines of buried channels. The River Tees cuts into solid rock throughout its length within this area, and in sections boulder clay is up to 20 ft thick. Shap Granite erratics are fairly common.

South of the River Tees there are a number of west to east buried channels, the chief of which lies along the Hutton Beck. Little is known about the boulder clay of this area; on the interfluves the type of distribution is broadly similar to that between Staindrop and Winston.

West Auckland, North of Bolam, South Church, Middridge and Redworth. Except along the lines of buried channels the boulder clay on the interfluves between the rivers Gaunless and Wear ranges generally between 10 and 30 ft, thickening gradually towards the east. It is generally a stiff bluish grey or yellow clay with local material, sometimes with lenses and bands of gravel. The following section was proved in a borehole [1985 2745] north-west of Tindale Crescent: Soil 1 ft, on mottled grey stony clay with coal and shale fragments 3½ ft, blue and grey clay with coal and shale traces 3 ft, dark grey clay with large boulders and coal and shale 3 ft 7 in and then gravel. A wide area east of West Auckland is covered with thick superficial deposits which include boulder clay. Deep buried channels filled with boulder clay enter this area from the west and south, following the approximate present-day course of the River Gaunless, the Oakley Cross and Hummer becks and associated tributaries. Between Evenwood and West Auckland the boulder clay is generally 25 to 35 ft thick, and also in places gravelly, as for example on the north side of Oakley Cross Beck. Near Evenwood Gate and Staindrop Field House opencast prospecting boreholes proved thicknesses in excess of 80 ft. In Hilton No. 55 Bore* [1702 2427] nearly ½ mile E of Evenwood Gate, the drift section was as follows: Soil 6 in, on sandy clay and boulders 10½ ft, clay and whinstone 14 ft, sand and boulders 15 ft, clay and boulders 24 ft, sandstone 2 ft, and boulders and clay 17½ ft. Between Bildershaw, Hummerbeck and Trunnelmire Plantation north-west of Bolam, the boulder clay, containing irregular lenses of sand and gravel, ranges from 10 to 40 ft in thickness increasing generally from east to west with the fall in elevation.

On the southern side of the quarry [1844 2315], 1300 yd WNW of Bolam, up to 12 ft of grey, brown and yellow stiff boulder clay contains, in addition to locally derived material, boulders of Shap granite and tuff.

South of Shildon, parts of the high ground between Brusselton, Bolam and Heighington are drift free, or covered with only a residual veneer of boulder clay. North-west of Royal Oak [2070 2369] the boulder clay generally ranges from 5 to 18 ft in thickness. At the New House opencast site [2037 2498] trenching operations carried out by the National Coal Board (Opencast Executive) have located the outcrops of the Harvey and Brockwell seams beneath the boulder clay. Between Heighington, Redworth and School Aycliffe the drift is generally thin, sandy and stony. It is much thicker, however, along the line of the railway east of Shildon railway station and along Red House Beck and its associated tributaries, the greatest proved thickness being 63 ft.

East of an approximate line from School Aycliffe to Spring Well [2645 2257] boulder clay thickens steadily eastwards, and in the eastern part of Newton Aycliffe Industrial Estate thick sandy stony boulder clays are present throughout, the Bakelite Water Bore* [2680 2329] proving 43 ft. Up to 20 ft of dull yellow or grey stiff clay are exposed along the railway cutting south of the former junction [2692 2411] containing a variety of erratics.

Heighington, Houghton-le-Side, Walworth, Summerhouse, Denton, High Coniscliffe. The steeper parts of the scarp which runs from west to east across the northern part of the area, for instance, around Leg's Cross, near Houghton-le-Side, and south of Heighington are drift-free but elsewhere the scarp is overlain by a thin covering of drift ranging from 5 to 15 ft in thickness. Locally moundy drift accumulations are present along the base of the feature. West of Twinsburn [2400 2200] and north of Halliwell House [2365 2243] the drift is thicker, along the line of a buried channel which follows Halliwell Beck. South of the scarp the boulder clay forms a sheet, 10 to 30 ft thick, generally consisting of a brown sandy stony clay. No good sections are visible, but in the stream 490 yd [2225 2039] north-north-east of Denton Grange East, 8 ft of sandy boulder clay, including large angular blocks of sandstone, are exposed. From there to the River Tees, between Piercebridge and High Coniscliffe, the drift deposits are absent or consist of a gently undulating sheet of boulder clay up to 15 ft thick. The River Tees itself has cut down into solid rock notably on the south bank immediately west of Piercebridge and on the north bank between Piercebridge and High Coniscliffe. In the north bank of the River Tees [2235 1540], 300 yd NNW of High Coniscliffe Church, boulder clay fills what is probably a buried channel (see p. 189).

Caldwell, Eppleby, Manfield and Brettanby. Apart from small relatively drift-free areas south and south-east of Aldbrough and south-west of Cleasby, the whole of this area is occupied by thick drift. South-west of

Cleasby, near High Farm [2467 1229] the boulder clay thins over the crest of a rockhead 'high' but even here solid has been exposed only by quarrying. Over the remainder of the area the drift, dominantly boulder clay but also including irregular lenses and layers of sand, gravel and laminated clay probably has a general thickness of 40 to 50 ft, though it is possibly more than 100 ft locally.

On the south bank of the River Tees [2092 1542], 200 yd WSW of Pierce Bridge, over 20 ft of grey and brown stiff stony boulder clay are exposed. At the top of the section is an ice-transported block of shale with sandstone ribs, 12 yd long.

The few records of boulder clay thicknesses from boreholes and wells in this area are shown on Plate XVIII.

Boulder clay and associated deposits are exposed locally on the south side of the River Tees from south of Holme House [2210 1497] to Cleasby and beyond. In most cases exposure is not good, and only partial sections are seen. At Manfield Scar [2335 1353] however, on the south bank of the River Tees, 1190 yd ENE of Manfield, one of the best drift exposures in the district is seen as follows:

	ft
Soil and sub-soil	2
Boulder clay, yellow and reddish grey sandy stony	8
Boulder clay, grey and blue stiff silty stony; many striated rounded or subrounded pebbles increasing in frequency to base; sporadic irregular lenses, up to 6 ft thick, of sand and gravel	10
Clay, sandy stiff dull red and brown; angular and subangular, predominantly small, stones; sporadic layers and lenses, often irregular and contorted, of silty sand, clayey silt, laminated clay and, more rarely, small gravel ..	30–35
Clay, stiff reddish brown; abundant small pebbles	8–10
Boulder clay, stiff reddish brown; subrounded to subangular striated stones up to small boulder size	8–9

At the north-west limit [2335 1367] of the section are 17 ft of pale brown and reddish stiff silty boulder clay equivalent to the 35-ft division above. The stone content increases markedly towards the base. In addition to material of local origin, erratics of Stainmore type are abundant throughout the section.

North and west of Darlington. Most of this area is covered by a thick sheet of drift which, in addition to boulder clay, includes sand and gravel, morainic drift and laminated clay. Few boreholes record thicknesses of less than 30 ft. The minimum recorded is 18 ft in a borehole [2535 1892] north of Cold Sides while the maximum is about 126 ft in TVCWB No. 1 Bore* [2552 1420] at Broken Scar Pumping Station on the western outskirts of Darlington. In contrast to areas farther west, the boulder clay is often discoloured dull reddish brown or greyish brown and is more silty. The stone content is of the familiar 'Stainmore' or local suite type, but there are in general fewer stones, and these are typically small subangular to subrounded pebbles or chips, though scattered larger stones occur throughout. Sand and gravel partings or irregular lenses also occur, as do small deposits of laminated or leafy clay which are too thin to map. Few sections are normally visible, but during construction of the Durham motorway a road cutting [2670 1868], 200 yd W of Burtree House exposed 12 ft of dull red, silty and gravelly clay passing down into dull red silty clay with laminated clay partings and irregular lenses of sand and gravel. On the west side of the motorway between a quarter and half a mile south-south-west of Burtree House the following section was noted: Clay, red and grey-greenish patches, with erratics of Carboniferous limestone, quartzite, ?diorite and dolerite 15 ft, sand 4 ft, on clay, sandy, red and brown. A slip-road cutting [2655 1762] west of Bottom House Farm exposed 12 ft of dull red and grey locally stony boulder clay. Silty boulder clay is exposed sporadically along the motorway south-west of Coniscliffe Grange, while just north of the bridge carrying the road between Darlington and Barnard Castle grey and dull reddish brown stony boulder clay underlies contorted sand, silt and gravel.

Typical of the drift of this area is the following section proved in a borehole [2617 1868] at Burtree Gate: Soil 1 ft, brown stony clay 14½ ft, red clay 6 ft, red sandy clay ½ ft, red and brown stony clay 14 ft, loamy sand with clay partings 2 ft, leafy clay with sand partings at base 8 ft, brown stony clay with sand partings 21½ ft, red loamy sand with clay partings 1½ ft, red stony clay with sand partings 3 ft, dark brown stony clay 19 ft, reddish brown stony clay 1½ ft, dark yellow stony clay 3 ft, on marl.

South-west of Darlington, Stapleton and Newton Morrell, East of Durham motorway. Except around High Farm, Cleasby, the area is covered by an extensive sheet of drift. The boulder clay is of a broadly similar character to the area immediately north of the Tees except that erratics of Stainmore type are considerably more abundant, suggesting that the main stream of ice moved south of the river. Along the western and south-western margins of this area the boulder clay is often a dull brownish red or grey very stiff strong clay containing many striated stones and boulders. Sand and gravel and laminated clay, although present, are less in evidence. On the east side of the Durham motorway [2470 1267] about 100 yd S of Manfield road bridge, 20 ft of brown, dull red and dark grey boulder clay including a bed of contorted laminated clay were exposed during construction of the road. The boulder clay contained local material in addition to Carboniferous limestone, sandstone, quartzite, dolerite and andesite. Broadly similar sections were seen elsewhere during construction of the motorway in cuttings for slip roads and in adjacent borrow pits. A cutting [2319 0971] about ½ mile SW of Clow Beck bridge exposed 15 ft of grey, dull yellow and reddish brown stiff boulder clay with erratics up to 2 ft in diameter. Boulders of Magnesian Limestone and Shap Granite are common on the ground surface but not in the clay itself.

In the bank edging the terraces of the River Tees below Cleasby dull red sandy and silty boulder clay is overlain by glacial sand and gravel. The maximum recorded thickness of boulder clay in this area, 90 ft, was proved in a borehole [2641 1214]. A borehole [2404 1064] at Clow Beck Farm proved 73 ft of boulder clay the upper part of which contained large boulders of highly polished Carboniferous limestone.

D.A.C.M.

DRUMLINS

The surface of the boulder clay in this district varies considerably, ranging from a comparatively flat to an irregular hummocky or moundy form with drumlins. The majority of the drumlins are assumed to be depositional accumulations mainly of lodgement till.

A belt of drumlins straddles the western margin of the district north-west of Romaldkirk, lying between a spread of glacial sand and gravel beside the River Tees and the foot of the scarp, which forms the southern side of the Tees valley. The largest drumlin [9844 2277] is approximately 70 ft high and 770 yd long, its long axis trending E20°S parallel to the valley side. The disposition and orientation of drumlins in this area indicates an ice movement down the Tees valley.

There is a small drumlin [1177 2175] 20 ft high and 210 yd long at the base of the scarp in Raby Park west of Raby Castle, and a number of small drumlin-like subrounded or oval mounds, composed predominantly of stiff clay with scattered boulders, are present nearly 1 mile WNW of Bolam. These mounds are margined on the west and north by higher ground and probably represent deposition of till, etc., from stagnant ice. A small drumlin [2112 1007] near Micklow Farm, 1 mile SE of Aldbrough, is sub-circular and 20 to 25 ft high, the long axis trending north-west. It is composed of bouldery gravel and clay, and rests on a flat rock platform.

The surface of the thick deposit of stiff grey boulder clay, with some sand and gravel, between Shotton Moor and Crook Beck is hummocky, with drumlin-like features having an amplitude of rarely more than 30 ft. This type of land-form is well displayed north-east of Keverstone Grange [144 230], at Cockshaw Hill [1503 2385].

D.A.C.M.

MORAINIC DRIFT

Morainic drift in this district takes three forms: a form of ground moraine, other than boulder clay, best described as ablation moraine, ice-margin (end or lateral) moraine and kame-type moraine. The ablation and ice-margin moraines are broadly similar, being composed of mounds or hummocks of disorientated, usually angular, blocks of rock debris, within a matrix of comminuted angular rock fragments. Their forms, however, differ in that the ablation moraines form undulating sheets unrelated to any ice margins, whereas the other morainic deposits form undulating ridges suggesting contacts with ice margins.

Ablation Moraine. Material which appears to have been deposited '*en masse*' during the decay of stagnant ice, either sub-glacially or englacially, is widely scattered throughout much of the district. On the western margin this type of moraine can be seen on Foggerthwaite Allotment [9800 2623] west of Eggleston Burn. A north-east to south-west belt of similar morainic drift occurs between Esperley Lane Ends and Evenwood, on the divide between the River Gaunless and Crook Beck. No natural sections are visible, but the deposit appears to be made up of sand, gravel and boulders with associated clay, and it has a hummocky surface. A small roughly circular shaped belt of hummocky topography around Carr House [1431 2069], east of Staindrop Church, is composed dominantly of small gravel, and overlying boulder clay. It occupies the almost imperceptible divide between the headwaters of the Langton Beck and Langley Beck and was probably instrumental in causing the waters of Langley Beck to be wholly diverted south-eastward from their original course along the Langton Beck valley.

Ice-Margin Moraine. An example of such a moraine, south-south-east of Romaldkirk near Hard Ings [9970 2096] trends east–west across the Tees valley. A lateral drainage channel which runs south from Romaldkirk against the western valley scarp passes round the western edge of the moraine and then issues into a flat sheet of fluvio-glacial sand and gravel cut into glacial outwash gravel. The most likely explanation is that the moraine marks a halt stage in a valley glacier retreating northwards in the present Tees valley. Alternatively, as the ice-contact slope of the moraine is apparently on the south side, the deposit may be the lateral moraine of a glacier in the Balder valley.

In the south-western part of the district small lateral moraines were apparently formed during the retreat of the Stainmore ice. The most notable forms a well-defined ridge running east-south-east from Eller Beck, for about three-quarters of a mile. It is well exposed to a thickness of 75 ft in Thwaite Beck [0102 1033], a small tributary of the River Greta.

Kame Moraine. An ill-defined belt of hummocky topography trends from the south of Houghton-le-Side to the western suburbs of Darlington, becoming broader in the same direction. The amplitude of the surface is rarely more than 30 ft where it is best displayed south and west of Walworth Moor, near Cowfold [240 202] and around Mowden [266 153]. Exceptionally, Great Boldearn's Hill [2247 2120] is a knoll, 45 ft high, apparently made up of a heterogeneous mass of bouldery and gravelly clay with the steepest face to the north and west. A similar mass near Mowden forms a broad ridge of stony gravelly clay, again with steep faces to north and west.

Within the morainic belt are elongate mounds or ridges, exemplified near Silver Hill [2397 1931], on the east side of Ling Beck, 380 to 1200 yd NNE of

Denton, and south of the Darlington–West Auckland road near Cross Lanes [2386 2113].

This suite of moraines is thought to have resulted from the dissolution and stagnation of an ice margin. Significantly, the belt apparently separates boulder clay deposits typical of most of the district from the more diverse deposits along part of the eastern margin and in the district to the east.

A small moraine [2497 2409] in the eastern part of the district, about 1 mile ENE of Redworth may also be regarded as kame moraine. It is composed of gravelly boulder clay, and is 700 yd long, the long axis trending a little north of east. D.A.C.M., J.H.H.

SAND AND GRAVEL

Sand and gravel occurs throughout the district either in sheets with gently undulating surfaces, in groups of closely associated mounds, as short irregular ridges, or, more rarely, as isolated mounds. It is difficult in the field to separate entirely deposits of exclusively glacial origin from those in which fluviatile action has played a part. In this account, however, the term *glacial sand and gravel* refers to material formed either englacially, sub-glacially or from the *in situ* melting of stagnant ice and deposited in sheets or random mounds. Other deposits, where fluvial action can be demonstrated, for example by the presence of false-bedding and current-bedding, and where the associated landforms suggest an extra-glacial fluviatile environment, have been mapped as *fluvio-glacial sand and gravel*. These may be either contemporaneous or penecontemporaneous with the general sheet of glacial sand and gravel or have been laid down by the more permanent and well established rivers in late glacial and post-glacial times. Though the landforms associated with these latter deposits are similar to those of recent rivers, their field relationships show them to be older. J.H.H., D.A.C.M.

DETAILS

Glacial Sand and Gravel. The highest level at which glacial sand and gravel occurs in the district is just above 1250 ft OD [0378 2511] near the summit of Woodland Fell. The deposit is thought to be a remnant of a wider sheet of early outwash material now partly removed by erosion. The younger and better preserved deposits are, in general, restricted to valleys where streams were operative in late-glacial and post-glacial times. The most extensive tracts occur

1. in the Tees valley between the western margin of the district at about 750 ft OD, and the area around Cotherstone, at about 550 ft OD.
2. in the old valley around Newsham between about 450 and 500 ft OD.
3. in the area south of Hutton Beck and north of White House [1431 1160] between 400 and 450 ft OD.
4. around Scargill south of the Greta at about 750 to 800 ft OD.
5. in the Gaunless valley between West Auckland and Fieldon.
6. south of the Tees valley at and to the south-east of Stapleton at about 175 ft OD.

In areas 1 to 4 the deposits are known only from small natural exposures and disused quarries. In areas 5 and 6, however, they have been proved in boreholes and locally mine shafts and drifts.

J.H.H.

The extensive sheet of sand and gravel between West Auckland and Fieldon (area 5 above) has a surface ranging between 320 and 390 ft OD. Besides sand and gravel it includes boulder beds, gravelly clay and, more rarely, laminated clay layers. The following examples from a mine drift, and the remainder from boreholes are cited as typical:

West Auckland Back Drift* [1827 2683]: 12 ft of medium rounded or subrounded to subangular gravel and sand in a silty clay matrix, on boulder clay.

Fieldon Bridge Industrial Site Bore No. 5 [1955 2665]: Soil 1 ft, on sandy gravel 9½ ft silty sand 5 ft 4 in, silty clay with sand partings 6 ft, sand with gravel 5 ft 4 in, and medium to coarse sand and gravel 18 ft.

NCB Brusselton No. 182 Bore [1920 2585]: Soil, Alluvium etc., 15 ft, on sand, silt and gravel 30 ft, with a layer of clay about 20 ft from the top, on brown fine to medium sand 9½ ft on boulder clay 18¾ ft. Examination of the drift from the borehole by the National Coal Board gave the following results:

Sample between 15 and 34 ft: Grey, medium to coarse poorly sorted sand, with abundant cobbles, and layers of gravel (mainly coarse sandstone).

Sample between 34 and 35½ ft: Grey brown strong structureless clay.

Sample between 37 and 40 ft: Buff and brown silt, with interbedded medium to coarse clayey sand and pebbles.

Sample between 40 and 46 ft: Sand with boulders, cobbles and pebbles; bands of clayey pebbly sand.

NCB Brusselton No. 181 Bore [1939 2598]: Soil 1 ft, on sandy clay with sandstone and dolerite boulders 5 ft, sand 4 ft, brown plastic silty laminated clay 4 ft, brown sand and gravel 2 ft, grey clay 1 ft, gravel 1 ft, brown sand with beds of gravel and silty sand 23 ft, and pebbles and boulders (sandstone and dolerite) 2 ft.

One of the most extensive sheets of glacial sand and gravel in the eastern part of the district (area 6 above) occurs south and south-east of Stapleton and along the lower part of Clow Beck east of Jolby Manor. This sheet is co-extensive with sand and gravel in Darlington and the adjacent Stockton (33) district. The base of the deposit is marked by a strong spring line on the feature margining the terrace and alluvial flats south-east of Stapleton. Resting on an irregular surface of boulder clay, the deposit thins north-westwards and dies out in the neighbourhood of Bank Top Farm [2609 1212] west of Stapleton. The thickness of 25 ft established at Murder Hill [2630 1150] is probably exceptional for this area. The deposit is being worked at a pit [2659 1170], near Stapleton Manor. At the western face of the pit (Plate XXA) are up to 25 ft of gravel containing subrounded to subangular stones, rarely more than 1 ft in diameter and including locally derived rocks and erratics of reddened sandstone, greywacke, Borrowdale Volcanic rocks and dolerite; Shap Granite is not common. The deposit contains sporadic lenses and irregular layers of sand, often with pebbles, silt and more rarely clay, and locally shows a very crude type of bedding. The deposit was worked south-south-west of Bank Top Farm where a previous trial hole [2601 1187] proved 18 ft of gravel.

Small deposits of glacial sand and gravel either as sheets, isolated mounds or interleaved with the boulder clay occur throughout the district although chiefly in the east and most of these are noted on Plate XVIII. Although few sections are visible in these deposits the following details may be noted:

Two small deposits of glacial sand and gravel occur on either side of the Red House Beck, ½ mile NNW of Redworth, up to 3 ft 1 in of gravel being proved. A small ridge of gravel, containing angular masses of sandstone occurs north and north-west of Newbiggin Farm [2200 2401] and there is a circular mound of gravel [2280 2209] near Elm Grange at Bellow Banks Hill. Further small masses of coarse gravel occur on the south-facing scarp 800 and 1200 yd SW of Bolam. West of Ingleton, small amounts of sand were worked at a pit [1650 2084] near Ingleton Grange. A section is no longer visible, but NCB Hilton No. 114 Bore* [1654 2082] at the extreme eastern end of the deposit, encountered 3 ft of sand at a depth of 3 ft 6 in, and NCB Hilton No. 113 Bore* [1577 2145] south of High Mulberry Farm, proved 5 ft of sand and boulders at a depth of 12 ft, in sandy boulder clay.

A small irregular dome-shaped mass of glacial sand and gravel occurs [234 179] on the south side of Denton Beck, south-south-east of Walworth.

One of the larger deposits of this type is at Hillhouse Hill [1845 1805], immediately north of Hill House. It forms a sub-circular, predominantly steep-sided mound, up to 40 ft above the surrounding terrain, and consists of angular and subangular blocks set in a matrix of gravel and clay. The gravel is mostly found on the north side of the hill and contains fragments of locally-derived material, as well as dolerite, basalt, greywacke and Borrowdale Volcanic rocks. On the south side of the hill, there is much more clay in the deposit. A similar, although smaller, mound [1861 1777] occurs north-east of Hill House and north-east [1912 1721] of Greystone cottages.

South of the River Tees, 33 ft of coarse gravel were recorded at 90 ft in a borehole [2239 1331] at Abbey Farm, Manfield and ?15 ft of gravel in the Grunton Bore* [2239 1148].

In a small area of sand north-north-west of Aldbrough, W. Gunn (in manuscript) noted 12 ft of sand in a well [2051 1213]. At Brook Wood, on the east side of Aldbrough Beck, north-east of Wath Erne, a section [2186 1021] comprises 6 ft of small and medium rounded and subrounded gravel on boulder clay. All these deposits may be contemporaneous with similar spreads of sand and gravel at Carlton Park, Forcett and Carkin, and those near White House (area 4 above).

Along the eastern margin of the district small deposits of glacial sand and gravel occur 900 yd NNW of Burtree House, while over ½ mile NNE of Burtree House and just east of the margin of the district, sand with pebbles was formerly worked at Whiley Hill Sand Pit [2750 1888]. At the site of the motorway bridge [2674 1861] over Burtree Lane, 4 ft of silty sand with a few stones and pebbles overlies laminated clay and boulder clay, while up to 4 ft of fine yellow and red sand with gravel was formerly exposed in a cutting [2648 1825] on the Durham motorway A1 (M) between 330 and 700 yd farther to the south-south-west. This sand interbedded in the boulder clay contains occasional pebbles and coal chips near the top.

In excavations [2472 1434] for the motorway immediately south-west of the CWS nurseries at Merrybent, 9 ft of contorted fine to medium angular to subrounded gravel with irregular silt lenses overlay dull reddish brown and grey silty stony clay. Sand and gravel was formerly worked at a small pit [2677 1578] on Berry Bank north-north-east of Mowden Hall.

Sand and gravel has been proved in a number of boreholes in this area, mainly along the line of the motorway. It probably forms irregular layers and lenses in the boulder clay and is rarely more than 3 to 4 ft thick, although 13¼ ft of uniform medium to coarse sand, with fine gravel mainly near the top was proved in a borehole [2634 1797] north-west of Bottom House. Over ¼ mile SSE of Bottom House, 8 ft of dull yellow and grey pebbly laminated silty sand were formerly exposed in a roadside cutting. Sand and gravel in boulder clay was also proved in boreholes along the line of the Tees valley notably at Broken Scar Pumping Station* [2552 1420]; ½ mile ESE of Low Coniscliffe near the weir [2590 1372]; and at Blackwell Bridge [2702 1260] near Stapleton where the following was proved: Alluvium 9 ft 2 in, reddish boulder clay 13 ft 1 in, sand and clay 4½ ft, reddish boulder clay 16 ft 7 in, red stony clay with sand partings 2¾ ft, red boulder clay 4 ft 7 in, sand 1¾ ft, red boulder clay 4 ft 7 in, dark grey stony clay with sand partings 3¼ ft, gravel bed with clay 16¼ ft, on blue and grey clay with sandstone boulders 3 ft 2 in. D.A.C.M.

Fluvio-glacial Sand and Gravel. Deposits of this type are commonly, but not exclusively, found in the main valleys. They are seen to advantage in the western part of the district, where in a quarry [9894 2385] on the eastern bank of Eggleston Burn the sequence exposed is:

	ft	in
Sand, coarse-grained; layers of fine gravel	4	0
Gravel, coarse; boulders up to 8 in diameter	5	0
Gravel, fine; maximum diameter of pebbles less than ¼ in	0	9
Sand, yellow fine-grained	0	6
Gravel, medium-grained; maximum diameter of boulders up to 4 in; a clayey matrix	2	0
Gravel, coarse; boulders up to 8 in diameter .. .. 4 ft 6 in to	9	0
Gravel, fine	0	5
Sand, fine-grained strongly false-bedded	0	6
Sand, dark brown fine-grained clayey	0	6
Sand, yellowish fine-grained ..	1	7
Gravel, fine	0	6
Gravel, mainly coarse; an impersistent bed..	6	0
Gravel, fine	0	6
Sand, fine-grained	0	9
Gravel; alternations of fine and medium grade	3	6
Sand, fine-grained	1	3

	ft	in
Gravel, fine	1	6
Sand	0	9
Gravel, coarse; layers of fine gravel and sand	5	0

The dimensions of the boulders in these gravels refer to the majority, and in the coarse gravel there are occasional boulders up to 3 ft across. Thicknesses are variable and those shown above should be taken to indicate the relative proportions of gravel to sand in the deposit as a whole. The boulders in the gravel consist mainly of Carboniferous limestones, sandstones and, more rarely, shale, together with red sandstone (?Triassic), red granite (?Eskdale Granite), granodiorite (?Carrock Fell Complex), red breccia (Brockram) and dolerite (Whin Sill). A study of pebble orientation in the gravels and false-bedding directions in the sands suggests that these deposits were derived from a river associated with the Stainmore ice rather than the Teesdale ice. Other spreads of fluvio-glacial sand and gravel in the west and south-west of the district include one south of the Tees in the vicinity of Cotherstone, and small mounds of gravel about half a mile north of the Tees Viaduct, near Barnard Castle.

J.H.H.

Fluvio-glacial sand and gravel extends eastwards from Westholme Hall [1384 1791] Winston as far as Langley Beck there merging into the terrace system of the stream. Throughout this area the derived soils are sandy clay loams with much subangular to rounded medium to coarse gravel. Fragments include locally derived debris, Carboniferous limestone, red sandstone, Brockram and Borrowdale Volcanic rocks. Nearby, Westholme Colliery No. 1 Shaft* [1346 1785] encountered 18 ft 6 in of sand and gravel at a depth of 20 ft.

In a working gravel pit west-north-west of Piercebridge the advancing face [1999 1579] at the time of survey (1961) exposed 19 ft of fluvio-glacial sand, silt and gravel, the upper surface of which was truncated and overlain by 6 to 8 ft of River Terrace gravels. The fluvio-glacial sand and gravels were strongly contorted with steep folds of low amplitude. All grades from the finest sand up to coarse rounded to subangular gravel were intermixed, but lenses and layers, often irregular, of sand and laminated silt with sand partings also occurred. They were up to 4 ft thick and were mainly confined to the anticlinal cores at the base of the section. The sand was typically a dirty yellow and dull grey coarse deposit containing pebbles and small chips of rock and in many instances was strongly false bedded. Irregular layers, often contorted, of grey and brown silt occurred within the sand. More rarely, irregular lenses of sand with some silt occurred in the higher levels of the deposit. Fig. 23 represents the structure seen in one of the anticlinal flexures.

D.A.C.M.

In late glacial and post-glacial times the established rivers of the district became sufficiently mature for the deposition of alluvium and the subsequent building of terraces. These fluviatile forms and associated deposits are demonstrably older than the oldest terraces of the present-day systems. In the west central part of the district the most important of these deposits are along the River Tees between its confluence with Thorsgill Beck and the River Greta; further deposits are seen in the River Greta and the River Tees towards West Layton. The back edge of the oldest terrace of this suite is at about 500 ft OD in the west and 475 ft OD in the east. A similar set of terraces was deposited, probably penecontemporaneously, in the Tees valley between West Thorpe Wood and Wycliffe Hall. These are at present 65 ft above present river level, their rear edge being at 440 ft OD at their western extremity. In other areas the courses of abandoned rivers are marked by perched expanses of 'older alluvium' which were probably laid down simultaneously with the earliest terrace deposits. Examples of this 'older alluvium' can be seen south of Hutton Beck near White House [1431 1160] where their closeness in time to the glaciation is indicated by their intimate relationship with a deposit of glacial sand and gravel containing a peat-filled kettlehole. Other examples of 'older alluvium' are seen in the old valley systems around and to the east of Forcett Church [1756 1223]. The gravels forming these and the terrace deposits are essentially similar to those of the glacial and fluvio-glacial sands and gravels except that the finer material has often been removed.

An expanse of 'older alluvium' (marked on the map with the alluvium symbol) east of Greta Bridge, is older than the oldest recent Greta terrace, but younger than the gravel

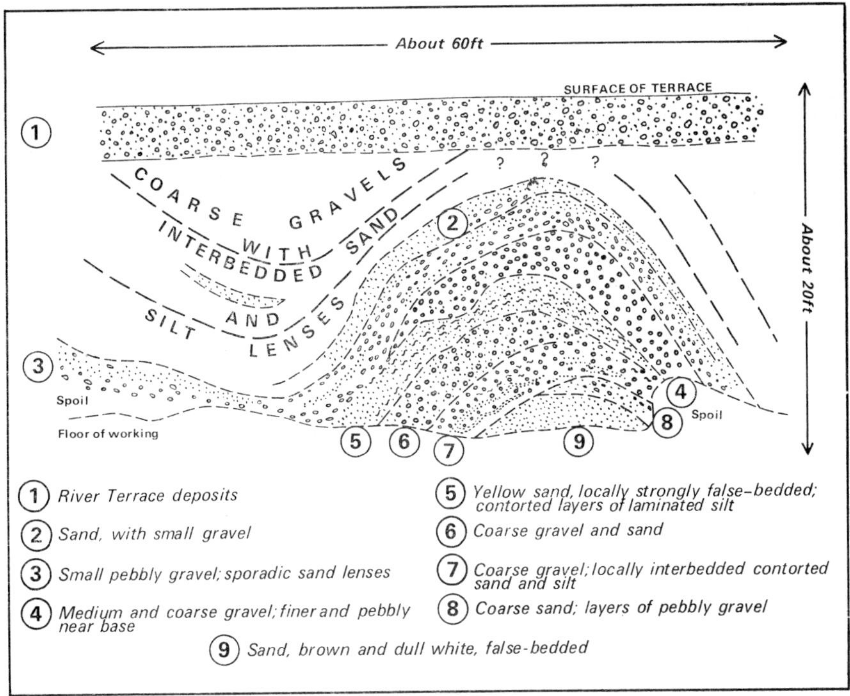

FIG. 23. *Diagrammatic representation of structure in drift deposits, formerly seen in gravel pit near Piercebridge*

terraces and associated river gravels described above. This deposit consists of a mixture of clay and peat covered by a thin film of gravel.

J.H.H.

On the Durham bank of the River Tees. [1619 1736] north-west of Gainford Church over 14 ft of coarse and bouldery gravel is exposed up to 75 ft above present river level. The deposit merges insensibly into boulder clays south of Gainford Great Wood. A series of small gravel terraces extend up the line of Langley Beck above the confluence of the Tees, the highest one of which merges into older river gravel at about 330 ft OD in the Westholme Hall area.

D.A.C.M.

GLACIAL LAKE DEPOSITS

The sites of former glacial lakes and associated deposits are shown on Plate XVIII. They are widely scattered throughout the district. They may perhaps be more accurately referred to as glacial or post-glacial meres which generally formed shallow sheets of water held up by low impounding walls of boulder clay or morainic drift. Most of them are infilled with laminated clay or brick earth, in many instances overlain by modern alluvium.

DETAILS

The sites of glacial lakes have been recognized in the southern and western parts of the district near Spanham Hill [004 102] and east of Greta Bridge. Little is known of the deposits, but at the latter [1050 1300] recent alluvium appears to overlie a peaty clay. Elsewhere in this area brick-earths, structureless clays with rare stones, have been worked west of Brignall [0511 1237], at Streatlam Tile Sheds [0705 2102] and south-east of East Layton [1688 0939]. These deposits are largely worked out, and sections are no longer visible.

Investigations into a peat-covered hollow near Romaldkirk show it to contain a sequence of calcareous marls and lake muds totalling 26 ft 8 in in thickness (Bellamy and others 1966, p. 429); an associated study of the fauna and flora suggests the presence of zones I to III of the late-glacial period and zones IV to VI of the post-glacial period.

D.A.C.M., J.H.H.

Brick-earth was formerly worked at a small pottery [1710 2400] 120 yd ESE of New Moors, Evenwood Gate. Very small amounts appear to have been removed, and the extent of the deposit is not known. A similar deposit was formerly dug at Barforth Tile Sheds [1615 1531] south of Barforth Hall. No sections are now visible but the deposit appears to extend farther south than the present pond limits. At the disused Hollymoor Brick and Tile Works [2087 1336] 1 mile W of Manfield, between 9 and 12 ft of laminated clay, dominantly yellowish brown in colour but pale blue at the base were formerly seen. There are other mapped occurrences west-south-west of Osmondcroft Farm [1293 1545], north-east of Wackerfield [159 229], north of Bolam [197 235] south-west of Walworth Hall [237 176] and in a series of disemminated irregular lake hollows east of Aldbrough and south of Manfield.

Laminated clays were proved in Winston No. 2 Bore [1415 1782] at a depth of 64 ft, while at a site [1262 1601] north-west of Osmondcroft Farm 9 ft of blue, stoneless, laminated clay was formerly worked.

Laminated clays are more widespread than has hitherto been suspected in the glacial deposits along the extreme eastern margin of the district. North of Burtree House [2690 1865] several lacustrine flats extend into the western part of the adjacent Stockton (33) district, the clay being overlain by a veneer of alluvium, locally peaty.

An investigation of the deposits in an arm of a one-time glacial lake [268 189] north-north-west of Burtree House recorded: sandy soil 50 cm, on peat 220 cm, black detritus mud 20 cm, calcareous marl 20 cm, muds and clays 30 cm (at least). The deposit includes material from post-glacial zones IV–VI (Bellamy and others 1966, p. 435). Laminated clay has also been noted in a number of boreholes along the line of the A1 (M) motorway.

A selection of shallow boreholes or excavations that have proved laminated clay or what is thought to be lacustrine clay are given below:

[2634 1799] at the site of the Darlington to West Auckland intersection with the Durham motorway: 7 ft 4 in of dark reddish brown laminated clay with sporadic sand laminae, at a depth of 22 ft.

[2580 1703] north-west of Stag House: laminated clay 10 ft 10 in.

[2526 1564] north-west of Coniscliffe Grange: laminated silty clay 16 ft 10 in.

[2582 1369] at the weir across the Tees, east-south-east of Low Coniscliffe: 8 ft of dark grey plastic clay with fine sand laminae at 29 ft depth.

[2469 1277] south-west of Cleasby, at a road bridge, the section proved was:

	ft	in
Top soil	0	10
Clay, pale brown silty; some stones	7	4
Clay, dark grey silty; small whinstones	12	1
Clay, dark red plastic; small gravel	4	3
Clay, dark red silty; small whinstones	2	9
Clay, dark red plastic; small gravel	5	11
Clay, sandy; limestone fragments	4	6

[2470 1267] on the south side of the motorway cutting, south of a bridge: a bed of contorted laminated clay was exposed at the base of the cut, underlying boulder clay.

D.A.C.M.

GLACIAL DRAINAGE CHANNELS

Following the glacial maximum a drainage system was established which almost certainly included sub-glacial, supra-glacial and extra glacial streams. Remnants of supra-glacial streams probably disappeared with the ice, but the other two categories are represented by landforms and associated deposits. The courses of some of these channels are shown in Plate XVIII.

At an early stage in the ice-retreat from the district it is probable that major sub-glacial rivers flowed along the approximate courses of the rivers Gaunless, Tees, Balder and the lower reaches of the proto-Greta, laying down the glacial sand and gravel deposits seen in the valleys. Other less erosive sub-glacial streams followed the grain of the country or the lines of pre-glacial streams and rivers, Langley Beck being one example. The major sub-glacial streams had tributaries, some of which were not exclusively sub-glacial, but sub-aerial in part. Channels near Folly House [0113 2309], south-east of Eggleston, exhibit the relationship of sub-glacial and sub-aerial channels. Here, one of the channels begins as a steep-sided flat-bottomed, alluvium-floored valley, with a maximum width of 75 yd. The valley trends westwards from its source for about 200 yd then north for a further 250 yd all the while running parallel to the base of a strong sandstone feature at about 1020 ft OD. The channel, having reached its northernmost extremity, then turns abruptly westwards through about 90°, and descends the present valley side towards the River Tees. In the latter part of its course the channel is V-shaped and incised. Other channels in the area (Plate XVIII), are similar except that after becoming incised at heights of up to 1020 ft they drain eventually into Langley Beck. The flat-bottomed upper reaches of these channels which have little or no gradient are interpreted as having been formed sub-aerially at the margin of a local ice sheet; those parts that are V-shaped and trend across the grain of the topography are thought to have been cut by much faster-flowing sub-glacial streams. If this hypothesis is correct it follows that the ice filling the Tees valley and relatively low-lying adjacent areas had an upper margin between 1000 and 1020 ft OD in the area south-east of Eggleston for sufficient time to allow the cutting of these channels.

All the major tributaries of the higher stretches of the River Tees in this district have been modified in some way or other by highly erosive streams closely associated with the ice, and the gorges of Deepdale and the River Balder are ascribed to this cause. The large quantities of gravel and sand east of West Auckland may be the outwash from sub-glacial and sub-aerial channels of this type.

Features interpreted as lateral drainage channels are assumed to have been sub-aerially formed along the margins of ice-sheets against abruptly rising ground. The best example is a steep-sided, flat-bottomed channel, up to 75 ft wide, which commences [9944 2188] south-west of Romaldkirk. It trends south between a strong boulder clay-covered scarp to the west and a lower boulder clay-covered plateau to the east, then changes course to round the western extremity of the Romaldkirk Moraine. Similar valleys near Barningham may also be lateral drainage channels of this type.

A suite of steep-sided overflow channels east of Eggleston Burn, near Knotts allotment [995 262] have been cut through the Hett Dyke at varying levels between 1220 and 1500 ft OD (Plate XIXB). These channels, which are approximately parallel and drain from north-west to south-east, were apparently formed when water, impounded in the upper reaches of the Eggleston Burn valley by ice and the Hett Dyke, overflowed at successive levels as the ice wasted or retreated.

An overflow channel in the Spanham area drained a glacial lake situated between Spanham West Hill [004 102] and Long Side [005 105]. The water drained east along the southern side of a moraine, ultimately entering the upper reaches of Gregory Beck. Below the old quarries [0172 1029] north of Spanham, the channel turns south-eastwards away from the present course of Gregory Beck for about 500 yd, then enters a fairly wide marshy hollow, possibly the site of another lake. The stream in the present Gregory Beck channel eventually cut back through thick boulder clay to capture the stream in the channel, leaving the lower part of it dry. Later still the moraine itself was breached by melt waters in a channel now occupied by Seavy Sike, a tributary of Thwaite Beck.

Immediately post-glacial drainage channels are generally cut into outwash deposits of sand and gravel, their presence often being recognized by the alluvial form of the re-worked sand and gravel deposits which fill the now dry valleys. Examples of these broad shallow dry valleys can be seen about 1 mile E of Hutton Magna, south of Forcett Hall [1728 1235], and about ½ mile S of Eppleby. East of Greta Bridge remnants of older river gravel are present in the extensive area of superficial fluvial deposits which occupy the broad valley north and east of Newsham and which trends south-eastwards towards Ravensworth. The river which formerly flowed in this valley was subsequently captured, and the drainage reversed at least in the area north of Smallways. This river drainage system with its associated alluvial deposits has, for the most part, masked or removed the older river gravels. J.H.H.

Recent Deposits

Introduction

The glacial and immediately post-glacial deposits of the district are overlain locally by deposits of recent gravel, sand and silt which form terraces or sheets of flood-plain alluvium. Peat occurs mainly as hill peat on the uplands, there being only small patches of lowland or basin peat. Landslips, mostly consisting of slipped drift material, occur as a minor deposit in the more deeply incised valleys. Hillwash, Creep, and Head, although known in the district have not been mapped.

River Terraces

The most extensive system of river terraces is in the valley of the River Tees, and below Winston they are virtually continuous, in many places on both sides of the river. Since only part of the Tees passes through this district no overall correlation of its terrace system has been attempted. In other valleys both the terraces and their relationship to those in the Tees valley are often obscure.

J.H.H., D.A.C.M.

Details

From Eggleston Burn to the confluence with the River Balder there is an almost continuous terrace, too narrow to be shown on the one-inch map, between 5 and 10 ft above the present level of the River Tees. This terrace consists predominantly of gravel with associated sand lenses. Near the mouth of Raygill Beck it is represented by a platform cut into solid rock on which virtually no superficial deposits remain. Along this same stretch of river older terraces are restricted to the vicinity of the confluence of Eggleston Burn and around the confluence with the River Balder. In the area near the Eggleston Burn confluence two terraces with a maximum vertical separation of 10 to 12 ft converge towards a knickpoint where the Lunedale Fault crosses the river. The lower of the two terraces falls from 20 to 12 ft above the present river level from the western

ɔf the country around Barnard

PLATE XVIII

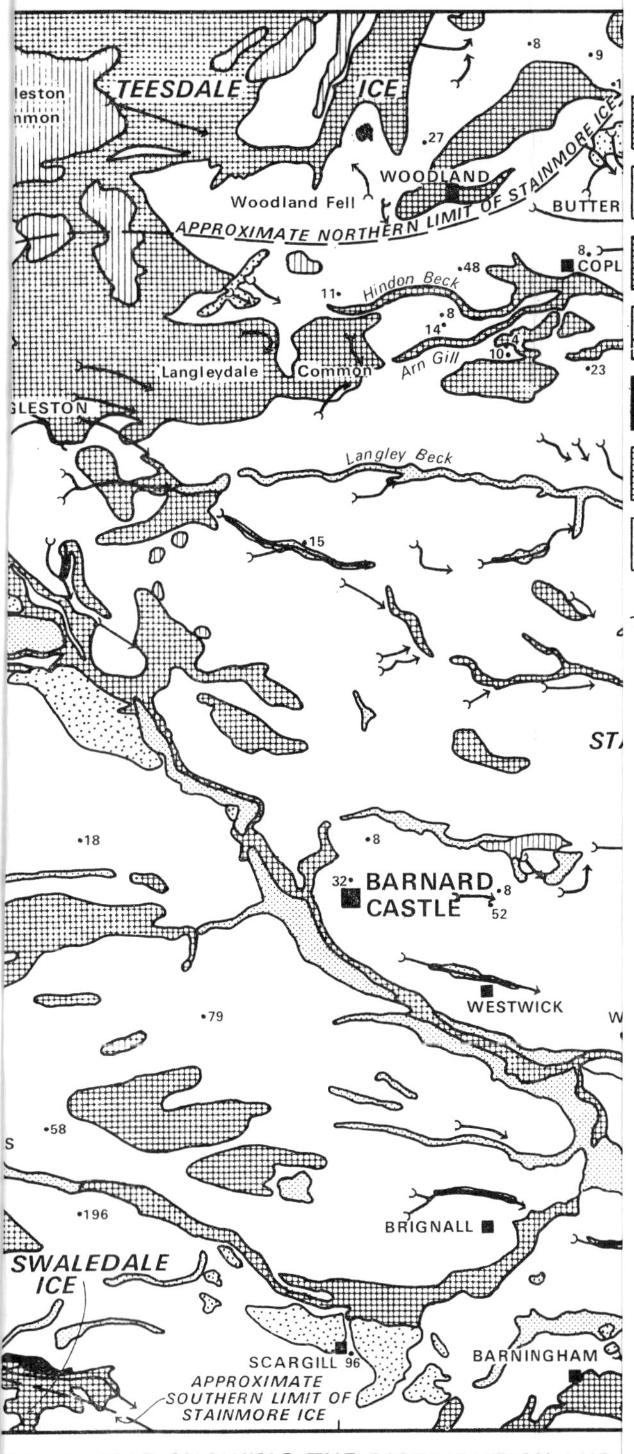

EXPLANATION

Solid rocks at surface

Peat

Alluvium; Terraces; Fluvio-glacial Sand and Gravel Terraces; Older River Gravels

Glacial Sand and Gravel

Morainic Drift

Glacial Lake Deposits

Boulder Clay

Glacial overflow channels

.56 *Thickness in feet of drift at point indicated*

KETCH-MAP SHOWING THE DISTRIBUTION AND

margin of the district to the line of the Lunedale Fault. The terraces near the River Balder confluence are approximately parallel to the base line of the Tees and on the western (Yorkshire) side of the river are at two levels 35 and 60 ft above the present river. On the eastern (Durham) bank there is, in addition to the flood plain terrace, a corresponding meander terrace about 40 ft above present river level. At both the above localities the terraces consist essentially of well-sorted gravel with subordinate sand lenses.

Between Cotherstone and Barnard Castle, both sides of the Tees are flanked by discontinuous flood-plain and older terraces. The deposits themselves, at all levels, are essentially of gravel, with subordinate amounts of sand. Some of the terraces are cut in solid rocks.

Older terrace deposits occur in two main areas. The first is between Towler Hill Farm [0364 1788] and the Yorkshire bank of the Tees, where five terraces are seen at heights of 25, 30, 50, 60 and 75 ft above the river. The other area is between the north bank of Deepdale Beck and Sour Beck, where terraces stand 25 and 75 ft above the river. South of Deepdale Beck the same two terraces are seen respectively at 20 and 70 ft above the river.

Between Barnard Castle and Ovington Weir [1314 1497] a series of immediately post-glacial terraces (as distinct from recent terraces, have been recognized. The base-line of these post-glacial terraces is broadly parallel to that of the present river but some 40 to 70 ft above it, the two levels being convergent downstream. Between Eggleston Abbey [0616 1503] and the confluence with the River Greta two divergent post-glacial terraces have been recognized, the lower being from 0 to 30 ft above base level and the higher a further 20 to 25 ft above in the area of maximum divergence. Downstream near West Thorpe [0996 1432] the two terraces are parallel to, or slightly convergent to, the uplifted base level; both terraces and the base level at this locality show the effect of a knickpoint caused by the outcrop of the Little Limestone in the present river [0975 1459]. Near this outcrop the terraces consist of coarse gravel whereas in the stretch of the river above the River Greta confluence they are mainly benches cut into solid rock which only have a thin gravelly veneer. In this same general area, at about 0 to 20 ft and 20 to 30 ft above present river level, there are younger convergent terraces which appear to be controlled by outcrop of the Bottom Little Limestone [0553 1557] near the Lendings Mill, Barnard Castle.

Between the eastern end of Abbey Bridge gorge and Ovington Weir a flood plain terrace stands 5 to 10 ft above present river level, except in Rokeby Gorge and in the vicinity of the knickpoint formed by the Little Limestone. Hereabouts, too, are isolated groups of meander terraces which are broadly divergent and which stand about 25 and 45 ft respectively above present river level. The recent terrace deposits are distinguished from the older post-glacial terraces by their relative positions and by the presence of sand and silt in the gravel.

J.H.H.

Along the Tees below Winston (Plate XXB) up to four gravel terraces have been recognized. The terraces of the River Tees probably constitute the most important source of gravel in the district.

Around and to west of Barforth Hall [1638 1761] river terrace deposits are at three levels notably 8 to 10 ft, 18 to 25 ft and 35 to 45 ft above the river level. The highest (third) terrace consists of coarse gravel with sand and silt lenses. South-east [2047 1636] of Piercebridge Grange up to 14 ft 6 in of gravel have been removed from the fourth and highest river terrace whose near edge is about 50 ft above river level. South [1999 1579] of White Cross Farm 8 ft of gravel forming the second river terrace (about 35 ft above river level), was formerly seen to overlie 19 ft of contorted fluvio-glacial gravel. Farther west, near Gainford this terrace forms a rock-cut platform covered with gravel up to 6 ft thick. South-east of Piercebridge, on the south side of the River Tees, extensive river terrace flats partially composed of gravel are present around Holme House and these are now being worked east of Cliffe. In the first and second river terraces, at and to east of Glebe House [2284 1421] considerable quantities of sand with lenses and irregular pockets of gravel occur.

Terraces are also extensive to south and south-east of High Coniscliffe. The first river terrace, which extends over a wide area, includes gravels, but the sand and silt fraction is very variable. The surface of the terrace is irregular and has three indistinct sub-levels but at the rear it is about 170 ft OD. Small quantities of gravel are present in the second and third river terraces in this area. The third terrace whose rear edge lies at about 190 ft OD some 35 to 40 ft above river level has a slipped leading edge and appears to consist of coarse gravel, but the amount of gravel is unknown as the rockhead surface here rises steeply.

West of Low Coniscliffe [249 139] large quantities of gravel have been removed from the first and second river terraces. At a locality [2440 1393] between 300 and 400 yd WNW of Low Coniscliffe the following section was formerly visible on the northern margin of the pit:

	ft	in
Soil, brown and grey sandy silty; sporadic pebbles	1	6
Clay, dull red and brown sandy silty; thins to west	2	0
Gravel with sporadic boulders; sand and silt lenses	4	0

Two hundred yards [2425 1403] farther to west-north-west the coarse gravel is 8 ft thick while at a further 120 yd [2414 1402] up to 12 ft of gravel is present, the overlying clay and soil being comparatively thin. At a borehole [2475 1427], 280 yd SW of the CWS Nurseries at Merrybent, 9 ft 2 in of gravel with some boulders, and underlying 5 in of soil correspond to the third terrace at about 160 ft OD. Part of this deposit was formerly seen in a cutting [2472 1434] on the Durham motorway adjacent to the CWS Nurseries, Merrybent, where 8 ft of mainly fine gravel, with irregular sand and silt lenses overlay contorted fluvio-glacial sand and gravel. South-west of the road bridge [2467 1418] 15 ft of gravel is wholly referable to river terrace deposits.

Around and to east of Cleasby [251 130], the first and second river terraces are prominent and gravel up to 15 ft thick has been worked north-west of Stapleton. The section was variable with lenses of grey and brown clay occurring both near the surface and at the base. Tree trunks, some well preserved, were occasionally encountered in the workings, together with horns, bones and antlers in a variable condition. At a locality [2612 1275] north of Stapleton, a tree trunk, 14 ft × 1½ ft in diameter was found 8 ft below surface, and underlying 3 ft of gravel; in addition a curved horn was found beneath 8 ft of gravel, though the exact locality was not recorded.

The following sequence was recorded in this pit [2611 1255] north-east of Stapleton:

	ft	in
Top soil, sandy silty	1	0
Gravel, fine; sandy and silty matrix; sporadic sand lenses	2	0
Gravel, fine and medium; sandy and silty matrix; gravel blackened at top and base and locally fused up to	3	6
Gravel, coarse; bouldery at base ..	6	0

The layers of blackened gravel may represent the temporary stabilization of an older upper surface by vegetation.

At a borehole [2627 1277] at the eastern limit of the pit 24 ft 2 in of terrace deposits were proved including 5 ft 1 in of sand 15 ft 8 in from the surface. Some distance to the south-west, in a borehole [2585 1240] at the site of the road bridge between Stapleton and Cleasby, the terrace deposits comprised only 13 ft of sand with coarse gravel in the basal 5 ft 4 in. At and to the east of Stapleton, first and second terrace deposits are predominantly formed of gravel overlying sand and silt.

In the smaller valleys in the district, terrace deposits are occasionally found but are very small. Small terraces occur in Eggleston Burn, in rivers Balder and Greta, and in Langley and Deepdale becks; in the Gaunless valley between Butterknowle and West Auckland; along Sudburn Beck east of Snotterton Hall, and along Clow Beck near Aldbrough and again near the eastern margin of the district.

D.A.C.M.

ALLUVIUM

In this account alluvium is defined as the most recent deposit of a river and is found bordering the stream where it is still liable to flooding. It is present along most of the rivers and streams in this district and borders the River Tees along much of its length, except in gorges such as those at Rokeby and Abbey Bridge and near the various waterfalls or "weirs" caused by the outcrop of solid resistant rocks. In the Tees valley the widest spreads of alluvium are east of Piercebridge, notably on the Yorkshire side of the river near High Coniscliffe, and Stapleton, and on the Durham side between Low Coniscliffe and Blackwell, south of Darlington.

In the west bank of the River Tees [2227 1473], south-east of Holme House, 3½ ft of interbedded grey and yellow sand and silt with occasional rounded pebbles overlie 5 ft of muddy silt. At the northern end of the exposure a bed of gravel wedges out to the south, and farther north along the river bank [2231 1492] 6 ft of dull grey and brown sand, bedded in 1-in units are exposed. Scattered small rounded or subrounded pebbles occur throughout, commonly as short discontinuous lenticles. In the alluvial flat on the north side of the River Tees, 1 mile ESE of Low Coniscliffe, the deposit consists mainly of fine interbedded sands and silts with occasional lenses of gravel. A trench section [2604 1364] exposed 10 ft of sand and interbedded silt.

On the Yorkshire side of the river [2643 1278] north of Stapleton, 5 ft of interbedded pebbly sand and silt, with occasional layers of fine gravel, rest on up to 2 ft of gravel.

Among the larger areas of alluvium associated with other streams in the district are those occurring along the River Gaunless east of Low Lands; the Red House Beck, north of Heighington; the Langley Beck below Staindrop; the Langton (Dyance) and Summerhouse becks; the Cocker Beck, south-east of Denton, and the Clow (or Hutton, Caldwell and Aldbrough) Beck, more especially in the neighbourhood of Aldbrough. In the Gaunless valley above West Auckland the alluvium is rather variable, and at a locality [1925 2637] south-east of St Helens Church, West Auckland, over 6 ft of gravel is recorded containing fragments of limestone, sandstone, quartz and tuff. Considerable spreads of gravelly alluvium flank Langley Beck, south-east of Staindrop.

Patches of older alluvium, varying from coarse to finer clayey gravel and elevated above the present drainage system, are generally associated with post-glacial drainage channels. D.A.C.M., J.H.H.

PEAT

In the western part of the district, hill peat forms a discontinuous cover on much of the high ground. On scarp slopes this peat is generally present only above 1500 ft, but on dip slopes it in places extends down to 1250 ft OD. The hill peat, usually heavily dissected and eroded into 'hags', appears everywhere to be wasting. Thicknesses in excess of 6 ft are rare in the north-western part of the district, but farther south 12 ft are locally seen. The amount of wood in the peat varies considerably from place to place; silver birch (*Betula sp.*) is common but oak (*Quercus sp.*) also occurs.

In the north-west there are scattered patches of basin peat on shale slacks above 900 ft OD and, in contrast to the hill peat, it does not appear to be wasting. Below 800 ft peat occurs either as a covering or infilling of kettleholes, as for example at Romaldkirk (Bellamy and others 1966, fig. 3) and in the group of five south of Howegill Plantation [029 123] or as a veneer to other superficial deposits, notably in glacial or post-glacial drainage channels. A good example occurs [0782 1750] north-east of Barnard Castle near Westwick. J.H.H.

Small areas of peat are present north of Westside House [1486 2000] in an area of impeded drainage on the shallow divide between the Langley Beck and the headwaters of the Langton Beck, and mainly on the southern side of the hummocky topography at Carr House [143 207]. An arm of the one-time glacial lake 300 yd NNW [268 189] of Burtree House has shown the presence of 220 cm of peat beneath 50 cm of soil (Bellamy and others 1966, p. 435). The peat is underlain by mud, on calcareous marl, muds and clays.

A borehole [2534 1586] north-north-west of Coniscliffe Grange recorded 8 ft of silty clay and peat at a depth of 9 ft 3 in, and a borehole [2464 1314], 350 yd ENE of Howden Hill, at the back edge of a terrace proved 4 ft of black peat at a depth of 8 ft. Similarly, in road excavations [2614 1229] 330 yd NW of Stapleton, peat with silt was reported under 2 ft of silt and silty clay near the rear edge of a terrace deposit. D.A.C.M.

CALCAREOUS MARL

Calcareous marl 20 cm thick has been recorded (Bellamy and others 1966, p. 435) in a lake deposit [268 189] north-north-west of Burtree House, at a depth of 290 cm, overlain by mud and peat.

TUFA

Small deposits of tufa occur on the banks of the River Tees where springs of calcareous water issue from the base of outcropping limestones. An example occurs on the Yorkshire bank of the River Tees [1434 1633] north-east of Winston Bridge.

LANDSLIPS

Extensive landslipping has occurred in the western bank of Eggleston Burn [985 264] where stream erosion has undercut thick deposits of boulder clay (Plate XIXA). The result of undercutting, and subsequent landslip, has been to displace the present stream to the east, where it has cut a new valley in solid rock. Elsewhere in the western part of the district landslips are common both in deeply incised valleys, where they generally occur below outcrop of gently dipping sandstones overlying shales, and on steep scarp slopes such as below Crag Top [0736 2351] south-west of Copley. Small landslips occur on the north bank of the Tees south of Osmondcroft Farm [1293 1545] and on the south bank of the Tees at Manfield Scar 2333 1363].

REFERENCES

BELLAMY, D. J., BRADSHAW, M. E., MILLINGTON, G. R. and SIMMONS, I. G. 1966. Two Quaternary deposits in the Lower Tees basin. *New Phyt.*, **65**, 429–52.

CARRUTHERS, R. G. 1938. Alston Moor to Botany and Tanhill (an adventure in stratigraphy). *Proc. Yorks. geol. Soc.*, **23**, 236–53.

CATT, J. A. and PENNY, L. F. 1966. The Pleistocene deposits of Holderness, East Yorkshire. *Proc. Yorks. geol. Soc.*, **35**, 375–420.

DWERRYHOUSE, A. R. 1902. The Glaciation of Teesdale, Weardale and the Tyne Valley, and their Tributary Valleys. *Q. Jnl geol. Soc. Lond.*, **58**, 572–608.

FAWCETT, C. B. 1916. The Middle Tees and its tributaries: a study in river development. *Geogr. Jnl.*, **48**, 310–25.

(L 316)

A. Landslip of boulder clay, Eggleston Burn

PLATE XIX

B. Glacial drainage channels on skyline cutting through Hett Dyke. From near Hill Top, Eggleston–Stanhope road

(L 318)

(L 371

A. Contorted glacial sand and gravel in pit (1963). Stapleton Manor, Stapleton

PLATE XX

B. View east-north-east over River Tees Terraces, Primrose Hill Farm, near Winston

(L 335

FRANCIS, E. A. 1970. The Quaternary: *in* Geology of Durham County. G. A. L. Johnson, Ed. *Trans. Nat. Hist. Soc. Northumb.*, **41,** 134–52.

GOODCHILD, J. G. 1875. The Glacial Phenomena of the Eden Valley and the Yorkshire Dale district. *Q. Jnl geol. Soc. Lond.*, **31,** 55–99.

HOWSE, R. 1864. On the glaciation of the counties of Durham and Northumberland. *Trans. N. England Min. Eng.*, **13,** (1863–64), 169–185.

KIRKBY, J. W. and DUFF, J. 1872. Notes on the Geology of part of South Durham. *Nat. Hist. Trans. Northumb.*, **4,** 150–98.

PENNY, L. F. 1964. A Review of the Last Glaciation in Great Britain. *Proc. Yorks. geol. Soc.*, **34,** 387–411.

——COOPE, G. R. and CATT, J. A. 1969. Age and Insect Fauna of the Dimlington Silts, East Yorkshire. *Nature, Lond.*, **224,** 65–67.

RAISTRICK, A. 1931. Glaciation. *In* Carruthers, R. G. and others. The geology of Northumberland and Durham. *Proc. Geol. Ass.*, **42,** 281–291.

SMITH, D. B. and FRANCIS, E. A. 1967. Geology of the Country between Durham and West Hartlepool. *Mem. geol. Surv. Gt Br.*

TAYLOR, B. J., BURGESS, I. C., LAND, D. H., MILLS, D. A. C., SMITH, D. B. and WARREN, P. T. 1971. Northern England. *Brit. reg. Geol.* x + 121 pp.

TROTTER, F. M. 1929a. The Tertiary Uplift and Resultant Drainage of the Alston Block and Adjacent Areas. *Proc. Yorks. geol. Soc.*, **21,** 161–80.

——1929b. On the Glaciation of Eastern Edenside, Alston Block and the Carlisle Plain. *Q. Jnl geol. Soc. Lond.*, **85,** 549–612.

WELLS, A. J. 1957. The Stratigraphy and Structure of the Middleton Tyas–Sleightholme Anticline, North Yorkshire. *Proc. Geol. Ass.*, **68,** 231–254.

WOOLACOTT, D. 1905. The superficial dpeosits and Pre-Glacial Valleys of the Northumberland and Durham Coalfield. *Q. Jnl geol. Soc. Lond.*, **61,** 64–96.

——1921. The Interglacial problem, and the Glacial and Post Glacial Sequence in Northumberland and Durham. *Geol. Mag.*, **58,** 21–32, 60–69.

Chapter 10

ECONOMIC GEOLOGY

THE ECONOMY of the northern part of the district has been almost wholly based on the coal mining industry which has been of outstanding importance over the last 150 years. Other less important industries directly dependent on the mineral resources of the district include quarrying of dolerite, for use as road metal, of sandstone for building purposes and road fill, and of limestone for flux and the manufacture of lime and cement. Gravel is extracted in the Tees valley above Darlington.

COAL MINING

Following intensive mining in the northern part of the district most of the coal reserves both from deep mines and major opencast sources are now virtually exhausted, and the area offers little scope for further development.

The following account of the coals of the district has been contributed by Mr T. S. Tomlinson of the National Coal Board.

COAL: RANK, QUALITY AND UTILISATION

The Durham coalfield produces coals which range from the prime coking coals of west Durham, with 24 to 32 per cent of volatile matter on the dry mineral matter-free basis, through coking-gas types to the gas coals of east Durham and then to house and industrial coals in which the volatile matter sometimes exceeds 38 per cent.

Although most coals can be used successfully for different purposes the property which decides the most efficient primary use is the rank. This is a fundamental property of the coal substance, i.e. that part of the coal which is of purely vegetable origin, omitting all forms of associated mineral matter.

The quality of the coal—measured by the amount and nature of the mineral matter—then becomes important as a secondary means of selection.

Many of the properties of coal vary systematically with the rank and may be used as parameters for a rank classification. Volatile matter and caking power are examples and these are used in the National Coal Board Classification which divides coal into classes each of which is designated by a coal rank code.

The general range of rank exhibited by Durham coals and the range of volatile matter (on the dry, mineral matter-free basis) for each rank class are shown in Table 4.

Within the Durham coalfield the rank of the seams varies in three ways:

1. there is a continuous regional change in each seam whereby the rank, highest in west Durham, falls as the seam is traced into other parts of the coalfield.
2. there is a general, but not systematic, increase in the rank of seams with depth.

TABLE 4

Coal Rank Code	*Description*	*Volatile Matter* (d.m.m.f.)
200 H 303 H	Heat-altered: mainly non-caking	9·1–19·5 19·6–27·0*
302 H	Heat-altered: mainly weakly-caking to medium-caking	19·6–32·0
301	Strongly-caking: prime coking coals	24·0–32·0*
401 402	Very strongly-caking: coking and gas coals ..	32·1–36·0 over 36·0
501 502	Strongly-caking: mainly coking and gas coals	32·1–36·0 over 36·0
601	Medium-caking: mainly house and industrial coals	32·1–36·0 over 36·0
701 702	Weakly-caking: house and industrial coals ..	32·1–36·0 over 36·0

*National limits 303 H, 301 } 19·6–32·0.

3. where a seam is cut by an igneous intrusion, such as a dyke, the heat of the intrusion causes partial devolatilisation of the coal—together with partial or complete destruction of the caking power—so that a higher rank is induced in the areas flanking the dyke. These metamorphosed coals are described as "heat-altered" in Table 4 and are distinguished by bearing the suffix H.

Within the boundaries of the Barnard Castle district coal mining has been concentrated mainly in the north-eastern half of the Coal Measures area, to which the following comments on rank and quality are therefore essentially restricted but where virtually all the major seams of the Durham succession are represented and have been worked.

Rank. Since the district includes only the south-western fringe of the Durham Coal Measures the regional rank variation in each seam is small. Nor is the increase in rank of the seams with depth particularly well marked. In addition, the metamorphic effect of both the Cleveland and Wackerfield dykes is confined to short distances from the intrusion, and there are no extensive areas of heat-altered coal. It follows, therefore, that the seams of this area show in general a restricted range in rank when compared with some other parts of the coalfield and the great majority have rank codes of 501, 502, 601 and 602. This is especially the case in the seams above the Bottom Busty where, however, some coal of rank codes 701 and 702 is sometimes present in seams down to, and including, the Brass Thill.

From the Ganister Clay to the Bottom Busty seam the most frequently encountered codes are 501 and 502 but isolated examples of higher rank coal occur, e.g. 301 rank code in the Brockwell seam at some points in the old Ramshaw take and 401 in the Marshall Green at Emms Hill.

Of the properties associated with rank the volatile matter shows a general range of 33 to 40 per cent, on the dry, mineral matter-free basis, but falls to under 32 per cent, of course, in the few cases where the rank code is 301. It usually exceeds 36 per cent in the seams of the Middle Coal Measures.

A calorific value of 15 500 Btu/lb and above (dry mineral matter-free) is usual in the Victoria seam and is common in some samples of other seams up to and including the Harvey: otherwise, the usual range is 14 800 to 15 400 Btu/lb.

Quality. Many of the seams contain dirt bands of variable thickness so that the overall ash content, of coal and dirt together, often shows a considerable range within the colliery take. In the following, however, the values quoted relate to the "seam excluding dirt", i.e. to the clean coal and middlings combined and may be taken to represent the basic quality of the coal.

The Harvey has been one of the most consistent of the seams and in the several mines in which it has been worked the ash content has seldom exceeded 7 per cent, being under 4 per cent at one or two points. The Main and Durham Low Main were of similar quality.

On the other hand, many of the seams which in some parts of West Durham were of excellent reputation—such as the Busty, Brockwell and Victoria seams—are here of more variable quality. They contain more than 10 per cent of ash in places although with a general range of 4 per cent to 8 per cent, which is also the usual range for the Main, Durham Low Main and Bottom Hutton seams.

Apart from the Brockwell seam, with under 1·2 per cent, none of the seams is consistently low in sulphur and many are variable in this respect. In most cases samples have given minimum values of 1·0 to 1·2 per cent and maximum values which have exceeded 3·0 per cent in the Five-Quarter, Durham Low Main, Brass Thill, Bottom Hutton, Tilley and Bottom Busty seams although in the last named the sulphur does not usually exceed 2·0 per cent. Samples of the Harvey have contained 0·9 to 2·6 per cent of this impurity.

The chlorine content of the seams is very low and even the highest value recorded, 0·17 per cent, is low by national standards.

The phosphorus content of the Harvey and Victoria seams appears always to be low, i.e. under 0·010 per cent. In most of the remaining seams also the amount is of this order but local increases to 0·030 per cent have been found. Occasional samples of the Main and Brockwell seams have reached 0·100 per cent of phosphorus which, on the national scale, is very high.

The Yoredale seam, in the Millstone Grit Series, was worked at North Tees. It was a strongly-caking coal with rank code 502, mean volatile matter and calorific value (both on the dry, mineral matter-free basis) of about 42 per cent and 14 920 Btu/lb respectively, ash 13 to 15 per cent, sulphur 2·1 to 4·5 per cent, low chlorine and phosphorus up to 0·021 per cent.

Utilisation. Many of the mines within the district reached their maximum production during the 1950's. New Shildon, for example, raised over 290 000 tons of coal in 1956 from the Brass Thill and Hutton seams: Brusselton produced over 171 000 tons in 1957 from the Durham Low Main, Brass Thill, Tilley and Busty seams. Other mines with smaller outputs must have numbered at least twenty.

On the basis of rank most of the coal, being of rank code 501 to 602, was of general purpose type and was used for coke and gas manufacture, for steam-raising and as a house fuel—depending on the size grading, the cleaning facilities and the available markets. A proportion of the output was also sent out for ships' bunkers.

No National Coal Board mines are now working in this part of the coalfield. There is a small output from privately owned drifts and small opencast sites.

T.S.T.

Limestone

All the limestones of Carboniferous age below the Crag Limestone have been worked locally in small quarries either as building stone or for making lime both for agriculture and building. Of these limestones the Great Limestone has been the most extensive source and at the time of the survey three quarries were in operation. Two of these, Kilmond Wood Quarry [0335 1370] and Hulands Quarry [0155 1383] are in the south-west of the district and the output from both was crushed to produce chippings for road-surfacing. In the south central part of the district limestone is worked in East Layton Quarry [1579 1127] for use both as a flux and a roadstone.

Dolomite and Dolomitic Limestones

Permian dolomites and dolomitic limestones have been quarried in many places north of the River Tees along the western margin of the formation where the superficial cover is relatively thin. Among the larger disused workings are those of Thickley Quarry [2408 2564] at East Thickley, High Bank Quarry [2400 2273] near Heighington, Denton Quarry [2100 1966], Summerhouse Quarry [2093 1875], Morton Limekilns [190 205] and Hobgate Quarry [1646 1888]. The only working quarries are Old Towns Quarry [257 246] near Middridge Grange re-opened after many years' disuse, and a new quarry [249 252] east of the old Middridge Quarry, which has now been filled in. Good reserves exist and the material extracted is used for flux and to produce agricultural lime. In the past this rock has been used as building stone, roadstone and for making lime for building. Summerhouse Quarry was in use well over a century ago. Browell and Kirkby (1866, p. 210) quote the following analysis for material from this quarry.

	per cent
$CaCO_3$	56.40
$MgCO_3$	38.88
Oxide of Fe_3Al	0.80
Sand and Clay	1.36
H_2O	2.56
	100.00

South of the River Tees small quarries around Eppleby and near Cleasby have been disused for many years; those near Cleasby were almost entirely filled in during construction of the Durham motorway. D.A.C.M., J.H.H.

Chert

The only workable deposits of chert occur in the "Main Chert" above the Great Limestone. These are restricted, within the district, to the area around

Bowes where there are numerous small disused quarries. The rock locally provided building stone, but its principal use was as a source of chippings for road surfaces. J.H.H.

SANDSTONE

All the Carboniferous sandstones, but more especially the sandstones of Millstone Grit Series and Lower Coal Measures age, have been quarried on a small scale for building stones, roofing slates, walling and flagstones. In the past the accessibility of a sandstone rather than its nature has governed where and for what purpose a bed has been quarried. Some of these sandstones are locally cemented with silica giving the rock an overall composition approaching that of ganister and at some of these localities the beds were formerly quarried for use as refractories.

Three quarries within the district are still working and all are in the same bed of sandstone, which lies about 150 ft above the Upper Felltop Limestone. In Shipley Bank Quarries [0182 2075], about $2\frac{1}{4}$ miles SE of Eggleston, the sandstone, which is generally evenly and regularly bedded and capable of being split into thin slabs, is now quarried on a small scale for walling and stone fireplaces. Formerly it was extensively utilized throughout the area as roofing material. In Stainton Quarry [0699 1884] north-north-east of Barnard Castle and Dunn House Quarry [1136 1938] near Staindrop, the bed is quarried for monumental and building stone, flags, crazy paving, rockery stone and for repair and restoration purposes. At all these localities there appear to be adequate reserves to supply these markets. D.A.C.M.

SHALE

A few small quarries were formerly worked in the siliceous shales which generally overlie the limestones of the lower part of the Millstone Grit Series. These beds appear to have been used exclusively for walling. J.H.H.

DOLERITE

The primary use of dolerite within the district has been for roadstone. The Hett Dyke was formerly worked extensively to the east of Eggleston Burn, especially in the vicinity of Moor House [9898 2599], economically workable reserves being now exhausted. This also applies to the Wackerfield Dyke, which was quarried in the area between north-east [1580 2266] of Wackerfield and east-north-east [1682 2307] of Hilton Moor Farm.

The Cleveland Dyke was formerly worked in small quarries along its strike in the north-western part of the district west of Hindon Beck [0325 2510]. Farther east in the Cockfield Fell area between The Slack [1154 2549] and a locality [1439 2443] near Buck Head Farm the dyke was extensively worked from about 1780 onwards, and where readily accessible the rock has been worked out. In the area north-west of Bolam the dyke was in the form of a laccolith, and although some dolerite remains, reserves of good stone are near exhaustion and quarrying has ceased. Other small disused quarries in this dyke occur to the east between Bolam and Leg's Cross House [210 226]. D.A.C.M., J.H.H.

Fireclay

At Eldon Hill Quarry [242 272] opencast operations have taken place for the extraction of the Main and Maudlin coals. The site was left open, and at the north-east end of the workings, fireclay underlying the Main Coal was subsequently worked. Now abandoned, the site is being used as a waste tip. E.A.F.

Sand and Gravel

Deposits of sand and gravel, including those of both glacial and fluvio-glacial origin, are relatively common within the district, but because they are generally associated with subordinate layers of clay, and are often of limited areal extent, these deposits have not been extensively exploited.

In the western part of the district fluvio-glacial sands and gravels were worked until 1960 in a quarry [9894 2385] on the east bank of Eggleston Burn near its confluence with the Tees. At the time of writing the only quarry still working in this area was south of the Tees, astride the margin of this district and the adjacent Brough-under-Stainmore (31) district. In the south-eastern part of the district a pit [2659 1170] about 300 yd ESE of Stapleton Manor exploits glacial sand and gravel. The remaining small deposits of glacial sand and gravel in the eastern part of the district are not economically attractive since many are coarse and intermixed with a fairly high proportion of clay.

River terrace gravels are currently being worked south of the Tees [215 153] east of Cliffe and at a pit [261 127] north-west of Stapleton. The material extracted near Piercebridge is used for ready-mixed concrete.

Possible sites for future exploration of economically workable gravel include an area near Barforth Hall [160 170] near Gainford; and further extension of reserves is possible in the Tees valley at and to the east of Piercebridge. Elsewhere no economically workable deposits are known. D.A.C.M., J.H.H.

Boulder Clay

Though boulder clay is a widespread deposit within the district it is of only limited economic use. In the eastern part of the district it was used as a constructional fill for an embankment on the Darlington spur of the Durham motorway [2593 1221] $\frac{1}{2}$ mile W of Stapleton. In the adjacent Brough-under-Stainmore (31) district more extensive use of boulder clay was made in the construction of the Selset Dam and to a lesser extent the Balderhead Dam. D.A.C.M., J.H.H.

Brick-clays

In the past glacial clays, including laminated clays, have been worked, where they are almost free of stones, for the manufacture of bricks and tiles. For the most part these deposits are now worked out and currently none is being exploited. In the western half of the district sites at which these clays were worked include a pit [0511 1237] west of Brignall and another [0705 2102] on Streatlam Moor. Amongst the larger workings in the remainder of the district are sites at Fieldon Bridge [201 269] near Tindale Crescent, at Osmondcroft [1262 1601] near Winston, at Barforth Tile Sheds [1615 1531] about 1 mile SSW of Gainford, near East Layton [1688 0939], and at Hollymoor [2087 1336] west of Manfield. J.H.H., D.A.C.M.

PEAT

Though peat occurs on the high ground there is no evidence to suggest that any has been cut recently for burning although much has been removed in the past for this purpose. J.H.H.

LEAD AND CALCITE

In the north-western part of the district a number of mineral veins were formerly worked for lead, in the form of galena, in the area between Spanham and Eller Beck. These lead veins, which are associated with the North Spanham Fault, appear to have been most productive where they cut the Great Limestone as for example in Eller Beck Hush [992 103] and Spanham Hush [009 099]. At these localities calcite, in a variety of habits, is the principal gangue mineral and it is particularly abundant in Spanham Hush. Outside the Spanham area galena has been recorded in small veins in the Coal Measures of the area between Woodland and Copley.

Apart from its association with lead veins calcite occurs in joints and veins in most limestones within the district. One notable section in calcite-veined limestone is seen in a quarry [0230 1353] in the Great Limestone south-west of Boldron; in these veins, up to 3 ft wide, the calcite is generally massive and pink or white in colour. The only known use to which the calcite has been put is as an ornamental stone. J.H.H.

BARYTE

Baryte is relatively rare within the district, and it has only been recorded from two small quarries [0781 2371 and 0794 2362] north of Copley and near the Wigglesworth Fault. These quarries are in a sandstone high in the Millstone Grit Series and pink baryte is associated, in veins and on joints, with calcite and some limonite. Both quarries were probably worked for sandstone rather than the vein minerals. Small amounts of baryte are occasionally found in the Permian rocks. J.H.H., D.A.C.M.

COPPER

Copper ores were formerly worked along a vein associated with a small fault in the Great Limestone at Sorrowful Hill [1531 1051], near East Layton. Recent workings in an adjacent limestone quarry have exposed the fault [1556 1066] in the floor of the quarry and it can be seen that the mineralization is almost wholly confined to the beds beneath the limestone, where nodules of bornite and covellite coated with malachite occur. At this locality the base of the limestone is ferruginous and dolomitic with very little copper. At an adjacent locality [1575 1067], however, a mineralized shatter-belt in the limestone is exposed which contains chalcocite, chalcopyrite, and malachite. J.H.H.

WATER SUPPLY

The hydrology of the Hydrometric Area of which this district forms a part has been described (Mills and Hull 1961). Subsequent data may be obtained from the annual "Surface Water Survey" of the Department of the Environment and from the annual "Water Services Handbook". The district lies within the Northumbrian Water Authority's region.

The greater part of the district receives a mains water supply provided by the Durham County Water Board in the north and the Tees Valley and Cleveland Water Board in the central and southern parts of the district. This supply is derived mainly from storage reservoirs and springs, and by abstraction from rivers. The Durham County Water Board serves part of the district with some water from the Wear Catchment whereas the Tees Valley and Cleveland Water Board obtains 36·5 million gallons per day (m.g.d.) of surface water from reservoirs in Baldersdale and Lunedale to the west of the district. These reservoirs, which are sited on rocks of Viséan or Namurian age, yield water in which carbonate hardness varies from 20 to 35, non-carbonate hardness from 12 to 17 and a total hardness from 37 to 47 parts per million. In addition to these supplies, between 8·5 and 15 m.g.d. is abstracted from the River Tees at Broken Scar [255 140] near Darlington. This water has a carbonate hardness of 63, a non-carbonate hardness of 45, and a total hardness of 108 parts per million.

To augment surface water supplies described above, or where mains water is not available, boreholes have been drilled to abstract underground water, the quality and quantity of which is dependent on the structure and permeability of the strata drilled. Water derived from the Carboniferous comes mainly from joints and to a limited extent from granular porosity in the sandstones; water from the Permian is derived both from joints and from the often considerable granular porosity in the limestone.

Viséan and lower Namurian strata in the district have yields of up to 130 000 g.p.d. In the North Tees Drift Mine [122 162] west of Winston strata were dewatered at the rate of 0·13 m.g.d.

Sandstones from the upper beds of the Namurian and the lower part of the Coal Measures (Westphalian) are thought to be capable of yielding up to about 120 000 g.p.d. depending on the degree of fracturing of the rock. Relatively few wells have been drilled in these strata and no detailed abstraction figures are available.

Quantities of between 0·027 and 1·825 m.g.d. were pumped in certain years from mine shafts in the north-east of the district. This water, which was derived from Lower and Middle Coal Measures and the overlying Permian, was discharged mainly into surface streams. Only a small fraction was used for industrial processes as, for example, at Randolph Colliery [158 249], Evenwood, whilst some of it was taken by the Durham County Water Board in time of drought.

Few records are available of the quality of water derived from boreholes but the following are quoted as examples.

A borehole [9846 1393] near Bowes into Namurian and Viséan strata:

	p.p.m.
Carbonate hardness	307·0
Non-carbonate hardness	370·0
Chloride	1·3
Nitrates	0·57
Albuminoid ammonia	0·01
Oxygen absorption	0·52
Total solids dried at 100°C	430·0

A borehole* [1228 1328] near Wycliffe Hall into Namurian and Viséan strata:

	p.p.m.
Carbonate hardness	4·0
Non-carbonate hardness	Nil
Chlorides	29·0
Alkalinity as $CaCO_3$	514·0
Nitric nitrogen	0·10
Oxygen absorption	1·2
Free ammonia	0·32
Iron	0·24
Lead, copper, zinc	None
pH	8·8

A borehole* [2281 2336] near Shackleton Beacon, Heighington, into basal Westphalian and upper Namurian strata:

	p.p.m.
Carbonate hardness	277·0
Non-carbonate hardness	106·0
Chloride	24·6
Nitrate	0·26
Ammonia	0·18
Albuminoid ammonia	0·07
Oxygen absorption	0·03
Total solids at 100°C	420·0

The carbonate hardness of this sample appears to be higher than one might have expected from this part of the Carboniferous sequence.

A spring [1620 1731], Gainford Spa Well, which issues from Namurian sandstones about $\frac{3}{4}$ mile NW of Gainford Church is rich in H_2S.

The Permian is an important aquifer throughout northern England and has great potential in this district, though at the time of writing it remains relatively untapped (Anon 1961, p. 21 and Map No. 111). A recent series of exploratory boreholes have indicated potential yields of up to 1 m.g.d. The greater part of the water issues from the Middle Magnesian Limestone, lower beds being relatively impervious. The quality of the water derived from the Permian is indicated by the following analysis of a sample from a borehole [2618 1868] at Burtree Gate north of Darlington:

	p.p.m.
Carbonate hardness	277·0
Non-carbonate hardness	40·0
Chloride	22·2
Nitrate	0·02
Ammonia	0·05
Albuminoid ammonia	0·04
Oxygen absorption	0·8
Total solids, dried at 100°C	370·0

In the south-eastern part of the district, in the general area between Manfield, Aldbrough and Newton Morrell, a number of wells obtain water from both the lower part of the Namurian and the overlying Permian limestones. In this area a borehole at Grunton Farm* [2239 1148] yields water with the following analysis:

	p.p.m.
Carbonate hardness	254·0
Non-carbonate hardness	106·0
Chloride	24·0
Alkalinity as $CaCO_3$	254·0
Free ammonia	0·05
Albuminoid ammonia	0·01
Iron	8·0
Total solids	400·0
pH	7

In former years yields of a few hundred gallons per day for domestic and farm use were obtained from superficial deposits within this district. As sources of this nature are prone to pollution the abstraction of water from them has, for the most part, been discontinued. D.A.C.M., J.H.H.

References

ANON. Annual. *The Surface Water year-book of Great Britain.* London: H.M.S.O.

BROWELL, E. J. and KIRKBY, J. W. 1866. On the chemical composition of various beds of Magnesian Limestone and associated Permian rock of Durham. *Nat. Hist. Trans. Northumb.*, **1,** 204–30.

MILLS, D. A. C. and HULL, J. H. 1961. pp. 19–23 and Map 3 in *Wear and Tees hydrological survey. Hydrometric areas 24 and 25.* London: H.M.S.O.

WILKINSON, D. and SQUIRES, N. 1964. *The Water Engineers Handbook.* London.

Appendix 1

SECTIONS MEASURED AT OUTCROP

Crag Gill

6-in NZ 02 SW. Section exposed in stream adjacent to White House [0268 2362]. Measured by J. H. Hull and G. Richardson; collected by G. Richardson and M. D. Traynor. Shown graphically in Fig. 9(3).

	Thickness ft	in	Total ft	in
CARBONIFEROUS				
NAMURIAN (MILLSTONE GRIT SERIES)				
FIRST GRIT (lower part of lower leaf)				
Sandstone, grey thick- bedded and massive medium- to coarse-grained feldspathic; small quartz pebbles [0262 2539] ..	20	0	20	0
Sandstone, grey silty thin-bedded carbonaceous micaceous laminae ..	9	0	29	0
Mudstone, grey sandy micaceous; fauna from basal 10 in includes zaphrentoid coral indet., *Archaeocidaris sp.* [spine], crinoid debris, *Fenestella sp.*, *Rhombopora sp. nov*, *Crurithyris sp.*, orthotetoid indet., *Productus carbonarius*, *Rugosochonetes sp. nov.*, *Spirifer ?*, *Coleolus* cf. *namurcensis*, *Euphemites sp.*, turreted gastropod indet., *Aviculopecten sp.*, *Euchondria sp. nov.*, *Palaeoneilo sp.*, *Paleyoldia macgregori*, *Catastroboceras* cf. *neilsoni*, *Metacoceras ?*, coiled nautiloid indet., *Reticuloceras stubblefieldi*, *Dithyrocaris sp.*, *Hollinella sp.*, ostracods indet., fish teeth (loc. 20a) ..	7	0	36	0
WHITEHOUSE LIMESTONE				
Limestone, grey impure; productoid fragments indet.; crinoid debris (loc. 19a)	0	11	36	11
Sandstone, grey silty thin-bedded; irregular micaceous carbonaceous laminae	11	0	47	11
Mudstone, dark grey shaly silty; micaceous and carbonaceous	0	2	48	1
Sandstone, grey calcareous fine-grained ..	1	0	49	1
Sandstone, grey silty cross-bedded; abundant micaceous carbonaceous laminae; silty mudstone bands	8	0	57	1
Mudstone, grey silty; sandy at top ..	2	0	59	1

	Thickness ft	Thickness in	Total ft	Total in
Mudstone, grey silty calcareous; *Rhombopora sp. Orbiculoidea sp., Productus carbonarius, Schizophoria sp., Coleolus* cf. *namurcensis* (loc. 18b)	4	6	63	7
Limestone, grey to dark grey argillaceous and arenaceous; crinoidal; *Rugosochonetes sp., Sanguinolites sp.* (loc. 18b)	1	2	64	9
Mudstone, grey sandy micaceous; rare carbonized plant fragments	0	6	65	3
Sandstone, grey thin irregularly bedded; micaceous carbonaceous laminae ..	2	9	68	0
Seatearth-sandstone ..	0	2	68	2
Mudstone, grey sandy; very micaceous; slightly carbonaceous	0	8	68	10
Sandstone, buff to pale grey thick-bedded micaceous irregular carbonaceous laminae; sporadic 'worm markings'	4	0	72	10
Mudstone, grey silty calcareous at top; *Dictyoclostus sp., Lingula sp., Productus carbonarius, Rugosochonetes sp., Coleolus sp., Retispira sp., Sanguinolites* aff. *v-scriptus, Streblochondria sp.* (loc. 18b)	16	6	89	4
Mudstone, grey silty calcareous; limestone nodules in basal 2 in; brachiopods	0	4	89	8
Mudstone, grey silty micaceous	0	4	90	0
Mudstone, grey silty micaceous; rare plant fragments ..	0	10	90	10
Seatearth-mudstone, black; coal fragments	0	1½	90	11½
Seatearth-mudstone, pale grey; abundant slipped surfaces ..	2	6	93	5½
Ganister, pale grey fine-grained	1	0	94	5½
Mudstone, grey silty calcareous; silty bands; crinoid columnals, *Cornulitella sp., Rhombopora sp., Dictyoclostus sp., Eomarginifera sp.*, orthotetoid indet., *Productus carbonarius, Rugosochonetes sp., Donaldina sp., Platyceras sp.* [juv.], *Streblochondria sp.* (loc. 18b) ..	16	0	110	5½
Horizon of GRINDSTONE LIMESTONE	—	—	—	—
Ganister, grey compact fine-grained.. ..	1	8	112	1½
Not exposed	2	0	114	1½
GRINDSTONE SILL [0258 2343]				
Sandstone, grey silty micaceous ..	3	0	117	1½
Seatearth-mudstone, silty; sandy bands; very carbonaceous at top; small coal fragments at base	4	9½	121	11
Sandstone, massive at top; thin-bedded at base carbonaceous and micaceous	16	0	137	11
Mudstone, grey silty micaceous; sandstone ribs; sporadic roots ..	6	0	143	11
Seatearth-sandstone, grey silty micaceous; slightly carbonaceous ..	2	0	145	11

	Thickness ft	in	Total ft	in
Ganister, grey fine-grained	4	6	150	5
Not exposed	3	0	153	5
Limestone, grey argillaceous fine-grained; iron-cemented calcareous mudstone bands; rare brachiopods including *Spirifer* cf. *trigonalis*	2	0	155	5
Seatearth-mudstone, grey silty	2	6	157	11
Mudstone, sandy; carbonaceous; rare carbonaceous rooty markings	4	0	161	11

Durham Motorway near Cleasby

6-in NZ 21 SW. Section [2458 1231] on east side of cutting on Durham motorway (A1(M)) near High Farm, Cleasby. Measured by D. A. C. Mills. Shown diagrammatically in Fig. 15, Plate XIVA.

	Thickness ft	in	Total ft	in
PERMIAN STRATA (see p. 133)				
CARBONIFEROUS				
NAMURIAN (MILLSTONE GRIT SERIES)				
Sandstone, reddened shaly ripple-marked micaceous; siltstone ribs and interlaminae	12	0	12	0
Sandstone, pale grey and brown; sporadic ripple-marked siltstone bands	7	0	19	0
Siltstone, grey	2	0	21	0
Sandstone, grey and green streaked; flaggy and shaly	3	0	24	0
Mudstone, grey and green finely micaceous clayey	1	0	25	0
Mudstone, purple and red shaly sparingly micaceous	10	0	35	0
Sandstone, grey and red well-bedded siliceous; sporadic shaly bands	7	0	42	0

Harthorn Quarry, Aldbrough

6-in NZ 21 SW. Measures exposed in old quarry [2060 1100] 350 yd SW of St Paul's Church, Aldbrough. Measured by D. A. C. Mills; collected by J. Pattison.

	Thickness ft	in	Total ft	in
CARBONIFEROUS				
NAMURIAN (MILLSTONE GRIT SERIES)				
TOP LITTLE LIMESTONE				
Limestone, flaggy argillaceous; calcareous shale partings especially at top; limestone sandy towards base; *Hyalostelia smithii*, *Alitaria panderi*, turreted gastropod indet.	12	3	12	3
Mudstone and shale, calcareous; thin bedded argillaceous limestone bands except at base; *Hyalostelia smithii*, *Fenestella* cf. *frutex*, *Fenestella sp.*, *Fistulipora sp.*, *Cleiothyridina sp.*, pectinoid indet.	16	8	28	11

	Thickness ft	in	Total ft	in
BOTTOM LITTLE LIMESTONE				
Limestone, dark grey hard massive impure fine-grained; calcareous shale band at top; *Alitaria* cf. *panderi*, *Antiquatonia sp.*, *Martinia sp.*, *Phricodothyris sp.*, *Rhipidomella michelinia*, *Schuchertella sp.*, *Spirifer* cf. *trigonalis* ..	2	8	31	7

Hedrick Gill

6-in NZ 01 NE and NZ 02 SE. Section exposed in Forthburn Beck [0569 2115], Hedrick Gill and Old Mill Gill [0736 1966]. Measured by G. Richardson and J. H. Hull. Shown graphically in Fig. 9(5).

	Thickness ft	in	Total ft	in
CARBONIFEROUS				
NAMURIAN (MILLSTONE GRIT SERIES)				
Sandstone, buff to brown coarse-grained [0569 2115] ..	20	0	20	0
Not exposed	11	0	31	0
Sandstone, grey thin locally cross-bedded, micaceous carbonaceous laminae; sporadic siltstone bands	45	0	76	0
Mudstone, grey silty micaceous; sandy laminae; plant fragments often carbonized becoming very abundant towards top	9	2	85	2
Mudstone, grey silty micaceous ?calcareous; blocky iron-cemented nodular bands; flaggy sandstone band 4 ft from base; plant fragments and worm tubes	4	0	89	2
Not exposed	4	0	93	2
Mudstone, grey silty; micaceous carbonaceous laminae; sandstone band at base	9	0	102	2
Siltstone, grey micaceous carbonaceous cross-bedded; ½-in fine-grained sandstone bands ..	5	0	107	2
Mudstone, grey silty calcareous micaceous; shaly bands; fauna of crinoids, bryozoa, brachiopods including productoids and spiriferids; bivalves ..	9	0	116	2
Mudstone, grey tinged green silty calcareous micaceous; abundant carbonaceous patches; iron-cemented bands; fauna as above	30	6	146	8
Not exposed .. about	1	0	147	8
Mudstone, grey silty micaceous; pyritous towards base; rare bivalves, gastropods and ostracods, ?worm tubes at base	8	0	155	8
Not exposed	3	0	158	8
FOSSIL SANDSTONE [0642 2009]				
Sandstone, grey thin-bedded medium-grained calcareous; abundant carbonaceous patches; locally rooty; 'cauda-galli' markings; sporadic crinoid ossicles and brachiopods	19	0	177	8

	Thickness ft	in	Total ft	in
Sandstone, grey fine-grained calcareous; irregular carbonaceous laminae; pyrite and calcite veinlets	1	4	179	0
Mudstone (partly exposed); rare brachiopods and gastropods at top ..	27	0	206	0
Sandstone; silty bands; abundant micaceous carbonaceous laminae	5	0	211	0
Sandstone, grey hard slightly calcareous; (assumed equivalent of UPPER FELLTOP LIMESTONE) ..	1	6	212	6
Mudstone, grey silty; silty sandstone bands	5	0	217	6
Mudstone, grey silty micaceous	5	0	222	6
Sandstone, pale grey thin-bedded fine grained; carbonaceous and micaceous	4	6	227	0
Sandstone, grey to dark grey nodular weathering fine-grained calcareous irony; plant fragments ..	1	6	228	6
Sandstone, grey silty thin-bedded; carbonaceous micaceous laminae	5	0	233	6
Sandstone, as at 228 ft 6 in	1	6	235	0
Sandstone, grey silty micaceous; siltstone and silty mudstone bands	6	0	241	0
Not exposed	8	0	249	0
Sandstone, grey thin-bedded fine-grained; micaceous carbonaceous; siltstone bands	22	0	271	0
Mudstone, grey silty micaceous carbonaceous; siltstone bands [0736 1966] ..	7	0	278	0

How Beck

(composite section)

6-in NY 91 NE. (31-Brough under Stainmore district). Measured by C. R. Burch; collected by J. Pattison. Shown graphically in Fig. 9(4).

	Thickness ft	in	Total ft	in
CARBONIFEROUS				
NAMURIAN (MILLSTONE GRIT SERIES)				
Sandstone, white thin-bedded fine-grained limonitic; jointed; coal streaks and plant debris in lower 1 ft	4	6	4	6
Sandstone, [9626 1755] silicified; phosphate nodules; fish teeth and scales ..	0	3	4	9
Mudstone, calcareous; limestone rib; *Fenestella sp.*, orthotetoid fragments, *Productus carbonarius*, *Quasiavonia sp.*, *Rugosochonetes sp.*, *Schizophoria sp.*, *Spirifer* cf. *bisulcatus*, *Aviculopecten sp.*, *Weberides sp.* [pygidium] (loc. 18a)	2	11	7	8
BOTANY LIMESTONE				
Not exposed (limestone elsewhere)	6	7	14	3
Limestone, bluish grey thin-bedded fine-grained; *Dibunophyllum bipartitum biparti-*				

	Thickness ft	in	Total ft	in
tum, *Buxtonia sp. nov.*, *Dictyoclostus sp.*, *Reticularia sp.*, *Rugosochonetes sp.*, *Schizophoria sp.*, cochliodont fish tooth (loc. 17a)	2	5	16	8
Not exposed: probably limestone at top and shale with sandstone ribs in lower half ..	22	0	38	8
Sandstone, grey thin-bedded fine-grained; plants and carbonaceous streaks ..	1	6	40	2
Shale, grey silty; sandstone ribs at top; pyrite and ironstone nodules at base ..	12	5	52	7
Coal [9630 1751] ..	0	5	53	0
Seatearth-sandstone grey fine-grained ..	0	3	53	3
Seatearth-mudstone, pale grey	1	0	54	3
BOTANY GRIT				
Sandstone, grey fine-grained, becoming coarser downwards	3	0	57	3
Fault (cutting out lower 20 ft of BOTANY GRIT and 30 ft of underlying beds: mainly shales)	50	0	107	3
Shales, grey; siltstone and thin sandstone ribs	26	0	133	3
Sandstone, white thin-bedded fine-grained micaceous limonitic	1	0	134	3
Shales, grey silty; siltstone and sandstone ribs common in top 4 ft	25	0	159	3
FOSSIL SANDSTONE [9700 1840]				
Sandstone, brown fine-grained siliceous limonitic; crinoids and shell casts	7	6	166	9
Not exposed	8	0	174	9
Sandstone, white and brown thin-bedded fine-grained micaceous carbonaceous; shaly at base ..	30	0	204	9
Shales, grey silty micaceous; siltstone and sandstone ribs ..	33	0	237	9
Shales, dark grey crinoidal; *Crurithyris sp.*, *Martinia sp.*, productoid fragments, *Euchondria?* [incomplete]; (loc. 16b)	2	0	239	9
UPPER FELLTOP LIMESTONE [9740 1907] ..				
Limestone, grey fine-grained slightly arenaceous; small shell fragments ..	1	3	241	0

Percy Beck

6-in NZ 01 NW. Measured by C. R. Burch and I. C. Burgess, and collected by J. Pattison. Shown graphically in Fig. 8(6).

	Thickness ft	in	Total ft	in
CARBONIFEROUS				
NAMURIAN (MILLSTONE GRIT SERIES)				
Shale	2	0	2	0
Sandstone, white thin-bedded fine-grained	30	0	32	0
Not exposed	14	0	46	0
Siltstone, grey sandy	1	0	47	0
Coal [0491 1762] ..	0	3	47	3
Seatearth-mudstone, grey	0	2	47	5
Sandstone, white and				

	Thickness ft	in	Total ft	in
brown thin-bedded fine-grained micaceous (including gaps)	56	0	103	5
Not exposed: (probably with a thin coal at base)	30	0	133	5
Sandstone-seatearth, white fine-grained micaceous	0	2	133	7
Not exposed	22	0	155	7
Sandstone, brown and dark grey thin-bedded micaceous carbonaceous ..	3	8	159	3
Siltstone, grey sandy; thin sandstone ribs..	17	0	176	3
Not exposed: (includes ?horizon of COALCLEUGH SHELL-BED)	2	0	178	3
Fault, small	—	—	—	—
Sandstone, white thin-bedded fine-grained	1	6	179	9
Shale, dark grey silty..	0	3	180	0
Sandstone, rubbly thin-bedded fine-grained feldspathic micaceous; some false-bedding	26	0	206	0
Coal .. ..1 in to	0	2	206	2
Sandstone, white thick-bedded false-bedded fine-grained.. ..	25	0	231	2
Shales and siltstones; sandstone ribs at top	26	0	257	2
Fault	—	—	—	—
Shales, grey silty; ironstone nodules and discontinuous conglomeratic sandstone ribs in top 4 to 5 ft together with the horizon of the Lower Felltop Limestone (loc. 14a)	34	0	291	2
Shales, black at base darker grey above; ironstone nodules above; fossiliferous especially at base; *Eomarginifera sp.*, *Orbiculoidea nitida*, *Rugosochonetes sp.*, *Spirifer* cf. *trigonalis*, *Straparollus sp.*, *Aviculopecten sp.*, *Palaeoneilo sp.*, *Polidevcia attenuata*, *Weberides sp.*, (loc. 15a)	20	0	311	2
ROOKHOPE SHELL-BEDS LIMESTONE [0471 1700]				
Limestone, grey fine-grained	0	4	311	6
Shale, black ..	0	4	311	10
Limestone, grey fine-grained crinoidal; *Lingula sp.*, orthotetoid indet., *Spirifer sp.*, smooth spiriferoid indet., *Edmondia sp.* (loc. 14a)	1	6	313	4
Shale, brown soft; calcareous; shell debris	0	6	313	10
Shale, dark grey sandy	0	8	314	6
Sandstone, white and brown fine-grained; slightly calcareous; siliceous wisps and plant debris at base	1	10	316	4
Shale, grey sandy; plants	0	2	316	6
Shale, black planty	0	8	317	2
Shale, coaly ..	0	1	317	3
Seatearth-mudstone, pale grey rooty at top	3	0	320	3
Not exposed ..	14	0	334	3
Limestone, grey fine-grained thin-bedded crinoidal; hematite at top [0469 1694] ..	2	4	336	7
Sandstone, brown fine-grained thick-bedded siliceous feldspathic; leached with shell casts in top 2 ft ..	5	0	341	7
Not exposed [including small fault] ..	5	6	347	1

	Thickness ft	in	Total ft	in
Sandstone, white fine-grained thin-bedded micaceous; some sandy shale partings	16	10	363	11
Shale, grey sandy micaceous; fine-grained white sandstone ribs	3	0	366	11
Not exposed	15	0	381	11
KNUCTON SHELL-BEDS LIMESTONE [0456 1679]				
Limestone, grey, fine-grained very thin-bedded siliceous; argillaceous partings .. 12 to	14	0	395	11

Rigg Farm

6-in NZ 01 NW. Section exposed in a tributary to Deepdale Beck about 480 yd S of Rigg Farm [0082 1660]. Measured by C. R. Burch.

	Thickness ft	in	Total ft	in
CARBONIFEROUS				
NAMURIAN (MILLSTONE GRIT SERIES)				
Shales	—	—	—	—
ROOKHOPE SHELL-BEDS LIMESTONE [0070 1616]				
Limestone	1	7	1	7
Sandstone, grey thin- and thick-bedded fine- to medium-grained calcareous; shelly and rooty in top 1 ft	7	6	9	1
Shale, grey; sandy at top; ironstone nodules in basal 1 ft ..	26	0	35	1
KNUCTON SHELL-BEDS LIMESTONE [0081 1615]				
Limestone, grey thin-bedded fine-grained siliceous; shells	17	6	52	7
Shale, dark grey; in part sandy	10	0	62	7
Sandstone, white and brown thin-bedded at base fine-grained siliceous	13	0	75	7
Not exposed	3	0	78	7
Sandstone, thin-bedded fine-grained micaceous; shell fragments in top 6 in ..	3	0	81	7
Shale, grey sandy; sandstone ribs ..	3	0	84	7
Sandstone, white thin-bedded fine-grained micaceous	6	0	90	7

River Gaunless near Butterknowle

6-in NZ 12 SW. Measures exposed on south bank of River Gaunless between a point [1082 2444] 250 yd NNE of Peathrow West Farm to a point [1032 2430] upstream 450 yd N of W of Peathrow West Farm. Measured by G. Richardson and D. A. C. Mills.

	Thickness ft	in	Total ft	in
CARBONIFEROUS				
WESTPHALIAN (LOWER COAL MEASURES)				
Sandstone, grey yellow weathering predominantly massive, locally false-bedded medium to coarse-grained; sporadic flaggy or shaly micaceous sandstone partings especially near top [1082 2444]	25	0	25	0
BOTTOM MARSHALL GREEN				
Coal, bright ..	1	0	26	0
Sandstone, ganisteroid hard rooty ..	1	0	27	0
Sandstone, flaggy predominantly fine-grained; carbonaceous micaceous; sporadic plant fragments and debris (*poorly exposed*) about	12	0	39	0
Shale, carbonaceous micaceous	1	0	40	0
Sandstone, massive fine- to medium-grained	5	0	45	0
Sandstone, shaly flaggy false-bedded carbonaceous micaceous; occasional more massive well-bedded bands (with gaps) about	6	0	51	0
Shales; thin mudstone and sandstone ribs (*only partially exposed*) .. about	30	0	81	0
GANISTER CLAY				
Coal (*not exposed*) about	1	0	82	0
Shale; sandstone ribs.. about	10	0	92	0
Sandstone, predominantly flaggy shaly thin-bedded fine- to medium-grained carbonaceous and micaceous; sandy shale bands	13	0	105	0
Shale, carbonaceous; and shaly mudstone	5	0	110	0
Shale, grey micaceous	20	0	130	0
Mudstone, sandy shaly carbonaceous and micaceous	1	6	131	6
Shale, grey; shaly mudstone bands at top with ironstone nodules; sporadic plant fragments at top ..	2	6	134	0
Shale, grey locally ferruginous; sporadic nodules about	12	0	146	0
Fireclay mudstone ..	0	2	146	2
Shale, grey barren [1032 2430] ..	2	6	148	8

River Tees, Baldersdale to Barnard Castle

(composite section)

6-in 01 NW. Measured by C. R. Burch. Shown graphically in Fig. 8(5).

	Thickness ft	in	Total ft	in
CARBONIFEROUS				
NAMURIAN (MILLSTONE GRIT SERIES)				
HORIZON OF UPPER FELLTOP LIMESTONE				
Coal [0254 1949] ..	0	2	0	2
Seatearth - mudstone, grey	0	4	0	6
Sandstone, white and brown thin- and thick- false-bedded fine-grained siliceous micaceous feldspathic	60	0	60	6

	Thickness ft	in	Total ft	in
Shales, grey silty; siltstone ribs and thin sandstone partings at base ..	27	5	87	11
Mudstone, grey silty sulphurous; planty ..	1	2	89	1
Coal [0272 1893] ..	0	10	89	11
Shale, black; coal films; ironstone nodules; sporadic slickensides; planty ..	4	4	94	3
Mudstone, grey silty; rooty; slickensides; siltstone and sandstone ribs; shell debris in basal 6 in ..	3	7	97	10
Not exposed .. about	6	0	103	10
Sandstone, white thin-bedded fine-grained feldspathic ..	14	0	117	10
Not exposed; shale debris evident ..	12	0	129	10
Coal	1	0	130	10
Shale, grey silty ..	2	0	132	10
Sandstone, white thin- and thick-bedded fine-grained micaceous carbonaceous; siltstone bands and coarse-grained lenses 37 ft to	50	0	182	10
?TANHILL COAL [0312 1824]				
Coal, shaly 12 in to	2	0	184	10
Shale, grey sandy ..	1	0	185	10
Not exposed	10	0	195	10
Sandstone, white and brown, thin-bedded fine-grained	2	0	197	10
Not exposed	25	0	222	10
Sandstone, white thin-bedded fine-grained carbonaceous feldspathic	9	0	231	10
Not exposed	6	6	238	4
COALCLEUGH COAL				
Coal, shaly; roots ..	1	4	239	8
Sandstone, grey fine-grained, roots at top	2	5	242	1
Siltstone, grey; sandstone ribs	1	10	243	11
Shale, coaly	0	4	244	3
Siltstone, grey sandy micaceous rooty ..	0	8	244	11
Sandstone, white and brown thin- to thick-bedded fine-grained micaceous; shale fragments and sporadic large feldspars	5	0	249	11
Siltstones, grey sandy	7	0	256	11
Sandstone, white thick- and false-bedded medium- to coarse-grained; small ironstone nodules in basal 2 ft 2 in ..	7	7	264	6
Siltstone, grey sandy; sporadic ironstone nodules	1	8	266	2
Shale, black coaly; roots	1	4	267	6
Not exposed	15	4	282	10
Sandstone, [0411 1739] white and brown massive coarse-grained; wedge of siltstone at top and disorientated ironstone nodules at base 12 ft to	15	0	297	10
Shale, black; ironstone nodules	4	0	301	10
Not exposed	12	0	313	10
HORIZON OF LOWER FELLTOP LIMESTONE ..	—	—	—	—

River Tees, Demesne Mill to Abbey Bridge

6-in NZ 01 NE. North bank of River Tees from Desmesne Mill [0533 1588] to Abbey Bridge [0663 1496]. Measured by G. Richardson and J. H. Hull; collected by G. Richardson. Shown graphically in Fig. 6(4).

	Thickness ft	in	Total ft	in
CARBONIFEROUS				
NAMURIAN (MILLSTONE GRIT SERIES)				
TOP LITTLE LIMESTONE				
Limestone, grey and buff predominantly fine-grained silicified; cherty laminae and patches; crinoids and some brachiopods	11	10	11	10
Limestone, grey thin-bedded cherty; 4-in crinoidal limestone band 11 in from top; abundant brachiopods	1	10	13	8
Limestone, pale grey silicified fine-grained; abundant irregular cherty laminae; ?sponge spicules, *Crurithyris sp.*, *Avonia sp.*, costate productoid indet. (loc. 5a) ..	2	0	15	8
Mudstone, grey silty finely micaceous..	0	2	15	10
Limestone, grey argillaceous silicified; cherty laminae; *Avonia* cf. *youngiana*, '*Camarotoechia*' *sp.*, *Cleiothyridina* cf. *fimbriata*, *Dielasma sp.*, *Eomarginifera* cf. *setosa*, *Martinothyris* cf. *lineata*, orthotetoid fragments, *Pugilis ?*, *Spirifer* cf. *trigonalis* (loc. 5a) 1 ft 4 in to	1	8	17	6
Mudstone, grey calcareous silty micaceous; 10 in ironstone band 3 ft from base; *Avonia sp.*, *Crurithyris sp.*, *Dielasma sp.*, *Eomarginifera sp.*, *Martinia sp.*, *Spirifer* cf. *trigonalis* (loc. 4a)	10	6	28	0
Limestone, grey argillaceous silty hard; *Phricodothyris sp.*, *Spirifer* cf. *trigonalis* (loc. 4a)	0	9	28	9
Mudstone, grey silty calcareous at top and base; *Hyalostelia parallela*, *Crurithyris sp.*, '*Camarotoechia*' *sp.*, *Dielasma sp.*, *Eomarginifera sp.*, *Echinoconchus* cf. *elegans*, orthotetoids indet., *Pugilis ?*, *Rugosochonetes sp.*, *Spirifer* cf. *trigonalis*, cf. *Edmondia laminata*, *Limipecten dissimilis* (loc. 4a) ..	10	10	39	7
Not exposed	0	11	40	6
Mudstone, grey silty calcareous micaceous; carbonaceous and pyritic patches; 'cauda-galli' structures..	0	9	41	3
Limestone, grey argillaceous fine-grained; pyritic patches; '*Camarotoechia*' *sp.*, orthotetoid fragments, *Phricodothyris sp.*, productoid fragments (loc. 4a) ..	0	10	42	1
Siltstone, grey calcareous micaceous; iron-rich bands and pyritic patches; silty mudstone bands at base; worm bores, productoid fragment	3	9	45	10
Mudstone, grey silty calcareous micaceous; pyritic patches; 'cauda-galli' structures; 'worm' markings and *Spirifer sp.*	2	0	47	10
BOTTOM LITTLE LIMESTONE				
Limestone, grey massive argilla-				

	Thickness ft	in	Total ft	in
ceous; semiconchoidal fracture; *Lingula sp.*, smooth spiriferoid indet. (loc. 3a)	1	1	48	11
Limestone, grey thin-bedded argillaceous silicified and iron cemented; 'cauda-galli' structures; 'worm' markings, indet. shell fragments ..	2	0	50	11
Limestone, grey argillaceous silicified and iron cemented; *Avonia sp.*, *Eomarginifera sp.*, *Orbiculoidea* cf. *nitida*, *Schizophoria sp.*, *Spirifer* cf. *trigonalis*, ostracods indet. (loc. 3a) ..	0	8	51	7
Limestone, grey thin-bedded argillaceous silicified; 'cauda-galli' structures; 'worm' markings; *Tornquistia* cf. *polita*, *Paraparchites sp.* (loc. 3a)	1	2	52	9
Limestone, grey fine-grained; *Avonia ?*, smooth spiriferoid indet., ostracods, crinoids and fish debris	1	6	54	3
Limestone, grey and brown fine-grained siliceous; sporadic pipe-like structures at base; rare brachiopods, ostracods and fish fragments; grades into	1	0	55	3
WHITE HAZLE				
Sandstone, brown calcareous; brachiopods	0	8	55	11
Sandstone, grey and buff, thick- to massive-bedded; current-bedded at base; predominantly fine- to medium-grained, micaceous and carbonaceous ..	13	0	68	11
Mudstone, dark grey sandy micaceous carbonaceous 0 in to	0	4	69	3
Sandstone, predominantly pale grey and buff thick often current-bedded; fine-grained micaceous carbonaceous laminae; becomes progressively more silty to base with mudstone bands; 'worm' tubes ..	19	6	88	9
Mudstone, grey micaceous carbonaceous; sandstone laminae at top ..	3	6	92	3
Mudstone, grey silty micaceous carbonaceous; ironstone nodules; rare fossils at top; plant fragments ..	6	0	98	3
Mudstone, grey silty calcareous and micaceous; abundant ironstone nodules except at top; crinoid columnals, *Fenestella frutex*, *F.* cf. *oblongata*, *Fistulipora sp.*, *Penniretepora sp.*, *Buxtonia ?*, '*Camarotoechia*' *sp.*, *Echinoconchus* cf. *elegans*, *Eomarginifera lobata*, *E.* cf. *setosa*, *Productus sp.*, *Aclisina* cf. *elongata*, *Ianthinopsis sp.*, *Pseudozygopleura* cf.				

	Thick-ness ft	in	Total ft	in
rugifera, *Retispira* sp., *Aviculopecten* cf. *clathratus*, *A.* cf. *plicatus*, *Edmondia laminata*, *Palaeoneilo* cf. *laevirostrum*, *Paleyoldia macgregori*, *Streblochondria* cf. *anisota*, ostracods, fish tooth (loc. 2a) ..	18	0	116	3
COAL SILLS				
Sandstone, grey and buff thin-bedded fine- to medium-grained; irregular micaceous carbonaceous laminae; rooty at top ..	14	0	130	3
Mudstone, grey sandy micaceous and carbonaceous; sandstone bands ..	1	6	131	9
Sandstone, pale grey thin-bedded ripple-marked fine-grained micaceous 4 ft to	8	6	140	3
Mudstone, grey silty micaceous carbonaceous; bands of siltstone and sandstone	6	0	146	3
Sandstone; buff ganisteroid in topmost 1 ft; otherwise cross-bedded micaceous and carbonaceous fine-grained	4	0	150	3
Mudstone, grey silty micaceous; pale grey sandstone bands and laminae	2	10	153	1
Sandstone, grey silty fine-grained micaceous and carbonaceous; silty mudstone bands; rare worm tubes about	12	0	165	1
Chert, black and white banded; irregular base 6 in to	0	10	165	11
GREAT LIMESTONE				
Limestone, grey thin-bedded medium-grained argillaceous bioclastic; crinoids and rare solitary corals ..	15	0	180	11
Limestone, grey thick-bedded massive medium-grained; rare zaphrentoids in upper 11 ft and rare *Dibunophyllum sp.* from 9 to 36 ft from top; *Calcifolium bruntonense* 25 ft from top	49	0	229	11
Limestone, grey to dark grey thin-bedded fine-grained argillaceous pyritous; sporadic zaphrentoids 4 to 8 ft above base at presumed horizon of *Chaetetes* Band; relatively abundant large productoids in basal 4 ft ..	8	0	237	11
VISÉAN (CARBONIFEROUS LIMESTONE SERIES)				
Sandstone-seatearth	—	—	—	—

River Tees near Gainford

6-in NZ 11 NE. Measures exposed on the south bank of the River Tees between a point [1927 1570] at Ashes Wood 490 NNE of Low Field to a point [1889 1588] 460 yd upstream and 640 yd NNW of Low Field. Measured by G. Richardson and D. A. C. Mills.

	Thickness ft	Thickness in	Total ft	Total in
CARBONIFEROUS				
NAMURIAN (MILLSTONE GRIT SERIES)				
Sandstone, grey irregularly thin-bedded silty; abundant micaceous carbonaceous patches; seatearth-siltstone bands in upper 4 ft [1927 1570] ..	9	0	9	0
Sandstone, grey hard fine- to medium-grained carbonaceous; plant fragments	4	0	13	0
Sandstone, grey and buff thin irregularly bedded sparingly micaceous; plant fragments	4	0	17	0
Siltstone, grey micaceous carbonaceous; irregularly interbedded with grey fine-grained sandstone and silty mudstone bands containing plant debris	40	0	57	0
Sandstone, grey and buff, false-, locally wedge-bedded predominantly fine-grained; sporadic thin bands with quartz pebbles; occasional coal scars 10 ft to	8	0	65	0
Sandstone, grey silty; false-bedded sandstone bands; occasional siltstone bands and coal scars ..	25	0	90	0
Mudstone, grey silty micaceous and carbonaceous	0	6	90	6
Sandstone, light grey compact fine-grained	0	8	91	2
Sandstone, buff thin-bedded abundant micaceous laminae; plant fragments ..	0	10	92	0
Sandstone, buff calcareous ganisteroid; sporadic carbonaceous patches ..	0	7	92	7
Sandstone, buff to grey false-bedded fine-grained; worm tracks and burrows ..	1	4	93	11
Sandstone, buff to grey massive at top thin-bedded at base; fine- to medium-grained; carbonaceous micaceous laminae; fine siltstone bands; occasional worm bores and tracks ..	6	0	99	11
Siltstone, grey micaceous	0	3	100	2
Sandstone, buff calcareous fine- to medium-grained; fragments of brachiopods, gastropods and bivalves	0	6	100	8
Mudstone, sandy micaceous	0	2	100	10
Mudstone, dark grey shaly carbonaceous sparingly micaceous	0	2	101	0
Sandstone, dark grey silty carbonaceous [1889 1588] ..	0	2	101	2

River Tees near Hedgeholme, Winston

6-in NZ 11 NW. Measures exposed [1474 1692] in east bank of River Tees 360 yd NE of Hedgeholme. Measured by D. A. C. Mills and G. Richardson.

	Thickness ft	in	Total ft	in
CARBONIFEROUS				
NAMURIAN (MILLSTONE GRIT SERIES)				
Sandstone, rusty brown ferruginous thin-bedded fine-grained; shaly partings ..	3	4	3	4
Sandstone, shaly micaceous and carbonaceous; shale bands	3	0	6	4
Sandstone, ferruginous	0	9	7	1
Sandstone, shaly; grades down into ..	2	0	9	1
Shale; sporadic ironstone ribs	8	0	17	1

River Tees, Winston Bridge

6-in NZ 11 NW. Measures exposed in west bank of River Tees upstream from a point [1431 1643] 180 yd below Winston Bridge to a point [1429 1629] under bridge. Measured by D. A. C. Mills and G. Richardson. Shown graphically in Fig. 8(8).

	Thickness ft	in	Total ft	in
CARBONIFEROUS				
NAMURIAN (MILLSTONE GRIT SERIES)				
YOREDALE				
Coal; *not exposed* [1431 1643] .. ?	2	6	2	6
Not exposed	10	0	12	6
Shale, dark	1	0	13	6
Fault	—	—	—	—
Sandstone, predominantly grey fine-grained; ganisteroid at top and base; sporadic plant fragments	13	6	27	0
Mudstone, blue and grey; sandy ironstone nodules	14	0	41	0
Ganister, white ..	2	0	43	0
Sandstone, ganisteroid	1	7	44	7
Mudstone, grey silty; thin ironstone ribs and iron-cemented sandstone bands ..	5	0	49	7
Sandstone, silty ganisteroid; white sandstone patches ..	1	7	51	2
Ganister	2	0	53	2
Mudstone, grey silty micaceous; sporadic ironstone ribs and sandy and silty bands	7	0	60	2
Fault (*minor*)	—	—	—	—
Mudstone, grey silty; sporadic sandy bands; iron-cemented at top	8	0	68	2
Mudstone, grey silty slightly micaceous; brachiopods and gastropods ..	24	0	92	2
LOWER FELLTOP LIMESTONE				
Limestone, grey argillaceous fine- to medium-grained; sporadic crinoids; rare brachiopods	3	0	95	2
Sandstone, grey fine-grained compact ..	2	4	97	6
Spiculite, light grey very hard cherty sandy; very fine-grained to sub-porcellanous (Sliced Rock Collection E. 32023, p. 56)	1	0	98	6
Sandstone, grey very fine-grained; chloritic patches ..	0	6	99	0
Mudstone, grey to green silty thin-bedded ..	0	3	99	3
Mudstone, mottled grey and green calcareous; rare small black nodules	4	6	103	9
Fault	—	—	—	—
Mudstone, grey slightly silty and shaly; sporadic pyritous bands and ironstone nodules; worm markings	5	6	109	3

	Thickness ft	in	Total ft	in
Fault	—	—	—	—
ROOKHOPE SHELL-BEDS LIMESTONE (base of)				
Limestone, sandy argillaceous crinoidal; shell fragments	0	6	109	9
Limestone, grey shaly argillaceous crinoidal; silty partings; sporadic bryozoa and brachiopods	1	6	111	3
Limestone, grey medium- to coarse-grained crinoidal; rare shell fragments	0	9	112	0
Limestone, grey argillaceous medium-grained: argillaceous partings; rare shell fragments	1	0	113	0
Sandstone [1424 1629]	11	0	124	0

River Tees, Ovington to Winston

6-in NZ 11 NW. Measures exposed on south bank of River Tees between a point [1428 1620] 70 yd above Winston Bridge and a point [1308 1489] 1 mile upstream and below the village of Ovington. Measured by G. Richardson and D. A. C. Mills. Collected by G. Richardson. Shown graphically in Fig. 7(8).

	Thickness ft	in	Total ft	in
CARBONIFEROUS				
NAMURIAN (MILLSTONE GRIT SERIES)				
ROOKHOPE SHELL-BEDS LIMESTONE				
Limestone, yellow to brown medium-grained; argillaceous laminae; irregular top and base; sporadic shell fragments ..	1	6	1	6
Not exposed ..	2	0	3	6
Mudstone, grey and brown calcareous silty earthy; crinoidal	1	0	4	6
Limestone, as at 1 ft 6 in	0	8	5	2
Mudstone, as at 4 ft 6 in; less crinoidal	0	6	5	8
Limestone, as at 1 ft 6 in; grey to yellow and more compact	0	10	6	6
Mudstone as at 4 ft 6 in	0	3	6	9
Limestone, argillaceous predominantly thick- to massive-bedded medium- to coarse-grained; sporadic 3 in calcareous mudstone bands; abundant crinoid debris	3	2	9	11
Mudstone, grey calcareous earthy; band of argillaceous limestone at top; *Plicochonetes sp.*	0	10	10	9
Mudstone, green mottled yellow and brown; laminated; rare crinoidal debris	0	10	11	7
Limestone, grey and green rubbly argillaceous irregularly bedded; bands of crinoidal limestone up to 4 in thick ..	2	3	13	10
Limestone, yellow and grey argillaceous predominantly medium-grained; *Semi-*				

	Thickness ft	in	Total ft	in
planus cf. *latissimus*	3	2	17	0
Mudstone, grey calcareous irregularly bedded; thin bands of coarsely crinoidal limestone at base; *Fenestella sp.*, *Eomarginifera sp.*, *Martinia sp.*, *Semiplanus* cf. *latissimus*, *Spirifer sp.*, *Tornquistia* cf. *polita* (loc. 13a)	5	9	22	9
Limestone, grey compact argillaceous medium-grained crinoidal	1	8	24	5
Limestone, grey argillaceous; crinoids and brachiopods	0	8	25	1
Mudstone, grey calcareous; brachiopods	3	8	28	9
Limestone, grey argillaceous fine- to medium-grained; muddy laminae throughout, but especially at base	4	4	33	1
Sandstone, yellow and brown massive predominantly medium-grained but with coarser patches and sporadic large quartz grains; ganisteroid with irregular carbonaceous surface at base; worm tubes and borings throughout	11	0	44	1
Sandstone, grey silty fine- to medium-grained; micaceous partings	6	6	50	7
Mudstone, grey silty sparingly micaceous; septarian ironstone nodules; *Pleuropugnoides sp.*, *Rugosochonetes sp.*, nuculoids indet. (loc. 12a) ..	15	0	65	7
Mudstone, grey and green calcareous; *Hyalostelia sp.*, *Fenestella* cf. *frutex*, productoid fragments, *Rugosochonetes sp.*, smooth spiriferoids indet., *Aviculopecten sp.*, nautiloid indet. (loc. 12a)	0	3	65	10
KNUCTON SHELL-BEDS LIMESTONE				
Limestone, grey compact fine-grained; caninoid indet. [crushed]; brachiopods including productoids	0	6	66	4
Limestone, grey argillaceous thin-irregularly bedded fine- to medium-grained; bands of more compact limestone throughout; pyritous nodules in upper 6 ft; grey very fine limestone nodules 6 ft 6 in from top; some calcite veining near top; fossiliferous throughout, but especially near base; *Hyalostelia smithii*, *Caninia sp.*, *Fenestella sp.*, *Antiquatonia sp.*, *Cleiothyridina* cf. *fimbriata*, *Echinoconchus punctatus*, *Rugosochonetes sp.*, *Spirifer* cf. *bisulcatus*, *Aviculopecten* aff. *interstitialis*, *Cypricardella sp.*, *Parallelodon sp.*, *Sulcatopinna costata*, *Catastro-*				

	Thickness ft	in	Total ft	in
boceras sp., Rayonnoceras sp., Paraparchites sp., fish tooth [1409 1591] (loc. 11a) ..	32	0	98	4
Mudstone, grey shaly silty carbonaceous and micaceous ..	0	5	98	9
Not exposed	3	3	102	0
Mudstone, grey silty calcareous; sparingly micaceous	9	0	111	0
Seatearth-sandstone, grey silty	4	0	115	0
Sandstone, dull yellow calcareous locally ochreous fine-grained; grey shaly micaceous carbonaceous	10	0	125	0
Shale and mudstone interbedded; *poorly exposed*	10	0	135	0
Mudstone, grey and yellow micaceous; sporadic shale partings	3	0	138	0
Shale, dark grey carbonaceous micaceous; thin fine-grained sandstone bands; ironstone nodules	3	3	141	3
Shale, flaky with interbedded carbonaceous micaceous mudstone; thin sandstone bands at top ..	2	6	143	9
Shale, dark grey carbonaceous micaceous	0	6	144	3
Shaly mudstone; fine shale and sandstone bands	2	6	146	9
Sandstone, muddy grey calcareous; fine shale bands ..	2	0	148	9
Sandstone, grey carbonaceous micaceous; calcareous in basal 1 ft	1	9	150	6
Mudstone, dark grey shaly carbonaceous micaceous ..	0	3	150	9
Limestone, sandy impure	1	0	151	9
Mudstone, calcareous shaly carbonaceous micaceous; ironstone nodules; plant fragments	1	6	153	3
Sandstone, grey carbonaceous and micaceous; calcareous at base with plant fragments	1	2	154	5
Shale, dark grey ..	0	3	154	8
Sandstone, calcareous; crinoidal debris ..	0	3	154	11
Shale, dark grey carbonaceous micaceous; thin often irregular bands and partings of fine-grained sandstone; ironstone ribs and nodules especially at top	10	0	164	11
Shale, grey; sporadic ironstone nodules	5	0	169	11
Shale, dark grey or black sparingly micaceous and carbonaceous; thin ironstone ribs and small ironstone nodules; plant fragments; *Fenestella sp., Rugosochonetes sp., Straparollus sp., Sanguinolites sp.* [1328 1500] (loc. 10a) about	30	0	199	11
TOP CRAG (upper leaf)				
Limestone, dark grey-brown weathering impure hard splintery fine-grained	2	4	202	3
Shale and mudstone, calcareous hard ..	1	6	203	9
TOP CRAG (lower leaf)				
Limestone, as at 202 ft 3 in [1322 1500]	2	9	206	6

	Thickness ft in	Total ft in
Shale, grey	2 3	208 9
Mudstone, calcareous shaly	0 9	209 6
Shale, calcareous; argillaceous limestone ribs	13 0	222 6
BOTTOM CRAG (upper leaf)		
Limestone, sandy hard	1 2	223 8
Shale, hard grey calcareous	1 1	224 9
BOTTOM CRAG (lower leaf)		
Limestone, grey massive impure ..	4 0	228 9
Fault	— —	— —
Sandstone, calcareous at top ferruginous at base fine-grained ..	1 8	230 5
Shale, carbonaceous micaceous	0 2	230 7
Sandstone, rubbly fine-grained	2 0	232 7
Shale, grey carbonaceous micaceous ..	0 10	233 5
Sandstone, ferruginous thin-bedded fine-grained; shale partings with sandstone blebs	3 0	236 5
Sandstone, ferruginous fine- to medium-grained; carbonaceous micaceous; coal trace [1308 1489]	2 11	239 4

Spurlswood Beck and Quarter Burn

6-in NZ 02 NW. Measured by J. H. Hull; collected by W. G. E. Graham and G. Richardson. Shown graphically in Fig. 10(4) and Plate VI.

	Thickness ft in	Total ft in
CARBONIFEROUS		
WESTPHALIAN (LOWER COAL MEASURES)		
Sandstone, grey and brown thin-bedded [0174 2655] ..	15 0	15 0
Mudstone, grey and green silty	12 0	27 0
Horizon of GANISTER CLAY COAL ..	— —	— —
Sandstone, grey silty thin-bedded ..	3 6	30 6
Mudstone, dark grey and grey slightly silty fissile	30 0	60 6
?Coal (outcrop obscured)	— —	— —
Seatearth-sandstone, grey and brown ..	1 0	61 6
Mudstone, dark grey sandy; carbonaceous and micaceous laminae; *Planolites sp.* and fish fragments	2 10	64 4
Horizon of RODDYMOOR MARINE BAND	— —	— —
Seatearth-mudstone, grey and green silty at top	1 3	65 7
THIRD GRIT		
Sandstone, grey to white medium- to fine-grained feldspathic	42 0	107 7
?Coal horizon ..	— —	— —
Sandstone, grey coarse-grained 1 ft to	2 6	110 1
Sandstone, grey weathering yellow coarse-grained cross-bedded; quartz pebbles 6 ft to	12 0	122 1
Horizon of KAYS LEA MARINE BAND	— —	— —
Seatearth-mudstone, grey and green; sphaerosiderite 1 ft to	3 0	125 1

	Thickness ft	in	Total ft	in
Seatearth-sandstone, grey silty argillaceous	1	6	126	7
Sandstone, grey to pale grey fine-grained thin-bedded; carbonaceous micaceous laminae ..	11	0	137	7
Mudstone, grey silty slightly micaceous; sandstone bands at top	10	6	148	1
QUARTERBURN MARINE BAND [0170 2676]				
Mudstone, grey shaly with sporadic limestone nodules; *Lingula mytilloides*, *Productus carbonarius*	1	7	149	8
NAMURIAN (MILLSTONE GRIT SERIES)				
Sandstone, grey to dark grey calcareous ..	0	2	149	10
Sandstone-seatearth, grey fine-grained compact siliceous ..	2	3	152	1
Sandstone, pale grey to green fine-grained; sphaerosiderite ..	1	3	153	4
Mudstone, grey; sphaerosiderite; carbonized plant fragments and fish scales at base ..	5	6	158	10
Mudstone-seatearth, grey to dark grey; coal laminae at base	1	3	160	1
Coal	0	6	160	7
Mudstone-seatearth, grey silty	2	0	162	7
SECOND GRIT (upper leaf)				
Sandstone-seatearth, grey slightly micaceous	3	0	165	7
Sandstone, pale grey to white thin- and false-bedded medium-grained ..	8	0	173	7
Sandstone, grey yellow and white massive and cross-bedded coarse-grained; quartz pebbles in lower part	52	0	225	7
Mudstone, grey silty	9	6	235	1
Sandstone, brown fine- to medium-grained thick-bedded ..	3	6	238	7
Mudstone, grey silty; sandy laminae ..	2	0	240	7
Sandstone, grey silty thin-bedded; sandy carbonaceous mudstone laminae ..	5	0	245	7
Mudstone, grey silty; carbonaceous laminae	8	0	253	7
SPURLSWOOD SHELL-BEDS				
Mudstone, grey; *Lingula mytilloides*, orthotetoid indet., *Productus carbonarius*, *Rugosochonetes sp.*, *Retispira sp.*, *Sanguinolites ?*, *Serpuloides sp.*	0	6	254	1
Mudstone, dark grey	12	0	266	1
Siltstone, grey ..	2	6	268	7
Mudstone, dark grey silty	2	0	270	7
Mudstone, dark grey; *Lingula mytilloides*, *Orbiculoidea* cf. *nitida*, orthotetoid indet., *Productus carbonarius*, *Rugosochonetes* cf. *hindi*, *Retispira sp.*, *Actinopteria regularis*, *Edmondia ?*, *Palaeoneilo ?*, *Serpuloides sp.*, *Paraconularia sp.*	0	6	271	1
Mudstone, grey to pale grey micaceous ..	8	0	279	1
Mudstone, grey silty, pale grey sandy laminae [0226 2686] ..	6	6	285	7

Stobgreen Sike

6-in NZ 02 SW. Measured by J. H. Hull; collected by G. Richardson. Shown graphically in Fig. 6 (3).

	Thickness ft	in	Total ft	in
CARBONIFEROUS				
NAMURIAN (MILLSTONE GRIT SERIES)				
Sandstone, white coarse-grained; mineralized [0056 2410]	31	0	31	0
Shale (including small gaps)	8	0	39	0
Mudstone	2	0	41	0
?WHITEHOUSE LIMESTONE				
Limestone, dark grey impure; crinoidal	5	0	46	0
Fault	—	—	—	—
CRAG LIMESTONE				
Limestone, argillaceous coarse-grained; mineralized	2	0	48	0
Fault	—	—	—	—
Mudstone, grey silty	4	6	52	6
Sandstone, pale grey medium- to coarse-grained	2	0	54	6
Seatearth-mudstone	1	0	55	6
Not exposed	8	0	63	6
Sandstone	1	0	64	6
Not exposed	16	0	80	6
Mudstone, grey calcareous; shell fragments; poorly exposed	13	0	93	6
CRAG LIMESTONE				
Limestone, grey fine- to medium-grained argillaceous; crinoid and brachiopod debris	2	6	96	0
FIRESTONE SILL				
Sandstone, brown thick-bedded medium-grained micaceous; jointed ..	40	0	136	0
Not exposed including Fault	18	0	154	0
LITTLE LIMESTONE				
Limestone, grey fine-grained; calcite veins; brachiopod fragments ..	10	0	164	0
Not exposed (including WHITE HAZLE)	7	0	171	0
Sandstone, grey thin-bedded carbonaceous micaceous; rooty	2	0	173	0
Not exposed	10	0	183	0
Mudstone, grey calcareous; brachiopods	7	0	190	0
Limestone, grey argillaceous; brachiopod fragments	2	0	192	0
Mudstone, grey calcareous; brachiopods	0	11	192	11
Limestone, grey argillaceous fine-grained; brachiopod fragments	0	8	193	7
Mudstone, grey calcareous; crinoid debris; trepostomatous bryozoa, '*Camarotoechia*' *sp.*, *Dictyoclostus sp.*, *Eomarginifera sp.*, orthotetoid indet., *Martinia sp.*, *Rugosochonetes sp.*, *Spirifer* cf. *trigonalis* (loc. 2b)	14	6	208	1
Mudstone, grey calcareous; '*Camarotoechia*' *sp.*, *Coleolus sp.*, bellerophontoid indet., *Palaeoneilo sp.* (loc. 2b)	8	0	216	1
Not exposed	7	0	223	1
Mudstone, grey, calcareous; '*Camarotoechia*' *sp.*, *Lingula sp.* [large], *Spirifer sp.* (loc. 2b)	5	4	228	5

	Thickness ft	in	Total ft	in
COAL SILLS				
Sandstone, thin-bedded fine-grained	12	6	240	11
Mudstone, grey silty micaceous; ironstone bands	5	0	245	11
Mudstone, grey silty; *Fenestella sp.*, *Eomarginifera sp.*, *Euphemites sp.* (loc. 2b)	2	6	248	5
Seatearth-mudstone; sandy laminae . .	0	3	248	8
Sandstone (including 2 ft gap); siliceous at top	8	3	256	11
Not exposed . .	2	0	258	11
Mudstone, grey silty micaceous; *Eomarginifera sp.* . .	0	6	259	5
Sandstone (including small gaps) . .	35	0	294	5
Mudstone, grey silty micaceous; *Paleyoldia macgregori* (loc. 2b)	4	0	298	5
Sandstone, grey cross-bedded fine- to medium-grained . .	4	6	302	11
Shale, dark grey; sandstone ribs [0019 2378]	2	0	304	11
Horizon of GREAT LIMESTONE	—	—	—	—

Whorlton Beck

6-in NZ 11 NW and NZ 11 SW. Measures exposed in Whorlton Beck from a point [1067 1519] 230 yd ENE of Whorlton Grange downstream to the confluence [1089 1456] with the River Tees. Measured by D. A. C. Mills and J. H. Hull. Collected by W. G. E. Graham. Shown graphically in Fig. 6(5).

	Thickness ft	in	Total ft	in
CARBONIFEROUS				
NAMURIAN (MILLSTONE GRIT SERIES)				
Shale, dark grey calcareous micaceous; calcareous mudstone bands; *Crurithyris sp.*, *Echinoconchus sp.*, *Eomarginifera sp.*	20	0	20	0
TOP CRAG LIMESTONE				
Limestone, ochreous earthy; sporadic shaly ribs . .	2	8	22	8
Mudstone, shaly calcareous; more massive at base; *Antiquatonia sp.*, *Eomarginifera sp.*, *Brachythyris sp.*, *Martinia sp.*, *Orbiculoidea nitida*, *Schuchertella sp.*, *Spirifer* cf. *trigonalis*, *Straparollus* cf. *carbonarius*, *Catastroboceras sp.*, (loc. 9a) . .	2	0	24	8
Limestone, fine- to medium-grained; locally ochreous; platy at top; sporadic flaggy ribs and calcareous mudstone partings; *Crurithyris sp.*, *Brachythyris sp.*, *Linoprotonia sp.*, *Martinia sp.*, (loc. 9a)	5	0	29	8
Shale, dark grey and black calcareous; bryozoa indet., *Antiquatonia* cf. *insculpta*, *Crurithyris sp.*, *Linoprotonia sp.*, *Tornquistia* cf. *polita*, *Cypricardella*				

	Thickness ft	in	Total ft	in
sp., ostracods (loc. 8a)	10	0	39	8
Limestone, flaggy irregularly-bedded; grades into calcareous shaly mudstone at base 6 in to	0	10	40	6
Mudstone, shaly calcareous; some poorly preserved brachiopods	1	2	41	8
BOTTOM CRAG LIMESTONE				
Limestone, grey siliceous fine-grained; sporadic cherty streaks and carbonaceous patches [1067 1500] *Cleiothyridina sp.*, *Spirifer* cf. *trigonalis* (loc. 7a)	9	0	50	8
FIRESTONE SILL				
Sandstone, dark grey fine-grained; abundant cauda-galli markings at top	2	4	53	0
Mudstone, grey silty micaceous carbonaceous	0	8	53	8
Not exposed ..	2	0	55	8
Sandstone, grey fine-grained; becoming medium-grained downwards; micaceous with sporadic carbonaceous streaks	2	6	58	2
Mudstone, dark grey silty; sandstone bands	1	2	59	4
Mudstone, dark grey; pyritized bivalves	1	6	60	10
Sandstone, grey thinly-bedded fine-grained siliceous; carbonaceous scars	4	6	65	4
Mudstone, dark grey silty micaceous; sandstone bands at top; sporadic bivalve fragments; indet. shell fragments at base ..	6	0	71	4
Not exposed ..	2	6	73	10
Sandstone, grey to brown thin-bedded fine-grained limonitic ..	3	6	77	4
Sandstone, grey massive false-bedded coarse-grained ..	5	4	82	8
Mudstone, grey silty micaceous; planty	1	0	83	8
Sandstone, grey thin-bedded fine-grained limonitic; coal and carbonaceous scars; rooty including *Cordaites sp.* ..	4	6	88	2
Siltstone; alternates with grey fine-grained sandstone	2	4	90	6
Shale, dark grey; *Euphemites sp.*	2	2	92	8
Not exposed	6	6	99	2
FARADAY HOUSE SHELL-BED				
Limestone, dark grey fine-grained ..	0	8	99	10
Shale, dark grey; *Cleiothyridina sp.*, productoid indet., *Spirifer sp.* (loc. 6a)	0	2	100	0
Limestone, dark grey fine-grained; brachiopods ..	0	10	100	10
Shale, dark grey ..	0	6	101	4
Limestone, grey rubbly fine-grained; shelly	1	0	102	4
Sandstone, grey flaggy and thinly-bedded ripple-marked fine-grained micaceous	10	0	112	4
Sandstone, grey shaly; abundant silty laminae	8	6	120	10
Shale, grey silty ..	0	2	121	0
Ironstone rib ..	0	3	121	3
Shale, dark grey micaceous; small ironstone nodules ..	1	9	123	0

	Thickness ft	in	Total ft	in
Ironstone nodule band	0	2	123	2
Shale, dark grey; ironstone nodules ..	7	0	130	2
Shale, dark grey; bryozoa, brachiopods, gastropods, crinoids	4	0	134	2
Ironstone; jointed ..	0	4	134	6
Shale, dark grey; septarian nodules; fossiliferous	7	0	141	6
Limestone, dark grey, argillaceous fine-grained; shelly ..	0	6	142	0
Ironstone, dark grey fine-grained.. ..	0	6	142	6
Shale, grey calcareous; *Fenestella* cf. *frutex*, trepostomatous bryozoa, *Brachythyris sp.*, *Cleiothyridina* cf. *fimbriata*, *Dielasma sp.*, *Echinoconchus sp.*, *Pleuropugnoides sp.*, *Rugosochonetes sp.*, *Palaeoneilo sp.*, *Schizodus sp.* ..	1	2	143	8
Limestone, grey siliceous fine-grained; shelly	0	3	143	11
Shale, dark grey; *Fenestella* cf. *frutex*, trepostomatous bryozoan indet., *Buxtonia sp.*, *Dielasma sp.*, *Echinoconchus* cf. *elegans*, *E. punctatus*, *Eomarginifera sp.*, *Lingula* cf. *squamiformis*, *Pleuropugnoides* cf. *pleurodon*, *Rugosochonetes sp.*, *Spirifer* cf. *trigonalis*, *Aviculopecten sp.*, *Limipecten dissimilis*, *Pernopecten sowerbii*, *Schizodus sp.* (loc. 6a)	2	4	146	3
Limestone, dark grey siliceous, fine-grained; argillaceous and shaly at base; *Brachythyris sp.*, *Pleuropugnoides sp.*, productoid fragment indet.	0	6	146	9
Shale, dark grey irony; *Antiquatonia sp.*, *Cleiothyridina sp.*	3	0	149	9
Limestone, dark grey fine-grained siliceous; argillaceous at base; orthotetoid indet., *Spirifer* cf. *bisulcatus*	0	6	150	3
Shale, dark grey; *Antiquatonia sp.*, ..	2	6	152	9
Limestone, dark grey argillaceous fine-grained; shelly ..	0	9	153	6
Shale, grey calcareous; shelly	3	9	157	3
Limestone, grey well-bedded, hard siliceous fine-grained; argillaceous and thinly-bedded at base; productoid fragments 11 in to	1	2	158	5
Shale, dark grey fossiliferous	1	2	159	7
Not exposed	6	0	165	7
Shale, dark grey; *Fenestella* cf. *frutex*, *Cleiothyridina* cf. *fimbriata*, *Dictyoclostus sp.*, *Echinoconchus* cf. *punctatus*, *Lingula sp.*, *Schizophoria resupinata*, *Spirifer sp.* (loc. 6a)	3	6	169	1
Not exposed	9	0	178	1
Shale, dark grey; *Catastroboceras sp.* (loc. 6a)	2	6	180	7
Limestone, dark grey fine-grained; weathers to dark olive colour	0	8	181	3
Shale, dark grey; *Rugosochonetes sp.*, *Edmondia* cf. *laminata*, *Parallelodon* cf. *semicostatus;* orthocone nautiloid (loc. 6a)..	2	6	183	9
Shale, calcareous; *Fenestella sp.*, tre-				

	Thickness ft	Thickness in	Total ft	Total in
postomatous bryozoan indet., *Cleiothyridina* cf. *fimbriata*, *Dielasma sp.*, *Aviculopecten sp.*, *Limipecten dissimilis*, *Parallelodon* cf. *semicostatus*, *Pernopecten sowerbii* (loc. 6a) ..	4	0	187	9
TOP LITTLE LIMESTONE				
Limestone, grey crinoidal; *Composita sp.*, *Echinoconchus sp.*, *Eomarginifera sp.*	0	9	188	6
Shale, grey calcareous; *Cleiothyridina sp.*, *Spirifer* cf. *bisulcatus* ..	2	7	191	1
Limestone, fawn to grey well-bedded fine-grained; *Buxtonia sp.*, *Composita sp.*, orthotetoid indet. ..	1	2	192	3
Shale	0	3	192	6
Limestone, grey fine-grained; crinoidal and shelly ..	0	9	193	3
Shale, dark grey calcareous; shelly ..	0	6	193	9

Appendix 2

RECORDS OF SELECTED SHAFT AND BOREHOLE SECTIONS

SECTIONS of some of the more important boreholes and shafts are given below. Most of them are abridged to a greater or lesser extent.

The logs of the majority of boreholes and shafts sunk before 1910 in this district were published in "An account of the strata of Northumberland and Durham as proved by borings and sinkings" issued in seven volumes (but usually bound in four volumes) between 1878 and 1910 by the North of England Institute of Mining and Mechanical Engineers. In these volumes the records are grouped according to locality (usually parishes), and in the text below a selection of them are given in summary form referred to by the abbreviation "B & S" followed by the serial number of the entry. In a small number of cases the depths to coal seams in the "B & S" logs differ slightly from more recent calculations by the National Coal Board.

In those records that are based on old logs local miners' terms are often used. A glossary of these is as follows:

Balls	Ironstone nodules
Bastard	Generally applied to intermediate rock types not sufficiently described under a single term
Bind	Silty mudstone
Black stone	Shale
Blaes	Mudstone
Brass	Iron pyrites
Fakey	Silty
Girdle	Thin band or rib
Grey Beds	Alternations of mudstone, siltstone and sandstone; sandstone fraction often predominates
Metal	Mudstone or Shale
Metal Stone	Alternations of mudstone, siltstone and sandstone
Post	Sandstone
Post girdles	Thin bands of sandstone in metal stone
Ramble	Sandstone
Rash	Irregular streaks or films of coal, often highly disturbed
Scares	Irregular streaks or discontinuous bands
Seggar	Fireclay, seatearth
Splint	Poor quality coal
Thill	Seatearth
Whin	A hard stone e.g. ganister, ironstone, dolerite or basalt. In this context almost invariably a ganister or ironstone.

The abbreviation AOD refers to levels above Ordnance Datum.

n

Arnghyll No. 1 Bore

Surface about 960 ft AOD. 6-in NZ 02 SE. [0668 2450] 650 yd WSW of Burfoot Leazes. Drilled 1950 by NCB. Cores examined by R. H. Price. Plate VII (9).

	Thickness ft	in	Depth ft	in
SUPERFICIAL DEPOSITS				
Soil and Boulder Clay	8	3	8	3
CARBONIFEROUS				
WESTPHALIAN (LOWER COAL MEASURES)				
Sandstone, predominantly brown medium-grained; grey flaggy sandstone bands ..	55	11	64	2
VICTORIA				
Coal	0	11	65	1
Sandstone, grey ganisteroid	2	0	67	1
Sandstone, flaggy curly-bedded; ironstone pellets	1	4	68	5
Sandstone, brown medium-grained; flaggy partings	3	9	72	2
Shale, grey sandy ..	2	7	74	9
Sandstone; shale partings towards base ..	20	5	95	2
Mudstone, blue sandy	0	9	95	11
Mudstone, dark grey and black sandy micaceous; ?fish scale at 97 ft ..	6	1	102	0
Mudstone, dark grey sandy	8	0	110	0
Shale, dark grey; high dip 102 ft to 111 ft 6 in	1	6	111	6
Mudstone, sandy ..	1	8	113	2
Sandstone, ankeritic micaceous	0	10	114	0
Mudstone, dark sandy	22	5	136	5
Sandstone	1	0	137	5
Mudstone, blue sandy; irony at base ..	6	4	143	9
BOTTOM MARSHALL GREEN				
Coal	0	6	144	3
Fireclay, sandy partings	7	9	152	0
Sandstone, grey medium- to coarse-grained	23	0	175	0
GANISTER CLAY				
Coal	2	4	177	4
Fireclay; sandy at base	2	8	180	0

Arnghyll No. 3 Bore

Surface about 975 ft AOD. 6-in NZ 02 SE [0667 2455], about 300 yd ESE of Hill House, Copley. Drilled 1950 by NCB.

	Thickness ft	in	Depth ft	in
SUPERFICIAL DEPOSITS				
Soil and Boulder Clay	8	1	8	1
CARBONIFEROUS				
WESTPHALIAN (LOWER COAL MEASURES)				
TOP MARSHALL GREEN				
Coal	1	11	10	0
Seggar	4	0	14	0
Shale; sandstone partings	8	10	22	10
BOTTOM MARSHALL GREEN				
Coal	1	3	24	1
Band	0	2	24	3
Coal	0	5	24	8
Seggar	2	1	26	9
Shale, blue; sandstone partings	12	3	39	0
Sandstone	4	0	43	0
Sandstone, grey; thin shale partings ..	1	11	44	11
Sandstone, brown ..	24	11	69	10
Sandstone, grey; coal pipings	0	2	70	0
Shale, blue	3	0	73	0
Shale, grey sandy ..	6	4	79	4
Post, grey	0	6	79	10

	Thickness ft	in	Depth ft	in
GANISTER CLAY				
Coal	1	0	80	10
Band	0	2	81	0
Coal, foul	0	6	81	6
Seggar	0	6	82	0
Sandstone, grey; black metal streaks ..	6	4	88	4
Shale, blue; sandstone band	0	2	88	6
Sandstone, brown ..	0	6	89	0
Shale, blue	0	2	89	2
Sandstone, grey; shaly bands	1	6	90	8
Sandstone, brown ..	3	3	93	11
Sandstone, grey ..	3	1	97	0
Shale, grey sandy ..	6	8	103	8
Shale, sandy	4	3	107	11

Bakelite Water Bore, Aycliffe

Surface about 330 ft AOD. 6-in NZ 22 SE. Site [2680 2329] Aycliffe Trading Estate. Drilled 1956 for Bakelite Ltd. Cores examined by W. Anderson.

	Thickness ft	in	Depth ft	in
SUPERFICIAL DEPOSITS				
Boulder clay	43	0	43	0
PERMIAN				
MIDDLE MAGNESIAN LIMESTONE				
Limestone, broken ..	8	0	51	0
Limestone, hard about	30	0	81	0
LOWER MAGNESIAN LIMESTONE				
Limestone, hard about	53	0	134	0
MARL SLATE				
Shale, black	13	0	147	0
CARBONIFEROUS				
NAMURIAN (MILLSTONE GRIT SERIES)				
?Sandy shale ..	24	0	171	0
Shale, black; clay layers	44	0	215	0
Sandstone, white coarse	30	0	245	0
Clay, black; shale layers	65	0	310	0
Grit, white coarse ..	4	0	314	0
Shale; *Spirifer sp.* on fragment of chippings at 368 ft ..	58	0	372	0

Bildershaw No. 2 Bore

Surface 359 ft AOD. 6-in NZ 12 NE. Site [1870 2550] 170 yd NW of Hummerbeck. Drilled 1954 by NCB. Cores examined by R. H. Price and G. Armstrong. Plate X(114).

	Thickness ft	in	Depth ft	in
SUPERFICIAL DEPOSITS				
Clay, sandy	50	0	50	0
CARBONIFEROUS				
WESTPHALIAN B (MIDDLE COAL MEASURES)				
Shale, grey sandy ..	2	6	52	6
Shale, grey; fragmentary mussels; *Spirorbis*	4	8	57	2
(Horizon of HARVEY MARINE BAND)				
WESTPHALIAN A (LOWER COAL MEASURES)				
Coal	0	5	57	7
Seggar	2	5	60	0
Shale, grey sandy ..	12	0	72	0
Sandstone, grey ankeritic	5	3	77	3
Sandstone, flaggy fine-grained argillaceous	11	0	88	3
Shale, grey; ankeritic nodules; plants ..	8	3	96	6
Shale, brown; ankeritic nodules; mussels	4	6	101	0
HARVEY (goaf) ..	1	3	102	3
Fireclay, grey and brown	2	3	104	6
Mudstone, grey sandy; rooty at top ..	5	6	110	0
Mudstone, grey sandy; sandstone partings ..	3	0	113	0

	Thickness ft	in	Depth ft	in
Shale, grey sandy micaceous; sandstone partings [apparent dip 30°] ..	7	3	120	3
Sandstone, grey fine-grained; argillaceous partings; ankeritic in basal 7 ft	11	9	132	0
Sandstone, light grey massive	8	0	140	0
Sandstone, light grey massive; predominantly medium- and coarse-grained; shale pellets and coal scars at base	64	6	204	6
(Horizon of TOP TILLEY)				
Fireclay-mudstone, dark grey	5	0	209	6
Shale, grey slightly sandy	7	0	216	6
Shale, grey; sandy partings; plant remains	5	0	221	6
Shale and siltstone; ironstone nodules ..	3	6	225	0
Shale, dark grey; large mussels ..	6	3	231	3
Sandstone	0	3	231	6
BOTTOM TILLEY				
Coal, shaly	0	4	231	10
Fireclay	0	11	232	9
Coal	0	11	233	8
Shale, grey sandy; ankeritic bands; roots	2	9	236	5
Fireclay-mudstone	0	4	236	9
Coal	1	8	238	5
Fireclay, shaly ..	1	7	240	0
Siltstone, grey; ironstone nodules; roots	1	3	241	3
Shale, grey silty; fine sandy partings ..	18	9	260	0
Shale, grey; ironstone nodules	23	11	283	11
TOP BUSTY				
Coal	2	1	286	0
Fireclay and Fireclay-Mudstone, grey; ironstone nodules at base	5	0	291	0
Shale, grey; rooty ..	4	6	295	6
Fireclay, dark grey; coaly plant debris at top; coal partings at base	5	2	300	8
Mudstone, coaly pyritic	0	3	300	11
BOTTOM BUSTY				
Coal	4	11	305	10
Fireclay	0	8	306	6
Coal	0	3	306	9
Fireclay, ankeritic ..	0	9	307	6

Brettanby Farm Bore

Surface about 240 ft AOD. 6-in NZ 21 SW. Site [2279 1038] at Brettanby Farm, Barton. Drilled 1956 by J. T. Hymas Ltd. Cores examined by W. Anderson. Fig. 6(10).

	Thickness ft	in	Depth ft	in
SUPERFICIAL DEPOSITS				
Boulder Clay ..	66	0	66	0
CARBONIFEROUS				
NAMURIAN (MILLSTONE GRIT SERIES)				
BOTTOM LITTLE LIMESTONE				
Limestone, dolomitized; shells and crinoids	14	0	80	0
Sandstone, brown massive coarse-grained ..	16	0	96	0
Shale, black; barren ..	2	0	98	0
Shale; *Productus sp.* [spines]	1	0	99	0
Shale, sandy grey; rootlets	2	0	101	0
Shale, calcareous grey; crinoidal	7	0	108	0
Mudstone, black fissile	2	0	110	0
Shale, grey; crinoidal ..	8	0	118	0
Mudstone, black fissile	6	0	124	0

	Thickness ft	in	Depth ft	in
Shale, grey	2	0	126	0
Limestone, earthy; spinose productoids and crinoids	2	0	128	0
Shale, grey barren ..	1	0	129	0
Shale, grey limy; shells and crinoids ..	5	0	134	0
Limestone, coarse crystalline; crinoids and productoids ..	2	0	136	0
Mudstone, grey; productoids, large crinoids	9	0	145	0
Limestone, grey coarse; shells; crinoids ..	2	0	147	0
Shale, grey limy; shells; crinoids	7	0	154	0
Shale, grey; *Fenestella sp.*, productoids, spiriferoids and bivalves	15	0	169	0
GREAT LIMESTONE				
Limestone, grey coarse crystalline; fossiliferous ..	2	0	171	0

Broken Scar Pumping Station, Darlington
Bores N 1, E 2 and E 1 (Composite section)

Surface about 145 ft AOD. 6-in NZ 21 SE. Sites: N 1 [2552 1420]; E 2 [2550 1403]; E 1 [2551 1402], Broken Scar Pumping Station, Darlington. Drilled 1966 by J. T. Hymas and Son Ltd for Tees Valley and Cleveland Water Board. Cores examined by D. A. C. Mills and D. B. Smith.

	Thickness ft	in	Depth ft	in
SUPERFICIAL DEPOSITS				
[Thicknesses of beds in Superficial Deposits are approximate]				
Clay, greyish brown silty; scattered small pebbles of Magnesian Limestone and Carboniferous mudstone	5	0	5	0
Clay, reddish brown silty gritty; small rounded pebbles and scattered larger fragments of Magnesian and Carboniferous limestone; abundant small pebbles including porphyritic andesite at base ..	50	0	55	0
Gravel; Carboniferous Limestone fragments 60%, Magnesian Limestone 20%, remainder includes mudstone and sandstone	5	0	60	0
Clay, dull grey gritty; pebbles both Carboniferous and Permian increasing to base ..	66	0	126	0
PERMIAN				
MIDDLE MAGNESIAN LIMESTONE (shelf/ lagoon facies)				
Dolomite, pale buff oolitic porous; scattered stromatolite films; rock becomes coarser to base with pisoliths often grotesque; total core recovery 5 ft; sporadic forms of bivalves including *Bakevellia* ..	39	0	165	0
Dolomite, grey-cream to buff saccharoidal; commonly oolitic and more rarely pisolitic; sporadic shell fragments including a productoid and bivalves	27	9	192	9
Dolomite, buff oolitic; fine to medium sand grade below 193 ft 9 in; below 202 ft rock more altered with scarcely recognizable ooliths	21	9	214	6
Dolomite, calcitic saccharoidal highly por-				

	Thickness ft	in	Depth ft	in
rous; calcite-lined cavities	2	0	216	6
Dolomite, creamy-buff saccharoidal porous; scattered ooliths ..	4	3	220	9
Dolomite, cream soft fine-grained; scattered ooliths ..	19	3	240	0
MIDDLE MAGNESIAN LIMESTONE (transitional beds)				
Dolomite, cream finely granular soft porous; scattered ooliths and calcite-lined cavities; small circular calcite crystals	34	6	274	6
[Exact position of boundary between Middle and Lower Magnesian Limestone not established because of poor recovery between 245 and 290 ft]				
LOWER MAGNESIAN LIMESTONE				
Dolomite, ?calcitic grey buff mainly finely crystalline; slightly porous; scattered cavities; fine dolomitic siltstone laminae at 300 ft, 300 ft 6 in and 301 ft 7 in, and at intervals to base; scattered plant fragments; small coiled foraminifera and plant fragments	37	0	311	6
Siltstone, grey dolomitic; flecked bedding planes: abundant carbonaceous plant fragments [Marl Slate lithology]	0	3	311	9
Dolomite, grey crystalline; thin dolomitic siltstone band at base	0	5	312	2
Limestone, ?dolomitic pale grey cavernous; shelly; grey siltstone band at 313 ft 8 in; sporadic forms of shells including productoids	2	4	314	6
Dolomite, grey finely crystalline; irregular films and laminae of siltstone; wispy stylolites; scattered plant fragments ..	1	5	315	11
BASAL PERMIAN DEPOSITS				
Sandstone, grey medium-grained hard; small angular fragments of Carboniferous Limestone and sandstone, some pyritic	0	9	316	8
Breccia; angular fragments of Carboniferous sandstone, limestone and siltstone in a matrix of fine- to medium-grained sandstone, with sporadic coarse frosted grains; basal 1 in composed of limestone in matrix of grey silty sandstone	5	0	321	8
CARBONIFEROUS				
NAMURIAN (MILLSTONE GRIT SERIES)				
Seatearth-mudstone ..	0	1	321	9
Sandstone, brownish grey fine-grained micaceous; carbonaceous at top and base; pyritic blebs and ironstone patches; irregular fine network of ramifying calcite veins and calcite-lined cavities throughout; rock becomes reddened to base	0	9	322	6

	Thickness ft	in	Depth ft	in
Mudstone, grey micaceous and carbonaceous; silty laminae	1	11	324	5
Sandstone, mainly grey or reddened fine-grained micaceous and carbonaceous; sporadic muddy interlaminae	3	2	327	7
Mudstone, grey or reddened finely micaceous	2	5	330	0
Mudstone, dark grey or reddened micaceous; silty at base ..	2	0	332	0

Brusselton No. 179 Bore

Surface 345.76 ft. 6-in NZ 12 NE. Site [1985 2582] 1210 yd ENE of Hummerbeck. Drilled 1958 by NCB. Cores examined by A. M. J. Clarke. Plate XI (123).

	Thickness ft	in	Depth ft	in
SUPERFICIAL DEPOSITS				
Boulder clay	35	2	35	2
CARBONIFEROUS				
WESTPHALIAN B (MIDDLE COAL MEASURES)				
DURHAM LOW MAIN POST				
Sandstone, predominantly light grey fine-grained thin and wispily bedded; micaceous carbonaceous partings especially at top	17	10	53	0
Siltstone, grey argillaceous; sandy bands below 55 ft; sporadic plant fronds and stems	11	0	64	0
Sandstone, predominantly light grey medium - grained strongly false-bedded: siltstone and breccia-conglomerate bands between 74½ and 77 ft; 1½ ft siltstone band at 78½ ft; black micaceous partings from 84 to 87 ft, 97½ to 101 ft, 108 to 110 ft; grey siltstone inclusions between 110 and 113 ft; coal scars between 136 and 137 ft at base	78	3	142	3
DURHAM LOW MAIN				
Coal	2	9	145	0
Shale, black carbonaceous	0	3	145	3
Seatearth-siltstone ..	1	3	146	6
Seatearth-mudstone ..	4	6	151	0
Sandstone and siltstone interlaminated; sandstone light grey, fine-grained irregular; siltstone, dark grey micaceous	16	6	167	6
Sandstone, light grey fine-grained wispy-bedded; dark micaeous partings at top; massive at base ..	3	6	171	0
Siltstone, dark grey finely micaceous; sandy bands ..	17	0	188	0
BRASS THILL SHELL-BED				
Mudstone, grey shaly silty at top; mussels from 193 ft to base; pyritized *Naiadites* 195 ft 6 in to 196½ ft	9	0	197	0

	Thickness ft	in	Depth ft	in
BRASS THILL				
Coal	4	0	201	0
Seatearth - siltstone, black; sandy partings	3	0	204	0
Siltstone, dark grey micaceous; fine sandstone bands; comminuted plant debris below 207½ ft ..	9	0	213	0
Shale, black carbonaceous; mussel fragments	0	6	213	6
Siltstone, black carbonaceous micaceous; rootlets	2	6	216	0
Mudstone, dark grey shaly; sporadic silty bands; mussel beds at 215, 217½ and 219 ft; scattered mussels below; pyritized *Naiadites* from 222 ft to base	9	0	225	0
Siltstone, dark grey micaceous; sandy bands; comminuted plant debris.. ..	4	0	229	0
TOP HUTTON (upper leaf)				
Coal	0	5	229	5
Seatearth-mudstone; ironstone nodules at base	4	7	234	0
Sandstone, grey fine-grained; irregularly interbedded with dark grey micaceous siltstone	6	6	240	6
(Horizon of TOP HUTTON, lower leaf)				
Seatearth-siltstone, grey; ironstone nodules near base ..	2	6	243	0
Mudstone, grey shaly; sporadic clay ironstone bands.. ..	9	0	252	0
Sandstone, light grey fine-grained; siltstone laminae; argillaceous below 256 ft	17	0	269	0
Mudstone, grey shaly; sporadic clay ironstone bands below 280 ft; sporadic plant fragments below 280 ft; sporadic mussels between 278 and 280 ft; fragments of *Naiadites* in basal 6 in ..	16	6	285	6
HUTTON (BOTTOM HUTTON)				
Coal	3	6	289	0
Seggar	1	0	290	0

Burfoot Leazes Bore

Surface about 950 ft AOD. 6-in NZ 02 SE. [0690 2458] 400 yd WSW of Burfoot Leazes. From the notebook of H. H. Howell. Plate VII (12).

	Thickness ft	in	Depth ft	in
SUPERFICIAL DEPOSITS				
Clay	15	3	15	3
CARBONIFEROUS				
WESTPHALIAN A (LOWER COAL MEASURES)				
Metal, grey	26	0	41	3
Metal stone, grey ..	10	9	52	0
Post, grey and white ..	48	1	100	1
Metal, black	2	11	103	0
?TOP MARSHALL GREEN				
Coal	1	2	104	2
Metal, grey	1	2	105	4
Post, grey and white ..	18	3	123	7
BOTTOM MARSHALL GREEN ("COWLEY")				
Coal	1	7	125	2
Metal	4	11	130	1
Post, white; median 2 ft 7 inches in 'whin				

	Thickness ft	in	Depth ft	in
band'	27	3	157	4
Metal, grey and black	14	10	172	2
GANISTER CLAY				
Coal	1	0	173	2
Post, grey and white	22	11	196	1
Metal, grey	21	9	217	10
Whin	1	6	219	4
?THIRD GRIT				
Post, white	16	9	236	1
Metal, grey and black	17	6	253	7
Coal	0	6	254	1
Post, grey and white	10	10	264	11
Metal	0	2	265	1
Coal	0	9	265	10
Metal, grey	27	8	293	6
Post, white (into)	2	1	295	7

Charles Pit, Middridge Colliery

Surface about 400 ft. 6-in NZ 22 NE. Site [2503 2634], Middridge. Information from B & S Nos. 1357 and 2852. Plates VII and X (159).

	Thickness ft	in	Depth ft	in
SUPERFICIAL DEPOSITS				
Outset	12	0	12	0
Soil and gravel	12	0	24	0
PERMIAN				
LOWER MAGNESIAN LIMESTONE				
Limestone	112	0	136	0
CARBONIFEROUS				
WESTPHALIAN A (LOWER COAL MEASURES)				
Post, yellow; grey at base	52	0	188	0
?BOTTOM TILLEY				
Coal	2	4	190	4
Seggar	1	6	191	10
Metal, blue	36	0	227	10
BUSTY				
Coal	3	3	231	1
Seggar	2	0	233	1
Coal	3	3	236	4
Seggar	2	0	238	4
Post, light brown	52	0	290	4
Coal	0	3	290	7
Post, white	12	0	302	7
THREE QUARTER				
Coal	0	6½	303	1½
Fireclay	4	0	307	1½
Metal, blue	3	4	310	5½
Coal	0	8	311	1½
Fireclay, coarse	3	0	314	1½
Post girdle	1	0	315	1½
Metal, blue	20	0	335	1½
Post, white strong	31	2½	366	4
Post, grey; partings	13	6	379	10
Metal, blue; girdles	4	0	383	10
BROCKWELL				
Coal	5	6	389	4
Boring from base of shaft				
Seggar clay	4	0	393	4
Post, grey and white	8	11½	402	3½
Metal, blue and grey	28	2	430	5½
Post, grey and white	13	2	443	7½
Metal, grey	5	0	448	7½
Metal, blue	6	1	454	8½
Seggar clay	4	2	458	10½
Metal, blue; post and ironstone girdles	4	1	462	11½
Metal, grey	16	10	479	9½
Shale, blue and black	0	9	480	6½
Coal	0	3	480	9½
Metal, blue and grey; post girdles	26	6	507	3½
Post, white strong	14	2	521	5½
Metal, blue; post girdles	6	1	527	6½
Post, white; whin and metal girdles	24	1	551	7½
Metal, grey	5	4	556	11½
Shale, black	0	2	557	1½
Metal, grey	4	4	561	5½
Metal, blue	1	10	563	3½
GANISTER CLAY				
Coal	0	2	563	5½
Stone band	0	0¾	563	6¼

	Thickness ft	in	Depth ft	in
Coal	0	4½	563	10¾
Post, white	0	4	564	2¾
Metal, grey	1	2	565	4¾
Coal	0	3	565	7¾
Metal, blue; post panels	3	2	568	9¾
Post, white	17	8	586	5¾
Metal, blue	1	0	587	5¾
Post, white; lower part mixed with whin	2	8¾	590	2½

Cold Sides Bore

Surface about 260 ft AOD. 6-in NZ 21 NE. Approximate site [254 188] near Cold Sides Farm, Burtree Gate. Date of drilling unknown. Information based on record in B & S No. 1760 and other sources.

	Thickness ft	in	Depth ft	in
SUPERFICIAL DEPOSITS				
Clay	3	0	3	0
Clay, gravel and stones	12	0	15	0
Sand, partly stony	8	0	23	0
Sand and clay	5	0	28	0
Clay; loose stones	2	0	30	0
Sand and clay	4	0	34	0
PERMIAN				
MIDDLE MAGNESIAN LIMESTONE				
Limestone, soft	13	0	47	0
Clay	0	8	47	8
Limestone	2	9	50	5
Conglomerate	0	3	50	8
Sandstone, soft	7	1	57	9
Limestone, hard	0	3	58	0
Sandstone, soft	3	0	61	0
Limestone	1	0	62	0
"Sandstone"	2	6	64	6
Limestone and gravel	12	3	76	9
Limestone, yellow [described as "bastard"]	40	3	117	0
LOWER MAGNESIAN LIMESTONE				
Limestone, mainly yellow [described as "bastard"]	77	0	194	0
Limestone, blue; cherty	17	5	211	5
CARBONIFEROUS				
NAMURIAN (MILLSTONE GRIT SERIES)				
Shale; clay parting 14 ft 6 in from top	44	9½	256	2½
Grey whin	0	6	256	8½
Blue shale	10	7	267	3½
Grey whin	14	8½	282	0
Blue shale	4	4	286	4
Grey whin	2	6	288	10
Black shale	1	6	290	4
Grey whin	5	0	295	4
?LOWER FELLTOP LIMESTONE				
Limestone	1	1	296	5
Grey whin	22	2½	318	7½
Black shale	28	11	347	6½
?ROOKHOPE SHELL-BEDS				
Grey limestone	4	4	351	10½
Black shale	2	4	354	2½
Grey limestone	1	5	355	7½
Black shale	1	8	357	3½
Grey limestone	0	7	357	10½
Black shale	14	0	371	10½
Grey limestone	1	8	373	6½
Black shale	2	10½	376	5
Grey sandstone	4	1	380	6
Limestone	1	4	381	10
Grey sandstone	8	5	390	3
Grey sandstone	3	9	394	0
Blue shale	14	9	408	9
Grey sandstone	11	1½	419	10½
Blue shale	3	4	423	2½
Calcareous shale	1	7	424	9½
Black shale	9	6	434	3½
Grey rock	9	11	444	2½
Calcareous shale (?KNUCTON SHELL-BEDS)	13	11½	458	2
Black shale	9	2	467	4
White sandstone	28	9	496	1
Shale	0	10	496	11
Black shale	2	10	499	9
White sandstone	7	6	507	3

	Thickness ft	in	Depth ft	in
Black shale	64	7	571	10
CRAG LIMESTONE				
Grey Limestone	7	1	578	11
Limestone	3	9	582	8

Note: According to another record the total depth of the borehole is 602 ft 0½ in the lowest entries being "Mountain Limestone" 20 ft, on white post 2 ft 6 in.

Cowley Bore

Surface about 1030 ft AOD. 6-in NZ 02 NE. Site [0629 2532] about 130 yd W of Cowley, near Woodland. Sunk 1899; B & S No. 3088.

	Thickness ft	in	Depth ft	in
SUPERFICIAL DEPOSITS				
Boulder Clay	6	0	6	0
CARBONIFEROUS				
WESTPHALIAN A (LOWER COAL MEASURES)				
Sandstone, grey hard	4	0	10	0
Shale, grey sandy	2	0	12	0
Sandstone, grey	9	0	21	0
Shale	5	9	26	9
VICTORIA				
Coal	2	1	28	10
Fireclay	3	6	32	4
Sandstone, grey	3	6	35	10
Shale, blue	12	10	48	8
Sandstone, grey	1	6	50	2
Shale	1	8	51	10
Coal, slaty	0	3	52	1
Shale, blue	2	4	54	5
Sandstone, grey	2	9	57	2
Shale, grey sandy	2	0	59	2
Shale, blue	6	10	66	0
Shale, grey sandy	13	6	79	6
Shale, grey and black; ironstone balls	29	5	108	11
?STOBSWOOD				
Coal	0	4	109	3
Fireclay	7	0	116	3
Sandstone, grey; hard jointy	1	9	118	0
Shale, grey sandy	3	9	121	9
Sandstone, grey hard	0	9	122	6
Shale, blue	7	6	130	0
Whin, hard and jointy	3	0	133	0
Sandstone, grey	1	6	134	6
Shale, blue	6	7	141	1
TOP MARSHALL GREEN				
Coal	1	4	142	5
Fireclay	6	1	148	6
Shale, grey sandy	6	3	154	9
Sandstone, grey; occasional shale partings	43	6	198	3
Shale, blue	3	2	201	5
Sandstone, grey	3	1	204	6
Shale, blue	7	2	211	8
BOTTOM MARSHALL GREEN ("COWLEY")				
Coal (thickness suspect)	5	7	217	3
Fireclay	4	8	221	11
Sandstone, grey	0	6	222	5

Diamond Pit, Butterknowle

Surface about 600 ft AOD. 6-in NZ 12 NW. Site [1094 2556], Butterknowle. Date of sinking unknown. B & S No. 369; Plates IX and XI (24).

	Thickness ft	Thickness in	Depth ft	Depth in
SUPERFICIAL DEPOSITS				
Soil and Gravel	5	0	5	0
CARBONIFEROUS				
WESTPHALIAN B (MIDDLE COAL MEASURES)				
Metal, grey	5	0	10	0
HUTTON (BOTTOM HUTTON)				
Coal	4	0	14	0
Clay	0	5	14	5
Post, grey	2	2	16	7
Metal, grey	18	0	34	7
Blackstone	2	0	36	7
Metal, blue; post girdles	5	0	41	7
Post, black	1	6	43	1
RULER (JUBILEE)				
Coal	1	6	44	7
Thill stone	5	0	49	7
Coal	1	0	50	7
Stone band	0	4	50	11
Coal	0	4	51	3
Thill stone	8	0	59	3
Post, white	2	1½	61	4½
Black slag stone	0	6	61	10½
Post, white	3	2½	65	1
Whin	0	6	65	7
Post, white strong	6	2	71	9
Metal, blue	4	9	76	6
Post girdles	0	6	77	0
Metal, blue	8	3¾	85	3¾
Post, white strong	2	11¾	88	3½
Whin girdle	0	11½	89	3
Post, white	2	6½	91	9½
Metal, grey	9	5½	101	3
Post, grey	7	10¾	109	1¾
Whin	0	7	109	8¾
Metal, blue	25	6½	135	3¼
Ironstone girdle	0	3	135	6¼
Metal, blue	8	5¾	144	0
(Assumed horizon of HARVEY MARINE BAND)				
WESTPHALIAN A (LOWER COAL MEASURES)				
Coal	1	0	145	0
Thill stone	1	0	146	0
Post, grey	4	4	150	4
Metal, blue	2	4	152	8
Post, white strong	6	1½	158	9½
Metal, grey	1	7½	160	5
Post, grey	6	2½	166	7½
Whin, brown	1	0	167	7½
Post, white	0	9	168	4½
Metal, grey	1	5¼	169	9¾
Whin	0	7	170	4¾
Metal, strong	8	2½	178	7¼
Post, strong	0	11	179	6¼
Metal, grey	4	2¾	183	9
Metal, black	1	7	185	4
HARVEY (YARD)				
Coal	3	6	188	10
Thill stone	7	9½	196	7½
Post, grey	1	3	197	10½
Whin	0	6	198	4½
Metal, strong; post girdles	10	5	208	9½
Post, white	2	6	211	3½
Metal, grey	8	0	219	3½
Metal, blue	28	0	247	3½
TOP TILLEY				
Coal, danty	0	10	248	1½
Thill stone	1	7	249	8½
Metal, blue	2	8	252	4½
Post, strong grey	1	10	254	2½
Metal, blue	1	10	256	0½
Metal, strong grey	13	6	269	6½
Metal, strong grey; post girdles	10	3	279	9½
Metal, blue	8	11	288	8½
Metal, grey	10	4	299	0½
Metal, black	1	6	300	6½
BOTTOM TILLEY				
Coal	0	2	300	8½
Thill stone	1	4	302	0½
Metal, strong grey; post girdles	4	5	306	5½
Metal, grey; post girdles	12	10	319	3½
Post, strong	28	11	348	2½
Metal, blue	0	3	348	5½
Metal, black; a little coal	4	2½	352	8
Thill stone, strong	1	1	353	9
BUSTY (FIVE-QUARTER OR CROW)				
Coal	1	4	355	1

	Thickness ft	in	Depth ft	in
Thill stone	2	0	357	1
Metal, grey strong; post girdles ..	21	1	378	2
Post, brown strong ..	3	10	382	0
Metal, blue	4	0	386	0
Post, white strong ..	3	2	389	2
Metal, grey strong ..	2	3	391	5
Post, white strong ..	5	4	396	9
Metal, grey	9	4	406	1
Post girdle	1	0	407	1
Metal, dark blue ..	5	9	412	10
BROCKWELL (MAIN)				
Coal	5	10	418	8
Thill	0	6	419	2

Eggleston No. 3 Bore

Surface about 1460 ft AOD. 6-in NZ 02 SW. Site [0153 2483], Woodland Fell. Date of drilling unknown. From field notes of W. Gunn. Fig. 10(2).

	Thickness ft	in	Depth ft	in
SUPERFICIAL DEPOSITS				
Soil and Gravel ..	1	6	1	6
CARBONIFEROUS				
WESTPHALIAN A (LOWER COAL MEASURES)				
Sandstone, brown ..	28	0	29	6
Metal, dark	13	2	42	8
Metal, blue	25	6	68	2
Metal, black	3	6	71	8
Metal, dark	8	9	80	5
Metal, black; some coal	5	7	86	0
Sandstone, hard ..	0	2	86	2
Metal; mixed with grey post	6	0	92	2
THIRD GRIT				
Sandstone, grey ..	1	6	93	8
Metal, grey ..	1	11	95	7
Metal, grey; sandstone ribs	7	0	102	7
Sandstone, grey hard; partings	12	11	115	6
Metal, grey and metal stone	11	0	126	6
Metal, dark	4	0	130	6
Coal	0	6	131	0
NAMURIAN (MILLSTONE GRIT SERIES)				
Metal stone, dark grey	6	6	137	6
Metal dark	2	11	140	5
SECOND GRIT (upper leaf)				
Sandstone, hard ..	5	10	146	3
Metal, grey ..	3	11	150	2
Sandstone, hard ..	3	4	153	6
Metal, grey and blue..	8	9	162	3
Hard girdle	0	5	162	8
Metal, blue and dark..	6	6	169	2
Metal, dark; mixed with coal	0	9	169	11
SECOND GRIT (lower leaf)				
Metal stone, grey ..	12	7	182	6
Metal, dark and grey	4	4	186	10
Sandstone, white ..	1	10	188	8
Metal stone, grey ..	6	8	195	4
Sandstone, white ..	20	7	215	11
Metal, blue	11	0	226	11
Metal stone, grey ..	2	7	229	6
Sandstone, grey (into)	3	2	232	8

Eggleston No. 4 Bore

Surface about 1475 ft AOD. 6-in NZ 02 NW. Site [0202 2518], Woodland Fell. Date of drilling unknown. From field notes of W. Gunn.

	Thickness ft	in	Depth ft	in
SUPERFICIAL DEPOSITS				
Soil and ramble ..	5	9	5	9
CARBONIFEROUS				
WESTPHALIAN A (LOWER COAL MEASURES)				
Sandstone, grey and brown	6	1	11	10
Metal, dark	20	8	32	6
Metal stone, dark ..	44	11	77	5
Sandstone, grey; ironstone rib	15	3	92	8
Whin	2	2	94	10
Metal stone, grey ..	8	3	103	1
Metal, grey and dark ..	5	9	108	10
Coal	0	6	109	4
Metal, grey; mixed with coal at top ..	2	0	111	4
THIRD GRIT				
Sandstone	5	9	117	1
Sandstone, white hard	14	8	131	9
Sandstone, brown ..	1	0	132	9
Metal stone, grey ..	13	8	146	5
NAMURIAN (MILLSTONE GRIT SERIES)				
Metal, grey; sandstone ribs	7	0	153	5
Metal, dark	2	2	155	7
?SECOND GRIT (upper leaf)				
Metal stone, grey ..	18	2	173	9
Metal, dark	1	2	174	11
Coal, foul	1	11	176	10
SECOND GRIT (lower leaf)				
Sandstone-seatearth	0	6	177	4
Sandstone, grey and white	26	4	203	8
Metal, grey ..	9	0	212	8
Sandstone, grey ..	10	6	223	2
Metal stone, grey ..	12	8	235	10
FIRST GRIT (upper leaf)				
Sandstone, hard ..	1	9	237	7
Metal stone, grey; sandstone ribs ..	17	9	255	4
Sandstone, white ..	6	5	261	9
Sandstone, grey ..	3	4	265	1
Metal, grey; (?including horizon of SHARNBERRY SHELL-BEDS)	14	3	279	4
FIRST GRIT (lower leaf)				
Sandstone, grey (into)	31	0	310	4

Eggleston No. 5 Bore

Surface about 1386 ft AOD. 6-in NZ 02 NW. Site [0073 2518], Woodland Fell. Date of drilling unknown. From notes of W. Gunn. Fig. 10(3).

	Thickness ft	in	Depth ft	in
SUPERFICIAL DEPOSITS				
Clay, gravel and ramble	7	6	7	6
CARBONIFEROUS				
NAMURIAN (MILLSTONE GRIT SERIES)				
SECOND GRIT (lower leaf)				
Sandstone, brown ..	28	0	35	6
Metal, dark; mixed with coal	0	4	35	10
Metal, grey	1	10	37	8
Sandstone, grey and brown	13	2	50	10
Metal, dark grey ..	17	4	68	2
Sandstone, grey ..	1	4	69	6
Metal, dark grey and grey	9	4	78	10
FIRST GRIT (upper leaf)				
Sandstone, white ..	4	10	83	8
Metal, grey ..	12	0	95	8
Sandstone, grey; metal partings ..	8	0	103	8
Metal, dark (?includ-				

	Thickness ft	in	Depth ft	in
ing horizon of SHARNBERRY SHELL-BEDS)	7	0	110	8
Metal stone, grey	5	9	116	5
FIRST GRIT (lower leaf)				
Sandstone, brown and grey	66	0	182	5
Metal stone, grey	24	0	206	5
Metal, dark grey	4	3	210	8
Sandstone girdle	0	9	211	5
Metal, grey	5	6	216	11
Metal, grey and metal stone (?including horizon of WHITE HOUSE LIMESTONE)	25	3	242	2
Sandstone, grey (into)	—	—	—	—

Eldon Colliery Shaft (Harry Pit)

Surface about 392 ft AOD. 6-in NZ 22 NW. Site [2392 2809] 700 yd NE of Eldon. Sunk 1864. Information from B & S No. 771 and Abandoned Mine Plans. Plates X, XI and XII (152).

	Thickness ft	in	Depth ft	in
SUPERFICIAL DEPOSITS				
Gravel, clay and sand	14	6	14	6
CARBONIFEROUS				
WESTPHALIAN B (MIDDLE COAL MEASURES)				
Metal, blue; iron girdles	27	6	42	0
Post	4	0	46	0
Black stone	0	6	46	6
HIGH MAIN				
Coal	2	0	48	6
Seggar	4	0	52	6
Post	4	0	56	6
Metal, blue	3	6	60	0
Metal, grey; post girdles	13	6	73	6
Black stone	15	0	88	6
Metal, blue; iron girdles	12	0	100	6
METAL (THREE-QUARTER)				
Coal, jet	3	0	103	6
Seggar	3	0	106	6
FIVE-QUARTER				
Coal	6	0	112	6
Seggar	2	0	114	6
Metal, blue and grey; 4 ft 7 in blue stone at 146 ft 8 in	34	4	148	10
Coal	0	10	149	8
Post; metal girdles	21	2	170	10
Metal, grey; ironstone bands	34	3	205	1
MAIN				
Coal	6	5	211	6
Seggar	0	10	212	4
Coal	1	5	213	9
Band	0	6	214	3
Coal	2	0	216	3
Band	1	2	217	5
Coal	0	11	218	4
Seggar	4	10	223	2
Metal grey; post girdles	69	10	293	0
Post (Durham Low Main Post)	48	0	341	0
DURHAM LOW MAIN				
Coal	2	8	343	8
Metal, grey	11	0	354	8
Metal, dark; ironstone bands	34	10	389	6
BRASS THILL				
Coal	2	0	391	6
Band	0	10	392	4
Coal	1	6	393	10
Metal, grey; post girdles	23	4	417	2
TOP HUTTON				
Coal	0	4	417	6
Metal, grey	0	8	418	2
Coal	0	9	418	11

	Thickness ft	in	Depth ft	in
Seggar	3	6	422	5
Metal, grey	2	0	424	5
Coal	0	4	424	9
Metal, grey; post girdles at top; ironstone girdles at base	35	11	460	8
BOTTOM HUTTON				
Coal	3	4	464	0
Metal, grey; post girdles	6	9	470	9
Coal	0	4	471	1
Metal, grey; post girdles	26	10	497	11
Post	12	6	510	5
Metal, grey	3	6	513	11
Coal	0	2	514	1
Metal, grey; post girdles	26	4	540	5
Black stone	1	7	542	0
Metal, grey; post girdles	15	0	557	0
Metal, dark, ironstone bands	19	0	576	0
(Assumed horizon of HARVEY MARINE BAND)				
WESTPHALIAN A (LOWER COAL MEASURES)				
Coal	0	3	576	3
Seggar	3	10	580	1
Metal, grey; post girdles	38	11	619	0
Black stone	0	10	619	10
HARVEY				
Coal	3	8	623	6
Seggar	2	0	625	6
Metal, grey	8	9	634	3
Coal	0	3	634	6
Seggar	2	2	636	8
Metal, grey	4	10	641	6
Post	13	0	654	6
Metal, grey; mixed with post	60	7	715	1
BOTTOM TILLEY				
Coal	2	2	717	3
Metal, grey	51	8	768	11
Post	3	6	772	5
BUSTY (?BOTTOM)				
Coal	2	8	775	1
Metal, grey	2	0	777	1
Post, white	46	0	823	1
Metal, grey	1	0	824	1
THREE-QUARTER				
Coal	0	5	824	6
Black stone	0	8	825	2
Post; metal partings	19	9	844	11
Coal	0	7	845	6
Post	5	3	850	9
Metal, grey	9	6	860	3
Coal	0	11	861	2
Seggar	3	0	864	2
Metal, grey	11	8	875	10
Post, metal partings	41	0	916	10
BROCKWELL				
Coal	5	2	922	0
Metal, grey; post and ironstone girdles; fireclay at top ..	24	0	946	0
Post	25	0	971	0
Metal, grey	5	0	976	0

Engine Pit, St Helen Auckland

Surface about 337 ft AOD. 6-in NZ 12 NE. Site [1970 2711] 1000 yd ENE of St Helen's Church. Sunk 1830. Sources: NCB, B & S 1650 and 1651 and Wm Coulson. Plates X and XI (122).

	Thickness ft	in	Depth ft	in
SUPERFICAL DEPOSITS				
Gravel, clay and sand; 14 ft of brown strong clay at base ..	37	8	37	8
CARBONIFEROUS				
WESTPHALIAN B (MIDDLE COAL MEASURES)				
Metal stone, blue ..	8	6	46	2

	Thickness ft in	Depth ft in
HUTTON (BOTTOM HUTTON)		
Coal	1 10	48 0
Thill	4 8	52 8
Metal, grey	4 6	57 2
Post girdle	0 11	58 1
Metal, blue	2 0	60 1
Post, white	9 2	69 3
Metal parting ..	0 6	69 9
Metal, grey	5 10	75 7
Metal grey; ironstone balls	3 6	79 1
?RULER (JUBILEE)		
Coal	0 6	79 7
Thill	1 8	81 3
Metal stone, grey ..	25 6	106 9
Post girdles, grey ..	1 0	107 9
Metal, grey jointy ..	0 9	108 6
Post girdles, grey ..	1 6	110 0
Metal, grey soft jointy	0 6	110 6
Post girdles, grey jointy	1 6	112 0
Metal stone, grey strong	0 10	112 10
Black stone, soft ..	1 6	114 4
Coal	0 6	114 10
Metal, grey	2 6	117 4
Dun Post; whin ..	0 6	117 10
Metal; dark grey ..	3 9	121 7
Post girdle	0 4	121 11
Metal, grey	1 3	123 2
Post girdle	0 4	123 6
Metal, grey strong ..	0 5	123 11
Post girdles, partings ..	0 11	124 10
Metal, grey	2 4	127 2
Coal	0 8	127 10
Black swad	0 3	128 1
Coal	0 4	128 5
Metal, grey strong ..	2 10	131 3
Post, white	0 10	132 1
Metal, grey strong ..	3 0	135 1
Post girdles	4 6	139 7
Post, white	2 6	142 1
Whin	0 10	142 11
Post, white and grey ..	3 6	146 5
Metal, black strong; ironstone girdles ..	10 7	157 0
Metal, grey jointy; large ironstone balls	9 6	166 6
Metal, dark; ironstone balls up to 3 in ..	6 11	173 5
(Assumed horizon of HARVEY MARINE BAND)		
WESTPHALIAN A (LOWER COAL MEASURES)		
Coal	0 2	173 7
Thill, grey strong ..	4 0	177 7
Metal, grey soft; ironstone balls	11 0	188 7
Post, grey	0 6	189 1
Dun whin	0 10	189 11
Post, grey; ironstone balls	7 0	196 11
Post, grey	3 5	200 4
Metal, black	1 4	201 8
Metal, grey	1 3	202 11
Blackstone	2 4	205 3
HARVEY		
Coal; $\frac{1}{2}$ in swad band near base	3 9	209 0
Thill, grey strong ..	1 0	210 0
Post, white strong ..	3 4	213 4
Post, white strong; metal partings ..	12 10	226 2
Metal, grey	0 5	226 7
Post, white; metal partings	6 10	233 5
Post, grey	1 3	234 8
Post, white; metal partings at base ..	2 6	237 2
Post, grey	6 3	243 5
Metal, grey	1 0	244 5
Post, grey shivery ..	10 2	254 7
TOP TILLEY		
Coal; 2 1-in bands	2 5	257 0
Thill, dark grey ..	2 9	259 9
Metal stone, dark grey; ironstone balls ..	8 10	268 7
Cockle and mussel shell bed	0 6	269 1
Black stone; coal and water	0 7	269 8
Metal, grey strong; post girdle at base ..	2 0	271 8
BOTTOM TILLEY		
Coal, good	0 6	272 2
Metal stone ..	1 5	273 7
Coal, coarse ..	0 5	274 0
Thill, grey	3 8	277 8
Metal, grey	7 3	284 11

	Thickness ft	in	Depth ft	in
Metal, dark blue; ironstone balls	18	3	303	2
Blackstone; ironstone balls	4	4	307	6
Metal, grey strong ..	12	8	320	2
Ironstone, grey ..	1	8	321	10
Post, white; black coal scars..	14	0	335	10
Metal stone, blue ..	0	8	336	6
Post, white; metal partings	24	2	360	8
Coal and stone ..	0	3	360	11
Post, white; metal partings	24	0	384	11
?THREE-QUARTER				
Coal	0	6	385	5
Metal, soft dark ..	3	6	388	11
Post, white; metal partings	12	11	401	10
Metal stone; post girdles	12	5	414	3
Coal	0	7	414	10
Thill, grey soft; metal, grey soft	1	8	416	6
Metal, grey strong; post girdles ..	8	11	425	5
Post girdles; metal partings	5	10	431	3
Metal, grey	1	3	432	6
Post, white strong ..	3	6	436	0
Metal, grey; post girdles	15	8	451	8
BROCKWELL (MAIN)				
Coal; 1½ in band 11 in from top ..	5	7½	457	3½
Thill, strong grey ..	0	11	458	2½

Conflicting records exist for details of strata below the Brockwell Coal, but the shaft was to a total depth of 685 ft with coal 14 in at about 546 ft and coal 16 in at about 582 ft.

Etherley No. 4 Bore

Surface about 580 ft AOD. 6-in NZ 12 NE. Site [1636 2765] Hunters' Hill House, Toft Hill. Drilled 1828. Information from B & S No. 813. Plate IX (79).

	Thickness ft	in	Depth ft	in
SUPERFICIAL DEPOSITS				
Soil and clay ..	9	9	9	9
CARBONIFEROUS				
WESTPHALIAN B (MIDDLE COAL MEASURES)				
Metal stone, dark grey; post girdles ..	30	3	40	0
Metal, black	9	5	49	5
(Assumed horizon of HARVEY MARINE BAND)				
WESTPHALIAN A (LOWER COAL MEASURES)				
Metal stone, grey; post girdles	10	0	59	5
Coal, foul	0	5	59	10
Metal, grey	1	10	61	8
Coal	0	10	62	6
Metal stone, grey ..	3	0	65	6
Post, brown and white	6	0	71	6
Metal stone, grey ..	5	0	76	6
Blackstone	3	6	80	0
HARVEY				
Coal	3	5	83	5
Metal, grey	4	7	88	0
Coal	0	6	88	6
Metal, grey	4	0	92	6
Post, white strong; grey and scarred at top	36	6	129	0
Metal stone, grey ..	6	0	135	0
Post, white strong ..	8	6	143	6
TOP TILLEY				
Coal	0	10	144	4
Metal, dark grey ..	3	10	148	2
Coal	1	4	149	6
Metal stone, grey; post girdles	24	6	174	0
Black metal	0	10	174	10

	Thickness ft	in	Depth ft	in
BOTTOM TILLEY				
Coal	1	1	175	11
Metal, grey ..	4	0	179	11
Metal and post, white	7	0	186	11
Metal, dark grey ..	2	0	188	11
Coal	0	8	189	7
Metal, grey ..	1	5	191	0
Coal	0	4	191	4
Metal stone, grey; post and whin girdles ..	51	8	243	0
Metal, dark grey; whin girdles	4	3	247	3
BUSTY				
Coal, splinty ..	0	10	248	1
Metal, mixed with foul coal	1	5	249	6
Coal	4	0	253	6
Metal stone, strong ..	21	0	274	6
THREE-QUARTER				
Coal, coarse ..	1	6	276	0
Metal stone, dark grey; post girdles ..	9	7	285	7
Post, white; whin ..	2	6	288	1
Metal stone; whin girdles	6	9	294	10
Coal	0	6	295	4
Metal, grey and dark grey; mixed with coal in top 4 in ..	16	1	311	5
Coal, foul; mixed with metal	1	0	312	5
Metal stone, dark grey	6	0	318	5
Post, white; whin girdles	9	2	327	7
Whin	1	6	329	1
Post, white	1	4	330	5
Metal, grey; thin girdles	19	1	349	6
BROCKWELL				
Coal	6	7	356	1
Metal, dark grey ..	1	0	357	1

Gordon Gill F Bore

Surface 569.76 ft AOD. 6-in NZ 12 NW. Site [1379 2691], Snape Foot 1500 yd NE of High Lands. Drilled 1956 by NCB. Cores examined by R. H. Price.

	Thickness ft	in	Depth ft	in
SUPERFICIAL DEPOSITS				
Soil and boulder clay..	19	0	19	0
CARBONIFEROUS				
WESTPHALIAN B (MIDDLE COAL MEASURES)				
Shale, grey soft; ironstone bands; vein ankerite	38	0	57	0
Shale, black; small mussels at top ..	5	8	62	8
HARVEY MARINE BAND				
Shale, black; *Lingula* in basal 4 in; *moderate dip*	0	4	63	0
WESTPHALIAN A (LOWER COAL MEASURES)				
Fireclay, dark grey; slickensided ..	1	10	64	10
Coal	0	8	65	6
Fireclay	1	6	67	0
Shale, dark grey ..	16	3	83	3
Coal	1	7	84	10
Shale, grey; ironstone nodules; mussels in basal 4 ft	11	2	96	0
Coal	0	7	96	7
Fireclay, dark grey ..	0	5	97	0
Shale, dark grey hard ankeritic	5	6	102	6
[*Fault*]	—	—	—	—
Fireclay-siltstone, sandy; ironstone nodules at base ..	9	0	111	6

	Thickness ft	in	Depth ft	in
Sandstone, dark grey fine-grained; roots	3	6	115	0
Mudstone, grey sandy; roots	2	0	117	0
Sandstone, argillaceous; micaceous carbonaceous partings	1	6	118	6
Shale, black; *high dip*	0	6	119	0
[*Fault*]	—	—	—	—
Sandstone, grey hard; ankeritic veins; ganisteroid at base; pyritic films	11	0	130	0
Shale, dark silty	4	0	134	0
Mudstone, shaly; ankeritic veins	1	8	135	8
Shale, dark grey sandy; thin shaly sandstone breccia at base	0	8	136	4
Sandstone, hard siliceous micaceous	1	0	137	4
Shale, dark grey sandy; silty at top	1	8	139	0
Sandstone, hard ganisteroid; basal 4½ ft argillaceous with rootlets	9	6	148	6
Shale, grey sandy; slumped argillaceous sandstone bands; sandstone pellets and ironstone nodules at top	9	0	157	6
Shale, grey silty; sporadic slumped sandy bands; ironstone nodules and bands; worm borings, plant remains	11	6	169	0
Sandstone, argillaceous at top; ganisteroid at base	2	5	171	5
Mudstone, grey shaly silty; micaceous bands at top	3	9	175	2
Mudstone, black silty carbonaceous; muscovite flakes; carbonised plant impressions (?horizon of RODDYMOOR MARINE BAND)	0	4	175	6
Coal	0	7	176	1
Shale, dark; sandy partings; roots	1	2	177	3
THIRD GRIT				
Sandstone, white and grey medium- to coarse-grained hard quartzo-feldspathic; argillaceous and rooty at top	26	9	204	0
Shale, grey sandy	3	0	207	0
Sandstone, white variably textured, predominantly medium- to coarse-grained	89	9	296	9
Mudstone, grey; roots	1	3	298	0
Seatearth-breccia; slickensided	1	0	299	0
NAMURIAN (MILLSTONE GRIT SERIES)				
Sandstone, medium-grained micaceous	2	7	301	7
Siltstone, blue-grey; roots	1	0	302	7
Sandstone, locally feldspathic predominantly fine- to medium-grained; argillaceous and micaceous at base	12	5	315	0
Sandstone, white-grey medium- to coarse-grained	9	6	324	6
Shale, grey silty	1	0	325	6

Gordon Gill K Bore

Surface 560.63 ft AOD. 6-in NZ 12 NW. Site [1380 2687] at Snape Foot. Drilled 1956 for NCB. Cores examined by R. H. Price. Plate IX (55).

	Thickness ft	in	Depth ft	in
SUPERFICIAL DEPOSITS				
Soil	1	0	1	0
Boulder Clay	10	0	11	0
Clay, grey	24	0	35	0
CARBONIFEROUS				
WESTPHALIAN A (LOWER COAL MEASURES)				
Clay, grey; ironstone nodules	6	0	41	0
?HARVEY				
Coal	2	3	43	3
Fireclay - mudstone; ironstone nodules ..	3	9	47	0
Mudstone, grey sandy irony	11	0	58	0
Shale, grey silty; sandy partings	3	0	61	0
Siltstone, locally sandy; sphaerosideritic; jointed	4	0	65	0
Shale, grey silty; sandy partings; ironstone bands	11	0	76	0
Mudstone, grey rooty; ironstone bands ..	8	0	84	0
Fireclay and fireclay-mudstone	11	0	95	0
Siltstone, pale grey; ironstone bands; sphaerosiderite bands at top ..	16	0	111	0
Fault breccia ..	1	0	112	0
Mudstone, silty; *high dip*	16	3	128	3
Siltstone, grey becoming more sandy towards base; plant remains at top ..	10	9	139	0
Sandstone, fine-grained; dark partings; ironstone bands; jointed ..	4	0	143	0
Shale, grey silty; sandy partings; slickensided	5	0	148	0
Siltstone, grey micaceous; jointed; a few rootlets	2	0	150	0

	Thickness ft	in	Depth ft	in
Fireclay - mudstone, dark grey; slickensided	2	9	152	9
Siltstone, dark grey micaceous and carbonaceous	1	9	154	6
TOP TILLEY				
Coal	0	8	155	2
Fireclay-siltstone, dark grey micaceous ..	1	10	157	0
Fireclay - mudstone, dark grey; rooty ..	1	0	158	0
Sandstone, argillaceous; rooty; fireclay siltstone bands	6	0	164	0
BOTTOM TILLEY				
Coal	0	3	164	3
Fireclay, grey ..	0	6	164	9
Sandstone, irony; rooty	1	3	166	0
Shale, silty; alternates with sandy shale; sandstone bands ..	10	0	176	0
Mudstone, grey silty; ironstone bands; slickensided ..	17	0	193	0
Fault breccia ..	1	0	194	0
Shale, dark grey; brecciated shale bands ..	3	0	197	0
Shale, black finely micaceous	2	0	199	0
Mudstone, dark; micaceous at base with sandy and ironstone bands	1	3	200	3
Coal	0	1	200	4
Fireclay and fireclay-mudstone; slickensided; broken ..	6	8	207	0
Shale, grey silty; rootlets	5	0	212	0
Sandstone, light grey ..	0	8	212	8
Shale, sandy; ankerite veins	3	4	216	0
Siltstone, shaly; plants	4	0	220	0
Shale, grey sandy; *dip* 45°	3	0	223	0
?Fireclay, soft; no recovery	6	0	229	0
[*Fault*]	—	—	—	—
Shale, dark grey; plants	2	4	231	4

	Thickness ft	in	Depth ft	in
?BUSTY				
Coal	4	3	235	7
Sandstone	0	11	236	6
Coal	1	0	237	6
Fireclay, brecciated ..	1	6	239	0
[*Fault* (?Butterknowle)]	—	—	—	—
Sandstone, grey ganisteroid	1	0	240	0
Shale, grey soft ..	9	0	249	0
NAMURIAN (MILLSTONE GRIT SERIES)				
SECOND GRIT (upper leaf)				
Sandstone, light grey micaceous; shaly bands	9	0	258	0
Sandstone, white medium- and coarse-grained	16	0	274	0
SPURLSWOOD SHELL-BEDS				
Siltstone, sandy; ironstone nodules	2	0	276	0
SECOND GRIT (lower leaf)				
Sandstone, medium- and coarse-grained gritty ..	20	3	296	3
Mudstone, grey silty; worm bores at base (including horizon of WOODLAND SHELL-BEDS AT BASE) ..	10	7	306	10
Shale, black leafy (including horizon of WOODLAND SHELL-BEDS	0	2	307	0
Sandstone, ganisteroid	2	0	309	0
Fireclay - mudstone, dark grey shaly ..	0	10	309	10
Sandstone, grey argillaceous micaceous..	2	2	312	0
Shale, grey sandy and silty; ankerite veins	2	9	314	9
Mudstone, dark grey shaly; sandy partings at base	5	3	320	0
FIRST GRIT (upper leaf)				
Sandstone, light grey; sporadic shaly silty partings	5	0	325	0
Sandstone, light grey and white; occasional shaly bands	12	0	337	0
Sandstone, light grey fine- to coarse-grained ..	35	0	372	0

Gordon House No. 1 Shaft

Surface 665.19 AOD. 6-in NZ 12 SW. Site [1334 2403] 520 yd ESE of St Mary's Church, Cockfield. Date of sinking not recorded. Information from B & S No. 2725 and NCB. Plate XI (48a).

	Thickness ft	in	Depth ft	in
SUPERFICIAL DEPOSITS				
Built-up ground, soil, and boulder clay ..	42	8	42	8
CARBONIFEROUS				
WESTPHALIAN B (MIDDLE COAL MEASURES)				
Shale, grey and blue ..	6	0	48	8
MAIN				
Coal	6	6	55	2
Fireclay	4	6	59	8
Shale, grey and blue; black at base ..	17	2	76	10
Fireclay	3	0	79	10
MAUDLIN				
Coal and band ..	6	5	86	3
Fireclay	6	9	93	0
Sandstone (Durham Low Main Post) ..	98	1	191	1
Shale, grey	5	0	196	1
Sandstone (Durham Low Main Post) ..	21	9	217	10

	Thickness ft	in	Depth ft	in
DURHAM LOW MAIN				
Coal, banded ..	5	2	223	0
Sandstone	2	10	225	10
Shale, blue	36	8	262	6
BRASS THILL				
Coal	3	0	265	6
Sandstone	2	9	268	3
Black band	18	10	287	1
(Assumed horizon of TOP HUTTON)				
Fireclay	5	0	292	1
Sandstone	6	9	298	10
Shale, blue	23	6	322	4
Sandstone	10	0	332	4
Shale, grey	7	8	340	0
Sandstone	4	0	344	0
Shale, grey	15	0	359	0
HUTTON (BOTTOM HUTTON)				
Coal	4	1	363	1
Fireclay	4	9	367	10
Shale, grey	4	2	372	0
Sandstone	27	6	399	6
Shale, grey	2	7	402	1
RULER (JUBILEE)				
Coal and band ..	3	5	405	6
Fireclay	1	8	407	2
Shale, blue and grey; sandstone ribs especially at top ..	23	7	430	9
Shale, grey; black at base	38	3	469	0
(Assumed horizon of HARVEY MARINE BAND)				
WESTPHALIAN A (LOWER COAL MEASURES)				
Coal	0	8	469	8
Fireclay	3	2	472	10
Sandstone	7	6	480	4
Shale; thin median sandstone rib ..	37	8	518	0
HARVEY				
Coal and splint ..	7	8	525	8
Fireclay	5	0	530	8
Shale, grey; sandstone girdles	18	10	549	6

Gordon House No. 1 Bore

(synopsis)

Surface about 610 ft AOD. 6-in NZ 12 SW. Site [133 243]. About 450 yd E by N of St Mary's Church, Cockfield. Drilled 1923 by Andrew Kyle Ltd.

	Thickness ft	in	Depth ft	in
SUPERFICIAL DEPOSITS				
Soil, gravel and boulders	16	6	16	6
CARBONIFEROUS				
WESTPHALIAN A (LOWER COAL MEASURES)				
Sandstone, grey and brown	25	0	41	6
Metal, dark at base ..	19	6	61	0
?BOTTOM TILLEY				
Coal	0	3	61	3
Fireclay	5	0	66	3
Metal, grey	10	2	76	5
Sandstone, grey ..	64	7	141	0
BUSTY (old workings)	5	6	146	6
Strata; sandy at top, otherwise mainly argillaceous.. ..	61	2	207	8
BROCKWELL (old workings)	6	0	213	8
Strata; sandy at top, argillaceous at base.	37	0	250	8
(VICTORIA RIDER HORIZON)				
Fireclay	14	4	265	0
VICTORIA				
Coal	1	0	266	0

	Thickness ft	in	Depth ft	in
Strata; mainly argillaceous; occasional sandstone ribs ..	31	1	297	1
BOTTOM VICTORIA (NCB)				
Coal, dirty	1	2	298	3
Strata (horizon of STOBSWOOD at 328 ft 2 in, and of TOP MARSHALL GREEN at 341 ft 11 in) ..	58	2	356	5
MARSHALL GREEN				
Coal	1	8	358	1
Strata; mainly argillaceous; occasional sandstone bands ..	52	11	411	0
GANISTER CLAY				
Coal	1	3	412	3
Band	0	5	412	8
Coal	0	2	412	10
Strata; mainly argillaceous	53	2	466	0
Metal, blue (horizon of RODDYMOOR MARINE BAND)	8	5	474	5
Coal	0	7	475	0
Fireclay, strong fakey	2	6	477	6
?THIRD GRIT				
Sandstone	10	9	488	3
Metal, grey (horizon of KAYS LEA MARINE BAND)	11	11	500	2
Strata; mainly argillaceous	29	4	529	6
Metal, grey with balls (horizon of QUARTERBURN MARINE BAND)	5	6	535	0
NAMURIAN (MILLSTONE GRIT SERIES)				
Coal	0	3	535	3
Fireclay	9	11	545	2
Metal, sandy and sandstone	4	10	550	0

Gordon House No. 93 Bore

(synopsis)

Surface 630.31 ft AOD. 6-in NZ 12 SW. Site [1334 2419] 450 yd ESE of St Mary's Church, Cockfield. Drilled 1957 by Cementation Co. Ltd. for NCB. Cores examined by A. M. J. Clarke. Plates VII and IX (47).

	Thickness ft	in	Depth ft	in
SUPERFICIAL DEPOSITS				
Soil, sandy clay and boulders	19	0	19	0
CARBONIFEROUS				
WESTPHALIAN A (LOWER COAL MEASURES)				
Shale	15	0	34	0
Sandstone	8	9	42	9
TILLEY				
Coal	2	9	45	6
Shale	17	6	63	0
Shale; sandstone bands	14	0	77	0
Sandstone, partly coarse-grained ..	65	0	142	0
Shale, sandy	7	0	149	0
BUSTY				
Coal	2	0	151	0
Shale	1	4	152	4
Coal	1	2	153	6
Seggar	1	6	155	0
Shale, sandy; sandstone bands ..	24	0	179	0
Sandstone	6	0	185	0
Shale, sandy; sandstone bands ..	18	0	203	0
Sandstone	4	0	207	0
Shale, sandy	19	2	226	2
BROCKWELL				
Coal	4	8	230	10
Shale	0	3	231	1
Coal	1	0	232	1

	Thickness ft	in	Depth ft	in
Seggar	1	11	234	0
Shale	3	0	237	0
(Horizon of BOTTOM BROCKWELL)				
Seatearth, sandy grey-brown	1	0	238	0
Strata, mainly arenaceous	19	9	257	9
Strata, mainly argillaceous	9	11	267	8
VICTORIA SHELL-BED				
Shale, dark grey; sandy partings; sporadic mussel bands	3	4	271	0
Mudstone, dark grey; mussel impressions	8	11	279	11
VICTORIA RIDER				
Coal	0	2	280	1
Seatearth, ironstone ..	5	11	286	0
Siltstone, rooty; ironstone bands ..	15	3	301	3
Siltstone	1	10	303	1
VICTORIA				
Coal	0	1	303	2
Shale	0	1	303	3
Coal	0	10	304	1
Strata, mainly argillaceous, including 2 ft 3 in shale band with mussels at 321 ft	30	3	334	4
BOTTOM VICTORIA (NCB)				
Coal	1	0	335	4
Strata, mainly arenaceous	1	0	335	4
Mudstone, silty shaly at base; sparse mussel impressions at base	10	4	354	4
STOBSWOOD				
Coal	0	5	354	9
Seatearth mudstone ..	4	4	359	1
(Horizon of TOP MARSHALL GREEN)				
Seatearth	0	8	359	9
Strata, mainly argillaceous	29	7	389	4
BOTTOM MARSHALL GREEN				
Coal	1	10	391	2
Strata, mainly arenaceous at top; argillaceous at base ..	54	4½	445	6½
GANISTER CLAY				
Coal	1	5	446	11½
Seatearth-sandstone	0	3¾	447	3¼
Coal	0	4¾	447	8
Siltstone; roots ..	5	4	453	0

Gordon House No. 96 Bore

(synopsis)

Surface 647.37 ft AOD. 6-in NZ 12 SW. Site [1294 2425], St Mary's Church, Cockfield. Drilled 1957 by Cementation Co. Ltd. for NCB. Cores examined by A. M. J. Clarke. Plates IX (41).

	Thickness ft	in	Depth ft	in
SUPERFICIAL DEPOSITS				
Clay, boulders and top soil	10	6	10	6
CARBONIFEROUS				
WESTPHALIAN A (LOWER COAL MEASURES)				
Sandstone	62	0	72	6
BUSTY				
Coal workings ..	6	6	79	0
Strata; mainly shale..	24	6	103	6
THREE-QUARTER				
Coal	1	3	104	9
Strata, mainly shale; sandstone 11 ft 6 in at 135 ft 6 in	39	11	144	8

	Thickness ft	in	Depth ft	in
BROCKWELL				
Coal; shale bands ..	6	1	150	9
Seggar	1	3	152	0
(?Horizon of BOTTOM BROCKWELL)				
Seatearth-mudstone ..	4	0	156	0
Sandstone	16	0	172	0
VICTORIA SHELL-BED				
Shale, silty; small mussels at 173, 174 and 177 ft 6 in ..	9	0	181	0
VICTORIA FISH-BED				
Mudstone; small mussels at top; fish-scale at 189 ft	8	6	189	6
(VICTORIA RIDER HORIZON)				
Seatearth	2	6	192	0
Strata; mainly mudstone and siltstone..	22	1	214	1
VICTORIA				
Coal	1	0	215	1
Strata; mainly argillaceous; sandstone bands	21	8	236	9
BOTTOM VICTORIA (NCB)				
Coal	0	8	237	5
Strata; sandstone at top, argillaceous at base	33	11	271	4
(STOBSWOOD horizon)				
Seatearth-sandstone	2	8	274	0
Strata; predominantly sandstone; TOP MARSHALL GREEN washed out	30	10	304	10
BOTTOM MARSHALL GREEN				
Coal	1	7	306	5
Strata; predominantly arenaceous; silty at top	37	1	343	6
Strata; predominantly argillaceous; 3-inch mudstone band at 355 ft 9 in containing minute fish fragments	17	3	360	9
GANISTER CLAY				
Coal	1	4	362	1
Sandstone band ..	0	2½	362	3½
Coal	0	2½	362	6
Seatearth	3	6	366	0
Sandstone and siltstone	2	0	368	0

Gordon House No. 98 Bore

Surface 655.31 ft AOD. 6-in NZ 12 SW. [1190 2497] 1340 yd NW of St Mary's Church, Cockfield. Drilled 1957 by Cementation Co. Ltd. for NCB. Cores examined by A. M. J. Clarke. Plates VII and IX (31)

	Thickness ft	in	Depth ft	in
SUPERFICIAL DEPOSITS				
Top soil and sandy clay	5	0	5	0
CARBONIFEROUS				
WESTPHALIAN A (LOWER COAL MEASURES)				
Sandstone	54	0	59	0
BUSTY				
Coal	5	0	64	0
Shale, sandy; sandstone bands ..	6	0	70	0
Sandstone	7	0	77	0
Shale, sandy	3	0	80	0
Shale, black	9	0	89	0
THREE-QUARTER				
Coal	1	6	90	6
Shale, dark sandy ..	2	6	93	0
Shale, grey	5	0	98	0
Shale, sandy	2	0	100	0

	Thickness ft	in	Depth ft	in
Sandstone	7	0	107	0
Shale, sandy	15	0	122	0
Sandstone	2	0	124	0
Shale	9	0	133	0
BROCKWELL				
Old workings	8	0	141	0
Shale, grey	4	0	145	0
Sandstone	3	0	148	0
Siltstone, sandy micaceous	0	10	148	10
Sandstone, mainly light grey medium-grained; sporadic micaceous partings; coal scars at base	62	2	211	0
Siltstone, shaly micaceous; plant remains at base	0	8	211	8
Sandstone, light grey wispy-bedded micaceous	1	7	213	3
VICTORIA FISH-BED				
Mudstone, dark grey silty; one fish spine at 219 ft	7	1	220	4
Shale, grey finely micaceous; occasionally sandy especially at top with plant debris	2	6	222	10
VICTORIA RIDER				
Coal	0	7	223	5
Seatearth-mudstone	3	0	226	5
VICTORIA				
Coal	0	10	227	3
Mudstone, shaly carbonaceous	0	$2\frac{3}{4}$	227	$5\frac{3}{4}$
Coal, bright and dull	0	$6\frac{1}{4}$	228	0
Seatearth-mudstone, black carbonaceous micaceous; clay ironstone nodules	2	0	230	0
BOTTOM VICTORIA (NCB)				
Coal	0	9	230	9
Shale, carbonaceous	0	$0\frac{3}{4}$	230	$9\frac{3}{4}$
Coal	0	1	230	$10\frac{3}{4}$
Shale, carbonaceous	0	1	230	$11\frac{3}{4}$
Coal	0	$2\frac{1}{4}$	231	2
Seatearth - sandstone, ganisteroid	1	10	233	0
Sandstone, grey fine- to medium-grained; micaceous bands and silty partings; sporadic plant debris	43	8	276	8
Mudstone, dark grey shaly silty micaceous; sporadic worm burrows; *Planolites sp.* at base	3	4	280	0
(?STOBSWOOD horizon)				
Sandstone and siltstone interbedded, sandstone predominating; micaceous partings throughout; plant remains	10	0	290	0
Shale, dark grey; plant remains	0	4	290	4
BOTTOM MARSHALL GREEN				
Coal	1	2	291	6
Seatearth-sandstone, ganisteroid	2	6	294	0
Sandstone, light grey fine- to medium-grained; silty and micaceous bands and partings; plant remains; ironstone nodules at base	9	6	303	6
Siltstone, dark grey micaceous; sandstone laminae; plant debris	1	8	305	2
Mudstone, dark grey shaly silty finely micaceous; *Cochlichnus* cf. *kochi* at 306 ft; *Planolites* cf. *ophthalmoides* and mussels at 307 ft; *Planolites sp.* at 308½ ft; worm burrows at base	4	4	309	6
Shale, black finely micaceous; indet. organic debris including fish spine at 313 ft	4	0	313	6

	Thickness ft	in	Depth ft	in
Siltstone; worm burrows	0	6	314	0
Mudstone, shaly silty; sporadic *Planolites sp.*, worm burrows and plant debris ..	2	0	316	0
Sandstone and siltstone, interbedded.. ..	9	0	325	0
Sandstone, grey fine- to medium-grained; silty micaceous bands and partings ..	11	0	336	0
Shale, dark grey muddy; *Planolites* cf. *ophthalmoides*	7	9	343	9
Siltstone, black sandy micaceous; worm burrows at top ..	1	1	344	10
Sandstone, dark grey fine-grained; worm burrows at top ..	1	2	346	0
Shale, dark grey silty micaceous; fish debris	0	6	346	6
?GANISTER CLAY				
Coal	0	0½	346	6½
Seatearth-mudstone ..	2	5½	349	0
Sandstone, light grey fine-grained massive; sporadic micaceous partings	7	6	356	6
Siltstone, dark grey sandy partings; comminuted plant debris	3	0	359	6
Shale, sandy	3	6	363	0

Gordon House No. 131 Bore

Surface 466.10 ft AOD. 6-in NZ 12 NW. Site [1385 2499] 1000 yd NW of Buck Head Farm. Drilled 1957 by Cementation Co. Ltd for NCB. Cores examined by A. M. J. Clarke. Plate IX (57).

	Thickness ft	in	Depth ft	in
SUPERFICIAL DEPOSITS				
Top soil, boulders and boulder clay ..	31	6	31	6
CARBONIFEROUS				
WESTPHALIAN A (LOWER COAL MEASURES)				
Shale, grey soft ..	2	8	34	2
BOTTOM TILLEY (?lower leaf)				
Coal	0	4	34	6
Shale, grey	19	6	54	0
Shale, sandy	1	6	55	6
Sandstone	30	6	86	0
Shale, grey	2	8	88	8
Shale, black; coal traces	1	7	90	3
Shale, dark grey ..	3	11	94	2
BUSTY				
Coal	3	2	97	4
Shale band	0	8	98	0
Coal	1	3	99	3
Seggar	2	3	101	6
Sandstone, grey ..	7	6	109	0
Shale, grey sandy ..	19	10	128	10
THREE-QUARTER				
Coal	0	7	129	5
Seggar	1	1	130	6
Sandstone	1	6	132	0
Shale, grey sandy ..	16	0	148	0
Sandstone, broken ..	3	3	151	3
Shale, broken ..	10	3	161	6
BROCKWELL (goaf) ..	6	6	168	0
Seggar	2	0	170	0
Seatearth-mudstone, dark grey slickensided	0	3	170	3
Sandstone; slump structures; roots; carbonized plant remains at base	1	6	171	9
Shale, dark grey carbonaceous	0	2	171	11
Sandstone fine-grained; roots	0	6	172	5

	Thickness ft	Thickness in	Depth ft	Depth in
Shale, black carbonaceous	0	1	172	6
(BOTTOM BROCKWELL horizon)				
Seatearth-sandstone; roots	1	4	173	10
Sandstone, grey fine- to medium-grained wispy-bedded; bands of shale	22	6	196	4
Shale with siltstone and sandstone	6	2	202	6
VICTORIA SHELL-BED				
Mudstone, silty; small mussels; plant debris at base	3	10	206	4
Sandstone, medium-grained	1	2	207	6
Mudstone, shaly; sporadic mussels and mussel fragments	6	9	214	3
VICTORIA FISH-BED				
Shale, dark grey finely micaceous; sporadic fish scales at base	4	0	218	3
Siltstone, grey; clay-ironstone bands; mudstone with plant remains at base ..	1	6	219	9
(VICTORIA RIDER horizon)				
Seatearth - siltstone; sphaerosiderite bands and nodules ..	4	3	224	0
VICTORIA				
Coal	0	10	224	10
Seatearth - mudstone and siltstone ..	2	2	227	0
Siltstone; rooty at top, hard and ferruginous at base	9	0	236	0
Sandstone, fine-grained ferruginous; roots	3	4	239	4
Mudstone, shaly; clay ironstone nodules; rare poorly preserved mussel fragments ..	1	2	240	6
Seatearth - sandstone, hard ferruginous ..	0	2	240	8
Seatearth-mudstone ..	1	4	242	0
Siltstone	2	4	244	4
Seatearth - mudstone, black	1	2	245	6
Shale; carbonized plant remains	0	10	246	4
BOTTOM VICTORIA, (lower leaf) (NCB)				
Coal	1	1	247	5
Shale, black	0	1	247	6
Seatearth - mudstone and sandstone ..	2	6	250	0
Siltstone, sandy; roots	2	6	252	6
Sandstone, white fine-grained	4	6	257	0
Shale, dark grey silty; sandstone partings; sporadic plant debris	7	0	264	0
Shale; mussel fragments at top ..	4	2	268	2
STOBSWOOD				
Coal	0	1	268	3
Seatearth-mudstone and sandstone ..	3	9	272	0
Sandstone, grey white fine-grained; shaly partings	3	3	275	3
Siltstone. grey micaceous	2	9	278	0
Shale, grey silty ..	1	4	279	4
TOP MARSHALL GREEN				
Coal	0	1	279	5
Shale, black; coal films at base; ?fish scales	0	11	280	4
Coal	0	1	280	5
Shale, black; coal films	0	3	280	8
Seatearth - mudstone dark grey	0	9	281	5
Siltstone, grey; roots at top	2	1	283	6
Sandstone, predominantly white fine-grained; shaly partings; fine- to medium-grained in basal 6 ft				

	Thickness ft in	Depth ft in
3 in; brecciated with shale fragments at base	16 9	300 3
Shale, grey silty; plant fragments at base ..	1 0	301 3
Shale, dark grey and black; plant debris ..	0 7	301 10
BOTTOM MARSHALL GREEN		
Coal; fusain bands ..	1 4	303 2
Seatearth-sandstone ..	3 8	306 10
Mudstone; grey silty	1 2	308 0
Sandstone, white fine-grained	2 0	310 0

Grunton Bore

Surface about 260 ft AOD. 6-in NZ 21 SW. Site [2239 1148], Grunton Farm, Manfield. Drilled 1959 by J. T. Hymas Ltd. Cores examined by W. Anderson.

	Thickness ft in	Depth ft in
SUPERFICIAL DEPOSITS		
Boulders and clay ..	26 0	26 0
PERMIAN		
BASAL PERMIAN SANDSTONE AND BRECCIA		
Conglomerate; subrounded pebbles up to 1 in, of chert, grit and limestone in matrix of sandy dolomitic limestone ..	44 0	70 0
CARBONIFEROUS		
NAMURIAN (MILLSTONE GRIT SERIES)		
Shale, reddened ..	10 0	80 0
TOP LITTLE LIMESTONE		
Limestone, blue crinoidal; reddened at top	10 0	90 0
Shale, black; interbedded chert ..	4 0	94 0
BOTTOM LITTLE LIMESTONE		
Limestone, yellow crinoidal; dolomitized	17 0	111 0
Sandstone, red massive fine-grained porous	6 0	117 0
Shale, grey and black sandy	5 0	122 0
Sandstone, grey ganister-like	2 0	124 0
Shale, grey	26 0	150 0
Limestone, blue; crinoids and productoids	2 0	152 0
Shale	8 0	160 0
Limestone, blue ..	2 0	162 0
Shale, dark grey ..	20 0	182 0
Shale and limestone .	4 0	186 0
GREAT LIMESTONE		
Limestone, hard ..	6 0	192 0

High Wham No. 31 Bore

Surface 808.68 ft AOD. 6-in NZ 12 NW. Site [1065 2750] 1110 yd NW of High Wham. Drilled 1958 by NCB. Cores examined by A. M. J. Clarke.

	Thickness ft	in	Depth ft	in
SUPERFICIAL DEPOSITS				
Made-up ground and clay	8	6	8	6
CARBONIFEROUS				
WESTPHALIAN A (LOWER COAL MEASURES)				
Shale	0	6	9	0
BROCKWELL				
Coal	3	0	12	0
Shale, grey	4	0	16	0
Seatearth - sandstone grey fine-grained; micaceous at base ..	2	0	18	0
Sandstone, grey fine- to medium-grained predominantly thin-bedded; sporadic silty bands and micaceous partings ..	20	10	38	10
Siltstone, grey thin-bedded; sandy partings; *Planolites* in basal 6 in	3	5	42	3
VICTORIA SHELL-BED				
Mudstone, grey silty at top; shaly and finely micaceous below with ironstone layers; mussels 45 ft to 46 ft; indet. organic debris at 50 ft 4 in ..	10	9	53	0
VICTORIA FISH-BED				
Shale, black finely micaceous; abundant fish scales and spines ..	0	6	53	6
(Horizon of VICTORIA)				
Seatearth - mudstone, brownish grey silty ..	4	7	58	1
Sandstone, grey fine-grained false-bedded; dark micaceous partings; sporadic plant debris	15	11	74	0
Seatearth - mudstone, silty towards base ..	3	3	77	3
Sandstone, grey medium-grained false-bedded; dark micaceous partings; silty mudstone bands between 84 ft 6 in and 96 ft; thin coal scars between 106 and 111 ft; erosive base ..	41	1	118	4
(Horizon of STOBSWOOD)				
Seatearth - sandstone, grey fine-grained micaceous	1	6	119	10
Mudstone, grey silty and finely micaceous	3	2	123	0
Seatearth - sandstone, light grey; ironstone nodules	2	0	125	0
Mudstone, grey shaly	5	0	130	0
Sandstone, light grey medium-grained false-bedded ..	5	0	135	0
Mudstone, grey ..	1	9	136	9
TOP MARSHALL GREEN				
Coal, bright ..	0	7	137	4
Seatearth - mudstone and siltstone ..	4	5	141	9
Sandstone, grey fine-grained, wispy-bedded; dark grey silty micaceous bands ..	8	3	150	0
BOTTOM MARSHALL GREEN				
Coal	1	10	151	10
Seatearth-sandstone ..	2	2	154	0
Sandstone, light grey fine-grained	1	0	155	0

High Wham No. 33 Bore

Surface 716.85 ft AOD. 6-in NZ 12 NW. Site [1141 2694] 390 yd NE of High Wham. Drilled 1958 by Cementation Co. Ltd. for NCB. Cores examined by A. M. J. Clarke.

	Thickness ft	Thickness in	Depth ft	Depth in
SUPERFICIAL DEPOSITS				
Soil and clay ..	10	0	10	0
CARBONIFEROUS				
WESTPHALIAN A (LOWER COAL MEASURES)				
Sandstone, light grey medium-grained false-bedded; jointed	23	7	33	7
Mudstone, grey silty micaceous; sporadic plant debris ..	1	11	35	6
Sandstone, light grey fine-grained thin-bedded; silty bands; comminuted plant debris	7	4	42	10
VICTORIA SHELL-BED				
Mudstone, dark grey shaly micaceous; mussels abundant between 46 ft 6 in and 46 ft 11 in and 50 ft 3 in and 50 ft 4 in	11	0	53	10
VICTORIA FISH-BED				
Shale, black finely micaceous; fish scales at base ..	0	9	54	7
(Horizon of VICTORIA)				
Seatearth-mudstone, silty at base; iron-nodules in basal 3 ft	5	5	60	0
Siltstone, thin-bedded; sandy partings ..	11	2	71	2
Sandstone, grey fine-grained; siltstone bands and lenses ..	3	11	75	1
Seatearth-mudstone and sandstone ..	3	7	78	8
Sandstone, light grey predominantly medium-grained wispy-bedded; coalified plant stems in basal 3 in	27	8	106	4
Mudstone, grey shaly micaceous	2	0	108	4
Coal	0	4	108	8
Sandstone, grey medium-grained; rootlets at top, plant debris below	0	4	109	0
Siltstone, grey; sporadic plant debris ..	0	6	109	6
Sandstone, grey mainly fine-grained wispy bedded; argillaceous in places; sporadic worm burrows ..	12	3	121	9
Siltstone, grey micaceous; sandy bands and partings; layers of comminuted plant debris	5	1	126	10
Mudstone, grey; mussel fragments and *Gyrochorte carbonaria* above 128 ft; sporadic worm-filled burrows in basal 2 ft ..	6	4	133	2
STOBSWOOD				
Coal	0	5	133	7
Seatearth-sandstone ..	3	5	137	0
TOP MARSHALL GREEN				
Coal	0	7	137	7
Seatearth-mudstone	0	3	137	10
Coal	0	2	138	0
Seatearth-sandstone ..	2	0	140	0
Siltstone, grey massive; sporadic sandy bands	7	11	147	11
BOTTOM MARSHALL GREEN				
Coal	1	7	149	6
Seatearth-mudstone	0	1	149	7
Coal	0	3	149	10
Seatearth-sandstone, brownish-grey ..	1	4	151	2
Sandstone, light grey; sporadic rootlets ..	1	10	153	0

Hilton No. 1 Bore

(synopsis)

Surface about 470 ft AOD. 6-in NZ 12 SE. Site [1730 2342] about 400 yd SW of Bolton Garths. Drilled 1907 by Andrew Kyle Ltd. for North Bitchburn Coal Co. Ltd.

	Thickness ft	in	Depth ft	in
SUPERFICIAL DEPOSITS				
Surface clay and stones	11	4	11	4
CARBONIFEROUS				
WESTPHALIAN A (LOWER COAL MEASURES)				
Metal, broken ..	4	4	15	8
GANISTER CLAY				
Coal	0	9	16	5
Strata; mainly silty and shaly; several thick sandstones ..	148	10	165	3
?QUARTERBURN MARINE BAND				
Fakes, dark hard; sandstone ribs ..	18	6	183	9
NAMURIAN (MILLSTONE GRIT SERIES)				
Strata; mainly argillaceous	15	6	199	3
SECOND GRIT (upper leaf)				
Fakes, dark and fakey sandstone	22	6	221	9
Shale and seatearth ..	4	7	226	4
SPURLSWOOD SHELL-BEDS				
Fakes, coarse ..	2	3	228	7
Strata; mainly shale; 6-ft fireclay at 237 ft	46	10	275	5
SECOND GRIT (lower leaf)				
Sandstone	10	0	285	5
Sandstone, dark fakey	3	3	288	8
Strata; sandstone, fireclay and shale (including horizon of WOODLAND SHELL-BEDS near base) ..	25	7	314	3
FIRST GRIT (upper leaf)				
Sandstone, coarse-grained	35	9	350	0
Blaes	6	1	356	1
?SHARNBERRY SHELL-BEDS				
Blaes, light fakey ..	3	2	359	3
FIRST GRIT (lower leaf)				
Sandstone, fakey ..	2	7	361	10
Conglomerate ..	18	9	380	7
Sandstone, dark fakey	1	1	381	8
Strata; mainly argillaceous	31	7	413	3

Hilton No. 2 Bore

(synopsis)

Surface about 410 ft AOD. 6-in NZ 12 SE. Site [1663 2127] at Todwell House 820 yd NW of Ingleton Vicarage. Drilled 1908 by Andrew Kyle Ltd. Plate VI (86).

	Thickness ft	Thickness in	Depth ft	Depth in
SUPERFICIAL DEPOSITS				
Clay and stones	9	0	9	0
CARBONIFEROUS				
WESTPHALIAN A (LOWER COAL MEASURES)				
Sandstone	9	0	18	0
TOP MARSHALL GREEN				
Coal	0	9	18	9
Blaes and fakes	8	3	27	0
Sandstone	22	6	49	6
Fakes	7	7	57	1
BOTTOM MARSHALL GREEN				
Coal	1	3	58	4
Blaes, fakey coaly	0	6	58	10
Coal	0	6	59	4
Strata; mainly argillaceous; 5½ ft sandstone at 107 ft	63	7	122	11
GANISTER CLAY				
Coal	2	7	125	6
Blaes, hard fakey	2	4	127	10
Sandstone	35	2	163	0
Strata; mainly sandstone	30	0	193	0
Coal	0	6	193	6
Strata; sandy at top, shaly at base	18	0	211	6
Blaes, fakey (horizon of RODDYMOOR MARINE BAND)	3	8	215	2
Coal	1	10	217	0
Strata (including THIRD GRIT and horizon of KAYS LEA MARINE BAND)	34	0	251	0
Fakes	3	2	254	2
Coal, burnt	0	4	254	6
Sandstone, red and white	24	0	278	6
Blaes and hard ribs (approximate horizon of QUARTERBURN MARINE BAND)	10	6	289	0

Hilton No. 3 Bore

(synopsis)

Surface about 480 ft AOD. 6-in NZ 12 SE. Site [1679 2203] Hilton Grange Farm, Hilton. Drilled by Andrew Kyle Ltd. Fig. 10(9).

	Thickness ft	Thickness in	Depth ft	Depth in
SUPERFICIAL DEPOSITS				
Clay and stones	9	6	9	6
CARBONIFEROUS				
WESTPHALIAN A (LOWER COAL MEASURES)				
TOP MARSHALL GREEN				
Coal	0	6	10	0
Blaes, fakey	4	0	14	0
Sandstone	19	6	33	6
BOTTOM MARSHALL GREEN				
Coal	2	0	35	6
Blaes, fakey	6	0	41	6
Sandstone	12	0	53	6
Blaes	1	0	54	6
Sandstone	0	5	54	11
Blaes	5	0	59	11
Sandstone	5	2	65	1
Blaes	4	8	69	9
Sandstone, fakey	7	6	77	3
Blaes, fakey	8	6	85	9
GANISTER CLAY				
Coal	2	7	88	4
Blaes, coaly	1	2	89	6
Fakes, dark	3	0	92	6
Sandstone	28	0	120	6
Fakes	9	6	130	0
Sandstone	17	6	147	6
Blaes and sandstone ribs (horizon of				

	Thickness ft	in	Depth ft	in
RODDYMOOR MARINE BAND at base) ..	21	2	168	8
Coal	1	10	170	6
Strata (including horizon of KAYS LEA MARINE BAND at base)	23	6	194	0
Coal	0	4	194	4
Strata (including horizon of QUARTERBURN MARINE BAND at base)	28	4	222	8
NAMURIAN (MILLSTONE GRIT SERIES)				
Strata; mainly argillaceous	45	4	268	0
SECOND GRIT (upper leaf)				
Sandstone, grey fakey	31	6	299	6
Blaes, dark (including horizon of SPURLSWOOD SHELL-BEDS at base)	11	0	310	6
SECOND GRIT (lower leaf)				
Sandstone, fakey bands	37	0	347	6
Strata (including horizon of WOODLAND SHELL-BEDS) ..	22	0	369	6
FIRST GRIT (upper leaf)				
Sandstone, grey; locally fakey ..	22	4	391	10
Strata (including horizon of SHARNBERRY SHELL-BEDS at base)	12	8	404	6
FIRST GRIT (lower leaf)				
Sandstone, grey ..	38	6	443	0
Blaes	13	0	456	0
?WHITEHOUSE LIMESTONE				
Kingle	1	2	457	2
Sandstone, grey and dark grey	21	4	478	6

Hilton No. 5 Bore

(synopsis)

Surface about 495 ft AOD. 6-in NZ 12 SE. Site [1756 2266] at southern edge of Trunnelmire Plantation. Drilled 1908 by Andrew Kyle Ltd.

	Thickness ft	in	Depth ft	in
SUPERFICIAL DEPOSITS				
Clay, and stones ..	15	0	15	0
CARBONIFEROUS				
WESTPHALIAN A (LOWER COAL MEASURES)				
Strata, mainly argillaceous; 13-ft sandstone at 38 ft ..	50	6	65	6
Blaes, fakey (including horizon of RODDYMOOR MARINE BAND)	7	0	72	6
Strata; alternating sandstone and mudstone	45	0	117	6
Blaes (including horizon of KAYS LEA MARINE BAND) ..	6	0	123	6
Sandstone	19	6	143	0
Blaes (including horizon of QUARTERBURN MARINE BAND) ..	4	6	147	6
NAMURIAN (MILLSTONE GRIT SERIES)				
Strata, alternating sandstone and mudstone (coal 8 in at 163 ft 6 in)	17	0	164	6
SECOND GRIT (upper leaf)				
Sandstone, grey ..	44	6	209	0
Fakes, hard (including horizon of SPURLSWOOD SHELL-BEDS) ..	6	0	215	0
SECOND GRIT (lower leaf)				
Sandstone, grey; 4-in coal at 234 ft ..	30	0	245	0
Strata; mudstone at top, sandstone at base	17	0	262	0

	Thickness ft	in	Depth ft	in
Blaes; sporadic sandstone ribs (horizon of WOODLAND SHELL-BEDS near base) ..	24	0	286	0
FIRST GRIT (upper leaf)				
Sandstone, grey coarse	26	0	312	0

Hilton No. 51 Bore

Surface 543.44 ft AOD. 6-in NZ 12 SE. Site [1628 2314] 380 yd NW of Hilton Moor. Drilled 1957 by Cementation Co. Ltd. for NCB. Plate VII (76).

	Thickness ft	in	Depth ft	in
SUPERFICIAL DEPOSITS				
Clay and boulders ..	30	0	30	0
CARBONIFEROUS				
WESTPHALIAN A (LOWER COAL MEASURES)				
Sandstone, grey fine-grained wispy-bedded	13	0	43	0
VICTORIA SHELL-BED				
Shale; large mussels	5	4	48	4
VICTORIA				
Coal	1	0	49	4
Seatearth, sandy; ironstone nodules ..	2	6	51	10
Sandstone, grey wispy-bedded; ironstone nodules at top ..	2	8	54	6
Shale, grey sandy; sandstone partings..	1	1	55	7
Sandstone, light grey fine-grained, wispy-bedded	10	5	66	0
Shale, grey silty; sandy in basal 2 ft ..	18	0	84	0
Mudstone, dark grey shaly finely micaceous; appears to be marine near base ..	8	0	92	0
STOBSWOOD				
Coal	0	9	92	9
Seatearth	2	3	95	0
Siltstone, finely micaceous; roots ..	3	10	98	10
TOP MARSHALL GREEN				
Coal	1	2	100	0
Shale, black	0	3	100	3
Seatearth, grey ..	0	11	101	2
Shale, black; coal partings	0	3	101	5
Siltstone, grey rooty; ironstone nodules ..	3	10	105	3
BOTTOM MARSHALL GREEN				
Coal	1	3½	106	6½
Shale	0	2	106	8½
Coal	0	1	106	9½
Shale	0	3	107	0½
Coal	0	3	107	3½
Seatearth-siltstone ..	4	4½	111	8
Sandstone, argillaceous; shale laminae	5	6	117	2
Shale, grey silty; sandy partings; sporadic plant remains ..	15	4	132	6
Sandstone, grey flaggy; dark partings ..	7	9	140	3
Shale, grey silty; sandy bands; sporadic plant remains and roots ..	5	3	145	6
Sandstone, grey wispy-bedded; sandy shale bands	13	7	159	1
Sandstone, argillaceous micaceous	0	7	159	8
GANISTER CLAY				
Coal	1	2	160	10
Band	0	3	161	1
Coal	0	7	161	8
Seatearth, black shaly	3	4	165	0

Hilton No. 53 Bore

(synopsis)

Surface 448 ft AOD. 6-in NZ 12 SE. Site [1748 2383] 1390 yd ESE of Evenwood Gate. Drilled 1957 by Cementation Co. Ltd. for NCB. Cores examined by R. H. Price. Plate VI (97).

	Thickness ft	in	Depth ft	in
SUPERFICIAL DEPOSITS				
Soil and sandy clay ..	10	0	10	0
CARBONIFEROUS				
WESTPHALIAN A (LOWER COAL MEASURES)				
Strata; mainly arenaceous	57	6	67	6
Shale, dark grey sandy micaceous; *Curvirimula sp.*, plant debris	2	9	70	3
Strata; mudstone at base contains plants: *Alethopteris sp.* and *Mariopteris sp.* ..	7	3	77	6
GANISTER CLAY				
Coal, baked ..	0	1½	77	7½
Whinstone, blue-grey; 1 in white trap at top; jointed	9	4½	87	0
Seatearth-mudstone, shaly black and grey; baked	2	0	89	0
Strata; mainly sandstone but including mudstone with *Curvirimula sp.* at 118 ft	36	2	125	2
Shale, dark sandy; sandstone inclusions	3	4	128	6
Sandstone, grey wispy-bedded; jointed ..	1	0	129	6
Shale, dark grey sandy; sandstone partings ..	3	6	133	0
Strata; argillaceous ..	15	6	148	6
RODDYMOOR MARINE BAND				
Shale, black; ironstone nodules; basal 4 in pyritic; *Serpuloides stubblefieldi*, *Planolites ophthalmoides*, *Lingula mytilloides*, *Euphemites sp.*	4	6	153	0
Coal	0	8	153	8
Cannel	0	1	153	9
Strata (including THIRD GRIT)	14	9	168	6
KAYS LEA MARINE BAND				
Shale, black; foraminifera including *Ammodiscus sp.*, *Planolites ophthalmoides*, *Lingula mytilloides* ..	2	0	170	6
Shale, dark grey silty	1	6	172	0
Shale, black; *Ammodiscus sp.*, pyritized sponge spicules, *Lingula mytilloides*	4	6	176	6
Seatearth, dark grey ..	0	6	177	0
Sandstone, dark grey argillaceous; shale partings	9	0	186	0
Siltstone, dark grey; sandy bands ..	3	0	189	0
Mudstone, silty; *Serpuloides sp.*, *Lingula sp.*, platformed conodont	4	0	193	0

Hilton No. 54 Bore

(synopsis)

Surface 419 ft AOD. 6-in NZ 12 SE. Site [1758 2426] 1400 yd E of Evenwood Gate. Drilled 1957 by Cementation Co. Ltd., for NCB. Cores examined by R. H. Price.

	Thickness ft	in	Depth ft	in
SUPERFICIAL DEPOSITS ..	53	0	53	0
CARBONIFEROUS				
WESTPHALIAN A (LOWER COAL MEASURES)				
Strata; mainly sandy near top, shaly below	50	5	103	5
GANISTER CLAY				
Coal	0	9	104	2
Strata; mainly mudstone and shale; sandstone ribs near top and base ..	33	10	138	0
Shale, dark grey; sandy micaceous partings; plant remains, fish scales	1	7	139	7
Strata; sandstone near top, mudstone and shale below ..	22	5	162	0
RODDYMOOR MARINE BAND				
Mudstone, dark grey shaly	6	5	168	5
Coal	0	7	169	0
Strata; mainly mudstone and shale ..	9	0	178	0
KAYS LEA MARINE BAND				
Siltstone, grey ..	2	0	180	0
Sandstone, grey argillaceous ..	2	6	182	6
Shale, dark ..	3	6	186	0
Shale, dark grey ..	2	6	188	6
Shale, dark grey ..	3	6	192	0
Strata; mainly mudstone and shale on 10-ft sandstone at base	18	0	210	0

Hilton No. 55 Bore

Surface 453.35 ft AOD. 6-in NZ 12 SE. Site [1702 2427] 800 yd E of Evenwood Gate. Drilled 1957 by Cementation Co. Ltd., for NCB. Cores examined by A. M. J. Clarke. Plate VII (88).

	Thickness ft	in	Depth ft	in
SUPERFICIAL DEPOSITS				
Boulders and Boulder Clay; some sand ..	83	6	83	6
CARBONIFEROUS				
WESTPHALIAN A (LOWER COAL MEASURES)				
Strata; predominantly argillaceous; sporadic sandstone ribs; "mussels" at 92 ft (Victoria Shell Bed)	24	4	107	10
VICTORIA FISH-BED				
Shale, dark grey; sporadic fish scales	2	0	109	10
VICTORIA RIDER				
Coal	0	11	110	9
Seatearth	1	11	112	8
Shale, black carbonaceous; coal partings	0	4	113	0
VICTORIA				
Coal	0	5	113	5
Seatearth ..	0	4	113	9
Siltstone and mudstone; sporadic argillaceous sandstone partings..	9	3	123	0
Sandstone, grey predominantly flaggy fine-grained.. ..	14	7	137	7
Shale, dark grey; fish scales at base ..	1	10	139	5
Ironstone; *Carbonicola*	0	3	139	8
Sandstone	0	4	140	0
Shale; *Naiadites* and fish scales	1	2	141	2
Coal ('Bottom Victoria')	0	9	141	11

	Thickness ft	in	Depth ft	in
Seatearth-sandstone ..	1	10	143	9
Shale, grading to mudstone; fish scales at base	16	3	160	0
(Assumed horizon of STOBSWOOD)				
Sandstone	1	3	161	3
Mudstone; fish scales and spines in basal 3½ ft	7	3	168	6
Seatearth	2	0	170	6
Sandstone, hard rooty	0	9	171	3
Seatearth	0	3	171	6
TOP MARSHALL GREEN				
Coal	0	5½	171	11½
Strata; predominantly argillaceous; thin sandstone band towards base	18	4½	190	4
BOTTOM MARSHALL GREEN				
Coal	1	7	191	11
Strata; predominantly sandstone; sporadic shale partings ..	34	10	226	9
Strata; mainly shale and mudstone; sporadic sandstone bands and partings ..	25	5	252	2
GANISTER CLAY				
Coal	1	5	253	7
Seatearth - sandstone	0	3	253	10
Cannel	0	1½	253	11½
Coal	0	3½	254	3
Seatearth	2	0	256	3
Sandstone	3	9	260	0

Hilton No. 60 Bore

Surface 460.6 ft AOD. 6-in NZ 12 SE. Site [1754 2175] near Hilton. Drilled 1957 by Cementation Co. Ltd. for NCB. Cores examined by A. M. J. Clarke. Fig. 10(11).

	Thickness ft	in	Depth ft	in
SUPERFICIAL DEPOSITS				
Soil clay and boulders	4	0	4	0
CARBONIFEROUS				
NAMURIAN (MILLSTONE GRIT SERIES)				
SECOND GRIT (upper leaf)				
Sandstone, predominantly coarse-grained pebbly locally feldspathic; jointed	32	6	36	6
Sandstone, light grey fine-grained flaggy; ferruginous bands	2	9	39	3
SPURLSWOOD SHELL-BEDS				
Shale, grey; *Lingula*	6	3	45	6
Mudstone	2	9	48	3
SECOND GRIT (lower leaf)				
Sandstone, white coarse - grained kaolinitic; angular quartz fragments	6	9	55	0
Sandstone, predominantly medium- to coarse - grained thin-bedded; micaceous at top ..	20	9	75	9
Shale, black carbonaceous pyritic; coaly partings	0	4	76	1
Coal	0	10	76	11
Seatearth	1	7	78	6
Seatearth - siltstone; ferruginous ..	6	2	84	8
Siltstone, sandy; ferruginous	6	4	91	0
Mudstone, shaly; sandy bands; abundant plants	7	0	98	0
Sandstone, fine white massive	2	0	100	0

Hilton No. 113 Bore

Surface 404.14 ft AOD. 6-in NZ 12 SE. Site [1577 2145] 1 mile NW of Ingleton. Drilled 1957 by Cementation Co. Ltd. for NCB. Cores examined by R. H. Price.

	Thickness ft	Thickness in	Depth ft	Depth in
SUPERFICIAL DEPOSITS				
Boulder Clay, sandy ..	25	0	25	0
CARBONIFEROUS				
WESTPHALIAN A (LOWER COAL MEASURES)				
Shale, sandy	7	0	32	0
Shale, silty micaceous; sandy at base ..	4	0	36	0
Sandstone rib ..	0	6	36	6
Shale, dark grey silty micaceous	3	7	40	1
Seatearth-sandstone ..	0	4	40	5
Sandstone, hard rooty	1	7	42	0
Ganister, pencil; porcellanous at base ..	4	0	46	0
Ganister, bastard; ankerite and pyrite ..	0	6	46	6
Sandstone, hard siliceous	0	11	47	5
Sandstone, argillaceous micaceous; slumped	4	7	52	0
Shale, grey silty; sandy partings	10	6	62	6
Mudstone, grey silty; shale bands ..	2	6	65	0
RODDYMOOR MARINE BAND				
Mudstone, dark grey finely micaceous; *Planolites ophthalmoides, Serpuloides stubblefieldi, Paraconularia sp., Orbiculoidea* cf. *nitida*	5	1	70	1
Mudstone, hard ankeritic	0	2	70	3
Coal	0	4	70	7
Shale, black carbonaceous	0	1	70	8
Seatearth-mudstone, ankeritic	2	2	72	10
Siltstone, grey sandy; plants	3	8	76	6
THIRD GRIT				
Sandstone, light grey; argillaceous at top ..	6	9	83	3
KAYS LEA MARINE BAND				
Shale, grey silty; *Lingula mytilloides, Planolites ophthalmoides, Ammodiscus sp.* and other foraminifera ..	6	9	90	0
Seatearth-sandstone, grey	1	0	91	0
Sandstone, massive fine-grained	6	6	97	6
Sandstone, argillaceous	3	6	101	0
Shale, grey slightly sandy micaceous; *Lingula mytilloides*; fish remains including palaeoniscid scales, *Rhabdoderma sp.*, and *Rhizodopsis sp.*	0	6	101	6
Shale, black carbonaceous; coal partings	0	11	102	5
Coal, hard bright ..	0	7	103	0
Seatearth-sandstone, dark grey	0	2	103	2
Siltstone, grey sandy; rooty at top ..	4	4	107	6
Sandstone, fine-grained	2	6	110	0
Mudstone, dark grey; fish debris at base ..	5	3	115	3
Sandstone, argillaceous	1	2	116	5
QUARTERBURN MARINE BAND				
Siltstone and shale; shell fragments; small *Lingula* at base	5	10	122	3
NAMURIAN (MILLSTONE GRIT SERIES)				
Mudstone, dark finely micaceous	0	3	122	6
Sandstone, ganisteroid	1	1	123	7
Sandstone, hard argillaceous	2	5	126	0

Hilton No. 114 Bore

Surface 377.11 ft AOD. 6-in NZ 12 SE. Site [1654 2082] 580 yd WNW of Ingleton Vicarage. Drilled 1957 by Cementation Drilling Co. Ltd. for NCB. Cores examined by R. H. Price. Fig. 10 (12).

	Thickness ft	in	Depth ft	in
SUPERFICIAL DEPOSITS				
Soil	0	6	0	6
Sand	3	0	3	6
Clay and Boulders ..	10	6	14	0
CARBONIFEROUS				
WESTPHALIAN A (LOWER COAL MEASURES)				
Sandstone, jointed ..	1	6	15	6
Shale, grey and brown	1	6	17	0
Shale, black silty; *Lingula mytilloides*, *Rhadinichthys sp.* ..	2	6	19	6
Sandstone, flaggy fine-grained micaceous; rooty at top.. ..	10	3	29	9
Shale, dark grey silty; sandy bands; plant fragments	4	9	34	6
QUARTERBURN MARINE BAND				
Shale, dark grey silty; fauna includes *Lingula mytilloides*	0	6	35	0
NAMURIAN (MILLSTONE GRIT SERIES)				
Sandstone-seatearth, hard ferruginous ..	6	0	41	0
Shale, grey micaceous rooty; argillaceous sandstone bands ..	4	9	45	9
Sandstone, grey argillaceous; irregularly-bedded	5	3	51	0
Sandstone, pale grey massive	10	6	61	6
Siltstone, grey micaceous; plants ..	1	0	62	6
Shale, black brecciated [*small fault*] ..	0	4	62	10
Seatearth, mottled reddish-brown; crush breccia at top ..	1	2	64	0
Siltstone, grey to reddish-brown rooty ..	1	0	65	0
SECOND GRIT (upper leaf)				
Sandstone, rooty ..	0	6	65	6
Sandstone, pale reddish-brown to grey soft massive; plants at base	27	6	93	0
Shale, dark grey; fish scales at base ..	1	4	94	4
Sandstone, grey to reddish-brown	4	6	98	10
Siltstone, dark grey shaly	3	2	102	0
Shale, dark grey silty; small ironstone nodules	5	0	107	0
SPURLSWOOD SHELL-BEDS				
Mudstone, dark grey shaly; ironstone nodules; slickensided; scattered fauna of *Orbiculoidea sp.* and productoids ..	5	0	112	0
Shale, grey silty; *Orbiculoidea sp.* ..	1	6	113	6
Shale, grey silty ..	0	6	114	0
SECOND GRIT (lower leaf)				
Sandstone, mainly coarse-grained gritty feldspathic; sporadic mudstone bands	20	0	134	0
Marl, reddish-grey soft	0	4	134	4
Sandstone, as at 134 ft	4	6	138	10
Mudstone, dark grey; mottled red at top	1	3	140	1
Sandstone, dark grey argillaceous ..	0	3	140	4

	Thickness ft	in	Depth ft	in
Sandstone, fine-grained; rooty at top	1	8	142	0
Siltstone, grey-green rooty	3	7	145	7
Sandstone, pale grey fine-grained; sporadic micaceous partings and mudstone bands; ganister 1 ft 11 in at 148 ft 6 in ..	8	9	154	4
Seatearth, greenish grey	2	2	156	6
Mudstone, greenish grey	1	0	157	6
Shale, silty	3	6	161	0
WOODLAND SHELL-BEDS				
Mudstone; abundant fauna on partings including productoids and bivalves	0	2	161	2
Shale, dark grey ..	3	4	164	6
Sandstone, grey argillaceous ..	0	6	165	0
Sandstone, grey to green hard argillaceous; iron-rich	1	3	166	3
Shale, grey micaceous rooty; fine-grained sandstone laminae	0	9	167	0
Siltstone, greyish green	4	6	171	6
Shale, dark grey silty; sporadic ironstone nodules; *Serpuloides sp.*, *Lingula sp.*, *Orbiculoidea* cf. *nitida*, *Productus carbonarius*, *Rugosochonetes sp.*, *Euphemites sp.* [juv.] (loc. 22a)	6	6	178	0
Shale, black; fauna at top includes *Serpuloides sp.*, *Lingula sp.*, *Orbiculooidea* cf. *nitida*, orthotetoid indet., *Productus carbonarius*, *Rugosochonetes sp.*, *Retispira ?* (loc. 22a) ..	3	3	181	3
FIRST GRIT (upper leaf)				
Sandstone, grey fine-grained; mainly flaggy and micaceous; shaly partings at base ..	11	3	192	6
Siltstone, dark grey; sandstone partings; plant debris ..	5	0	197	6
SHARNBERRY SHELL-BEDS				
Shale, dark grey; 7-in ironstone at base; sporadic fossils including *Bucanopsis sp.*, *Retispira sp.*, (loc. 21a) ..	5	7	203	1
Siltstone, grey sandy; plant debris ..	6	1	209	2
Siltstone, sandy; plant debris; interbedded with flaggy micaceous sandstone	4	10	214	0
FIRST GRIT (lower leaf)				
Sandstone, pink green-mottled fine- to medium-grained; shale pellets at base; plant debris	13	4	227	4
Mudstone, crinoid debris; *Dictyoclostus sp.*, *Lingula mytilloides*, orthotetoid indet., *Rugosochonetes sp.*	1	8	229	0
Shale, dark grey ..	1	0	230	0
Shale, black; sandstone partings ..	0	6	230	6
WHITEHOUSE LIMESTONE				
Limestone, grey fine-grained	1	6	232	0
Sandstone, light grey flaggy rooty; shaly at base	4	6	236	6
Siltstone, grey micaceous; sandstone partings	0	6	237	0

	Thickness ft	in	Depth ft	in
Seatearth, dark grey ..	2	0	239	0
Mudstone, light grey ..	1	0	240	0
Siltstone, light grey ..	3	6	243	6
Sandstone, grey fine- to medium-grained micaceous	3	6	247	0
Siltstone, grey ..	0	6	247	6
Seatearth-mudstone ..	2	6	250	0

Hilton No. 115 Bore

Surface 399.22 ft AOD. 6-in NZ 12 SE. Site [1721 2072] Ingleton. Drilled 1957 by Cementation Co. Ltd. for NCB. Cores examined by A. M. J. Clarke. Fig. 10(13).

	Thickness ft	in	Depth ft	in
SUPERFICIAL DEPOSITS				
Soil and sandy clay ..	11	6	11	6
CARBONIFEROUS				
NAMURIAN (MILLSTONE GRIT SERIES)				
SECOND GRIT (upper leaf)				
Sandstone, mainly fine- to medium-grained current-bedded	21	6	33	0
Sandstone, mainly medium- to coarse-grained current-bedded	14	0	47	0
SPURLSWOOD SHELL-BEDS				
Shale, dark grey; ironstone nodules; crinoid columnals, *Serpuloides* sp., *Lingula* sp. [juv.], *Productus carbonarius*, *Donaldina* sp. (loc. 23c) ..	4	9	51	9
Seatearth	0	9	52	6
Siltstone, grey sandy; roots at top ..	4	6	57	0
SECOND GRIT (lower leaf)				
Sandstone, grey argillaceous; plant remains at base ..	7	4	64	4
Sandstone, light grey coarse-grained current-bedded; plant debris at base	35	7	99	11
Coal	0	5	100	4
Seatearth	2	3	102	7
Siltstone	3	2	105	9
Sandstone, argillaceous; roots ..	1	3	107	0
Seatearth-sandstone ..	0	3	107	3
Sandstone, light grey fine-grained shale partings; plants at base	9	6	116	9
Shale, micaceous sandy	0	3	117	0

Hilton Hall Bore

Surface about 400 ft AOD. 6-in NZ 12 SE. Site [1698 2143] 860 yd N of Ingleton Vicarage. Drilled 1942 by Messrs A. Kyle for Pease's West Collieries. Fig. 10(10) and Plate VI (91).

	Thickness ft	in	Depth ft	in
SUPERFICIAL DEPOSITS				
Clay, stones and boulders	13	6	13	6
CARBONIFEROUS				
WESTPHALIAN A (LOWER COAL MEASURES)				
Metal, blue and grey; hard ribs; coal traces	6	6	20	0
Sandstone, grey ..	1	0	21	0

	Thickness ft	in	Depth ft	in
Shale, sandy; coal traces	6	0	27	0
Sandstone, hard; shaly bands	2	6	29	6
Sandstone and sandy shale	6	0	35	6
Shale, blue; hard bands; *Palaeostachys sp.* at 36 ft 4 in	8	0	43	6

?GANISTER CLAY

	Thickness ft	in	Depth ft	in
Coal, soft	2	0	45	6
Coal, hard with pyrites	1	0	46	6
Fireclay, hard fakey; hard balls	3	6	50	0
Sandstone, hard broken	10	6	60	6
Bind, coaly	0	6	61	0
Sandstone, hard broken	5	9	66	9
Shale, blue sandy; plant debris	22	9	89	6
Sandstone; soft shale and grey metal bands; sandstone partings; plant debris ..	7	0	96	6
Shale, dark; ironstone balls; plant debris..	6	6	103	0
Post hard; spar joints	1	7	104	7
[Strata affected by faults between 66 ft 9 in and 104 ft 7 in]				
Shale, dark grey pyritic	0	10	105	5
Coal, inferior ..	0	7	106	0
Fireclay, pale grey ..	2	0	108	0
Fireclay, hard sandy..	5	0	113	0

RODDYMOOR MARINE BAND

	Thickness ft	in	Depth ft	in
Shale, ironstone balls; *Lingula sp*	2	0	115	0
Shale, dark jointy; plant fragments including *Lepidophloios laricinus* at 118 ft	4	0	119	0
Coal, inferior	1	6	120	6
Fireclay, sandy ..	2	9	123	3
Sandstone, grey hard shaly; plant debris..	1	0	124	3
Fireclay, hard sandy..	1	0	125	3
Sandstone, fakey partings	4	0	129	3
Shale, sandstone partings (base at horizon of KAYS LEA MARINE BAND)	4	3	133	6
Fakes and fakey sandstone	1	9	135	3
Shale, dark grey; jointed	5	0	140	3
Shale; ironstone balls	3	3	143	6
Shale; jointy	4	6	148	0
Coal	3	0	151	0
Fireclay, hard sandy..	2	0	153	0
Sandstone; hard pale; [faulted strata] ..	18	0	171	0

QUARTERBURN MARINE BAND

	Thickness ft	in	Depth ft	in
Shale; ironstone balls; *Lingula sp.*, *Orbiculoidea* cf. *nitida*, productoid fragments and palaeoniscid scale ..	9	0	180	0

NAMURIAN (MILLSTONE GRIT SERIES)

	Thickness ft	in	Depth ft	in
Metal, blue; ironstone balls; coal traces ..	3	0	183	0
Fireclay, hard sandy..	7	0	190	0
Sandstone, fakey ..	5	6	195	6
Shale, sandy; plant debris; *Lingula mytilloides;* and fish remains	13	0	208	6
Shale, broken; *Lingula sp.*	6	9	215	3
Metal, blue sandy; *Lingula sp.* [juv.] ..	2	0	217	3
Fakes	4	9	222	0
Fireclay	0	9	222	9
Sandstone	4	6	227	3
Shale, sandy	3	9	231	0
Shale, sandy; ironstone nodules; plant debris	1	6	232	6
Coal	0	9	233	3

SECOND GRIT (upper leaf)

	Thickness ft	in	Depth ft	in
Sandstone, coarse-grained, feldspathic	37	9	271	0
Shale; ironstone balls; includes SPURLSWOOD SHELL-BEDS; crinoid columnals, *Dictyoclostus sp.*, *Lingula*				

	Thickness ft	in	Depth ft	in
sp. [juv.], productoids indet., turreted gastropods indet., *Rhadinichthys sp.* [scale], fish fragments indet.	6	0	277	0
Shale, grey sandy ..	1	9	278	9
Fireclay, hard sandy; ironstone balls ..	3	9	282	6
Shale, grey sandy ..	3	0	285	6
SECOND GRIT (lower leaf)				
Sandstone, coarse ..	18	0	303	6

Hilton Moor No. 2 Bore

Surface 495 ft AOD. 6-in NZ 12 SE. Site [1644 2360] 750 yd SSE of Evenwood Gate. Drilled 1954 by NCB. Cores examined by G. Armstrong and R. H. Price. Plate VI (80).

	Thickness ft	in	Depth ft	in
SUPERFICIAL DEPOSITS				
Soil	2	10	2	10
Boulder clay	5	2	8	0
Clay, brown silty ..	9	0	17	0
Clay, grey	7	8	24	8
CARBONIFEROUS				
WESTPHALIAN A (LOWER COAL MEASURES)				
Shale, mainly sandy ..	13	0	37	8
VICTORIA/VICTORIA RIDER				
Coal	0	6	38	2
Fireclay, grey ..	2	10	41	0
Shale, grey silty; ironstone nodules ..	15	0	56	0
Sandstone, argillaceous; shale partings	11	9	67	9
Shale, dark grey; *Carbonicola pseudorobusta ?*, *Naiadites flexuosus ?*, *Carbonita humilis*	1	9	69	6
BOTTOM VICTORIA (NCB)				
Coal	1	0	70	6
Sandstone, grey argillaceous rooty ..	1	11	72	5
Shale, sandy micaceous rooty	1	6	73	11
Sandstone, grey fine-grained	6	0	79	11
Shale, grey micaceous	0	8	80	7
Sandstone, pale grey argillaceous; rooty at base	1	1	81	8
Shale, grey silty micaceous	12	7	94	3
Sandstone, flaggy; shale partings	1	6	95	9
Shale, dark grey micaceous; ironstone nodules at base ..	6	4	102	1
STOBSWOOD				
Coal	0	2	102	3
Fireclay, grey sandy ..	2	3	104	6
Fireclay, grey shaly; ironstone nodules ..	2	1	106	7
Shale, grey silty rooty	1	0	107	7
TOP MARSHALL GREEN				
Coal	0	7	108	2
Fireclay, dark grey; shaly at base ..	3	10	112	0
Shale, grey micaceous rooty	2	0	114	0
Sandstone, light grey fine- to medium-grained	23	3	137	3
Shale, grey sandy ..	0	3	137	6
BOTTOM MARSHALL GREEN				
Coal	1	10	139	4
Fireclay, grey ..	1	0	140	4
Sandstone, light grey fine-grained flaggy; alternating with grey sandy silty micaceous shale	30	7	170	11
Sandstone, light grey fine-grained; shaly below 172 ft 6 in ..	6	11	177	10

	Thickness ft	in	Depth ft	in
Shale, grey sandy; argillaceous flaggy sandstone partings especially below 181 ft 6 in	9	2	187	0
Sandstone, ganisteroid	1	0	188	0
Shale, grey micaceous rooty; plant debris	3	7	191	7
GANISTER CLAY				
Coal	1	7	193	2
Sandstone, black carbonaceous rooty; 1-in coal rib	0	5	193	7
Fireclay, dark grey and black slickensided ..	2	2	195	9
Shale, grey micaceous; rooty	2	2	197	11
Sandstone, grey fine-grained flaggy; rooty at top	1	0	198	11
Shale, grey finely micaceous; rooty ..	0	11	199	10
Sandstone, pale grey fine-grained thin-bedded; ironstone inclusions	10	6	210	4
Siltstone, grey micaceous; sandstone partings	0	11	211	3
Sandstone, grey and white fine-grained; flaggy at top massive at base; sporadic pyritic inclusions and some ankerite ..	22	9	234	0
Shale, grey finely micaceous; sandstone partings at top; silty at base	13	9	247	9
Sandstone, grey argillaceous micaceous ..	0	9	248	6
Shale, grey micaceous; silty at top; ironstone bands.. ..	14	6	263	0
Sandstone, hard ankeritic argillaceous ..	0	2	263	2
Shale, dark grey micaceous sandy; ironstone nodules; rootlets at top	5	4	268	6

	Thickness ft	in	Depth ft	in
RODDYMOOR MARINE BAND				
Shale, black; *Ammodiscus sp., Lingula mytilloides, Donaldina?, Elonichthys sp.* [scales]	2	3	270	9
Coal, canneloid ..	0	7	271	4
Fireclay, dark grey slickensided ..	0	7	271	11
THIRD GRIT				
Sandstone, dark grey rubbly argillaceous micaceous rooty; ironstone nodules	4	1	276	0
Sandstone, grey fine-grained flaggy rooty	0	10	276	10
Sandstone, pale grey fine-grained flaggy micaceous ..	9	5	286	3
Siltstone, grey sandy micaceous	0	9	287	0
Shale, dark grey; siltstone bands.. ..	2	0	289	0
KAYS LEA MARINE BAND				
Shale, dark grey and black; foraminifera including *Ammodiscus sp., Planolites ophthalmoides, Lingula mytilloides* ..	4	0	293	0
Shale, dark grey sandy	1	0	294	0
Siltstone, dark grey sandy	1	4	295	4
Mudstone, dark grey; ironstone nodules and sporadic shell fragments	6	10	302	2
Shale, black silty micaceous; *Ammodiscus sp.* [rare], *Lingula mytilloides* and *Orbiculoidea* cf. *nitida* at base	6	10	309	0
Shale, dark grey silty; rare *Lingula sp.* ..	3	9	312	9
Siltstone, dark micaceous pyritic ..	0	8	313	5
Coal, dirty	1	1	314	6

	Thickness ft	in	Depth ft	in
Mudstone, dark grey micaceous carbonaceous; coaly films and pyrite	0	3	314	9
Sandstone, grey argillaceous micaceous; rooty	0	6	315	3
Sandstone, pale grey rooty; micaceous partings; sporadic slickensides ..	1	9	317	0
Sandstone, grey and brown fine- to medium-grained rooty; irregular shaly partings	1	9	318	9
QUARTERBURN MARINE BAND				
Mudstone, grey silty; ironstone nodules; *Lingula mytilloides*, *Orbiculoidea sp.*, *Aviculopecten* cf. *delepinei* and *Donaldina sp.* ..	9	9	328	6
NAMURIAN (MILLSTONE GRIT SERIES)				
Sandstone, grey argillaceous rooty ..	1	6	330	0
Mudstone, grey sandy; coaly plant films ..	1	4	331	4
Shale, black carbonaceous pyritic; coal partings	0	4	331	8
Coal	0	2	331	10
Fireclay - mudstone, dark grey sandy micaceous	1	2	333	0
Sandstone, grey argillaceous micaceous rooty; ironstone nodules; ankeritic films at base	6	3	339	3
Mudstone, grey sandy	1	10	341	1
Sandstone, hard grey fine- to medium-grained; irregular micaceous partings; coaly plant debris near base	6	1	347	2
Mudstone, dark grey silty	0	3	347	5
Fireclay-mudstone ..	0	7	348	0
Sandstone, grey fine-grained	3	11	351	11
Fireclay, grey	2	4	354	3
Sandstone, light grey coarse; kaolinitic at base; 2-in shale band at 356 ft 6 in ..	7	4	361	7
Shale, dark grey silty	1	0	362	7
Coal	0	2	362	9
Fireclay, grey	0	10	363	7
Sandstone, grey fine; dark partings and bands; argillaceous at base; roots at top	11	1	374	8
Shale, grey silty; sandy partings	1	4	376	0
Mudstone, dark grey sandy micaceous; sporadic plants ..	2	6	378	6
Sandstone, grey argillaceous	1	0	379	6
Shale, grey sandy micaceous	0	6	380	0
Shale, grey silty micaceous; fine sandstone partings; sporadic plant impressions ..	8	0	388	0

Jane Pit, Adelaide Colliery

Surface about 410 ft AOD. 6-in NZ 22 NW. Site [2218 2744] 750 yd SW of Coundon Grange. Date of sinking 1884. Information from B & S No. 7 and NCB. Plates X, XI and XII (142).

	Thickness ft	in	Depth ft	in
SUPERFICIAL DEPOSITS				
Soil, Clay, Sand and Gravel	8	10	8	10
CARBONIFEROUS				
WESTPHALIAN B (MIDDLE COAL MEASURES)				
Freestone, brown soft	8	0	16	10
Metal, blue and white; post girdles at base..	24	0	40	10
Coal	0	2	41	0
Thill, grey	4	0	45	0
Freestone, grey strong	1	2	46	2
Metal, grey strong; rare post girdles ..	17	10	64	0
Metal, blue soft; ironstone girdles ..	4	4	68	4
Coal	0	8	69	0
Thill, grey	6	5	75	5
Post girdles, grey jointy	11	9	87	2
Metal, blue jointy ..	5	6	92	8
Metal, blue strong; post girdles at top ..	10	0	102	8
Metal, black; ironstone girdles	2	8	105	4
TOP HIGH MAIN				
Coal, splint and black jet	2	4	107	8
Thill, grey; post girdles	14	0	121	8
Metal, grey with ironstone	4	0	125	8
BOTTOM HIGH MAIN (YARD)				
Coal, coarse; splinty bands	2	4	128	0
Thill, grey tough ..	5	0	133	0
Metal, grey; post girdles	1	9	134	9
Post, grey; strong at top shivery at base..	4	10	139	7
Metal blue; ironstone balls; black metal in basal 2 ft	11	5	151	0
Coal and jet	2	2	153	2
Thill, grey	2	3	155	5
Metal, grey and blue; ironstone girdles and balls; 10-in post girdle at 169 ft 10 in	35	2	190	7

	Thickness ft	in	Depth ft	in
METAL				
Jet, black	4	0	194	7
Ironstone band ..	0	2	194	9
Jet, black	1	4	196	1
Thill, grey	2	4	198	5
Metal, black	4	2	202	7
FIVE-QUARTER				
Coal; coarse splint in basal 20 in ..	5	4	207	11
Thill, grey	4	3	212	2
Post, white; grey at base	18	10	231	0
Stone, blue	13	1	244	1
Coal, foul or jet ..	2	9	246	10
Blue stone; ironstone balls	6	0	252	10
Jet	0	8	253	6
Post and whin, white and grey; metal and grey beds	20	11	274	5
Metal blue; ironstone girdles and balls ..	42	1½	316	6½
MAIN				
Coal; 1-in slaty band at 318 ft 0½ in ..	5	5	321	11½
Splint, black stone and thill	2	11½	324	11
Coal, strong; bands every 3 to 4 in ..	3	6	328	5
Black stone ..	1	2	329	7
Coal, strong ..	0	10	330	5
Post, grey and white; occasional metal and whin partings (DURHAM LOW MAIN POST)	101	9	432	2
Metal, grey strong ..	6	0	438	2
(Assumed horizon of DURHAM LOW MAIN)				
Post, white	4	0	442	2
Metal, blue soft; ironstone nodules ..	12	0	454	2
Metal, grey coarse ..	28	6	482	8
BRASS THILL				
Coal	3	5	486	1
Thill, pasty	5	1	491	2
Metal, grey and black; ironstone girdles at top	72	5	563	7

	Thickness ft	in	Depth ft	in
BOTTOM HUTTON (HUTTON)				
Coal	2	0	565	7
Seggar clay	3	6	569	1
Post, grey strong ..	15	6	584	7
Metal, grey; post girdles	45	0	629	7
Coal, good	0	3	629	10
Metal, dark grey ..	7	0	636	10
Post, white; mixed with whin..	9	0	645	10
Metal, grey	2	0	647	10
Coal, coarse	0	9	648	7
Post, white; metal partings	18	0	666	7
Metal, black; ironstone girdles	24	0	690	7
(Assumed horizon of HARVEY MARINE BAND)				
WESTPHALIAN A (LOWER COAL MEASURES)				
Thill stone	3	0	693	7
Post and whin, grey ..	15	0	708	7
Metal, black	1	2	709	9
Post, dark strong ..	1	8	711	5
Coal, coarse	0	2	711	7
Thill stone, soft ..	4	0	715	7
Metal, grey; post girdles	28	5	744	0
Post, white strong ..	11	11	755	11
HARVEY				
Coal	4	1	760	0
Thill, black leafy ..	7	0	767	0
Post, white; whin and metal partings ..	63	0	830	0
TOP TILLEY				
Coal	0	3½	830	3½
Thill stone	2	0	832	3½

	Thickness ft	in	Depth ft	in
Metal, grey	10	8½	843	0
Post, grey strong ..	6	0	849	0
BOTTOM TILLEY				
Coal, coarse ..	0	3½	849	3½
Post, dark	4	4	853	7½
Coal, good	0	4	853	11½
Metal, dark.. ..	1	5	855	4½
Coal, coarse ..	2	0½	857	5
Metal, grey strong ..	43	3	900	8
BUSTY (BUSTY BANK)				
Coal	3	6	904	2
Metal, grey	5	0	909	2
Post, white; metal partings	33	0	942	2
Post, white; mixed with whin..	28	0	970	2
Metal, blue	4	0	974	2
THREE-QUARTER				
Coal	0	4	974	6
Metal, grey; post girdles	10	0	984	6
Post, white strong ..	3	0	987	6
Metal, grey; post girdles	4	2	991	8
Metal, dark blue ..	7	0	998	8
Coal, good	1	1	999	9
Metal, grey; post girdles	23	10	1023	7
BROCKWELL				
Coal	5	0	1028	7
Post, white; whin ..	12	0	1040	7
Metal, dark grey; post girdles	3	0	1043	7
Metal, strong; thick post girdles ..	42	0	1085	7
Coal, coarse cannel ..	1	0	1086	7
Thill, white; iron balls	2	0	1088	7

John Pit, New Copley Colliery

Surface about 774 ft AOD. 6-in NZ 12 SW. Site [1156 2397] 910 yd NW of Burnt Houses. Sunk 1868. Information from B & S No. 2557 and NCB. Plates IX and XI (29).

	Thickness ft	in	Depth ft	in
SUPERFICIAL DEPOSITS				
Boulder Clay ..	17	0	17	0
CARBONIFEROUS				

	Thickness ft	in	Depth ft	in
WESTPHALIAN B (MIDDLE COAL MEASURES)				
Shale, blue; sandstone band at top.. ..	15	0	32	0

	Thickness ft	in	Depth ft	in
HUTTON (BOTTOM HUTTON)				
Coal	4	0	36	0
Fireclay	2	8	38	8
Shale, blue	6	2	44	10
Sandstone, grey and white	11	11	56	9
Shale, blue	1	6	58	3
TOP RULER (JUBILEE)				
Coal	0	11	59	2
Band	0	1	59	3
Coal	1	4	60	7
Fireclay	0	7	61	2
Sandstone, grey and white	8	7	69	9
Shale, blue	1	1	70	10
BOTTOM RULER (JUBILEE)				
Coal	0	4	71	2
Fireclay	0	8	71	10
Sandstone, grey and white	12	7	84	5
Shale, blue	0	10	85	3
Sandstone, white and grey	1	5	86	8
Shale; 3-in ironstone rib	7	8	94	4
Coal	0	4	94	8
Sandstone	0	5	95	1
Shale, blue and black	2	1	97	2
Sandstone, white	0	10	98	0
Shale, grey and blue; black at base	43	6	141	6
(Assumed horizon of HARVEY MARINE BAND)				
WESTPHALIAN A (LOWER COAL MEASURES)				
Coal, cannel	0	3	141	9
Fireclay	1	6	143	3
Sandstone	5	9	149	0
Shale, blue	1	7	150	7
Sandstone, grey	2	6	153	1
Shale, blue	4	7	157	8
Sandstone, grey	4	7	162	3
Shale, blue; black at base	6	9	169	0
HARVEY				
Coal and band; 1 ft 4 in fireclay at 174 ft 8 in	6	5	175	5
Fireclay	3	0	178	5
Coal and band	1	4	179	9
Fireclay	3	3	183	0
Shale, blue; white sandstone bands	23	4	206	4
Sandstone	10	6	216	10
TOP TILLEY				
Coal	1	5	218	3
Fireclay	2	6	220	9
Sandstone	5	0	225	9
Coal	1	0	226	9
Fireclay	2	8	229	5
Coal	0	4	229	9
Fireclay	4	10	234	7
Shale, blue	29	4	263	11
Sandstone	5	6	269	5
BOTTOM TILLEY				
Coal, splint	1	0	270	5
Fireclay	3	2	273	7
Sandstone; predominantly white	56	6	330	1
BUSTY				
Coal; two thin bands	7	0	337	1
Sandstone	16	9	353	10
Shale, blue; sandstone rib at 362 ft 5 in	9	11	363	9
THREE-QUARTER				
Coal	0	4	364	1
Fireclay	1	8	365	9
Shale; sporadic sandstone and ironstone girdles	16	6	382	3
Sandstone, white	4	5	386	8
Brass band	0	4	387	0
Sandstone, white	5	11	392	11
Shale, blue; ironstone girdles	24	10	417	9
Shale, coaly	0	2	417	11
BROCKWELL				
Coal; cannel in top 7 in	4	$8\frac{1}{4}$	422	$7\frac{1}{4}$
Band	0	$0\frac{1}{4}$	422	$7\frac{1}{2}$
Coal	1	$2\frac{1}{2}$	423	10
Sump (no details)	26	2	450	0

Keverstone No. 44 Bore

Surface 762.91 ft AOD. 6-in NZ 12 SW. Site [1220 2329] 140 yd S of Burnt Houses. Drilled 1957 by Cementation Co. Ltd. for NCB. Cores examined by A. M. J. Clarke. Plate VI (34).

	Thickness ft	in	Depth ft	in
SUPERFICIAL DEPOSITS				
Soil, clay and boulders	30	0	30	0
CARBONIFEROUS				
WESTPHALIAN A (LOWER COAL MEASURES)				
Siltstone, grey thin-bedded sandy; micaceous partings; sporadic sandstone bands; abundant plant debris ..	8	10	38	10
Mudstone, dark grey silty micaceous; plants, worm tracks and fish scales in basal 1 ft ..	8	2	47	0
Siltstone, dark grey micaceous; sporadic sandy bands; comminuted plant debris and rootlets at top ..	10	2	57	2
Sandstone, pale to dark grey wispy-bedded micaceous; silty and shaly in basal 22 ft; sporadic rootlets; fish scales at 77½, 84½ and 89 ft 10 in; worm burrows near base ..	33	10	91	0
Shale, black; fish scales	2	6	93	6
Mudstone, dark grey sandy micaceous; plant fragments; mussels and fish scales at 96¼ ft; roots in basal 2 ft ..	5	6	99	0
Sandstone, grey hard massive fine- to medium-grained; roots at top; coal scars and silty laminae in basal 2 ft ..	11	6	110	6
Mudstone, dark grey; pyrites in basal 2 in; worm tubes ..	2	0	112	6
Coal, bright	1	4	113	10
Sandstone-seatearth ..	0	5	114	3
Siltstone-seatearth ..	0	2	114	5
Mudstone, dark grey; slickensided; roots	4	7	119	0
Siltstone, dark grey hard massive ..	2	6	121	6
Mudstone, dark grey silty shaly; fish scale at 122 ft 10 in ..	5	6	127	0
Mudstone, dark grey; silty in upper 5 ft; mussels and fish fragments (?horizon of RODDYMOOR MARINE BAND)	7	11	134	11
Coal, dull	0	1	135	0
THIRD GRIT				
Sandstone-seatearth, dark grey massive fine-grained; amorphous white mineralization on joints	1	2	136	2
Sandstone, pale grey very hard massive fine- to medium-grained; sporadic coal scars and pyrite in basal 2 ft	9	10	146	0
Sandstone, pale grey ganisteroid rooty	0	8	146	8
Mudstone, grey shaly; slickensided; plants	5	4	152	0
Sandstone, pale grey medium-grained irregularly bedded; mineralized subvertical joints ..	2	10	154	10
KAYS LEA MARINE BAND				
Siltstone, grey sandy; fragments of *Lingula*; fish scales and worm tubes ..	2	10	157	8
Sandstone, dark grey; roots	0	6	158	2
Sandstone, pale grey				

	Thickness ft	in	Depth ft	in
medium-grained; rootlets	1	4	159	6
Siltstone, pale grey massive ferruginous	1	6	161	0
Mudstone, grey slickensided; ironstone nodules; rootlets ..	2	0	163	0
Mudstone, grey shaly; mussel fragments and *Lingula* at 165 and 166½ ft; 4-in arenaceous and micaceous band at base ..	5	0	168	0
Sandstone-seatearth ..	2	0	170	0
Siltstone, dark grey micaceous; sandstone lenticles ..	1	0	171	0
Siltstone, dark grey micaceous shaly; abundant *Lingula* in basal 6 in	1	0	172	0
Sandstone-seatearth, brownish grey; ferruginous in basal 6 in ..	4	6	176	6
Sandstone, pale grey massive medium-grained micaceous; silty micaceous laminae in basal 2½ ft ..	9	0	185	6
Siltstone, grey shaly; sand partings; plant fragments	9	6	195	0
Mudstone, grey silty; *Planolites* 5½ ft from base	8	6	203	6
QUARTERBURN MARINE BAND				
Mudstone, grey; *Lingula*	1	6	205	0

Keverstone No. 133 Bore

(synopsis below Bottom Brockwell Seam)

Surface 611.26 ft AOD. 6-in NZ 12 SW. Site [1455 2318] 970 yd NE of Keverstone Grange. Drilled 1957 by Cementation Co. Ltd. for NCB. Cores examined by A. M. J. Clarke. Plates VII and IX (63).

	Thickness ft	in	Depth ft	in
SUPERFICIAL DEPOSITS				
Overburden and drift ..	38	6	38	6
CARBONIFEROUS				
WESTPHALIAN A (LOWER COAL MEASURES)				
Sandstone	0	6	39	0
BUSTY				
Coal	1	3	40	3
Band	0	1	40	4
Coal	0	8	41	0
Seatearth-mudstone ..	2	6	43	6
Siltstone, light grey; rooty at top ..	5	6	49	0
Mudstone, grey silty; ironstone bands; sandy at base with sporadic plant remains	8	6	57	6
THREE-QUARTER				
Coal	1	4	58	10
Seatearth-sandstone ..	3	8	62	6
Sandstone, light grey ..	3	0	65	6
Siltstone, dark grey sandy and shaly; micaceous bands and partings	5	0	70	6
Mudstone, dark grey silty shaly; ironstone bands	2	0	72	6
Mudstone; abundant poorly preserved mussels	1	2	73	8
Siltstone, dark grey finely micaceous; shale bands and sandstone partings; plant debris	3	4	77	0
Sandstone, fine-grained; silty bands; 10-in ironstone band at 81				

	Thickness ft	in	Depth ft	in
ft; comminuted plant debris	4	10	81	10
Mudstone, dark grey silty shaly; clay ironstone bands and comminuted plant debris; mussels from 89 to 90 ft with abundant small mussels from 91 ft 2 in to 91 ft 4 in	10	2	92	0
Seatearth-mudstone ..	1	2	93	2
Sandstone, light grey fine-grained wispy-bedded	20	0	113	2
Mudstone, silty at top	3	10	117	0
BROCKWELL (goaf) ..	4	0	121	0
Shale, black; vitrain partings	0	5	121	5
Seatearth-mudstone; coalified rootlets ..	1	10	123	3
Mudstone, dark grey silty; roots; plant fragments at base ..	3	9	127	0
Siltstone, grey shaly ..	1	0	128	0
[Fault]	—	—	—	—
Sandstone; siltstone laminae	4	0	132	0
BOTTOM BROCKWELL				
Coal, canneloid at top	0	10	132	10
Seatearth-mudstone, dark grey ..	1	8	134	6
Mudstone, dark grey shaly; clay ironstone bands ..	2	6	137	0
Coal, dull inferior	1	11	138	11
Sandstone, carbonaceous; slumped ..	0	0½	138	11½
Coal, bright fusainous	0	2½	139	2
Strata, arenaceous ..	24	4	163	6
Mudstone, silty shaly; plant remains ..	9	6	173	0
VICTORIA SHELL- AND FISH-BED				
Mudstone, black and shaly at base; sporadic ironstone nodules; mussels at 174½ ft, fish scale fragments below 180 ft	8	0	181	0
(Horizon of VICTORIA RIDER)				
Strata, argillaceous ..	17	0	198	0
Seatearth-mudstone; coalified plant remains	2	4	200	4
Shale, black	0	2	200	6
Seatearth-mudstone ..	1	2	201	8
VICTORIA				
Coal	0	11	202	7
Strata; variable sequence of siltstone, mudstone, sandstone and shale, including fault breccia at 218 ft and seatearth at 223½ ft	57	5	260	0
Mudstone, dark grey; shaly below 263 ft with fish remains ..	5	0	265	0
Siltstone, black; sandstone lenses ..	1	0	266	0
Seatearth-mudstone, sandy; clay ironstone nodules at base	5	3	271	3
Siltstone; clay ironstone nodules ..	1	7	272	10
TOP MARSHALL GREEN				
Coal	0	10	273	8
Strata, including 38 ft of mainly medium-grained sandstone at base	48	4	322	0
BOTTOM MARSHALL GREEN				
Coal	1	10	323	10
Strata; mainly mudstone with small mussels at base	21	11	345	9
Strata; mainly arenaceous	19	3	365	0
Mudstone, shaly; plant remains at top ..	3	1	368	1
Coal, inferior	0	2	368	3
Shale, dark grey; plant remains and rootlets	3	1	371	4
GANISTER CLAY				
Coal	1	2	372	6
Seatearth-sandstone	0	7½	373	1½
Coal	0	1	373	2½
Seatearth-mudstone and siltstone ..	4	9½	378	0

Ladysmith Shaft, Brusselton Colliery

Surface about 376 ft AOD. 6-in NZ 12 NE. Site [1937 2553] 580 yd ENE of Hummerbeck. Date of sinking unknown. Information from B & S No. 2466 and NCB. Plate X (118).

	Thickness ft	Thickness in	Depth ft	Depth in
SUPERFICIAL DEPOSITS				
Soil and Boulder Clay	6	0	6	0
CARBONIFEROUS				
WESTPHALIAN B (MIDDLE COAL MEASURES)				
Sandstone	21	0	27	0
Shale, black	2	11	29	11
Coal (?RULER)	0	3	30	2
Metal, blue and grey; post girdles	20	2	50	4
Coal, coarse splint	0	9	51	1
Seggar	1	9	52	10
Post, leafy; metal bands	12	6	65	4
Metal, blue; soft at base	42	6	107	10
(Assumed horizon of HARVEY MARINE BAND)				
WESTPHALIAN A (LOWER COAL MEASURES)				
Coal	0	7	108	5
Seggar, bastard	5	7	114	0
Coal	0	3	114	3
Seggar	4	3	118	6
Metal, grey; whin and post girdles	20	0	138	6
Shale, black; sandstone rib at base	4	6	143	0
HARVEY				
Coal	1	6	144	6
Band	0	3	144	9
Coal	3	8	148	5
Seggar	1	7	150	0
Post; rare metal partings; coal pipings at base	109	0	259	0
TOP TILLEY				
Coal	0	8	259	8
Seggar, bastard	11	4	271	0
Metal, blue	8	0	279	0
BOTTOM TILLEY (CONSTANTINE)				
Coal	0	6	279	6
Seggar	1	9	281	3
Coal	0	6	281	9
Post, white	2	6	284	3
Metal, grey	1	0	285	3
Coal	1	3	286	6
Seggar	5	0	291	6
Metal, grey and blue	38	0	329	6
TOP BUSTY (BEAUMONT)				
Coal	3	0	332	6
Seggar	11	6	344	0
Post, grey; metal parting	4	0	348	0
Shale, black	3	6	351	6
BOTTOM BUSTY				
Coal	3	2	354	8
Seggar	0	4	355	0
Coal	1	6	356	6
Seggar	5	6	362	0
Post, grey	3	0	365	0
Metal, blue; post girdle	14	0	379	0
THREE-QUARTER				
Coal	0	6	379	6
Seggar	1	6	381	0
Post; metal partings	9	6	390	6
Metal, blue; post rib	18	11	409	5
Coal	0	6	409	11
Seggar	6	0	415	11
Post; sporadic metal partings	36	6	452	5
Metal, blue	13	3	465	8
BROCKWELL				
Coal	4	2	469	10
Band	0	3	470	1
Coal	2	1	472	2
Post	13	6	485	8
Metal, blue	7	6	493	2

Lutterington Estate Bore

Surface about 450 ft AOD. 6-in NZ 12 SE. Site [1875 2447] Lutterington Hall, Bildershaw. Drilled 1834. For detailed section see B & S No. 2833. Fig 10 (7) and Plate VI (115).

	Thickness ft	Thickness in	Depth ft	Depth in
SUPERFICIAL DEPOSITS				
Soil and stony clay ..	14	6	14	6
CARBONIFEROUS				
WESTPHALIAN A (LOWER COAL MEASURES)				
Post, brown	18	6	33	0
Metal, blue; girdles ..	13	2	46	2
GANISTER CLAY				
Coal	1	4	47	6
Strata; predominantly siltstones, mudstones and shales; sandstone 14 ft 4 in at 98 ft 10 in; sandstone 12 ft at 125 ft 6 in; sandstone 10 ft at 175 ft 4 in; approximate horizon of QUARTERBURN MARINE BAND at about 186 ft	138	6	186	0
NAMURIAN (MILLSTONE GRIT SERIES)				
Strata; predominantly sandy siltstones and mudstones	37	6	223	6
SECOND GRIT (upper leaf)				
Post, white	48	0	271	6
Coal	0	7	272	1
Post, white	8	0	280	1
?SPURLSWOOD SHELL-BEDS				
Metal, grey; post girdles in basal 8 ft	13	5	293	6
SECOND GRIT (lower leaf)				
Post, brown white and grey	62	4	355	10
?WOODLAND SHELL-BEDS				
Metal, dark grey; post girdles ..	10	2	366	0
FIRST GRIT (upper leaf)				
Post, white; mixed with whin at top ..	36	0	402	0
?SHARNBERRY SHELL-BEDS				
Metal, grey ..	25	0	427	0
FIRST GRIT (lower leaf)				
Post, grey and white	7	0	434	0
Metal, dark grey ..	14	4	448	4
(?Horizon of WHITEHOUSE LIMESTONE)				
Metal stone, grey; ironstone girdles ..	15	5	463	9
Post, grey strong; metal partings	7	1	470	10
Whin into	0	5	471	3

Machine Pit, Auckland Park Colliery

Surface about 338 ft AOD. 6-in NZ 22 NW. Site [2270 2846] 560 yd NNE of Coundon Grange. Information from B & S No. 29 and NCB. Plates X, XI and XII (145).

	Thickness ft	Thickness in	Depth ft	Depth in
SUPERFICIAL DEPOSITS				
Clay and Sand ..	16	0	16	0
CARBONIFEROUS				
WESTPHALIAN B (MIDDLE COAL MEASURES)				
Metal, soft blue ..	7	0	23	0
Metal, blue; ironstone girdles	16	0	39	0

	Thickness ft	in	Depth ft	in
HIGH MAIN				
Coal	2	0	41	0
Post	8	8	49	8
Metal, blue; black at base	14	10	64	6
Metal, grey strong ..	16	0	80	6
METAL [DICKY DANT: THREE-QUARTER]				
Coal; slaty bands ..	2	5	82	11
Thillstone	4	0	86	11
Post, grey	25	0	111	11
Metal, blue	20	1	132	0
FIVE-QUARTER				
Coal	5	0	137	0
Post, grey strong ..	7	0	144	0
Metal, grey strong ..	14	0	158	0
Metal, blue and black	20	6	178	6
Post, grey strong ..	9	0	187	6
Whin	4	0	191	6
Post, grey strong ..	8	0	199	6
Metal, blue; ironstone girdles	22	6	222	0
MAIN				
Coal	7	0	229	0
Thillstone, white strong	2	10	231	10
?MAUDLIN				
Coal; slaty bands ..	3	4	235	2
Thillstone, dark ..	1	3	236	5
Coal	0	10	237	3
Post, white strong ..	2	0	239	3
Metal, grey strong ..	39	0	278	3
Post, grey strong ..	5	0	283	3
Metal, grey strong; post girdles	22	0	305	3
Post, white	23	0	328	3
Metal, grey; post girdles	5	0	333	3
Post, white	22	0	355	3
(Assumed horizon of DURHAM LOW MAIN)				
Metal, grey; ironstone girdles	34	10	390	1
BRASS THILL				
Coal	3	3	393	4
Metal, grey; post girdles	5	1	398	5
Metal, black	2	0	400	5

	Thickness ft	in	Depth ft	in
Post, white with whin	7	3	407	8
Metal grey; black at base	10	3	417	11
TOP HUTTON				
Coal	0	4	418	3
Thill	5	6	423	9
Coal	0	4	424	1
Thill	1	4	425	5
Post, grey and white strong	19	7	445	0
Metal, grey strong; ironstone girdles ..	14	6	459	6
BOTTOM HUTTON				
Coal	2	3	461	9
Metal, grey; ironstone girdles	15	5	477	2
Post, white strong ..	15	0	492	2
Metal, black	2	0	494	2
Post, grey strong ..	16	3	510	5
Coal	0	4	510	9
Post, grey strong ..	19	2	529	11
Metal, dark grey ..	1	5	531	4
Coal	0	8	532	0
Post, grey and white ..	14	3	546	3
Metal, grey; post girdles	7	0	553	3
Metal, black	20	0	573	3
(Assumed horizon of HARVEY MARINE BAND)				
WESTPHALIAN A (LOWER COAL MEASURES)				
Thillstone, grey strong	3	0	576	3
Post, grey strong ..	8	3	584	6
Metal, grey strong ..	13	9	598	3
Post, grey and white; with whin	36	1	634	4
HARVEY				
Coal	3	5	637	9
Black stone	0	2	637	11
Metal thill, grey soft ..	4	0	641	11
Metal, grey	8	10	650	9
Ironstone girdle ..	1	5	652	2
Metal, grey; post girdles	7	1	659	3
Post, white; and whin	2	6	661	9
Metal, grey	2	4	664	1
Post grey; metal partings	4	0	668	1

	Thickness ft	in	Depth ft	in
Coal	0	5	668	6
Metal, grey; ironstone balls	5	6	674	0
Post, grey	10	3	684	3
Metal, grey	14	0	698	3
TOP TILLEY				
Coal	0	10	699	1
Fireclay, strong	5	4	704	5
BOTTOM TILLEY				
Coal, coarse	0	7	705	0
Metal band, grey	1	9	706	9
Coal	0	4	707	1
Metal, grey	1	8	708	9
Coal; soft scar bands	0	9	709	6
Post, grey	6	0	715	6
Metal grey; ironstone girdles	38	0	753	6
TOP BUSTY				
Coal	1	0	754	6
Fireclay	3	3	757	9
BOTTOM BUSTY				
Coal	0	9	758	6
Post girdle	0	4	758	10
Coal	2	1	760	11
Metal, grey	4	2	765	1
Post, white	40	0	805	1
Whin	8	0	813	1
Post, white	14	8	827	9
Ironstone	1	5	829	2
THREE-QUARTER				
Coal	2	2	831	4
Blackstone	0	8	832	0
Fireclay	7	5	839	5
Coal	0	9	840	2
Fireclay	5	0	845	2
Post and metal, grey	7	0	852	2
Post, white	4	6	856	8
Post, white flimby	32	0	888	8
Metal, grey strong	8	4	897	0
BROCKWELL (MAIN)				
Coal	5	0	902	0
Fireclay	8	0	910	0
Metal, dark grey	5	0	915	0
Post, white strong	6	6	921	6

Moorhill No. 1 Bore

(synopsis)

Surface 831.38 ft AOD. 6-in NZ 12 NW. Site [1027 2801] 1 mile NW of High Wham. Drilled 1957 by Boldon Drilling Co. for NCB. Cores examined by R. H. Price. Plates VI and VII (21).

	Thickness ft	in	Depth ft	in
SUPERFICIAL DEPOSITS				
Sand and boulders	12	0	12	0
CARBONIFEROUS				
WESTPHALIAN A (LOWER COAL MEASURES)				
Strata; sandy at top	30	9	42	9
Shale, dark grey; mussels throughout, abundant between 44 and 45 ft	11	3	54	0
VICTORIA RIDER				
Coal	0	2	54	2
Seatearth, grey brown; calcareous base; slickensided	1	7	55	9
VICTORIA				
Coal, inferior; shaly partings; slickensided	1	0	56	9
Strata; argillaceous in upper 22 ft; remainder sandstone; STOBSWOOD COAL absent	59	5	116	2
TOP MARSHALL GREEN				
Coal	0	4	116	6

	Thickness ft	in	Depth ft	in
Strata; predominantly sandstone but silty and shaly at top ..	23	10	140	4
BOTTOM MARSHALL GREEN				
Coal	1	7	141	11
Strata; sandstone at top and base; shales and siltstones in middle part of sequence	44	1	186	0
GANISTER CLAY				
Coal	0	8	186	8
Seatearth, sandy micaceous	2	5	189	1
Sandstone, grey wispy-bedded; irony bands	18	5	207	6
Shale, grey and dark grey sandy silty micaceous; worm bores below 235 ft; black at base ..	43	8	251	2
Coal, inferior	0	4	251	6
Shale, black carbonaceous	0	2	251	8
Seatearth	0	4	252	0
Siltstone, argillaceous (approximate horizon of RODDYMOOR MARINE BAND) ..	7	6	259	6

	Thickness ft	in	Depth ft	in
THIRD GRIT				
Sandstone, grey and white medium-grained; 3 ft 8 in shale band at 281 ft 10 in	23	4	282	10
Shale, grey micaceous	0	11	283	9
Sandstone, grey fine-grained wispy-bedded	0	10	284	7
Shale, dark grey; occasional micaceous partings; sporadic worm bores below 288 ft 6 in ..	5	4	289	11
KAYS LEA MARINE BAND				
Shale, dark grey finely micaceous; *Ammodiscus sp.*, *Lingula mytilloides*, *Euphemites sp.*, ostracods indet., palaeoniscid scales including *Rhadinichthys sp.* ..	5	9	295	8
Coal, inferior	0	3	295	11
Seatearth, sandy; carbonaceous films ..	0	4	296	3
Seatearth, brown shaly	3	9	300	0

Moorhill No. 118 Bore

Surface 869.18 ft AOD. 6-in NZ 02 NE. Site [0978 2817] 1550 yd SSW of Softley. Drilled 1957 by NCB. Cores examined by A. M. J. Clarke. Plate VII (18).

	Thickness ft	in	Depth ft	in
SUPERFICIAL DEPOSITS				
Top Soil, Boulder Clay and Clay	20	8	20	8
CARBONIFEROUS				
WESTPHALIAN A (LOWER COAL MEASURES)				
BROCKWELL				
Old Coal Workings	7	4	28	0
Shale, grey; rooty at base	3	0	31	0

	Thickness ft	in	Depth ft	in
Seatearth, shaly; ironstone nodules ..	2	0	33	0
Sandstone, light grey false-bedded fine-grained	15	1	48	1
Mudstone, sandy; alternating with fine-grained sandstone ..	1	11	50	0
Sandstone, light grey massive false-bedded fine-grained	7	8	57	8
Siltstone, dark grey micaceous; plant re-				

	Thickness ft	in	Depth ft	in
mains	1	10	59	6
Sandstone, light grey false-bedded fine-grained; dark micaceous partings; jointed	3	2	62	8
VICTORIA SHELL-BED				
Mudstone, dark grey; sandy partings at top; plants and poorly preserved mussels in lower 5½ ft	7	4	70	0
Mudstone, dark grey sandy; a few plant remains	4	0	74	0
VICTORIA FISH-BED				
Shale, black micaceous; fish scales at base	2	11	76	11
VICTORIA RIDER				
Coal	0	3	77	2
Seatearth, shaly greyish brown	2	1	79	3
VICTORIA				
Coal, poor ..	0	1	79	4
Shale, rooty	0	2	79	6
Siltstone, grey rooty; ironstone nodules ..	3	2	82	8
Sandstone, grey argillaceous; rootlets ..	3	4	86	0
Siltstone, grey; predominantly sandy; sandstone partings at base	8	0	94	0
Siltstone-seatearth, grey sandy; ironstone nodules ..	0	6	94	6
Siltstone, dark grey sandy; irony bands; a few plants ..	17	8	112	2
Sandstone, white fine-grained variably bedded; shaly bands at 123 ft; brecciated with shale fragments in basal 3 ft ..	35	10	148	0
Shale, grey silty ..	1	0	149	0
TOP MARSHALL GREEN				
Coal	0	10	149	10
Seatearth-mudstone, dark rooty; coal films	1	8	151	6
Shale, dark carbonaceous	0	6	152	0
Seatearth-mudstone, grey sandy	0	6	152	6
Sandstone, white fine-grained predominantly massive ..	12	4	164	10
BOTTOM MARSHALL GREEN				
Coal	1	7	166	5
Seatearth-mudstone, dark shaly	0	4	166	9
Sandstone, grey fine rooty	1	0	167	9
Sandstone, grey fine-grained; shaly at top	3	3	171	0

Mount Pleasant Bore

see Appendix 3

New Shildon No. 117 Bore

Surface 343 ft AOD. 6-in NZ 22 NW. Site [2114 2719] 870 yd NE of Fieldon Bridge. Drilled 1957 by Cementation Co. Ltd. for NCB. Plates X, XI and XII (134).

	Thickness ft	in	Depth ft	in
SUPERFICIAL DEPOSITS				
[Open holed to 172 ft]				
Top Soil	0	9	0	9
Sandy Clay, Sand and Boulders	22	3	23	0
CARBONIFEROUS				
WESTPHALIAN B (MIDDLE COAL MEASURES)				
Shale, grey sandy ..	22	0	45	0
Shale, sandy ..	9	0	54	0
FIVE-QUARTER				
Goaf..	3	6	57	6
Shale, grey; soft at top	41	6	99	0
Shale, black	4	0	103	0
Shale, grey	9	0	112	0
Sandstone	12	0	124	0
Shale, grey	28	3	152	3
Coal	0	9	153	0
Shale, grey sandy ..	3	0	156	0
Shale, sandy	8	0	164	0
MAIN				
Coal	5	3	169	3
Fireclay	0	9	170	0
Fireclay, sandy ..	2	0	172	0
[Cored to base]				
Siltstone, grey irony; rooty at top ..	7	0	179	0
Mudstone, grey silty..	9	4	188	4
Shale, grey sandy; sandstone bands ..	8	8	197	0
Shale, grey silty; banded with sandstone	33	0	230	0
Siltstone; sandy partings	8	0	238	0
Siltstone; alternates with flaggy sandstone	4	0	242	0
Sandstone, light grey fine-grained; occasional coaly plant debris; carbonaceous shale films in basal 6 in [DURHAM LOW MAIN POST]	30	6	272	6
DURHAM LOW MAIN				
Coal	0	9	273	3
Seatearth, grey silty ..	1	3	274	6
Siltstone, grey rooty; hard and sandy at top	4	6	279	0
Siltstone, grey sandy; rootlets and ironstone bands.. ..	4	0	283	0
Shale, grey; *Naiadites*	4	0	287	0
Siltstone, grey sandy; sandstone bands ..	4	0	291	0
Shale, grey sandy micaceous; sandstone partings and irony bands; occasional plants	3	6	294	6
Siltstone, grey; sandy partings; plant remains	3	6	298	0
Mudstone, grey silty; occasional irony bands; plant remains, small mussels	14	0	312	0
Mudstone, dark grey; ironstone nodules; abundant mussels; *Spirorbis*	7	9	319	9
BRASS THILL				
Coal	3	0	322	9
Shale, carbonaceous; coaly films	0	1	322	10
Seatearth-mudstone, sandy; coaly films ..	0	3	323	1
Sandstone, grey current-bedded fine-grained	0	11	324	0
Mudstone, grey; plants and mussel fragments	7	0	331	0
Shale, dark; sandy in basal 2 ft with irony bands; mussels ..	3	0	334	0
Shale, grey sandy rooty; sandstone bands towards base; mussels	5	0	339	0
Mudstone, dark grey sandy; irony bands and ironstone nodules; plant debris ..	11	0	350	0
Mudstone, light grey sandy and silty; plant debris ..	23	6	373	6

	Thickness ft	in	Depth ft	in
(Assumed horizon of TOP HUTTON)				
Sandstone, whitish fine-grained.. ..	10	6	384	0
Shale, grey sandy ..	13	6	397	6
Shale, dark grey; ironstone bands; abundant mussels at intervals	11	5	408	11
BOTTOM HUTTON				
Coal	2	4	411	3
Seatearth-mudstone, grey sandy ..	4	9	416	0
Sandstone, fine-grained wispy bedded	11	6	427	6
Mudstone, grey sandy; shaly at top; sandstone bands; scattered plants.. ..	29	6	457	0
Mudstone, grey; ironstone bands; scattered plants; mussels from 459 ft ..	3	1	460	1
Coal	0	2	460	3
Seatearth-mudstone, grey sandy; ironstone nodules ..	1	6	461	9
Mudstone, grey sandy rooty; ironstone nodules and bands ..	6	9	468	6
Sandstone, brownish wispy bedded fine-grained; shale partings	5	6	474	0
Shale, grey; ironstone nodules and bands; mussels	2	2	476	2
Sandstone	0	6	476	8
Shale, dark grey; ironstone nodules ..	0	8	477	4
Coal, cannel	0	4	477	8
Seatearth-mudstone, grey	0	3	477	11
Sandstone, whitish fine-grained micaceous; sandy shale bands	22	7	500	6
Shale grey; ironstone nodules and bands; sandy at top; mussels at base	7	6	508	0
Shale, dark; ironstone nodules and bands; abundant mussels especially below 511 ft; black shale with fish scales in basal 1 ft	11	6	519	6
(Assumed horizon of HARVEY MARINE BAND)				
WESTPHALIAN A (LOWER COAL MEASURES)				
Mudstone, grey sandy; ironstone nodules; rooty	1	6	521	0
Sandstone, light grey wispy bedded ..	0	9	521	9
Siltstone, grey and light grey; often irony and sandy; 4-ft mudstone band at 533 ft 6 in..	46	3	568	0
Sandstone, white and light grey massive; sporadic micaceous partings; shale breccia at base	80	2	648	2
Siltstone, rooty finely micaceous	1	4	649	6
Sandstone, grey fine-grained argillaceous	4	6	654	0
Siltstone, grey shaly; sandy bands ..	14	0	668	0
TOP TILLEY				
Coal	0	2	668	2
Seatearth, dark grey and grey-brown, irony	2	10	671	0
Siltstone, grey; shaly and finely micaceous; irony and rooty at top; sandy partings	10	0	681	0
Mudstone, grey shaly; irony bands; plant remains	2	0	683	0
Shale, dark grey; abundant large *Carbonicola* and other mussel fragments ..	2	0	685	0
Shale, black carbonaceous; abundant mussels; *Carbonicola*	1	0	686	0

	Thickness ft	in	Depth ft	in
Siltstone, finely micaceous	0	3	686	3
Sandstone, argillaceous micaceous slumped	0	9	687	0
Shale, grey	0	9	687	9
BOTTOM TILLEY				
Coal	0	6	688	3
Seatearth	1	2	689	5
Coal	0	2	689	7
Shale, grey sandy rooty	0	10	690	5
Coal	1	4	691	9
Seatearth, dark grey sandy	1	3	693	0
Siltstone, grey finely micaceous; plant remains; coaly at base	2	1	695	1
Shale, dark grey carbonaceous; coaly plant remains ..	0	2	695	3
Shale, dark grey carbonaceous rooty ..	1	7	696	10
Seatearth, dark grey shaly; thin coal partings	0	2	697	0
Mudstone, grey and dark grey; irony; silty partings ..	30	3	727	3
Shale, dark grey finely micaceous; coalified plant remains and carbonaceous films ..	2	2	729	5
TOP BUSTY				
Coal	0	9½	730	2½
Seatearth, dark grey carbonaceous and silty at top; otherwise sandy	4	9½	735	0
Sandstone, grey argillaceous	7	0	742	0
Shale, dark grey; plant remains	2	2	744	2
BOTTOM BUSTY				
Coal	3	2	747	4
Seatearth, grey ..	0	10	748	2
Sandstone, grey wispy bedded fine-grained argillaceous; rooty at top	1	10	750	0
Sandstone, light grey and white massive fine-grained; occasional kaolinitic medium- and coarse-grained bands ..	36	6	786	6
Seatearth, grey-brown; sandy at base ..	2	0	788	6
Sandstone, grey argillaceous	1	6	790	0
Shale, grey sandy micaceous	0	6	790	6
Sandstone, wispy bedded fine-grained ..	6	2	796	8
Siltstone, grey micaceous shaly; fragmentary mussels ..	0	8	797	4
Mudstone, dark grey shaly; plant debris..	7	8	805	0
Siltstone, dark grey shaly; finely micaceous	7	0	812	0
THREE-QUARTER				
Coal	0	7	812	7
Seatearth, grey-brown; grades to seatearth-siltstone	3	10	816	5
Siltstone, grey; sandy bands	3	7	820	0

New Shildon No. 146 Bore

Surface 441.90 ft AOD. 6-in NZ 22 NW. Site [2172 2736] 1200 yd SW of Coundon Grange. Drilled 1958 by Cementation Co. Ltd. for NCB. Plates X, XI and XII (139).

	Thickness ft	in	Depth ft	in
SUPERFICIAL DEPOSITS				
[Open holed to 829 ft]				
Top Soil, Sandy Clay with sandstone boulders	18	0	18	0
Coal (Drift)	1	6	19	6
Clay and Boulders ..	8	6	28	0
CARBONIFEROUS				
WESTPHALIAN B (MIDDLE COAL MEASURES)				
Shale, predominantly grey sandy	48	0	76	0
Sandstone	5	0	81	0
Shale, grey	1	0	82	0
Coal	1	0	83	0
Shale, grey sandy ..	23	0	106	0
Sandstone, brown (broken)	11	0	117	0
Sandstone; shale bands	8	0	125	0
Shale, sandy	4	0	129	0
HIGH MAIN				
Coal	2	0	131	0
Shale, grey	2	0	133	0
Sandstone	6	0	139	0
Shale, sandy; sandstone blebs	9	0	148	0
Shale, grey sandy ..	15	0	163	0
Sandstone	6	0	169	0
Shale, sandy; sandstone bands.. ..	9	0	178	0
Shale, grey sandy ..	12	0	190	0
Sandstone	2	0	192	0
Shale, grey	6	0	198	0
Sandstone, hard ..	2	0	200	0
Shale, grey sandy ..	2	0	202	0
Shale, sandy	13	0	215	0
Shale, dark grey; coal bands	4	0	219	0
(Assumed horizon of METAL)				
Shale, grey sandy ..	13	0	232	0
(Assumed horizon of FIVE-QUARTER)				
Shale, sandy; sandstone bands.. ..	17	0	249	0
Shale, grey and sandy	11	0	260	0
Shale, black	9	0	269	0
Shale, dark grey ..	3	0	272	0
Shale, black	2	0	274	0
Shale, grey sandy ..	14	0	288	0
Sandstone, grey hard..	10	0	298	0
Shale, sandy	11	0	309	0
Sandstone	6	0	315	0
Shale, grey sandy ..	23	0	338	0
Shale, soft dark grey; coal bands	4	0	342	0
MAIN				
Coal	3	0	345	0
Shale, dark grey ..	7	0	352	0
MAUDLIN				
Coal; mixed with shale	3	0	355	0
Sandstone, grey ..	47	0	402	0
Shale, sandy; sandstone patches ..	20	0	422	0
Sandstone	5	0	427	0
Shale	2	0	429	0
Sandstone, grey ..	3	0	432	0
Sandstone	5	0	437	0
Shale, grey sandy ..	19	0	456	0
DURHAM LOW MAIN				
Coal	2	1	458	1
Sandstone	2	11	461	0
Shale, sandy	8	0	469	0
Sandstone	3	0	472	0
Shale, sandy	19	0	491	0
BRASS THILL				
Old Workings ..	?17	0	508	0
Shale, grey	4	0	512	0
Shale; ironstone bands	6	0	518	0
Shale, grey sandy ..	66	0	584	0
BOTTOM HUTTON				
Goaf	7	0	591	0
Shale, sandy; sandstone bands.. ..	33	0	624	0
Sandstone	7	0	631	0
Shale, sandy	4	0	635	0
Sandstone	5	0	640	0
Shale, soft	19	0	659	0
Shale, sandy	9	0	668	0
Shale; ironstone bands	18	0	686	0

	Thickness ft	in	Depth ft	in
(Assumed horizon of HARVEY MARINE BAND)				
WESTPHALIAN A (LOWER COAL MEASURES)				
Shale, sandy	6	0	692	0
Sandstone	1	0	693	0
Shale, sandy	45	0	738	0
Shale, sandy; sandstone bands.. ..	35	0	773	0
Shale, sandy	16	0	789	0
Sandstone; broken at base	16	0	805	0
HARVEY				
Old Workings ..	3	6	808	6
Shale, soft	9	6	818	0
Shale, sandy	5	0	823	0
Shale; soft patches ..	6	0	829	0
[Cored from 829 ft]				
(Assumed horizon of TOP TILLEY)				
Seatearth-siltstone, grey	0	6	829	6
Siltstone, grey micaceous; argillaceous at base	2	0	831	6
Mudstone, grey; clay-ironstone nodules; rootlets	2	2	833	8
Shale, black canneloid; mussels and mussel impressions ..	0	4	834	0
Siltstone, grey micaceous	0	8	834	8
Sandstone, light grey wispy bedded fine-grained; silty bands	2	2	836	10
Siltstone, grey finely micaceous; plant debris	3	3	840	1
Sandstone, whitish grey wispy bedded fine-grained micaceous and carbonaceous; silty at base..	1	3	841	4
Seatearth-mudstone, dark grey silty ..	1	6	842	10
BOTTOM TILLEY				
Coal	1	8	844	6
Seatearth-siltstone, grey; sandy partings at base	1	0	845	6
Siltstone, grey thin irregularly bedded; sandy partings; scattered rootlets; a few clay ironstone nodules; becomes argillaceous to base ..	8	6	854	0
Mudstone, grey silty shaly; clay ironstone bands below 865 ft; *Naiadites* at 880 ft; ironstone nodules below 881 ft; scattered small mussels between 883 ft 6 in and 884 ft 6 in ..	31	0	885	0
Shale, black splintery; fibrous plant debris and fish scales at 885 ft 1 in; very carbonaceous at base with abundant vitrain partings	1	7	886	7
TOP BUSTY				
Coal	0	9	887	4
Seatearth-mudstone, dark grey silty ..	1	0	888	4
Mudstone, grey silty; clay ironstone nodules; abundant vitrain partings to base	4	8	893	0
BOTTOM BUSTY				
Coal	0	3	893	3
Mudstone, dark grey carbonaceous ..	0	1	893	4
Coal, dull durainous	2	8	896	0
Seatearth-siltstone, grey argillaceous ..	3	0	899	0
Sandstone, grey fine-grained	0	6	899	6
Sandstone, grey massive medium-grained locally coarse-grained; coal scar at 915 ft; coal scar and shale breccia between 968 ft and 970 ft ..	72	2	971	8

	Thickness ft	in	Depth ft	in
THREE-QUARTER				
Coal	0	10	972	6
Seatearth-mudstone, brownish grey ..	2	6	975	0
Mudstone, grey; some silty bands; scattered rootlets	3	0	978	0

New Shildon No. 147 Bore

Surface 478.85 ft AOD. 6-in NZ 22 NW. Site [2178 2631] 430 yd ENE of Blue House. Drilled 1957–58 by Cementation Co. Ltd. for NCB. Plates X, XI and XII (141).

	Thickness ft	in	Depth ft	in
SUPERFICIAL DEPOSITS				
[Open holed to 208 ft]				
Top soil; sandy clay with sandstone boulders	20	0	20	0
CARBONIFEROUS				
WESTPHALIAN B (MIDDLE COAL MEASURES)				
Shale, grey; sandy at base	35	0	55	0
Sandstone	1	0	56	0
Shale, sandy	3	6	59	6
FIVE-QUARTER				
Coal	5	0	64	6
Shale, grey	10	6	75	0
Sandstone	3	0	78	0
Shale, sandy	12	0	90	0
Sandstone, hard ..	1	0	91	0
Shale, sandy; sandstone patches ..	17	0	108	0
Shale, grey sandy ..	16	6	124	6
MAIN				
Coal	2	6	127	0
Shale, dark grey ..	21	0	148	0
Sandstone	3	0	151	0
Shale, sandy; patches of sandstone ..	9	0	160	0
(?Horizon of MAUDLIN)				
Sandstone	14	0	174	0
Shale, sandy; patches of sandstone ..	8	0	182	0
Sandstone	9	0	191	0
Shale, grey sandy ..	7	0	198	0
Sandstone	10	0	208	0
[Cored between 208 ft and 280 ft]				
Siltstone, grey; sandy partings	8	3	216	3
Sandstone, massive false-bedded fine- to medium-grained scattered micaceous partings; coal scars at 235 ft; shale pellets in basal 5 ft (DURHAM LOW MAIN POST) ..	47	8	263	11
(Horizon of DURHAM LOW MAIN)				
Siltstone, dark grey micaceous; sandstone interlaminae ..	4	1	268	0
Shale, grey silty; sandy partings; scattered plant debris	6	0	274	0
Mudstone, grey shaly; silty at top with clay ironstone bands ..	6	0	280	0
[Open holed from 280 ft to 575 ft]				
Shale, grey	4	0	284	0
BRASS THILL				
Old workings ..	6	0	290	0
Shale, grey; often sandy	81	0	371	0
BOTTOM HUTTON (HUTTON)				
Coal	2	0	373	0
Shale, grey; often sandy	26	0	399	0
Shale, grey; locally sandy	11	0	410	0
Sandstone	6	0	416	0
Shale, sandy	3	0	419	0
Sandstone	3	0	422	0

	Thickness ft	in	Depth ft	in
Shale, sandy; sandstone bands	10	0	432	0
Shale, sandy	1	6	433	6
Coal	1	4	434	10
Shale, grey sandy	28	2	463	0
Sandstone	9	0	472	0
Shale, sandy	1	0	473	0
Sandstone	7	0	480	0
Shale, grey; sandstone bands	9	0	489	0
Shale, grey sandy	32	0	521	0
(Assumed horizon of HARVEY MARINE BAND)				
WESTPHALIAN A (LOWER COAL MEASURES)				
Shale, sandy; sandstone bands	8	0	529	0
Sandstone; broken at base	31	0	560	0
HARVEY				
Old workings	4	0	564	0
Shale, grey	11	0	575	0
[Cored from 575 ft to 796 ft]				
Sandstone, light grey often false flaggy and wispy bedded medium-grained; siltstone intercalations in basal 6 in	42	6	617	6
Siltstone, grey	1	4	618	10
Sandstone, grey massive medium- to coarse-grained; coal scars from 627 to 630 ft	11	8	630	6
Breccio-conglomerate; shale sandstone and ironstone in sandstone matrix	1	6	632	0
Sandstone, grey false-bedded medium-grained; micaceous partings; conglomeratic with siltstone and ironstone in basal 2 ft	8	6	640	6
Seatearth-siltstone	2	6	643	0
Siltstone, grey thin-bedded; sandy partings; muddy at base	4	5	647	5
Shale, grey; *Spirorbis sp.*, *Carbonicola* cf. *cristagalli*, *C.* cf. *oslancis*, ostracods indet., fish debris	2	0	649	5
TOP TILLEY				
Coal, inferior; shale partings with mussels at top	0	5	649	10
Mudstone, grey silty shaly; comminuted plant leaves and stems	2	2	652	0
Sandstone, light grey medium-grained; micaceous partings	2	6	654	6
Siltstone, grey shaly; plant debris	3	0	657	6
Sandstone, grey medium-grained	0	8	658	2
Siltstone, grey fissile micaceous	2	5½	660	7½
BOTTOM TILLEY				
Coal, bright	0	5½	661	1
Seatearth-mudstone	0	9	661	10
Shale, black carbonaceous; coal partings	0	2	662	0
Coal, bright	1	7	663	7
Seatearth-siltstone, grey sandy	3	11	667	6
Sandstone, light grey thin-bedded fine-grained; silty partings	6	6	674	0
Siltstone, grey thin-bedded; fine sandy partings; argillaceous towards base	7	0	681	0
Mudstone, grey shaly; clay ironstone nodules	18	5	699	5
TOP BUSTY				
Coal, bright	3	2	702	7
Seatearth-mudstone, grey silty	1	5	704	0
Siltstone, grey; micaceous; scattered rootlets	4	5	708	5

	Thickness ft	in	Depth ft	in
BOTTOM BUSTY				
Coal	3	2	711	7
Seatearth-siltstone, grey; abundant rootlets in top 4 in ..	2	11	714	6
Siltstone, grey massive; scattered rootlets; occasional sandstone bands	9	6	724	0
Mudstone, grey silty shaly	6	0	730	0
Sandstone, light grey predominantly massive medium- to coarse-grained; coal scars between 743 ft 6 in and 744 ft; clay ironstone breccia in basal 10 in ..	21	0	751	0
Mudstone, grey shaly; scattered clay ironstone nodules ..	4	0	755	0
THREE-QUARTER				
Coal	1	2	756	2
Seatearth-mudstone, grey silty	4	10	761	0
Mudstone, grey silty shaly micaceous; plant remains in basal 2 ft 6 in	7	0	768	0
Siltstone, grey shaly micaceous; sandy bands; worm burrows	3	0	771	0
Sandstone, light grey wispy bedded micaceous	2	0	773	0
Seatearth-mudstone, brownish grey; silty at base	1	0	774	0
Siltstone, grey massive	2	6	776	6
Seatearth-siltstone, grey; abundant rootlets	3	6	780	0
Siltstone, grey; sandy partings	1	0	781	0
Mudstone, grey massive; clay ironstone layers; poorly preserved mussels at 784 ft, *Carbonicola* cf. *pseudorobusta* at base	5	0	786	0
Coal, bright	0	6	786	6
Seatearth-siltstone, micaceous	2	6	789	0
Siltstone, grey; sandy partings; sporadic roots	1	6	790	6
Sandstone, light grey fine-grained wispy bedded; micaceous partings	5	6	796	0
[Open holed to base]				
Sandstone; broken at base	26	6	822	6
BROCKWELL				
Cavity; old workings	3	6	826	0
Shale, sandy	5	0	831	0

North Tees No. 1 Bore

Surface 420 ft AOD. 6-in NZ 11 NW. Site [1212 1762] Little Newsham. Drilled 1965 by NCB.

	Thickness ft	in	Depth ft	in
SUPERFICIAL DEPOSITS				
Soil and sandy boulder clay	6	0	6	0
CARBONIFEROUS				
NAMURIAN (MILLSTONE GRIT SERIES)				
Shale, sandy	8	0	14	0
Mudstone, grey; *Lingula sp.* at 14½ ft ..	1	0	15	0
Sandstone, grey micaceous fine- to medium-grained; a few siltstone bands; plant debris and rootlets	6	0	21	0

	Thickness ft	in	Depth ft	in
Siltstone, grey micaceous; a few sandstone bands; plants	4	6	25	6
Sandstone, fine ..	1	0	26	6
Seatearth, silty ..	3	0	29	6
Sandstone, thin-bedded fine-grained micaceous	4	0	33	6
Siltstone, grey micaceous; sandy towards base; abundant plant fragments	3	6	37	0
Mudstone, grey; crinoid debris; *Crania sp.*, productoid fragments, pectinoid indet., ostracods including *Hollinella sp.* (loc. 16a)	5	0	42	0
Siltstone, sandy calcareous; gastropods and fragmentary shells at top	2	0	44	0
Siltstone, grey shaly; sporadic sandy bands; plants especially below 49 ft, crinoid columnals at 70½ ft, brachiopods from 72 to 74¼ ft including smooth spiriferoids indet. ..	34	6	78	6
Mudstone, grey; sporadic fossils including brachiopods and crinoids; bivalves abundant at base. *Fault* 81 ft to 81 ft 10 in ..	6	9	85	3
Mudstone, calcareous; abundant fossils ..	0	3	85	6
?UPPER FELLTOP LIMESTONE				
Limestone; light grey crystalline at top and base otherwise fine-grained to subporcellanous; crinoid debris ..	2	0	87	6
Mudstone, sandy grey micaceous; bands of fine sandstone; sporadic rootlets ..	2	6	90	0

	Thickness ft	in	Depth ft	in
Sandstone, light grey mainly coarse-grained; siltstone bands at top and base	16	0	106	0
Siltstone, grey; plant debris	3	0	109	0
Mudstone, silty; sandy bands; plant debris, sporadic small shells towards base. *Fault at about* 115 ft ..	6	0	115	0
Siltstone, grey micaceous; rootlets ..	2	0	117	0
Sandstone, light grey medium-grained ..	3	0	120	0
Mudstone, silty; sandstone partings; roots	2	9	122	9
Sandstone, light grey medium-grained; siltstone bands and partings towards base; *fault at* 133 ft ..	24	9	147	6
Siltstone, highly micaceous; carbonaceous at top	11	6	159	0
Mudstone, slightly silty; sporadic ironstone nodules; *Planolites ophthalmoides* at 167 and 177 ft, goniatite at 171½ ft, marine bivalves below ..	24	4	183	4
Mudstone, grey; crinoid debris	1	2	184	6
Sandstone, fine-grained argillaceous; rootlets; small quartz pebbles	1	0	185	6
Seatearth, silty micaceous	1	6	187	0
Siltstone; sandstone beds	2	0	189	0
Siltstone, grey micaceous; nodules and strings of clay ironstone; 6-in sandstone at 212½ ft	34	0	223	0
Sandstone, grey fine-grained	1	0	224	0
Siltstone; lenticular sandstone partings and ironstone nod-				

	Thickness ft	in	Depth ft	in
ules at top	13	0	237	0
Mudstone, dark grey; sandy and carbonaceous at base; marine fossils at top ..	1	0	238	0
Ganister	1	6	239	6
Seatearth, sphaerosiderite	3	0	242	6
Siltstone, grey micaceous; ironstone nodules; rootlets near top	4	6	247	0
Mudstone, silty grey; siltstone bands; sporadic ironstone nodules	11	6	258	6
Sandstone, fine-grained; siltstone partings at base	5	0	263	6
Siltstone; sandstone bands	1	0	264	6
Mudstone, silty; crinoid debris; cf. *Planolites montanus* at 268 ft (loc. 15b), worm trails and burrows ..	5	6	270	0
Limestone, sandy; *Spirifer sp.*, smooth spiriferoid indet. ..	0	6	270	6
Sandstone, fine-grained bioturbated; silty bands; sporadic plant fragments; *Chondrites ?*, *Schellwienella sp.*, *Spirifer sp.* at 274 to 275 ft (loc. 15b) ..	5	6	276	0
Mudstone, silty; slumped siltstone bands; fauna between 283½ and 338 ft includes crinoid debris, *Antiquatonia sp.*, *Productus carbonarius*, *Rugosochonetes sp.*, *Tornquistia* cf. *polita*, *Streblochondria sp.*, *Weberides* cf. *mucronatus* (loc. 15b)	63	0	339	0
Shale, black	4	0	343	0

Old Eldon 'A' Bore

Surface 393.75 ft AOD. 6-in NZ 22 NW. Site [2323 2748] Eldon. Drilled 1958 by Cementation Co. Ltd. for NCB. Plates X, XI and XII (147).

	Thickness ft	in	Depth ft	in
SUPERFICIAL DEPOSITS				
[Open holed to 200 ft]				
Top Soil, Sandy Clay and Boulders ..	18	0	18	0
CARBONIFEROUS				
WESTPHALIAN B (MIDDLE COAL MEASURES)				
Shale, grey	7	6	25	6
FIVE-QUARTER				
Coal	2	0	27	6
Old Workings ..	7	6	35	0
Shale, grey sandy ..	26	0	61	0
Sandstone	4	0	65	0
Shale, grey	11	0	76	0
Sandstone	3	0	79	0
Shale; ironstone bands	6	0	85	0
Shale, grey sandy; 1 ft sandstone band at 108 ft	56	6	141	6
MAIN				
Coal	2	6	144	0
Shale	1	0	145	0
Coal	3	6	148	6
Fireclay	2	6	151	0
Shale, grey	5	0	156	0
Sandstone, grey ..	3	0	159	0
Shale, grey sandy ..	26	0	185	0
Sandstone, grey ..	6	0	191	0
Shale, sandy	4	0	195	0
Sandstone, grey ..	5	0	200	0
[Cored to base]				
Sandstone, light grey massive fine-grained	0	10	200	10

	Thickness ft	in	Depth ft	in
Siltstone, grey fissile micaceous	0	8	201	6
Sandstone, light grey wispy bedded fine-grained; silty partings	4	6	206	0
Siltstone, grey massive; plant debris ..	6	7	212	7
Breccio-conglomerate; siltstone fragments in sandstone matrix ..	5	5	218	0
Sandstone, light grey massive fine- to medium-grained; occasional coal scars; 2-in breccio-conglomerate at 243 ft 6 in and at base [DURHAM LOW MAIN POST]	27	3	245	3
DURHAM LOW MAIN				
Coal	2	6	247	9
Seatearth-mudstone, grey silty	1	3	249	0
Siltstone, grey massive; rootlets at top ..	5	0	254	0
Mudstone, grey silty shaly; thin clay ironstone bands; mussels from 259 ft	14	0	268	0
Mudstone, grey silty argillaceous; plant debris	10	0	278	0
Mudstone, grey shaly; mussels to 282 ft; plant debris from 284 to 287 ft; mussel fragments from 289 ft to base	20	0	298	0
BRASS THILL				
Coal	3	11	301	11
Seatearth-mudstone, dark grey micaceous	2	1	304	0
Mudstone, black silty finely micaceous; mussel impressions in top 2 in and at base	6	11½	310	11½
Coal	0	0½	311	0
Mudstone, black finely micaceous; mussels (mainly *Anthracosia*); rootlets below 312 ft	3	0	314	0
Mudstone, dark grey finely micaceous; mussels in top 3 in..	3	0	317	0
Mudstone, grey silty; plant fragments; sandy bands in basal 1 ft	5	0	322	0
Mudstone, black silty; mussels including *Naiadites*, *Spirorbis*	1	3	323	3
TOP HUTTON				
Coal	0	0½	323	3½
Seatearth - mudstone	1	11½	325	3
Coal	0	6	325	9
Seatearth-mudstone, dark grey	3	3	329	0
Mudstone, dark grey shaly; mussels (including *Anthracosia*) between 330 and 333 ft; plant debris at base	6	6	335	6
Siltstone and sandstone, grey argillaceous massive ..	9	6	345	0
Mudstone, grey shaly; mussels from 352 to 352 ft 8 in; clay ironstone bands in basal 3 ft	18	6	363	6
BOTTOM HUTTON				
Coal	2	5	365	11
Seatearth-sandstone, grey; clay ironstone nodules at base ..	5	1	371	0
Mudstone, black silty; micaceous	2	6	373	6
Sandstone, light grey wispy bedded fine-grained	1	6	375	0
Coal	0	1½	375	1½
Seatearth-siltstone, brownish-grey argillaceous	1	1½	376	3
Siltstone, grey; sandy at top and base, otherwise argillaceous	12	1	388	4

	Thickness ft	in	Depth ft	in
(Horizon of RULER coal)				
Sandstone, light grey thin-bedded, silty at top massive below; wispy bedded below 407 ft	30	5	418	9
Seatearth-mudstone, dark grey	1	0	419	9
Siltstone, grey fissile with micaceous partings; plant debris; sandy at top and base	9	1	428	10
Sandstone, whitish grey thin and wispy bedded	3	0	431	10
Siltstone, grey argillaceous; sandy intercalations; mussel impressions in basal 6 in	4	1	435	11
Coal, bright	0	6	436	5
Seatearth-siltstone, grey micaceous; sandy bands at base	1	7	438	0
Siltstone, dark grey argillaceous; 11-in sandstone band at 440 ft 1 in	4	4	442	4
Sandstone, light grey wispy bedded fine-grained; silty partings	8	8	451	0
Mudstone, dark grey shaly; silty in top 3 ft; black and shaly at base; mussels from 459 ft to 461 ft and from 463 ft to base	25	0	476	0
Shale, black fissile finely micaceous; poorly preserved mussel impressions at top; fish scales and spines in basal 6 in	2	0	478	0
(Assumed horizon of HARVEY MARINE BAND)				
WESTPHALIAN A (LOWER COAL MEASURES)				
Seatearth-sandstone, light grey fine-grained	2	0	480	0
Sandstone, light grey false and wispy bedded fine-grained; dark silty bands and partings	10	9	490	9
Mudstone, dark grey; plant debris at base	2	6	493	3
Seatearth-mudstone, light grey silty	2	9	496	0
Siltstone, grey massive; more argillaceous to base	11	5	507	5
Siltstone, grey thin-bedded; plant leaves and stems	8	7	516	0
Sandstone, light grey fine-grained; siltstone bands and partings	1	6	517	6
Siltstone, grey massive; occasional sandy partings; very sandy in bottom 1 ft	6	2	523	8
Sandstone, light grey false and wispily bedded fine and fine- to medium-grained; dark micaceous partings	15	7	539	3
Shale, black carbonaceous; pyritic; fish scales	0	4	539	7
Mudstone, grey silty shaly; abundant plant leaves and stems	0	7	540	2
HARVEY				
Coal; canneloid at top	0	7	540	9
Seatearth-mudstone	0	3	541	0
Coal	3	6	544	6
Seatearth-mudstone, dark grey silty; clay ironstone nodules	4	6	549	0
Siltstone, grey sandy and silty flaggy at top; light grey sandstone bands; sporadic rootlets at top	10	2	559	2
Sandstone, light grey fine- to medium-grained wispily bedded; silty bands and				

	Thickness ft	in	Depth ft	in
silty micaceous partings	24	2	583	4
Sandstone, grey massive medium-grained; breccio - conglomeratic with siltstone and mudstone fragments in upper 4 ft and again at base ..	13	9	597	1
Sandstone, light grey fine-grained; abundant silty bands and partings	17	0	614	1
(Horizon of TOP TILLEY)				
Seatearth-mudstone, grey silty	1	5	615	6
Siltstone, grey massive and flaggy; rootlets at top; plant stems at base	5	0	620	6
Sandstone, grey wispy bedded fine-grained; locally grades to siltstone with plant stems and leaves ..	5	9	626	3
Siltstone, grey thin-bedded; sandy bands and partings at top; plant debris	1	1	627	4
BOTTOM TILLEY				
Coal	0	6	627	10
Seatearth-mudstone	0	9	628	7
Shale, black; abundant coaly partings	0	1	628	8
Coal	1	1	629	9
Seatearth-siltstone, grey massive ..	2	3	632	0
Siltstone, grey thin-bedded; sandy bands and partings ..	4	0	636	0

Pioneer Shaft and Bore, Crake Scar Colliery

Surface about 1050 ft AOD. 6-in NZ 02 NE. Site [0792 2759] 1240 yd NNW of Woodland. Sunk and bored 1893. Information from B & S No. 2562 and NCB.

	Thickness ft	in	Depth ft	in
CARBONIFEROUS				
WESTPHALIAN A (LOWER COAL MEASURES)				
Superficial clay (thickness unknown) and sandstone	20	0	20	0
BROCKWELL				
Coal	6	0	26	0
Fireclay	13	6	39	6
Sandstone	12	0	51	6
Grey beds	9	0	60	6
Shale, blue	10	4	70	10
Coal	0	6	71	4
Fireclay	16	6	87	10
Sandstone	3	0	90	10
VICTORIA				
Coal	0	2	91	0
Ironstone band ..	0	2	91	2
Coal	0	6	91	8
Fireclay	3	0	94	8
Grey beds	15	9	110	5
Sandstone	2	3	112	8
Shale, blue	7	4	120	0
TOP MARSHALL GREEN				
Coal	0	2	120	2
Sandstone, grey hard ..	7	6	127	8
Sandstone	16	10	144	6
BOTTOM MARSHALL GREEN				
Coal	0	11	145	5
Shale, blue	8	6	153	11
Sandstone	34	6	188	5
[Boring from base of shaft to 299 ft 8 in]				
Shale, blue	1	10	190	3
Sandstone	28	9	219	0
Shale, blue; interbedded sandstone ..	16	6	235	6
Shale, black	19	0	254	6
Fireclay	2	0	256	6
Shale, blue	14	0	270	6
Sandstone	20	0	290	6
Shale, blue	1	8	292	2
Fireclay	7	6	299	8

Quarry West Pit, Butterknowle

Surface about 702 ft AOD. 6-in NZ 02 NE. Site [0982 2566] 1300 yd ENE of Copley. Sunk 1876. Information from B & S No. 2486. Plates IX and XI (19).

	Thickness ft	Thickness in	Depth ft	Depth in
SUPERFICIAL DEPOSITS				
Boulder clay	12	4	12	4
CARBONIFEROUS				
WESTPHALIAN B (MIDDLE COAL MEASURES)				
Seggar Clay; iron nodules	0	6	12	10
TOP BRASS THILL				
Coal	0	8	13	6
Metal or plate clay	1	6	15	0
Coal, splinty and rusty	3	0	18	0
Metal, black splinty and grey; iron girdles in basal 1 ft 6 in	4	3	22	3
BOTTOM BRASS THILL				
Coal, splinty and rusty	0	6	22	9
Post, grey jointy	1	10	24	7
Coal, rusty	0	9	25	4
Post, grey; metal girdles	16	11	42	3
Metal, blue; iron girdles	10	10	53	1
Seggar clay	0	2	53	3
TOP HUTTON				
Coal	0	3	53	6
Seggar clay	0	5	53	11
Coal	0	6	54	5
Metal, dark; iron nodules	6	7	61	0
Iron band	0	3	61	3
Metal grey	1	2	62	5
Post, grey and white; jointy at top	11	10	74	3
Metal, blue; bastard whin band	29	1	103	4
HUTTON (BOTTOM HUTTON) (FOUR FEET)				
Coal	3	10	107	2
Seggar Clay	2	6	109	8
Post; metal partings; mixed with whin at base	14	0	123	8
Metal, blue	8	6	132	2
RULER (JUBILEE)				
Coal	2	8	134	10
Seggar clay	3	0	137	10
Post; rare metal partings; 1 ft 9 in in seggar clay band at 161 ft 5 in	26	7	164	5
Metal, blue	5	0	169	5
Coal	0	3	169	8
Metal, blue	2	0	171	8
Post	2	0	173	8
Metal, blue	6	3	179	11
Metal, blue; grey post partings	5	0	184	11
Ironstone band	0	9	185	8
Metal, blue; iron girdles	37	5	223	1
(Assumed horizon of HARVEY MARINE BAND)				
WESTPHALIAN A (LOWER COAL MEASURES)				
Coal	0	3	223	4
Seggar clay	2	6	225	10
Metal, grey	6	0	231	10
Post; mixed with whin	21	0	252	10
Metal, blue	9	0	261	10
HARVEY				
Coal	2	2	264	0
Seggar clay	3	0	267	0
Metal, grey; iron girdles at top; post girdles at base	69	0	336	0
Post	6	0	342	0
TILLEY				
Coal	0	8	342	8
Band	0	3	342	11
Coal	0	7	343	6
Seggar Clay	2	6	346	0

	Thickness ft	in	Depth ft	in
Metal, grey; post partings	23	0	369	0
[Butterknowle Fault at about 369 ft—strata below said to be about 414 ft below the position of the Brockwell seam]				
NAMURIAN (MILLSTONE GRIT SERIES)				
Metal, grey	15	0	384	0
Coal	0	4	384	4
Seggar clay	2	8	387	0
Metal, grey; post girdles	9	0	396	0
Metal, blue	6	0	402	0
Metal, grey and blue; post girdles ..	10	3	412	3

Raby Castle Water Bore

Surface about 480 ft AOD. 6-in NZ 12 SW. Site [1288 2218] Raby Castle, Staindrop. Sunk 1943 by J. T. Hymas and Son Ltd. Samples examined by A. Fowler and C. J. Stubblefield. Fig. 10(6).

	Thickness ft	in	Depth ft	in
SUPERFICIAL DEPOSITS				
Boulder clay; shale debris .. about	65	0	65	0
CARBONIFEROUS				
NAMURIAN (MILLSTONE GRIT SERIES)				
Shale	21	0	86	0
SECOND GRIT				
Sandstone, pale grey medium- to coarse-grained; shale ribs	15	0	101	0
WOODLAND SHELL-BEDS				
Shale, silty dark grey micaceous; plants and marine fossils; *Crurithyris sp., Eomarginifera sp., Orbiculoidea nitida, Productus sp., Shansiella globosa* (loc. 22b) ..	15	0	116	0
Sandstone, pale grey silty medium-grained; greenish blue marly pellets	1	0	117	0
Shale	7	0	124	0
FIRST GRIT (upper leaf)				
Sandstone, predominantly coarse pebbly, quartzo-feldspathic	57	0	181	0

Randolph Colliery No. 2 Shaft, Evenwood

Surface about 548 ft AOD. 6-in NZ 12 SE. Site [1586 2489] Evenwood. Sunk 1893. Information from B & S No. 2697 and NCB. Plates IX and XI (72b).

	Thickness ft	in	Depth ft	in
SUPERFICIAL DEPOSITS				
Clay, blue and yellow; 1 ft sand bed at 12 ft	24	0	24	0
CARBONIFEROUS				
WESTPHALIAN B (MIDDLE COAL MEASURES)				
Ramble, jointy ..	12	0	36	0

	Thickness ft	in	Depth ft	in
Post, grey brown and white; jointed; metal partings at base [DURHAM LOW MAIN POST]	63	6	99	6
DURHAM LOW MAIN				
Coal	2	0	101	6
Seggar clay	0	6	102	0
Metal, dark; ironstone girdles; sandstone bands at top ..	41	0	143	0
BRASS THILL				
Coal	1	9	144	9
Band	0	6	145	3
Coal	0	9	146	0
Metal; post girdle near base	6	0	152	0
Coal	0	1½	152	1½
Metal, grey; sandstone bands and ironstone girdles	26	10	178	11½
TOP HUTTON				
Coal	0	3	179	2½
Band	0	3	179	5½
Coal	0	6	179	11½
Metal, blue; sandstone 8 ft at 197 ft 11 in; sandstone with metal partings, 11 ft 6 in at 225 ft	54	1	234	0½
HUTTON (BOTTOM HUTTON)				
Coal	3	9	237	9½
Seggar clay	3	7	241	4½
Post, grey hard ..	4	10	246	2½
Coal	0	4	246	6½
Seggar clay; iron balls	3	0	249	6½
Metal, grey	6	0	255	6½
Post, dark grey; iron girdles	6	0	261	6½
Post, grey hard ..	8	0	269	6½
RULER (JUBILEE)				
Coal	0	2	269	8½
Band	0	1	269	9½
Coal	0	6	270	3½
Seggar clay	1	0	271	3½
Post, white and grey; leafy near top ..	20	10	292	1½
Metal, blue; iron girdles at top ..	45	8	337	9½
(Assumed horizon of HARVEY MARINE BAND)				
WESTPHALIAN A (LOWER COAL MEASURES)				
Coal	0	3	338	0½
Seggar clay	4	0	342	0½
Post, dark grey ..	3	0	345	0½
Metal, hard; post girdles	5	0	350	0½
Post, dark grey; white spar	6	0	356	0½
Post, dark grey ..	6	0	362	0½
Metal, grey	4	0	366	0½
Post, white hard ..	11	0	377	0½
HARVEY				
Coal	3	0	380	0½
Post, white hard ..	48	3	428	3½
TILLEY				
Coal	0	9	429	0½
Seggar clay	2	0	431	0½
Shale, grey	25	0	456	0½
Post, grey hard ..	20	3	476	3½
BUSTY				
Coal	3	2	479	5½
Seggar clay	3	0	482	5½
Post, white and grey ..	10	0	492	5½
Metal, grey	3	0	495	5½
THREE-QUARTER				
Coal	1	0	496	5½
Strata; predominantly sandstone; thin metal bands and partings ..	32	3	528	8½
Coal	0	6	529	2½
Seggar clay	5	6	534	8½
Post, grey; metal partings	9	6	544	2½
Metal, grey hard ..	1	6	545	8½
Post, grey leafy ..	1	6	547	2½
Post, white hard; leafy and grey at base ..	15	0	562	2½
Metal, grey hard ..	6	0	568	2½
BROCKWELL				
Coal, cannel ..	4	0	572	2½
Coal	5	6	577	8½
Band	0	3	577	11½
Coal	0	6	578	5½
Sump (no details) ..	16	6½	595	0

Rush Pit, Old Etherley Colliery

Surface about 482 ft AOD. 6-in NZ 12 NE. Site [1804 2812] 1000 yd W of Woodhouses. Sunk 1869. Information from B & S No. 829 and NCB. Plate XI (100).

	Thickness ft	in	Depth ft	in
SUPERFICIAL DEPOSITS				
Soil	2	0	2	0
CARBONIFEROUS				
WESTPHALIAN B (MIDDLE COAL MEASURES)				
Post, laminated	6	0	8	0
Metal, grey and "grey beds"	28	2	36	2
Dun Post	2	6	38	8
Metal, grey	4	3	42	11
TOP BRASS THILL				
Coal	0	4	43	3
Black stone	1	3	44	6
Post, mild strong	5	0	49	6
Grey beds	3	0	52	6
Black metal	1	6	54	0
BOTTOM BRASS THILL				
Coal	0	5	54	5
Seggar	1	6	55	11
Black stone	1	9	57	8
Metal, grey and blue; post girdles	14	0	71	8
Black stone	1	3	72	11
Seggar	1	6	74	5
TOP HUTTON				
Coal	0	7	75	0
Grey beds with post	9	0	84	0
Metal, blue soft	2	0	86	0
Blackstone, coarse	1	3	87	3
Metal, blue and grey; ironstone nodules at top	24	0	111	3
HUTTON (BOTTOM HUTTON)				
Coal	1	6	112	9
Stone band	0	2	112	11
Coarse splint	1	6	114	5
Seggar, dark	0	8	115	1
Coal, coarse or splint	0	2	115	3
Thill, strong	6	3	121	6
Sandstone, silicated	1	8	123	2
Metal, strong; iron nodules	10	6	133	8
Black and blue stone	3	6	137	2
Metal, blue	0	3	137	5
Ironstone band	0	3	137	8
Metal, blue	3	0	140	8
Post, white	6	0	146	8
Black stone	0	2	146	10
TOP RULER (JUBILEE)				
Coal	0	2	147	0
Thill, dark strong; soft at top	11	3	158	3
Metal, blue	4	0	162	3
Black stone	1	2	163	5
BOTTOM RULER (JUBILEE)				
Coal	0	4	163	9
Thill, strong coarse	3	6	167	3
Sandstone, alternating with blue and grey metal and "grey-beds"	52	10	220	1
Dun whin	4	0	224	1
Metal, dark blue	10	6	234	7
Metal, black	16	0	250	7
(Assumed horizon of HARVEY MARINE BAND)				
WESTPHALIAN A (LOWER COAL MEASURES)				
Coal	0	4	250	11
Seggar and Thill	6	0	256	11
Metal, dark	0	6	257	5
Post, dun strong coarse	8	0	265	5
Post, strong silicated	12	0	277	5
Metal, black; post girdles	5	0	282	5
Seggar, coarse	4	0	286	5
Band, black	0	2	286	7
[?*Fault*]	—	—	—	—
HARVEY				
Coal	3	4	289	11
Seggar	1	0	290	11
Coal, coarse	0	6	291	5
Strata	45	9	337	2

	Thickness ft	in	Depth ft	in
TILLEY				
Coal	0	8	337	10
Band	1	0	338	10
Coal	1	2	340	0
Strata	103	6	443	6
TOP BUSTY				
Coal	2	6	446	0
Strata	5	6	451	6
BOTTOM BUSTY				
Coal	1	6	453	0
Drift to Brockwell seam				

Shackleton Beacon Bore

Surface about 560 ft AOD. 6-in NZ 22 SW. Site [2281 2336] Shackleton Beacon, 1½ miles WNW of Heighington. Drilled 1937 by Wm. Coulson Ltd. Fig. 10(8).

	Thickness ft	in	Depth ft	in
SUPERFICIAL DEPOSITS				
Sandy soil and stones ..	0	10	0	10
CARBONIFEROUS				
WESTPHALIAN A (LOWER COAL MEASURES)				
Sandstone, coarse ..	3	6	4	4
Shale, grey sandy; mudstone and ironstone bands	5	3	9	7
Mudstone, blue; ironstone bands	25	2	34	9
Sandstone, ganister-like	2	7	37	4
Shale, brown buff and grey sandy; ripple-marked shaly sandstone bands	14	8	52	0
Shale	6	0	58	0
QUARTERBURN MARINE BAND				
Shale; '*Productus*' ..	0	6	58	6
NAMURIAN (MILLSTONE GRIT SERIES)				
Sandstone, calcareous	1	0	59	6
Shale, sandy ..	2	0	61	6
Sandstone, grey medium-grained; some sandy shale ..	7	2	68	8
Shale, grey sandy; some ripple-marked shaly sandstone bands	5	0	73	8
Shale, dark grey; sporadic plant remains	21	4	95	0
Fireclay	5	0	100	0
Shale, dark grey; '*Productus*' ..	12	0	112	0
Fireclay, greenish grey	4	0	116	0
SECOND GRIT				
Sandstone, siliceous hard fine-grained	6	0	122	0

Shawbrow Hill No. 5 Bore

Surface 486.6 ft AOD. 6-in NZ 22 NW. Site [2143 2712] Shawbrow Hill 1120 yd ENE of Fieldon Bridge. Drilled 1965 by Boyles Bros. for NCB. Plate XII (137).

	Thickness ft in	Depth ft in
SUPERFICIAL DEPOSITS		
Soil	1 0	1 0
Clay, brown; often sandy; pebbles at top	7 10	8 10
CARBONIFEROUS		
WESTPHALIAN B (MIDDLE COAL MEASURES)		
Siltstone, brown micaceous; plant fragments; 4-in sandstone band at 16 ft 3 in	9 2	18 0
Sandstone, pale grey medium-grained; carbonaceous films	0 11	18 11
Mudstone, grey irony patches; coaly films at 20 ft 9 in; silty below with pale sandstone bands.. ..	3 7	22 6
Mudstone, grey ..	0 9	23 3
Sandstone, pale grey medium-grained carbonaceous; finer-grained below 54 ft..	33 10	57 1
?RYHOPE LITTLE		
Coal and coaly shale	0 1	57 2
Seatearth, grey ..	0 2	57 4
Coal	1 10	59 2
Sandstone, pale grey	0 1½	59 3½
Coaly shale ..	0 1	59 4½
Coal	0 1½	59 6
Seatearth, grey ..	0 9	60 3
Sandstone, pale grey fine- and fine- to medium-grained	11 0	71 3
Siltstone, grey; pale sandstone bands ..	1 3	72 6
Mudstone, grey; silty bands; occasional ironstone bands; occasional mussels ..	5 4	77 10
Mudstone, grey; below 82 ft occasional ironstone bands with mussels	13 6	91 4
Siltstone, grey rooty ..	0 10	92 2
Mudstone, dark grey; thin irony bands ..	2 8	94 10
Coal; coaly shale bands	0 6	95 4
Seatearth, grey and dark grey; silty below 96 ft 6 in ..	3 2	98 6
Mudstone, grey silty grading to siltstone, locally tinged green, below 100 ft ..	4 6	103 0
Mudstone, grey silty; siltstone bands below 108 ft	8 0	111 0
Siltstone, grey finely laminated	4 6	115 6
Mudstone, grey; occasional irony nodules	9 3	124 9
Coal	0 8	125 5
Seatearth, grey ..	4 4	129 9
Sandstone, grey laminated fine-grained; occasional slump structures	15 3	145 0
Siltstone, grey; pale sandstone laminae; grades to silty mudstone below 146 ft and mudstone below 150 ft	7 10	152 10
Siltstone, grey.. ..	0 7	153 5
Mudstone, dark grey; some ironstone bands; abundant mussels especially in basal 6 in	3 4	156 9
Mudstone, black; abundant mussels; some fish scales ..	0 6	157 3
TOP HIGH MAIN		
Coal	0 7	157 10
Seatearth, grey; slightly silty below 163 ft ..	9 2	167 0
Mudstone, grey; abundant plants	3 5	170 5
BOTTOM HIGH MAIN		
Coal	2 2	172 7
Pyrite band.. ..	0 1	172 8
Coal, shaly	0 2	172 10
Seatearth, grey ..	1 8	174 6
Sandstone, pale grey fine- to medium-grained	5 6	180 0

Staindrop Field House Drift

Surface level about 460 ft AOD. 6-in NZ 12 SE. Site [1725 2461] 390 yd ESE of Staindrop Field House, West Auckland. Examined by Mr. G. Richardson.

	Thickness ft	Thickness in	Depth ft	Depth in
CARBONIFEROUS				
WESTPHALIAN A (LOWER COAL MEASURES)				
Mudstone, grey silty carbonaceous; plants	0	5	0	5
BUSTY				
Coal	4	0	4	5
Seatearth-mudstone, grey	0	8	5	1
Coal	0	1	5	2
Seatearth-mudstone, silty; ironstone nodules	4	0	9	2
Seatearth-sandstone, grey silty; ironstone nodules at top ..	5	6	14	8
Sandstone, grey silty..	1	0	15	8
Mudstone, grey silty; sandy laminae; rare ironstone nodules; plant fragments ..	9	0	24	8
THREE-QUARTER				
Coal..	1	0	25	8
Seatearth-mudstone ..	1	0	26	8
Sandstone, light grey fine- to medium-grained	11	0	37	8
Mudstone, grey silty micaceous and carbonaceous	0	1	37	9
Sandstone, grey false-bedded micaceous carbonaceous laminae	1	11	39	8
Mudstone, grey silty micaceous; plants ..	4	9	44	5
Ironstone band ..	0	5	44	10
Mudstone, grey; silty bands; mussels ..	2	3	47	1
Sandstone, buff irregularly bedded fine-grained; silty mudstone bands with mussels	1	6	48	7
Mudstone, grey slightly silty; mussels ..	0	2	48	9
Sandstone, grey and buff fine-grained micaceous and carbonaceous	1	3	50	0
Mudstone, grey silty; mussels	0	11	50	11
Sandstone, buff, false-bedded silty micaceous and carbonaceous; iron-cemented nodular patches ..	2	0	52	11
Mudstone, grey silty slightly micaceous ..	0	2	53	1
Sandstone, buff ferruginous; worm tracks	0	2	53	3
Mudstone, grey; silty bands; mussels ..	1	10	55	1
Mudstone, dark grey very silty micaceous and carbonaceous abundant mussels ..	1	6	56	7
(Note: the faunal assemblage in the mudstone and silty mudstone fraction of the above $11\frac{3}{4}$ ft of strata includes *Carbonicola browni*, *C. declivis*, *C.* cf. *martini* [juv], *Curvirimula sp.* [frags.])				
Mudstone, dark grey micaceous carbonaceous; mussel fragments	0	$7\frac{1}{2}$	57	$2\frac{1}{2}$
Mudstone, grey shaly carbonaceous, ?mussel fragments ..	0	$1\frac{1}{2}$	57	4
Coal	0	2	57	6
Seatearth-mudstone; silty bands	4	0	61	6
Sandstone, grey silty micaceous carbonaceous	2	0	63	6
Sandstone, grey thin-bedded micaceous and carbonaceous ..	0	7	64	1
Sandstone, grey and buff false-bedded				

	Thickness ft	in	Depth ft	in
micaceous and carbonaceous; (minor fault 8 ft from base)	25	9	89	10
Seatearth-mudstone, dark grey carbonaceous; coaly laminae	0	6	90	4
Mudstone, grey carbonaceous; coaly at top	2	2	92	6
Seatearth-mudstone, grey	0	4	92	10
Sandstone, buff compact fine- to medium-grained; micaceous carbonaceous laminae	0	8	93	6
Mudstone, grey sandy; irregular sandstone bands; plant fragments	0	10	94	4
Sandstone, grey silty micaceous and carbonaceous	3	8	98	0
Mudstone, grey silty; sandy bands; plant fragments	7	6	105	6
Mudstone, grey, carbonaceous; *Carbonicola polmontensis*, *Curvirimula subovata*	3	6	109	0
BROCKWELL				
Cannel; fish fragments	1	0	110	0
Coal	4	0	114	0
Seatearth-mudstone	1	1	115	1
Coal	1	0	116	1
Seatearth-mudstone	0	4	116	5
Coal	0	8	117	1

TVCWB Bore No. 3

Surface 352.6 ft AOD. 6-in NZ 21 NE. Site [2520 1982] Swan House near Heighington. Drilled 1967–8 by J. T. Hymas and Son Ltd. for Tees Valley and Cleveland Water Board. Cores examined by D. A. C. Mills.

	Thickness ft	in	Depth ft	in
SUPERFICIAL DEPOSITS				
Soil	1	6	1	6
Clay, mottled	6	6	8	0
Silt; veins of sand ..	10	0	18	0
Boulder clay, brown ..	23	0	41	0
Silt, brown	4	0	45	0
Boulder clay	6	0	51	0
Sand, brown; silt and clay bands ..	9	0	60	0
Boulder clay, mainly brown; bands of silt at top	17	0	77	0
Broken rock	1	0	78	0
Gravel, fine and medium	2	0	80	0
Boulder clay, grey hard	4	0	84	0
Gravel; clay band ..	1	0	85	0
PERMIAN				
MIDDLE MAGNESIAN LIMESTONE				
No core	9	6	94	6
LOWER MAGNESIAN LIMESTONE				
Dolomite, buff-yellow hard finely granular; fine stylolites ..	1	1	95	7
Dolomite, grey to yellow hard finely granular microporous well-bedded; sporadic stylolites and carbonate-lined cavities and vugs; rare flaggy bands up to 2 in thick	11	2	106	9
Dolomite, grey to creamy-buff hard dense; finely granular; horizontal and sub-horizontal irregular stylolites commonly with films of residual clay; large empty or crystal-				

	Thickness ft	in	Depth ft	in
lined cavities and vertical or sub-vertical veinlets ..	29	9	136	6
Dolomite, grey and brownish yellow hard dense; occasional small wavy stylolites	5	4	141	10
Dolomite, as at 136 ft 6 in	4	0	145	10
Dolomite, as at 141 ft 10 in..	1	4	147	2
Dolomite, brown and yellow dense; poorly developed stylolites; plant debris.. ..	2	8	149	10
Dolomite, auto-brecciated; broken angular fragments of dolomite, set in a matrix of ?calcitic dolomite	1	0	150	10
Dolomite, calcitic; becoming progressively more dolomitic to base; abundant plant debris and filaments, scattered bryozoa, *Bakevellia ceratophaga* at 154 ft 10 in, *Schizodus sp.* at 157 ft	20	0	170	10
Dolomite, grey calcitic bedded; sporadic stylolitic partings and small cavities, mainly at top; grades down into limestone with little or no dolomite content but with sporadic dolomite bands; disseminated plant debris.. ..	10	7	181	5
Dolomite, buff-yellow soft less dense; irregular calcite veinlets and empty or carbonate-lined cavities; disseminated plant debris at base where rock becomes interbedded with fine laminae of Marl Slate lithology	6	2	187	7
MARL SLATE				
Dolomitic mudstone, grey argillaceous; alternates with beds of limestone lithology	2	1	189	8
Dolomite, argillaceous; fine laminae of dolomitic limestone ..	6	7	196	3

TVCWB Bore No. 7

Surface 172.40 ft AOD. 6-in NZ 20 NE. Site [2610 0993] 480 yd SE of Jolby Farm, Newton Morrell. Drilled 1968 by J. T. Hymas and Son Ltd. for Tees Valley and Cleveland Water Board. Cores (Permian section) examined by D. B. Smith and D. A. C. Mills.

	Thickness ft	in	Depth ft	in
SUPERFICIAL DEPOSITS				
Soil	2	6	2	6
Clay, brown	72	6	75	0
Boulder clay, grey ..	10	0	85	0
?Rock	5	0	90	0
Boulder clay	18	0	108	0
PERMIAN				
UPPER MAGNESIAN LIMESTONE				
No core (stated to be soft limestone) ..	14	0	122	0
Dolomite, grey fine-grained calcitic; coquinoid bands containing *Calcinema sp.*, *Schizodus sp.* and *Liebea sp.*	6	0	128	0
Dolomite, pale creamy-grey with faint purple laminae; hard finely crystalline calcitic; small carbonate-lined cavities and sporadic ¼ to ½-in bands of				

	Thickness ft	in	Depth ft	in
grey dolomitic shale; bedding planes slightly irregular and mottled with patches of red or orange material	6	0	134	0
Dolomite, creamy-grey, finely saccharoidal; thin beds of grey argillaceous dolomite; MnO_2 speckling ..	15	0	149	0
MIDDLE PERMIAN MARL				
Alternations of grey gypsum, slightly dolomitic, and mudstone and clay (poor recovery)	19	0	168	0

Thornton Hall Bore

Surface about 165 ft AOD. 6-in NZ 21 NW. Site [2480 1634] 1300 yd SE of Thornton Hall, High Coniscliffe. Date of drilling not known. Information from Wm. Coulson Ltd. and B & S No. 2581.

	Thickness ft	in	Depth ft	in
SUPERFICIAL DEPOSITS				
Common Earth ..	3	0	3	0
Clay	3	0	6	0
Sand and Loam ..	22	6	28	6
PERMIAN				
MIDDLE MAGNESIAN LIMESTONE				
Marl	9	6	38	0
Magnesian Limestone	13	4	51	4
LOWER MAGNESIAN LIMESTONE				
Broken ground ..	36	3	87	7
Limestone, brown ..	51	6	139	1
MARL SLATE				
Shale, brown	4	0	143	1
?BASAL PERMIAN DEPOSITS				
Post, white	3	5	146	6
CARBONIFEROUS				
NAMURIAN (MILLSTONE GRIT SERIES)				
Shale, red and white; ironstone rib ..	18	0	164	6
Sandstone, red ..	31	2½	195	8½
Shale, grey	29	7	225	3½
Shale, soft black; ironstone rib	32	3½	257	7
Sandstone, grey and white	12	8	270	3
Shale; sandstone ribs ..	60	9	331	0
Shale, grey	33	2	364	2
?COALCLEUGH SHELL-BED				
Limestone	0	4	364	6
Shale, grey	2	0	366	6
Limestone	0	4	366	10
Shale, grey	2	0	368	10
Sandstone, dark grey	0	8	369	6
Shale, grey; ironstone bands	14	10	384	4
Shale; sandstone ribs ..	16	2	400	6
Shale, black	0	3	400	9
?YOREDALE				
Coal	0	9½	401	6½
Sandstone; shale bands	6	7½	408	2
Sandstone, grey ..	18	0	426	2
Shale, grey	35	6	461	8
Sandstone, grey ..	1	0	462	8
Shale, grey	16	8	479	4
?ROOKHOPE SHELL-BEDS LIMESTONE				
Limestone, grey ..	17	7	496	11
Shale; ironstone rib ..	0	10	497	9
Shale, grey	6	11	504	8
Shale, black	0	9	505	5
Limestone, strong grey	6	0	511	5
Limestone	15	3	526	8

Toft Hill Underground B Bore

Surface 440.42 ft (Brockwell seam floor). 6-in NZ 12 NE. Site [1646 2800] 300 yd E of Hunter's Hill House. Drilled 1956 by NCB. Cores examined by G. Armstrong.

	Thickness ft	Thickness in	Depth ft	Depth in
CARBONIFEROUS				
WESTPHALIAN A (LOWER COAL MEASURES)				
BROCKWELL SEAM FLOOR (440.42 ft AOD)				
Fireclay	6	0	6	0
Shale, sandy	2	3	8	3
Sandstone, argillaceous; roots	1	7	9	10
Sandstone, light grey	18	0	27	10
VICTORIA SHELL-BED				
Shale, dark grey; ironstone bands; mussels and plant remains at base	11	7	39	5
Sandstone, light grey ferruginous	6	7	46	0
Shale, grey sandy; plant remains	7	1	53	1
Sandstone, argillaceous; sandy shale bands; sporadic plant remains	2	0	55	1
Shale, grey silty; sandy partings; plant remains	1	11	57	0
VICTORIA FISH-BED				
Shale, black; fish remains	2	0	59	0
(VICTORIA RIDER horizon)				
Mudstone, canneloid	0	3	59	3
Fireclay, grey-brown	4	6	63	9
Shale, black; coal films rootlets	1	0	64	9
Fireclay	0	4	65	1
VICTORIA				
Coal	0	6	65	7
Fireclay	1	0	66	7
Siltstone, grey shaly; sandy partings towards base	10	10	77	5
Sandstone	0	9	78	2
Shale, black	0	7	78	9
BOTTOM VICTORIA, TOP LEAF (NCB)				
Coal	0	5	79	2
Shale, black sandy micaceous	0	2	79	4
Seatearth-sandstone	0	7	79	11
Fireclay	2	0	81	11
BOTTOM VICTORIA, BOTTOM LEAF (NCB)				
Coal	0	11	82	10
Seatearth, sandy	1	0	83	10
Fireclay-mudstone	2	8	86	6
Shale, grey silty; sporadic plants and rootlets	8	6	95	0
Shale, dark grey; silty at base	5	6	100	6
(Assumed STOBSWOOD horizon)				
Sandstone, grey wispy-bedded	8	7	109	1
Mudstone, dark grey	0	6	109	7
Siltstone, grey; sporadic plant remains at base	3	2	112	9
TOP MARSHALL GREEN				
Coal	0	7	113	4
Fireclay	0	3	113	7
Seatearth-sandstone	0	5	114	0
Sandstone, argillaceous	6	0	120	0
Shale, grey sandy; silty at base	3	0	123	0
Shale, black micaceous; coal films, plant and root remains	3	0	126	0
Siltstone; plants	5	0	131	0
BOTTOM MARSHALL GREEN				
Coal	1	2	132	2
Sandstone	0	7	132	9
Coal	0	11	133	8
Fireclay	1	4	135	0
Shale, dark grey silty	5	0	140	0

	Thickness ft	in	Depth ft	in
Sandstone	6	0	146	0
Shale, grey silty; sandy bands and partings..	4	4	150	4
Sandstone, light grey wispy bedded ..	6	0	156	4
Shale, dark grey silty and sandy; plants ..	8	1	164	5
Sandstone, wispy-bedded	1	10	166	3
Shale, dark grey mainly sandy	9	2	175	5
GANISTER CLAY				
Coal	1	1½	176	6½
Sandstone, ganisteroid	0	11½	177	6
Coal	0	3	177	9
Fireclay, black and shaly at top; silty at base	2	8	180	5
Sandstone, light grey sporadic shaly partings; ganisteroid at top; massive at base	12	9	193	2
Shale, grey silty; ironstone bands except at top	12	7	205	9
Sandstone, grey-brown argillaceous.. ..	1	7	207	4
Shale, dark grey ..	1	0	208	4
Sandstone, light grey massive; ganisteroid and rooty at top ..	4	2	212	6
Shale, grey silty; rare plants	8	4	220	10
Sandstone, dark grey shaly	9	2	230	0
RODDYMOOR MARINE BAND				
Shale, dark grey; fish fragments at base	1	9	231	9
Coal	0	7	232	4
THIRD GRIT				
Sandstone, grey wispy-bedded; rooty and ganisteroid at top ..	4	7	236	11

Wath Erne Farm Water Bore

Surface about 270 ft AOD. 6-in NZ 21 SW. Site [2160 1001] Wath Erne. Drilled 1959 by J. T. Hymas and Son Ltd. Fig. 6 (9).

	Thickness ft	in	Depth ft	in
SUPERFICIAL DEPOSITS				
Clay and boulders ..	25	0	25	0
CARBONIFEROUS				
NAMURIAN (MILLSTONE GRIT SERIES)				
?CRAG LIMESTONE (BOTTOM)				
Limestone, hard yellow	21	0	46	0
Shale	24	0	70	0
TOP LITTLE LIMESTONE				
Limestone, blue ..	4	0	74	0
Shale	3	0	77	0
BOTTOM LITTLE LIMESTONE				
Limestone, blue ..	27	0	104	0
Shale	40	0	144	0
GREAT LIMESTONE				
Limestone	30	0	174	0

West Auckland Colliery, Back Drift

Surface level at drift mouth 360.23 ft AOD; at floor of Bottom Busty Seam 186.69 ft AOD. Inclination about 1 in 3. 6-in NZ 12 NE. Site [1827 2683] of drift mouth 620 yd N of W of St Helen's Church, West Auckland. Made 1957 for NCB. Examined by D. A. C. Mills and G. Richardson.

	Thickness ft	Thickness in	Depth ft	Depth in
SUPERFICIAL DEPOSITS				
Sand and gravel; silty clay matrix	12	0	12	0
Boulder clay; large sandstone fragments	18	0	30	0
CARBONIFEROUS				
WESTPHALIAN A (LOWER COAL MEASURES)				
Shale, blue	5	0	35	0
Sandstone, grey muddy flaggy micaceous; silty mudstone laminae	6	0	41	0
Mudstone, grey silty micaceous carbonaceous; basal 8 in shaly; *Spirorbis sp.*, *Anthraconaia* cf. *williamsoni*, *Naiadites sp. nov.*, *Carbonita humilis*, *Geisina arcuata*	3	0	44	0
HARVEY				
Coal	3	6	47	6
Seatearth-mudstone, grey silty	5	0	52	6
Sandstone, grey fine-grained micaceous rooty	1	0	53	6
Mudstone	5	0	58	6
Sandstone, grey and buff flaggy; mainly fine-grained; carbonaceous partings	36	6	95	0
TOP TILLEY				
Coal	0	10	95	10
Band	0	10	96	8
Coal	1	2	97	10
Seatearth-mudstone, grey silty	0	6	98	4
Seggar and sandstone	0	8	99	0
Sandstone, mainly light grey flaggy fine-grained; micaceous carbonaceous partings	6	6	105	6
Mudstone, grey silty micaceous; sporadic plant fragments . .	4	0	109	6
Mudstone	4	0	113	6
Mudstone, grey; *Carbonicola* cf. *cristagalli*, *C. decorata?*, *C. oslancis*, *C. rhomboidalis*, *Geisina arcuata*	0	6	114	0
BOTTOM TILLEY				
Coal	0	9	114	9
Seatearth-mudstone, sandy	0	8	115	5
Seatearth-sandstone, ganisteroid . .	0	3	115	8
Seatearth-sandstone; argillaceous nodules	3	6	119	2
Coal	0	7	119	9
Seatearth-sandstone, grey fine-grained . .	2	6	122	3
Shale, grey	8	5	130	8
Sandstone, grey silty . .	1	0	131	8
Mudstone, grey silty; sporadic ironstone nodules	5	0	136	8
Shale, grey	1	0	137	8
Mudstone, silty; ironstone laminae and nodules; *Carbonicola oslancis*, *Naiadites flexuosus*, *Geisina arcuata*	3	6	141	2
(Horizon of TOP BUSTY)				
Shale, grey	5	0	146	2
Mudstone, dark grey sandy flaggy; poorly preserved plant fragments	9	9	155	11

	Thickness ft	in	Depth ft	in
Sandstone, grey silty fine-grained micaceous; plant fragments; faulted at base	6	0	161	11
Shale, grey	4	0	165	11
Sandstone	4	0	169	11
Siltstone, grey micaceous; plant fragments at base ..	1	10	171	9
Sandstone, light grey and buff fine- to medium-grained; muddy at base ..	10	9	182	6
Sandstone, light grey and white medium-grained; rare muddy partings	2	9	185	3
BOTTOM BUSTY				
Coal	0	11	186	2
Band	0	0½	186	2½
Coal	1	11	188	1½
Seatearth-mudstone (into)				

West Auckland Colliery No. 168 Bore

Surface 445.03 ft AOD. 6-in NZ 12 NE. Site [1810 2744] 1090 yd NW of St Helen's Church, West Auckland. Drilled 1958 by NCB. Cores examined by A. M. J. Clarke. Plate IX (104).

	Thickness ft	in	Depth ft	in
SUPERFICIAL DEPOSITS				
Soil and boulder clay..	68	0	68	0
CARBONIFEROUS				
WESTPHALIAN A (LOWER COAL MEASURES)				
Sandstone, light grey fine-grained false-bedded	20	0	88	0
TOP TILLEY				
Coal	1	3	89	3
Seatearth-mudstone, light grey silty; sphaerosiderite band at 92½ ft	5	1	94	4
Siltstone, grey fissile; sandy partings at base; plant debris at top	6	3	100	7
Sandstone, light grey fine-grained; silty micaceous partings	7	9	108	4
Mudstone, grey shaly; silty bands; worm burrows, mussel fragments; plant debris	5	8	114	0
BOTTOM TILLEY				
Coal (Top Leaf) ..	0	6	114	6
Seatearth-siltstone ..	4	0	118	6
Coal (Bottom Leaf)	0	8	119	2
Seatearth-mudstone ..	0	4	119	6
Siltstone, grey; ironstone bands; roots..	2	0	121	6
Sandstone, grey fine-grained; thin irregular silty micaceous bands and partings..	5	6	127	0
Siltstone, grey; fine sandstone bands; worm burrows from 131½ to 134 ft ..	8	2	135	2
Mudstone, grey silty; shaly below 151½ ft; coaly partings in basal 6 in; ironstone layers; fine quartz veining from 142½ to 144½ ft	21	4	156	6
Seatearth-mudstone ..	0	6	157	0
Mudstone, dark grey; mussel fragments ..	2	6	159	6
Seatearth-mudstone ..	0	3	159	9
Mudstone, dark grey shaly	1	9	161	6
Seatearth-mudstone ..	2	8	164	2
Siltstone, grey micaceous; sandstone bands; mussel fragments	2	10	167	0

	Thickness ft	in	Depth ft	in
(?TOP BUSTY horizon)				
Siltstone, grey micaceous	1	4	168	4
Sandstone, light grey fine-grained; silty bands, especially at top	21	8	190	0
BOTTOM BUSTY				
Coal	1	7	191	7
Seatearth-mudstone	0	2	191	9
Coal	1	6	193	3
Seatearth-mudstone	2	8	195	11
Siltstone, light grey; thin sandstone bands	5	0	200	11
Sandstone, light grey fine-grained; silty micaceous partings	15	7	216	6
Mudstone, grey silty; clay ironstone bands	1	6	218	0
THREE-QUARTER				
Coal	0	8	218	8
Seatearth-mudstone; silty at base	2	10	221	6
Sandstone, light grey fine- to medium-grained	11	6	233	0
Siltstone, grey thin-bedded; sandy partings below 235 ft	4	0	237	0
Mudstone, grey silty; mussel fragments from 237 to 238 ft and 244 to 246 ft	12	0	249	0
Shale, black; fish scales and spines	4	0	253	0
Seatearth-mudstone	1	0	254	0
Mudstone, grey shaly; plant fronds and stems	4	4	258	4
Sandstone, light grey fine-grained; silty micaceous bands and partings	13	2	271	6
Mudstone, grey silty	7	6	279	0
BROCKWELL (old workings)	6	0	285	0
Seggar	0	3	285	3

West Auckland Colliery No. 177 Bore

(synopsis)

Surface 377.76 ft AOD. 6-in NZ 12 NE. Site [1862 2733] 660 yd W of N of St Helen's Church, West Auckland. Drilled 1958 by NCB. Cores examined by A. M. J. Clarke. Plate X (109).

	Thickness ft	in	Depth ft	in
SUPERFICIAL DEPOSITS				
Boulder clay	26	0	26	0
CARBONIFEROUS				
WESTPHALIAN B (MIDDLE COAL MEASURES)				
Fireclay	16	0	42	0
Shale, dark grey	4	0	46	0
Siltstone, grey sandy	8	6	54	6
(Assumed horizon of HARVEY MARINE BAND at 54 ft 6 in)				
WESTPHALIAN A (LOWER COAL MEASURES)				
Strata; mainly sandstone; siltstone bands; seatearth-mudstone at top	30	6	85	0
HARVEY				
Goaf and collapsed strata containing *Naiadites*, mussel debris and *Spirorbis*	4	0	89	0

	Thickness ft	in	Depth ft	in
Strata; mainly sandstone, medium-grained; 2 ft seatearth-siltstone at top	39	0	128	0
TOP TILLEY				
Coal	1	0	129	0
Seatearth-mudstone, black	0	9	129	9
Coal	1	9	131	6
Seatearth-mudstone	3	0	134	6
Mudstone, grey	3	3	137	9
Sandstone, light grey fine-grained; siltstone interlaminae at base with worm borings	2	6	140	3
Seatearth-siltstone	1	9	142	0
Mudstone, grey; *Gyrochorte* between 145 ft 1 in and 146 ft; ostracods at 146 ft and 146 ft 5 in; mussel fragments below 146 ft 3 in; *Spirorbis sp.* at base	4	8	146	8
BOTTOM TILLEY				
Coal, dull inferior	0	0½	146	8½
Siltstone, black carbonaceous micaceous; abundant mussel impressions in shale in basal ½ in	0	3½	147	0
Coal, bright	0	10	147	10
Seatearth-siltstone, black micaceous	2	4	150	2
Coal	0	7	150	9
Seatearth-mudstone, light grey	1	3	152	0
Siltstone, grey argillaceous	8	0	160	0
Mudstone, grey shaly silty at top; *Planolites montanus?* at 166 ft 6 in worm markings below 170 ft; plant debris below 173 ft	14	0	174	0
Mudstone, as above; worm markings in top 3 ft and from 180 to 182 ft; mussels at 184 ft; indet. organic debris at 185½ ft; fusainous plant fragments in basal 6 in	12	2	186	2
(Horizon of TOP BUSTY)				
Strata; mainly siltstone in upper 20 ft; mainly sandstone below	33	0	219	2
BOTTOM BUSTY				
Coal	2	1	221	3
Seatearth-siltstone	3	9	225	0
Sandstone	33	2	258	2
THREE-QUARTER				
Coal	0	8	258	10
Seatearth-mudstone	2	2	261	0
Sandstone	3	0	264	0

West Auckland Colliery, Spring Gardens No. 2 Bore

Surface 382 ft AOD. 6-in NZ 12 NE. [1824 2699] 700 yd WNW of St Helen's Church, West Auckland. Drilled 1953 by Andrew Kyle Ltd. for NCB. Cores examined by A. M. J. Clarke. Plate X(105).

	Thickness ft	in	Depth ft	in
SUPERFICIAL DEPOSITS				
Boulder clay and soil	50	0	50	0
CARBONIFEROUS				
WESTPHALIAN A (LOWER COAL MEASURES)				
Shale, sandy	2	0	52	0
Siltstone, sandy at base	11	9	63	9
Shale, grey sandy silty	2	0	65	9
HOPKINS BAND				
Shale, dark grey; *Spirorbis sp.*, *Naiadites sp.* [fragments], *Carbonita humilis*, *Geisina arcuata*	0	6	66	3

	Thickness ft	in	Depth ft	in
HARVEY (Goaf) ..	3	6	69	9
Fireclay	0	6	70	3
Shale, grey sandy silty; roots	1	9	72	0
Shale, grey sandy; plants	4	6	76	6
Sandstone, light grey mainly flaggy; some micaceous partings	41	6	118	0
TOP TILLEY				
Coal	0	10	118	10
Fireclay	0	11	119	9
Coal	1	3	121	0
Fireclay, greyish brown sandy	3	4	124	4
Siltstone	3	8	128	0
Sandstone, argillaceous	0	6	128	6
Shale, grey sandy; ankeritic at base; sandstone partings	1	2	129	8
Siltstone, grey; plant debris	1	6	131	2
Shale, grey; plant debris	0	4	131	6
Siltstone, grey.. ..	7	0	138	6
Shale, dark grey; *Carbonicola oslancis*, *Geisina arcuata*, fish scale; sporadic plant debris	2	3	140	9
BOTTOM TILLEY				
Coal	0	10	141	7
Fireclay	0	4	141	11
Shale; fireclay bands	2	5	144	4
Mudstone	0	2	144	6
Coal	0	3	144	9
Fireclay	0	6	145	3
Shale, grey silty; ironstone nodules; roots	4	3	149	6
Shale, grey sandy; sandstone partings; plants	4	3	153	9
Siltstone, grey finely micaceous; ironstone bands.. ..	16	3	170	0
Mudstone, grey; fissile at base	6	6	176	6
Shale, grey	2	0	178	6
Shale, dark grey; indet. mussels	1	6	180	0
(Horizon of TOP BUSTY)				
Sandstone, fine-grained ankeritic	0	3	180	3
Shale, grey silty micaceous; roots ..	1	0	181	3
Shale, grey sandy ..	3	0	184	3
Sandstone, argillaceous wispy-bedded ankeritic	1	0	185	3
Shale, grey silty; sandy bands; sporadic plant debris	5	9	191	0
Sandstone, argillaceous wispy-bedded ..	5	0	196	0
Shale, grey silty; sandy bands; bands with abundant plant debris including *Calamites* and *Neuropteris*	12	0	208	0
Sandstone, fine-grained wispy-bedded ..	1	0	209	0
Shale, sandy; sporadic plant debris including *Neuropteris* ..	6	6	215	6
Sandstone, ankeritic; small brown shale pellets at base ..	1	1	216	7
Shale, grey micaceous	0	10	217	5
Sandstone, light grey medium-grained ..	8	4	225	9
BUSTY				
Coal (1-in band at 227 ft 2 in) ..	2	10	228	7
Fireclay	2	11	231	6

West Auckland Colliery Underground No. 2 Bore

Up bore from Bottom Busty workings. 6-in NZ 12 NE. Site [1836 2732] 790 yd NW of St Helen's Church, West Auckland. Drilled 1961 by NCB. Cores examined by A. M. J. Clarke. Plate X (113).
Up Bore: measurements from floor of Bottom Busty at 192.98 AOD.

	Thickness ft	in	Height ft	in
CARBONIFEROUS				
WESTPHALIAN A (LOWER COAL MEASURES)			106	0
Sandstone, light grey fine-grained	5	7	100	5
TOP TILLEY				
Coal	1	4	99	1
Band	1	0	98	1
Coal	0	10	97	3
Seatearth-mudstone . .	4	3	93	0
Siltstone, grey laminated; sandstone interlaminae; rootlets, especially at top . .	8	0	85	0
Mudstone, grey shaly; sporadic mussel and plant fragments . .	5	0	80	0
BOTTOM TILLEY				
Coal	1	1	78	11
Seatearth - mudstone	2	11	76	0
Coal	0	10	75	2
Seatearth-mudstone . .	1	2	74	0
Mudstone, mainly grey shaly; fine sandstone laminae above 66 ft; clay ironstone band at 52 ft; mussels to 42 ft; plant fragments from 49 to 52 ft; worm trails from 58 to 63 ft; sand-filled worm burrows from 66 to 63 ft	34	4	39	8
(Horizon of TOP BUSTY)				
Siltstone, grey micaceous flaggy; abundant rootlets in top 1 ft 6 in; large plant fragments at base . .	8	8	31	0
Sandstone, light grey mainly medium-grained; siltstone bands from 2 in to 2 ft throughout . .	12	0	19	0
Open hole [post] . .	9	0	10	0
Excavation	10	0	0	0

Westholme Colliery No. 1 Shaft

Surface about 340 ft AOD. 6-in NZ 11 SW. Site [1346 1785] 400 yd WSW of Westholme Hall, Winston. Drilled 1910–12 by Hetton Colliery Co. Fig. 8(7).

	Thickness ft	in	Depth ft	in
SUPERFICIAL DEPOSITS				
Soil	1	6	1	6
Sand and gravel . .	18	6	20	0
Clay, brown	22	0	42	0
Sand, loamy; clay at base	7	6	49	6
Gravel	3	4	52	10
Clay	0	8	53	6
CARBONIFEROUS				
NAMURIAN (MILLSTONE GRIT SERIES)				
Metal, blue	54	2	107	8
Post, grey; sporadic blue metal bands . .	12	0	119	8
Post, white leafy . .	30	0	149	8
Metal, blue	3	6	153	2
Post, leafy	0	6	153	8
Metal, blue	5	0	158	8
Post, grey leafy; metal bands between 170 ft 6 in and 177 ft . .	33	4	192	0
Metal blue; ironstone bands	70	0	262	0
YOREDALE				
Coal	3	6	265	6
Coal, inferior . .	0	6	266	0
Sandstone and ganister	12	0	278	0
Shale, dark grey . .	33	6	311	6
Post; shale partings . .	51	0	362	6
Post, hard	36	6	399	0
Shale; ironstone balls; including assumed horizon of LOWER FELLTOP LIMESTONE 18 ft above base) . .	72	9	471	9
Coal	1	0	472	9
Shale	5	3	478	0

West Tees No. 1 Bore

Surface 452.19 ft AOD. 6-in NZ 12 NW. Site [1487 2621] 620 yd SSE of Bowes Close. Drilled 1953 by NCB. Cores examined by D. E. White.

	Thickness ft	in	Depth ft	in
SUPERFICIAL DEPOSITS				
Made ground and boulder clay	6	0	6	0
CARBONIFEROUS				
WESTPHALIAN A (LOWER COAL MEASURES)				
Shale	6	0	12	0
Sandstone	0	6	12	6
Shale, sandy	5	4	17	10
Sandstone	0	10	18	8
Shale	10	1	28	9
(BROCKWELL WORKINGS)	6	5	35	2
Fireclay	2	0	37	2
Shale, grey	9	2	46	4
Post, grey	0	8	47	0
Coal (?BOTTOM BROCKWELL)	0	4	47	4
Fireclay-mudstone ..	0	6	47	10
Sandstone, light grey massive fine-grained	1	2	49	0
Mudstone, sandy shaly	1	6	50	6
Sandstone, light grey fine-grained current-bedded	19	0	69	6
Mudstone, grey shaly; plant debris at top ..	21	7	91	1
VICTORIA				
Coal	0	9	91	10
Fireclay	2	2	94	0
Mudstone, sandy; plant debris	5	0	99	0
Siltstone; sporadic plant debris	12	8	111	8
BOTTOM VICTORIA (NCB)				
Coal	0	5	112	1
Fireclay	4	3	116	4
Coal	0	8	117	0
Ganister	2	10	119	10
Mudstone, sandy; plant debris	1	5	121	3
Sandstone, light grey current-bedded fine-grained	3	5	124	8
Mudstone, sandy; rare plant debris	4	10	129	6
Mudstone, shaly ..	6	1	135	7
Sandstone, ganisteroid	5	4	140	11
Mudstone, sandy; plant debris	3	4	144	3
Fireclay	2	9	147	0
Mudstone, shaly sandy; planty at base; fronds of *Alethopteris*, *Neuropteris*	16	11	163	11
Sandstone, ganisteroid	2	1	166	0
Sandstone, light grey massive fine-grained	8	9	174	9
(BOTTOM MARSHALL GREEN horizon)				
Shale, black carbonaceous	0	1	174	10
Sandstone, light grey massive fine-grained	15	7	190	5
Mudstone, sandy; traces of plant debris ..	15	3	205	8
Sandstone, light grey current-bedded fine- to medium-grained	31	10	237	6
GANISTER CLAY				
Coal	1	3	238	9
Sandstone	0	3	239	0
Fireclay-mudstone ..	2	0	241	0
Sandstone, argillaceous	6	0	247	0

Windlestone Supplementary B Bore

Surface 396.2 ft AOD. 6-in NZ 22 NE. Site [2664 2866] Windlestone Hall. Drilled 1959 by NCB. Plate XI (165).

	Thickness ft	in	Depth ft	in
SUPERFICIAL DEPOSITS				
[Open holed to 235 ft]				
Soil and subsoil ..	1	6	1	6
Boulder Clay; sand partings	76	6	78	0
PERMIAN				
MAGNESIAN LIMESTONE (Lower and Middle Divisions)				
Limestone	2	0	80	0
Limestone; sand panels (probably friable dolomite)	30	0	110	0
Limestone	125	0	235	0
[Cored from 235 ft]				
Limestone, grey flaggy porcellanous; thinner bedded towards base; thin bands of Marl Slate lithology in basal 4 ft ..	16	0	251	0
MARL SLATE				
Shale, dark grey thinly laminated; brown speckling on bedding planes	7	8	258	8
[*Fault*]				
Shale, as at 258 ft; ?plant leaf at 259 ft 3 in; fish scales at 260 ft; galena at 260 ft 6 in; hard dolomitic bands in basal 2 in..	3	4	262	0
BASAL PERMIAN SANDSTONE AND BRECCIA				
Sandstone, bluish grey hard breccia-conglomerate fine-grained pyritic; faceted and polished quartz pebbles and elongate flat siltstone breccia fragments; abundant 'millet seed' quartz grains	1	7	263	7

	Thickness ft	in	Depth ft	in
CARBONIFEROUS				
WESTPHALIAN B (MIDDLE COAL MEASURES)				
Shale, black carbonaceous locally canneloid	1	5	265	0
METAL				
Coal	0	6	265	6
Shale, black carbonaceous; reddened bands	2	6	268	0
Seatearth-mudstone ..	3	0	271	0
Mudstone, light grey; reddened bands; scattered rootlets and plant debris.. ..	10	3	281	3
FIVE-QUARTER				
Coal, shaly	0	10	282	1
Seatearth-mudstone, dark grey	1	2	283	3
Sandstone, light grey with reddened bands fine-grained; occasional wispy bedded bands; thin siltstone bands and partings; scattered rootlets ..	12	9	296	0
Mudstone, grey silty; shaly at top; poorly preserved mussel at 298 ft; scattered plant fragments	14	0	310	0
Shale, black carbonaceous; canneloid bands; fish scales at top; mussels (mainly *Anthracosia*) at base	2	10	312	10
Coal	0	6	313	4
Seatearth-mudstone, grey; carbonaceous at top	1	8	315	0
Mudstone, grey shaly; abundant plant debris	5	10	320	10
Sandstone, grey fine-grained wispy and turbulently bedded; siltstone intercalations..	3	2	324	0

	Thickness ft	in	Depth ft	in
Sandstone, light grey fine- to medium-grained false-bedded at top, massive below 336 ft; sporadic micaceous partings	23	6	347	6
Siltstone, grey thin-bedded; abundant sandy partings; comminuted plant debris	10	6	358	0
Sandstone, light grey fine-grained; siltstone interlaminae ..	5	0	363	0
Mudstone, grey massive to flaggy; *Gyrochorte* cf. *carbonaria* and *Cochlichnus kochi* from 373 to 374 ft; clay ironstone nodules and layers to base	15	2	378	2
Shale, black carbonaceous; abundant fish scales and spines ..	0	5	378	7
MAIN				
Coal	1	4	379	11
Seatearth-mudstone, grey	1	9	381	8
Coal	5	4	387	0
Seatearth	0	7	387	7
Coal	0	6	388	1
Seatearth-mudstone, grey; ironstone nodules at base ..	5	1	393	2
Mudstone, grey shaly; abundant plant leaves and stems; scattered rootlets	9	9	402	11
Mudstone, brownish grey shaly; large coalified portions of bark	1	1	404	0
Seatearth-mudstone, brownish grey ..	1	6	405	6
Mudstone, dark grey; abundant plant debris; coalified and scattered vitrain partings	5	3	410	9
Coal	0	2	410	11
Seatearth-mudstone, light grey	4	1	415	0
Mudstone, grey shaly; scattered rootlets ..	10	0	425	0
TOP MAUDLIN				
Coal	0	1	425	1
Seatearth	0	1	425	2
Coal	0	10	426	0
Seatearth - mudstone	2	0	428	0
Coal	0	10	428	10
Seatearth-mudstone, light grey	2	2	431	0
Mudstone, light grey silty flaggy; scattered rootlets	7	0	438	0
Siltstone, grey thin and wispily bedded; abundant thin sandstone bands; scattered plant fragments	8	1	446	1
BOTTOM MAUDLIN				
Coal	1	3	447	4
Seatearth-mudstone, light grey	3	5	450	9
Mudstone, grey silty shaly; scattered plant fragments becoming abundant at base ..	9	6	460	3
Sandstone, light grey fine-grained thin-bedded silty at top; irregular base ..	13	3	473	6
Siltstone, grey thin and wispily bedded; fine sandstone band from 480 to 483 ft and from 486 ft to base; plant debris ..	13	8	487	2
Mudstone, grey shaly; silty to 496 ft; clay ironstone bands towards base; mussel fragments in basal 1 ft	11	5	498	7
Shale, black carbonaceous; mussels and *Rhabdoderma* at base	0	4	498	11
DURHAM LOW MAIN				
Coal	1	8	500	7
Seatearth-siltstone, brownish grey; ironstone nodules at base	1	8	502	3

	Thickness ft	in	Depth ft	in
Siltstone, grey flaggy; sandy bands below 510 ft	23	4	525	7
Sandstone, whitish grey medium-grained false bedded; siltstone layers from 538 ft 3 in to 539 ft 6 in; clay ironstone conglomerate at 547 ft 6 in; shale pellets and fusainous plant debris in basal 1 ft ..	23	8	549	3
Mudstone, grey shaly; scattered mussels between 550 ft 6 in and 551 ft 6 in	3	4	552	7
TOP BRASS THILL				
Coal	2	5	555	0
Seatearth-mudstone ..	3	6	558	6
Sandstone, light grey thin-bedded fine- to medium-grained micaceous	1	10	560	4
Siltstone, dark grey micaceous and carbonaceous; fusainous plant fragments at base	1	1	561	5
BOTTOM BRASS THILL				
Coal	1	0	562	5
Seatearth-mudstone, dark grey	1	7	564	0
Mudstone, dark grey shaly; coalified plant debris in basal 4 in..	3	6	567	6
Seatearth-sandstone, whitish grey irregularly bedded ..	1	6	569	0
Sandstone, light grey wavily bedded fine-grained; dark micaceous partings; silty at base	9	0	578	0
Siltstone, medium grey thin irregularly bedded; dark micaceous partings; thin silty bands	8	5	586	5
Sandstone, light grey thin-bedded fine-grained; black micaceous partings ..	6	7	593	0
Mudstone, grey silty shaly finely micaceous; pyritised fragments of *Naiadites* at base	1	9	594	9
TOP HUTTON				
Coal	0	6	595	3
Shale, black carbonaceous	0	1	595	4
Coal	0	11	596	3
Seatearth-mudstone, black	3	4	599	7
Coal, inferior ..	0	1	599	8
Seatearth-mudstone, dark grey	0	10	600	6
Sandstone, light grey massive fine-grained; occasional siltstone interlaminae ..	11	7	612	1
Siltstone, grey flaggy micaceous; argillaceous at base ..	7	5	619	6
Mudstone, grey shaly; clay ironstone bands; mussels between 622 ft 6 in and 623 ft 6 in; shaly and pyritic in basal 6 in	16	1	635	7
BOTTOM HUTTON				
Coal	3	6	639	1
Seatearth-mudstone, dark grey; silty and sandy below 640 ft 4 in	3	2	642	3

Woodland Bore

(synopsis)

Surface 931.75 ft AOD. 6-in NZ 02 NE. Site [0910 2770] 1 mile 330 yd NE of Woodland village. Drilled 1962 by Boyles Bros. Ltd. for Geological Survey. Cores examined by D. A. C. Mills. For full details see

Mills and Hull 1968. The Geological Survey Borehole at Woodland, Co. Durham (1962). *Bull geol. Surv. Gt Br.*, No. 28, pp. 1–37. Figs. 4(3), 6(2), 7(2), 8(2), 9(2) and 10(5). Plates VI and VII (16).

	Thickness ft	in	Depth ft	in
SUPERFICIAL DEPOSITS				
Boulder clay	10	0	10	0
CARBONIFEROUS				
WESTPHALIAN A (LOWER COAL MEASURES)				
Strata	43	9	53	9
?VICTORIA RIDER				
Coal	0	3	54	0
Strata	82	9	136	9
BOTTOM MARSHALL GREEN				
Coal	1	6	138	3
Strata	45	5	183	8
GANISTER CLAY				
Coal	0	11	184	7
Strata	51	11	236	6
THIRD GRIT				
Sandstone	37	9	274	3
Strata	1	3	275	6
KAYS LEA MARINE BAND				
Shale	6	5	281	11
Strata	47	11	329	10
QUARTERBURN MARINE BAND				
Shale	2	10	332	8
NAMURIAN (MILLSTONE GRIT SERIES)				
Strata	9	0	341	8
SECOND GRIT (upper leaf)				
Sandstone	49	0	390	8
Strata	0	8	391	4
SECOND GRIT (lower leaf)				
Sandstone; shale and mudstone bands . .	64	5	455	9
WOODLAND SHELL-BEDS				
Shale; sandstone bands	15	1	470	10
Strata	3	4	474	2
FIRST GRIT				
Sandstone; sporadic shale bands . .	108	4	582	6
Strata	4	4	586	10
WHITEHOUSE LIMESTONE				
Limestone	1	1	587	11
Strata	127	1	715	0
GRINDSTONE LIMESTONE				
Limestone; sandstone and mudstone bands . .	14	1	729	1
GRINDSTONE SILL				
Sandstone	42	11	772	0
Strata	24	1	796	1
UPPER FELLTOP LIMESTONE				
Limestone; shale and sandstone bands . .	25	5	821	6
Strata	71	7	893	1
COALCLEUGH SHELL-BED				
Limestone	1	5	894	6
Strata	39	6	934	0
LOWER FELLTOP LIMESTONE				
Limestone	0	8	934	8
Strata	44	5	979	1
ROOKHOPE SHELL-BEDS LIMESTONE				
Limestone with median mudstone . .	12	10	991	11
Strata	75	6	1067	5
KNUCTON SHELL-BEDS				
Limestone and mudstone	10	1	1077	6
Strata	3	0	1080	6
CRAG LIMESTONE				
Limestone; mudstone band . .	23	6	1104	0
Strata	1	1	1105	1
FIRESTONE SILL				
Sandstone; shale and				

	Thickness ft	in	Depth ft	in
coal parting ..	7	5	1112	6
Strata	4	1	1116	7
FARADAY HOUSE SHELL-BED				
Limestone	0	11	1117	6
Strata	53	6	1171	0
LITTLE LIMESTONE				
Limestone	4	5	1175	5
WHITE HAZLE				
Sandstone	17	11	1193	4
Strata	9	6	1202	10
COAL SILLS				
Sandstone; shale bands	41	8	1244	6
Strata	22	7	1267	1
GREAT LIMESTONE				
Limestone	52	11	1320	0
VISÉAN (CARBONIFEROUS LIMESTONE SERIES)				
Strata	69	2	1389	2
FOUR-FATHOM LIMESTONE				
Limestone	32	1	1421	3
Strata	38	2½	1459	5½
LITTLE WHIN SILL				
Dolerite	7	2	1466	7½
Strata	66	2½	1532	10
THREE-YARD LIMESTONE (upper leaf)				
Limestone	6	3½	1539	1½
Strata	2	2½	1541	4
GREAT WHIN SILL (into)				
Quartz-dolerite ..	58	8	1600	0

Woodland Colliery No. 29 Bore

Surface about 1100 ft AOD. 6-in NZ 02 NE. [0644 2702] 1340 yd NW of Woodland. Sunk 1872. Information from B & S No. 3092.

	Thickness ft	in	Depth ft	in
SUPERFICIAL DEPOSITS				
Boulder Clay ..	27	0	27	0
CARBONIFEROUS				
WESTPHALIAN A (LOWER COAL MEASURES)				
Shale, black	7	3	34	3
Fireclay	4	10	39	1
VICTORIA				
Coal and black shale	1	0	40	1
Fireclay	4	0	44	1
Sandstone, leafy ..	2	6	46	7
Shale, blue	1	0	47	7
Sandstone, grey ..	53	6	101	1
Shale, blue	8	6	109	7
TOP MARSHALL GREEN				
Coal	1	4	110	11
Fireclay	3	0	113	11
Sandstone, grey ..	11	9	125	8
Shale, blue	1	4	127	0
BOTTOM MARSHALL GREEN				
Coal	1	0	128	0
Band	0	2	128	2
Coal	2	11	131	1
Fireclay	1	3	132	4
Sandstone; alternating grey and leafy ..	33	5	165	9
Shale, blue	5	0	170	9
Sandstone, grey ..	5	11	176	8
GANISTER CLAY				
Coal and blue metal	1	7	178	3
Fireclay	1	3	179	6
Sandstone, leafy ..	5	2	184	8
Shale, blue	2	6	187	2
Sandstone, grey and leafy	6	3	193	5
Shale, blue	4	10	198	3
Sandstone, leafy ..	1	2	199	5
Shale, blue and black ..	25	11	225	4
Sandstone, leafy ..	2	6	227	10
Shale, blue and black ..	23	3½	251	1½

	Thickness ft	in	Depth ft	in
THIRD GRIT				
Sandstone, grey; leafy at top	31	4½	282	6
Shale, blue	3	5	285	11
Sandstone, leafy ..	7	0	292	11
Shale, blue	11	7	304	6
Sandstone, leafy ..	3	0	307	6
Shale, blue (assumed horizon of QUARTERBURN MARINE BAND)	4	0	311	6
NAMURIAN (MILLSTONE GRIT SERIES)				
Sandstone, leafy ..	3	0	314	6
Coal	1	2	315	8
Fireclay	1	10	317	6
SECOND GRIT				
Sandstone, grey (into)	63	10	381	4

WT/4 Bore, Eggleston Moor

(synopsis)

Surface 1580 ft AOD. 6-in NY 92 NE. [9986 2750] Millstone Rigg, Eggleston Common. Drilled 1971 by Boyles Bros. for Northumbrian River Authority. Cores examined by D. A. C. Mills.

	Thickness ft	in	Depth ft	in
SUPERFICIAL DEPOSITS				
[Open holed to 623 ft 4 in]				
Head, wash, etc ..	3	2	3	2
CARBONIFEROUS				
NAMURIAN (MILLSTONE GRIT SERIES)				
Open hole; estimated base of First Grit at 345 ft; estimated position of Upper Felltop Limestone at 509 ft	620	2	623	4
Sandstone, grey moderate to thick-bedded fine-grained siliceous; sporadic micaceous carbonaceous partings and shaly laminae; coarse-grained towards base with pseudo-conglomerate band; grades to	40	11	664	3
Mudstone, silty dark grey finely micaceous and carbonaceous ..	23	11	688	2
Sandstone, grey and dark grey; abundant micaceous carbonaceous partings ..	5	2	693	4
Siltstone, dark grey laminated, sandstone and mudstone laminae	5	3	698	7
Sandstone and mudstone interbedded, grey fine- and very fine-grained ..	5	2	703	9
Sandstone, light grey to dark grey thin-to moderately-bedded very fine- to medium-grained; basal 25 ft interbedded and laminated with siltstone and mudstone ..	39	1	742	10
Sandstone, light grey moderate- to thick-bedded medium- to coarse-grained; occasional micaceous carbonaceous partings	53	1	795	11
Mudstone, dark grey laminated to thin-bedded finely micaceous and carbonaceous; scattered fossil debris	1	8	797	7
Sandstone, light grey moderately-bedded coarse- and very coarse-grained; mica-				

	Thickness ft	in	Depth ft	in
ceous carbonaceous patches and blebs ..	2	2	799	9
Mudstone, sandy and silty dark grey laminated to thin-bedded finely micaceous and carbonaceous; planty	4	10	804	7
Sandstone, dark grey muddy fine-grained calcareous; fossil debris	0	5	805	0
Sandstone, grey ganisteroid at top; muddy, micaceous and carbonaceous below ..	3	5	808	5
Sandstone, grey moderately-bedded coarse- and very coarse-grained	6	4	814	9
KNUCTON SHELL-BEDS				
Mudstone and siltstone, calcareous laminated to thin-bedded fossiliferous with ribbed shelly fossils at top	3	5	818	2
TOP CRAG LIMESTONE				
Limestone, dark grey muddy impure very fine; calcareous mudstone bands; sporadic stylolites and calcite-lined vugs; fossiliferous with much comminuted debris ..	9	8	827	10
Mudstone, dark grey calcareous; limestone nodules ..	0	5	828	3
BOTTOM CRAG LIMESTONE				
Limestone, predominantly as at 827 ft 10 in	8	8	836	11
Limestone-conglomerate; limestone fragments in matrix of dark calcareous mudstone ..	1	7	838	6
Limestone, predominantly as at 827 ft 10 in; grades at base to calcareous mudstone ..	8	10	847	4
Mudstone, dark grey laminated finely micaceous and carbonaceous	1	1	848	5
Coal	0	1	848	6
Mudstone, dark grey and black coaly ..	0	6	849	0
FIRESTONE SILL				
Sandstone, dark grey muddy thin-bedded micaceous and carbonaceous rooty	1	4	850	4
Sandstone, grey very hard ganisteroid ..	2	10	853	2
Mudstone, dark grey and black; coaly at top	3	1	856	3
FARADAY HOUSE SHELL-BED				
Sandstone, predominantly grey thin-bedded fine-grained calcareous; ribbed shelly fossils at top ..	6	7	862	10
Mudstone, dark grey laminated very fine micaceous and carbonaceous; occasional calcareous siltstone and impure limestone bands ..	34	10	897	8
LITTLE LIMESTONE				
Limestone, grey thin-bedded fine ..	4	0	901	8
WHITE HAZLE				
Sandstone, grey thin- to moderately-bedded fine- to medium-grained micaceous and carbonaceous; muddy and impure at base	14	1	915	9
Mudstone, dark grey laminated very fine				

	Thickness ft	Thickness in	Depth ft	Depth in
micaceous and carbonaceous	2	2	917	11
COAL SILLS				
Sandstone, predominantly grey and dark grey thin-bedded often muddy fine-grained; micaceous and carbonaceous; mudstone, silty mudstone and siltstone laminae and bands increasing in thickness to base to give 'striped' and banded appearance ..	28	8	946	7
Mudstone; striped and banded with siltstone and muddy sandstone; dark grey laminated and thin-bedded fine- and very fine-grained micaceous and carbonaceous; very fissile mudstone at base ..	9	10	956	5
Sandstone, grey and dark grey thin- to moderately-bedded fine-grained micaceous and carbonaceous; muddy at top; mudstone laminae and bands towards base ..	16	1	972	6
Mudstone, dark grey laminated very fine-grained micaceous and carbonaceous; sporadic sandstone and siltstone interlaminae; becomes progressively more fissile with depth with fossil content increasing; calcareous at base ..	11	1	983	7
GREAT LIMESTONE				
Limestone, grey and dark grey moderate- to thick-bedded fine and medium crystalline; muddy bands at top; stylolitic partings; calcite hair veins; vertically and sub-vertically jointed; pronounced coral band between 1001 ft 10 in and 1008 ft 7 in; pronounced coral-brachiopod band 1 ft 10 in at 1017 ft; comminuted fossil debris; sharp base	59	1	1042	8
VISÉAN (CARBONIFEROUS LIMESTONE SERIES)				
TUFT SANDSTONE (or SILL)				
Sandstone, grey and dark grey thin- to thick-bedded fine-grained calcareous; rooty at top underlain by ganisteroid band; becoming dark grey shaly thin-bedded micaceous and carbonaceous towards base with bands of mudstone	5	8	1048	4
Sandstone and mudstone alternating; grey and dark grey laminated and thin-bedded fine- and very fine-grained micaceous and carbonaceous; thin calcareous bioturbate fossiliferous sandstone and sandy limestone ribs at 1054 ft 1 in and 1059 ft 9 in ..	13	7	1061	11

Wycliffe Hall Bore

Surface about 450 ft AOD. 6-in NZ 11 SW. Site [1228 1328] 1000 yd NNW of Hutton Magna. Drilled 1953 by J. T. Hymas for Wycliffe Estate. Cores examined by A. J. Wells. Figs. 4(6) and 6(6).

	Thickness ft	Thickness in	Depth ft	Depth in
SUPERFICIAL DEPOSITS				
Boulder clay, with large boulders of mostly local origin; also Shap Granite and Borrowdale volcanics	18	0	18	0
Clay with cobbles ..	58	0	76	0
CARBONIFEROUS				
NAMURIAN (MILLSTONE GRIT SERIES)				
Shale, dark grey; sporadic ironstone nodules; rare septarian nodules; predominantly sparsely fossiliferous but occasional thin bands containing a fauna of productoids, rhynchonelloids, orthocone nautiloids and rare trilobites (*recovery poor*) ..	94	0	170	0
GREAT LIMESTONE				
Limestone, shaly; bryozoa, brachiopods and crinoids	1	6	171	6
Limestone, grey; median 4 in shaly band	4	10	176	4
Shale, calcareous; productoids and rhynchonelloids ..	0	10	177	2
Limestone, grey crinoidal; shaly at base with rhynchonelloids	3	7	180	9
Limestone, grey crinoidal; calcareous shale at base containing rhynchonelloids	1	11	182	8
Limestone, grey crinoidal; argillaceous streaks	38	0	220	8
Limestone, grey and brownish grey; crinoidal	6	2	226	10
VISÉAN (CARBONIFEROUS LIMESTONE SERIES)				
Shale, calcareous; sporadic productoids ..	0	5	227	3
Shale, dark grey sandy micaceous; pyrite ..	1	0	228	3
Shale, dark grey sandy; thin coal streaks and carbonized fragments	1	9	230	0
Sandstone, grey fine-grained micaceous; sandy shale streaks ..	2	10	232	10
Shale, sandy; alternates with shaly sandstone	3	3	236	1
Shale, predominantly sandy micaceous; pyrite; sporadic plant fragments	5	11	242	0
Sandstone, light grey medium-grained current-bedded; sporadic often irregular micaceous shale streaks and bands; pyrite nodules at base	13	0	255	0
Shale, sandy; alternates with shaly sandstone; sporadic pyrite nodules; productoids ..	4	0	259	0
Shale, sandy shale bands; sporadic ironstone and pyritic nodules	1	3	260	3
Shale, grey; sporadic sandy bands; small ironstone and pyritic nodules; brachiopods including productoids (especially in band 11 in from base), bivalves and crinoids	10	0	270	3
Shale, grey; ironstone nodules; rare pro-				

	Thickness ft	in	Depth ft	in
ductoids and plant fragments	13	0	283	3
Shale, grey sandy predominantly micaceous; sporadic brachiopods, including spinose forms, bivalves, gastropods and crinoids . .	18	0	301	3
Shale, dark grey; sporadic ironstone nodules; sporadic crinoids and plant remains	15	6	316	9
Shale, dark grey; ironstone nodules except at base; '*Chonetes*' *sp.*, at 325 ft 3 in; '*Productus*' *sp.*, at 326 ft 9 in; and a gastropod at 328 ft . .	30	9	347	6
Mudstone; productoids	7	6	355	0
Shale; brachiopods including *Lingula sp.*, at 356 ft and '*Chonetes*' *spp.* at 359 ft, nautiloids and crinoids	10	0	365	0
FOUR-FATHOM LIMESTONE				
Limestone, greyish brown; productoids and crinoids	18	0	383	0
Shale, calcareous, pyritic; crinoids . .	0	1	383	1
Ganister	3	0	386	1
Sandstone, grey fine-grained micaceous . .	2	8	388	9
Shale, sandy micaceous; pyritic nodules; plants	3	1	391	10
Sandstone, pale grey thick-bedded medium-grained micaceous; sandy shale parting towards base	12	8	404	6
Shale, sandy micaceous; shaly sandstone bands at top; plant remains	21	0	425	6
Sandstone, pale grey medium-grained; sandy micaceous shale bands . .	9	6	435	0
Shale, sandy; abundant plant remains . .	3	6	438	6
Shale, sandy; alternates with sandstone bands; sporadic plants	4	6	443	0
Shale, sandy; brachiopods and bivalves; plant remains . .	2	6	445	6
Sandstone, pale hard; sandy micaceous shale partings locally with sand-filled burrows	3	6	449	0
Shale, sandy; sandstone bands; sporadic plants	3	0	452	0
Sandstone, sandy micaceous carbonaceous; shale partings . .	6	0	458	0

Appendix 3

THE LITHOLOGICAL AND FAUNAL SUCCESSION OF THE MOUNT PLEASANT BORE NEAR BARNARD CASTLE, COUNTY DURHAM

by G. A. L. JOHNSON[1], D.SC.

Summary The Mount Pleasant Water Bore was drilled near Barnard Castle, County Durham between 1949 and 1951. After penetrating 79 ft of drift, Carboniferous strata encountered included 515 ft of low Namurian and high Viséan beds of Yoredale-type facies, and in particular the complete Four-Fathom Limestone and underlying Three-Yard Limestone cyclothems. A large collection of fossils has been made from the cores and detailed faunal lists have been prepared. An important top Viséan goniatite/bivalve fauna from above the Four-Fathom Limestone in the borehole has given critical evidence on the position between the Upper and Lower Carboniferous in northern England.

INTRODUCTION

THE Mount Pleasant Bore is situated on the south side of the River Tees near the crest of a rise overlooking Barnard Castle. Its exact site [0328 1508] is 500 yd S 7°E of Mount Pleasant Farm on six-inch map NZ 01 NW. The surface level of the borehole is about 765 ft above OD. The borehole, drilled between May 1949 and October 1951, was sunk by Messrs Isler of London for Barnard Castle Urban District Council. The greater part of the borehole was cored using a rotary shot drilling technique, to produce a large diameter hole. A chisel bit was used where the shot drilling proved too slow. No continuous cores were obtained through the superficial deposits and some of this ground was chiselled. Drilling to 84 ft produced some 24-in diameter cores but slow progress made a reduction in the size of the hole necessary. Boring continued to 173 ft producing 20-in diameter core but only broken cores were obtained which proved of little stratigraphical use. Owing to slow progress the diameter of the hole was again reduced at this depth. The 18-in diameter cores obtained after this were badly broken to the top of the Great Limestone at 192 ft but good core recovery was obtained below this. The upper four feet of chert at 320 ft proved too hard for shot drilling and were chiselled; no specimens were obtained. Good recovery of 18-in diameter cores continued to 395 ft when the size of the hole was again reduced. From this depth to the bottom of the boring at 594 ft almost continuous cores 14 inches in diameter were obtained. The total amount of core recovered from the boring was over 90 per cent.

[1]Department of Geological Sciences, University of Durham

The results of the borehole are summarized as follows:

	Thickness ft in	Depth ft in
Superficial Deposits	79 3	79 3
Namurian (Millstone Grit Series)	160 9	240 0
Viséan (Carboniferous Limestone Series)	354 0	594 0

The geology of the area is shown on the published one inch to one mile geological Sheet 32 (Barnard Castle), the description of which is included in the accompanying memoir (Mills and Hull 1976) of which this appendix forms part. Prior to the re-survey of the district, several workers had studied the area. The primary geological survey of the district was carried out by Gunn, the map being published in 1881. Very much later the stratigraphy and structures of the area were studied by Reading (1957) and Wells (1957).

Preliminary mention of the fossils found in the borehole, in particular the goniatites from below the Underset (= Four-Fathom) Chert was made in Rayner (1953) and Johnson, Hodge and Fairbairn (1962). The goniatite fauna is of considerable importance in tracing the base of the Upper Carboniferous, the Viséan–Namurian boundary, in northern England; this boundary is now recognized as lying between the Four-Fathom and the Great limestones.

Stratigraphical Succession

The stratigraphical succession in the surrounding area is clear, and the sequence as proved in the borehole closely agrees with this. Rockhead was proved at 79 ft 3 inches in sandstone thought to underlie the Bottom Little Limestone. The Great Limestone, 55 ft thick at a depth of 240 ft, the base of which forms the junction between the Millstone Grit and Carboniferous Limestone Series accords well with the thickness proved in exposures to the south where it forms a broad scarp dipping generally northwards at between 5 and 8°. In the Carboniferous Limestone sequence the sediments show well developed cyclic deposition and the Four-Fathom and Three-Yard cyclothems, each composed of the general sequence limestone, sandstone, shale, seatearth and coal, are well developed. The Underset (= Four-Fathom) Chert (Wells 1957) is 32 ft thick and is separated from the Four-Fathom Limestone, 25 ft thick, by 13 ft of mudstone and shale containing many fossils. *Girtyoceras*? *costatum* Ruprecht indicative of a P_{2c} age has been recorded. The measures between the Four-Fathom and Three-Yard limestone total 150 ft, but the thick porous sandstone often present below the Four-Fathom Limestone hereabouts was not proved. As a result of this, drilling was continued to below the Three-Yard Limestone. The Three-Yard Limestone was 14¼ ft thick including shale bands and a 4-in seatearth-sandstone 10 in from the base. The measures between the Three-Yard and Five-Yard limestones are only 31¼ ft thick, some 50 ft less than normal. This is attributed to faulting cutting out the lower part of the clastic sequence underlying the Three-Yard Limestone. Two feet of core were lost above the Five-Yard Limestone at the base of the borehole and it is therefore possible that the cores of the fault zone were not recovered. Supporting evidence for a fault is the fact that the limestone at the base of the borehole is lithologically and microfaunally similar to the Five-Yard Limestone proved elsewhere in the district. The borehole was drilled to 594 ft, 8½ ft of the Five-Yard Limestone having been proved.

Specimens collected from the borehole are held at the Department of Geological Sciences, University of Durham.

Acknowledgements. The advice of the late Dr W. S. Bisat, the late Professor O. M. B. Bulman, Dr H. M. Johnson, Mr M. Mitchell, Dr W. H. C. Ramsbottom and Sir James Stubblefield on the identification of certain fossils is gratefully acknowledged. I am also indebted to Barnard Castle UDC for granting permission to publish the geological information obtained from the borehole.

LOG OF THE BOREHOLE

Surface 765 ft AOD. 6-in NZ 01 NW. Site [0328 1508] 500 yd S7°E of Mount Pleasant Farm. Drilled 1949–51 by Messrs Isler of London for Barnard Castle UDC. Shown graphically in Fig. 4, No. 5.

	Thickness ft	in	Depth ft	in
SUPERFICIAL DEPOSITS				
Drift	79	3	79	3
CARBONIFEROUS				
NAMURIAN (MILLSTONE GRIT SERIES)				
Grit, light brown coarse	42	9	122	0
Shale, banded sandy	8	0	130	0
Shale, dark fine laminated	29	6	159	6
Shale, dark (*core lost*)	25	6	185	0
GREAT LIMESTONE (to 240 ft)				
Limestone and shale bands (*core lost*) (Tumbler Beds)	7	0	192	0
Limestone, grey compact; dolomitic at 236 ft; strong open joint from 207 ft to near base; *Calcifolium bruntonense*, *Garwoodia sp.*, *Girvanella sp.*, endothyrids, textulariids, *Dibunophyllum bipartitum*, trepostomatous bryozoa, *Dielasma sp.*, *Phricodothyris sp.*, productoids indet., *Semiplanus latissimus*, crinoid columnals, *Archaeocidaris sp.* [spines]	48	0	240	0
VISÉAN (CARBONIFEROUS LIMESTONE SERIES)				
Shale, dark soft; *Fenestella sp.*, trepostomatous bryozoa, *Actinoconchus planosulcatus*, *Antiquatonia muricata*, *Buxtonia scabricula*, *Eomarginifera lobata*, *Overtonia fimbriata*, *Productus concinnus*, *Orbiculoidea nitida*, *Rugosochonetes hardrensis* group, *R. laguessianus*?, *Schellwienella crenistria*, *Schizophoria sp.*, *Spirifer duplicicosta*	0	2	240	2
Shale, brown sandy micaceous; *Fenestella sp.*, *Actinoconchus planosulcatus*, *Eomarginifera lobata*, orthotetoids indet., *Productus concinnus*, *Rugosochonetes sp.*, *Spirifer sp.*	0	4	240	6
Coal, thin; on seatearth, sandy micaceous	1	6	242	0
Shale, grey and white sandy pyritic; many black nodules	1	6	243	6
Sandstone, yellow fine-grained; dark shaly carbonaceous partings	1	9	245	3
Sandstone, yellow medium-grained speckled	2	0	247	3
Sandstone and shale interbedded; fine grained	5	0	252	3
Shale dark, fissile; *Orbiculoidea nitida*, productoids indet., *Aclisina elongata*, *Euphemites urii*, *Ianthinopsis sp.*, '*Dentalium*' *sp.*, *Palaeoneilo luciniformis*, *P. undulata*, *Polidevcia* cf. *attenuata*, *Cycloceras sp.*, orthocone nautil-				

	Thickness ft	in	Depth ft	in
oids, *Tylonautilus sp. nov.* [=*T. nodiferus* early mut. Auctt.], ostracods, crinoid columnals [up to 2 mm diam.], palaeoniscid scales, *Pyritosphaera sp.*	2	0	254	3
Sandstone, white medium- and coarse-grained current-bedded; shaly at top with inclusions and roots	1	6	255	9
Mudstone, sandy micaceous carbonaceous	0	3	256	0
Sandstone, white	0	3	256	3
Shale, fissile; fish scales indet.	0	3	256	6
Sandstone, white medium-grained current-bedded	1	9	258	3
Mudstone, dark micaceous	0	6	258	9
Sandstone, yellow speckled medium-grained current-bedded	3	3	262	0
Sandstone, white; finely interbedded dark shale bands ..	5	0	267	0
Mudstone, sandy carbonaceous micaceous	8	0	275	0
Sandstone and shale finely interbedded; ironstone concretions; worm tracks; *Antiquatonia sp.*, *Buxtonia scabricula*, *Crurithyris sp.*, *Productus concinnus*, *Euphemites urii*, *Retispira sp.*, nuculoids indet.	8	0	283	0
Mudstone, dark micaceous; ironstone nodules; plants indet., *Buxtonia scabricula*, cf. *Fluctuaria undata*, *Orbiculoidea nitida*, *Pleuropugnoides pleurodon*, *Productus concinnus*, *Euphemites urii*, *Ianthinopsis sp.*, *Murchisonia sp.*, cf. *Naticopsis planispira*, *Retispira sp.*, *Soleniscus* cf. *ovalis*, '*Dentalium*' *sp.*, *Edmondia* cf. *arcuata*, *Koninckopecten scoticus*, *Palaeoneilo undulata*, *Polidevcia attenuata*, *Posidonia corrugata*, *Streblochondria anisota*, *Wilkingia variabilis*, *Cycloceras sp.*, *Paraparchites sp.*, crinoid columnals [up to 2 mm diam.]	18	8	301	8
Mudstone, dark; *Productus concinnus*, *Palaeoneilo luciniformis*, *Posidonia corrugata*	0	1	301	9
Mudstone, dark micaceous; ironstone nodules; *Cordaites*?, *Sphenopteris sp.*, *Pleuropugnoides pleurodon*, *Productus concinnus*, *Euphemites urii*, '*Dentalium*' *sp.*, *Palaeoneilo luciniformis*, *P. undulata*, *Posidonia corrugata*, orthocone nautiloids, crinoid columnals [up to 1 mm diam.] ..	3	0	304	9
Mudstone, dark micaceous; some ironstone nodules; plants indet., *Fenestella sp.*, trepostomatous bryozoa, *Chonetipustula plicata*, *Crurithyris urii*, *Orbiculoidea nitida*, *Phricodothyris sp.*, *Productus concinnus*, *Pustula sp.*, *Euphemites* aff. *pentonensis*, *E. urii*, '*Dentalium*' *sp.*, *Palaeoneilo undulata*, *Polidevcia attenuata*, *Nuculopsis gibbosa*, *Posidonia corrugata*, *Streblochondria anisota*, *Wilkingia variabilis*, *Catastroboceras* cf. *kilbridense*, orthocone nautiloids, *Pseudorthoceras sp.*, *Solenocheilus dorsalis*, ostracods indet., crinoid columnals [up to 16 mm diam.], *Pyritosphaera sp.*	4	9	309	6
Shale, dark fissile; *Fenestella sp.*, trepostomatous bryozoa, athyrids indet., *Buxtonia scabricula*, *Chonetipustula plicata*, *Crurithyris* cf. *urii*, *Lingula mytilloides*, *Orbiculoidea nitida*, *Productus concinnus*, *Euphemites* aff. *pentonensis*, *E. urii*, *Ianthinopsis sp.*, *Naticopsis sp.*, *Pseudozygopleura robroystonensis*, *Soleniscus* cf. *ovalis*, *S. turgidus*, *Straparollus sp.*, *Coleolus sp.*, '*Dentalium*' *sp.*, *Aviculopecten sp.*, *Palaeoneilo luciniformis*, *P. undulata*, *Polidevcia attenuata*, *Posidonia corrugata*, *Sanguinolites plicatus*, *Sulcatopinna*?, *Wilkingia*				

	Thickness ft	Thickness in	Depth ft	Depth in
variabilis [juv.], *Pseudorthoceras sp.*, orthocone nautiloids, ostracods, *Ureocrinus bockschii*, crinoid columnals [up to 16 mm diam.]	3	0	312	6
Shale, dark; large ironstone nodules; plants indet., *Fenestella sp.*, athyrids indet., *Orbiculoidea nitida*, productoids indet., *Euphemites urii*, '*Dentalium*' *sp.*, *Actinopteria persulcata*, *Palaeoneilo luciniformis*, *Posidonia corrugata*, cf. *Pseudaviculopecten rigidus*, *Schizodus sp.*, *Wilkingia variabilis*, *Tylonautilus sp. nov.* [=*T. nodiferus* early mut. Auctt.], orthocone nautiloids, *Catastroboceras kilbridense*, *Solenocheilus dorsalis*, *Coleolus namurcensis*, cf. *Poecilodus gibbosus* ..	2	0	314	6
Shale, dark; small ironstone nodules; *Buxtonia scabricula*, *Crurithyris sp.*, *Productus concinnus*, *Palaeoneilo undulata* ..	1	6	316	0
Shale (*core lost*)	4	0	320	0
UNDERSET (=FOUR FATHOM) CHERT (to 352 ft)				
Chert, very hard (*no core*)	4	0	324	0
Chert, limey; plants indet., sponge spicules indet., *Diploporaria sp.*, *Fenestella sp.*, *Rhabdomeson sp.*, trepostomatous bryozoa, *Brachythyris decora*, *Buxtonia scabricula*, ?*Crurithyris sp.*, *Echinoconchus sp.*, *Lingula mytilloides*, *Overtonia fimbriata*, *Phricodothyris sp.*, costate productoids, *Spirifer bisulcatus*, *Bucanopsis sp.*, *Donaldina* cf. *grantonensis*, *Euphemites urii*, *Hypergonia* cf. *acuticarinata*, *Murchisonia sp.*, *Pseudozyglopleura rugifera*, *Straparollus sp.*, '*Dentalium*' *sp.*, *Aviculopecten clathratus*, *Conocardium sp.*, *Palaeoneilo undulata*, *Nuculopsis gibbosa*, ?*Sanguinolites sp.*, *Schizodus axiniformis*, *Streblopteria concentrica*, *Catastroboceras quadratum*, *Cycloceras sp.*, orthocone nautiloids, *Coleolus sp.*, trilobites indet., ostracods, crinoid columnals [up to 6 mm diam.], palaeoniscid fish scales, *Pyritosphaera sp.*	3	3	327	3
Mudstone, dark; plants indet. [including bark preserved as vitrain], *Hyalostelia parallela*, *H. sp.* [very slender form], *Pleurodictyum dechenianum*, '*Zaphrentis*' *sp.*, *Fenestella sp.*, *Palaeocoryne sp.*, *Polypora sp.*, *Rhabdomeson sp.*, trepostomatous bryozoa, *Serpula sp.*, *Spirorbis sp.*, *Brachythyris decora*, *Chonetipustula sp.*, *Echinoconchus* cf. *elegans*, *Eomarginifera setosa*, *Girtyella sacculum*, *Lingula mytilloides*, cf. *Martinia sp.*, *Overtonia fimbriata*, *Pleuropugnoides pleurodon*, *Spirifer bisulcatus* group, *Tornquistia polita*, *Aclisina* cf. *elongata*, *Euphemites urii*, *Loxonema sulculosa*, *Murchisonia quadricarinata*, *Naticopsis sp.*, *Pseudozygopleura rugifera*, *P. scalarioidea*, *Retispira concinna*, *Straparollus* aff. *pentangulatus*, *Aviculopecten clathratus*, *Koninckopecten scoticus*, cf. *Nuculopsis gibbosa*, *Pernopecten sowerbii*, *Polidevcia attenuata*, *Schizodus* cf. *axiniformis*, *Catastroboceras quadratum*, *Cycloceras sp.*, '*Cyrtoceras*' *rugosum*, orthocone nautiloids, ?*Vestinautilus sp.*, *Weberides mucronatus*, ostracods, ?Phyllocarid crustaceans, '*Dictyonema*' *sp.*, *Archaeocidaris sp.*, crinoid columnals [up to 10 mm diam.], fish teeth and scales [including large cycloid scales]	3	9	331	0
Limestone, blue argillaceous; *Hyalostelia parallela*, '*Dentalium*' *sp.*, *Pleurodictyum dechenianum*, *Fenestella sp.*,				

	Thickness ft in	Depth ft in
Polypora sp., Cornulitella sp., productoids indet., *Spirifer bisulcatus, Aclisina sp., Euphemites urii, Pseudozygopleura sp., Aviculopecten* cf. *clathratus*, nuculoids indet., trilobite pygidia indet., ostracods, crinoid columnals [up to 10 mm diam.]	1 6	332 6
Mudstone, black calcareous; *Hyalostelia parallela, Claviradix ashi, Fenestella sp., Penniretepora sp., Rhabdomeson sp.*, scolecodonts [several jaw parts in association], *Eomarginifera setosa, Lingula mytilloides* [in position of growth], *Martinia sp., Pleuropugnoides pleurodon*, spiriferoids indet., *Tornquistia polita, Bellerophon sp., Euphemites urii*, cf. *Portlockiella parallela, Pseudozygopleura rugifera, Retispira sp., Polidevcia attenuata, 'Cyrtoceras' rugosum, 'Stroboceras' sp., Weberides mucronatus, Lophodus serratus* ..	0 6	333 0
Limestone, black argillaceous; ($7\frac{1}{2}$ *ft of core lost*); *Pleurodictyum dechenianum, Brachythyris decora, Lingula squamiformis, Spirifer bisulcatus*, pleurotomariids indet., *'Dentalium' sp., Catastroboceras kilbridense*, orthocone nautiloids, ostracods, crinoid columnals [up to 1 mm diam.], fish scales and bone fragments common, *Prioniodus sp.*, ..	19 0	352 0
Mudstone, black calcareous; *Buxtonia scabricula, Lingula mytilloides, L. squamiformis, Schizophoria sp.*, spiriferoids indet., *Euphemites sp., Porcellia sp.*, cf. *Phymatopleura sp., 'Dentalium' sp., Aviculopecten sp., Koninckopecten scoticus, Nuculopsis gibbosa, Polidevcia attenuata, Catastroboceras kilbridense*, orthocone nautiloids, goniatite indet. [large fragment], *Coleolus namurcensis*, palaeoniscid scales, *Prioniodus sp.*	1 0	353 0
Mudstone, black; rolled clisiophyllids including ?*Dibunophyllum sp., Pleurodictyum dechenianum*, zaphrentoids, *Eomarginifera* cf. *setosa, Lingula squamiformis, Orbiculoidea nitida*, costate productoids indet., *Schizophoria sp., Spirifer bisulcatus, Angyomphalus radians, Euphemites urii, E. urii hindi, Mourlonia expansa, Plagioglypta* cf. *meekiana, Aviculopecten sp., Nuculopsis gibbosa, Polidevcia attenuata, Schizodus sp., Cycloceras sp., Bistrialites bistrialis*, cf. *Mesochasmoceras latidorsatum*, orthocone nautiloids, *Pseudorthoceras sp.*, ?*'Stroboceras' sp., Coleolus namurcensis*	1 6	354 6
Mudstone, black pyritic; small ironstone nodules; *Paraconularia inaequicostata, Pleurodictyum dechenianum, Brachythyris sp., Echinoconchus sp., Rugosochonetes celticus, Spirifer bisulcatus, Euphemites* aff. *pentonensis, E. urii, Glabrocingulum armstrongi, Pseudozygopleura robroystonensis, P. rugifera, Retispira decussata, R. striata, Soleniscus intermedius, Plagioglypta* cf. *meekiana, Nuculopsis gibbosa, Palaeoneilo luciniformis, P. undulata, Schizodus sp., Catastroboceras kilbridense, Cycloceras sp.*, orthocone nautiloids, *Coleolus namurcensis*, crinoid columnals [up to 4mm diam.], fish remains [including ?operculum bone]	3 0	357 6
Mudstone, black; highly pyritic; *Rhombopora sp., Brachythyris sp., Crurithyris* cf. *magnispina, Eomarginifera setosa, Lingula mytilloides, Orbiculoidea nitida, Plicochonetes*		

	Thickness ft	in	Depth ft	in
waldschmidti auriculatus, Rhipidomella michelini, Rugosochonetes celticus, Spirifer bisulcatus, Tornquistia polita, Angyomphalus sp., Euphemites aff. *pentonensis, E. urii, Glabrocingulum armstrongi, G. atomarium, Pseudozygopleura sp., Retispira decussata, R. passeletensis, Straparollus* (*Euomphalus*) *pentangulatus, Aviculopecten sp., Nuculopsis gibbosa, Palaeoneilo luciniformis, P. undulata, Polidevcia attenuata, Bistrialites* cf. *bistrialis, Catastroboceras kilbridense, Cycloceras sp.*, orthocone nautiloids, *Pseudorthoceras sp.*, cf. *Vestinautilus paucicarinatus, Girtyoceras? costatum, G.* cf. *shorrocksi, G. sp.* [fragments], *Sudeticeras sp.* [juv.], *Coleolus namurcensis, Weberides mucronatus*, ostracods, crinoid columnals [up to 3 mm diameter], fish bones and scales	3	6	361	0
Mudstone, black; ironstone nodules; athyrids indet., *Composita sp., Echinoconchus elegans, Eomarginifera setosa*, orthotetoids, *Productus concinnus, Rugosochonetes sp., Schizophoria sp., Spirifer bisulcatus, Tornquistia polita, Euphemites* aff. *pentonensis, E. urii, Glabrocingulum armstrongi, Retispira roscobiensis, R. striata, Straparollus* (*Euomphalus*) aff. *pentangulatus, Plagioglypta meekiana, Aviculopecten sp., Caneyella membranacea, Nuculopsis gibbosa, Palaeoneilo luciniformis, Polidevcia attenuata, Sanguinolites* cf. *costellatus, Catastroboceras kilbridense, Cycloceras sp.*, orthocone nautiloids, *Coleolus namurcensis*, crinoid columnals [up to 5 mm diam.], *Lophodus serratus, Pristodus falcatus*	2	6	363	6
Shale, black (*broken core*); *Brachythyris sp., Buxtonia scabricula, ?Composita sp., Crurithyris urii, Eomarginifera setosa, Productus concinnus?, Spirifer bisulcatus, Parallelodon verneuiliana, Cycloceras sp.*, orthocone nautiloids, fish tooth indet	1	6	365	0
FOUR FATHOM LIMESTONE (to 390 ft)				
Limestone, grey; crinoidal; band of limestone conglomerate at 383 ft 3 in; (*3 ft of core lost*); *Calcisphaera sp., Calcifolium bruntonense* [rare]., *Garwoodia sp., Girvanella sp., Spongiostroma sp.*, endothyrids., *Howchinia bradyana*, tetrataxiids, bryozoans [abundant], crinoid columnals [abundant], *Archaeocidaris sp.*,	25	0	390	0
Shale, dark; simple corals indet. [crushed], *Fenestella sp., Antiquatonia muricata, Athyris lamellosa, ?Crurithyris sp., Productus concinnus?*, crinoid columnals cf. *Woodocrinus* type [up to 12 mm diam.], *Archaeocidaris sp.*	0	3	390	3
Coal, bright cleaty	0	1	390	4
Seatearth shale, roots in situ	3	7	393	11
Ganister, light coloured; rippled partings	1	6	395	5
Shale, dark micaceous; plant remains indet. [abundant], *Productus concinnus ?, Posidonia corrugata*	0	1	395	6
Sandstone and shale, finely interbedded; plant remains indet	2	0	397	6
Sandstone, fine-grained current-bedded	0	6	398	0
Sandstone and shale, finely interbedded micaceous; plant				

	Thickness ft	in	Depth ft	in
remains indet. . .	3	6	401	6
Shale, dark pyritic; plant remains including *?Cordaites sp.*, *Lingula squamiformis*, ?palaeoniscid scales [large] . .	1	3	402	9
Sandstone and shale interbedded, micaceous; plant remains indet.	1	6	404	3
Sandstone, flaggy medium-grained calcareous; micaceous and carbonaceous matter on bedding planes; plant remains indet. [abundant]	8	6	412	9
Shale, dark grey; sandy bands; ironstone nodules; plant remains indet., *Lingula squamiformis* [abundant in *Lingula* band at 418 ft 3 in]	7	0	419	9
Sandstone, medium-grained; carbonaceous partings	3	0	422	9
Sandstone, medium-grained speckled; many shale bands; plants	9	0	431	9
Shale, dark, a few sandy bands; plant remains indet. . .	3	0	434	9
Sandstone, grey rooty; carbonaceous parting at top . .	1	9	436	6
Sandstone, light grey, fine- to medium-grained; roots	2	6	439	0
Sandstone, fine-grained; irregular partings	2	0	441	0
Sandstone and flaggy sandstone, interbedded, fine-grained	9	6	450	6
Sandstone and shale, interbedded; plant remains including *Calamites sp.* and *Cordaites sp.*	7	6	458	0
Shale, grey and black; a few sandy bands; plant remains indet.	8	0	466	0
Shale, grey sandy; plant remains indet., *Eomarginifera setosa*, *Euphemites sp.*	6	0	472	0
Shale, dark micaceous; sandy bands; ironstone nodules; plant remains indet., *Rugosochonetes sp.* [in position of growth], *Lingula squamiformis* [some in position of growth], *Productus concinnus?*, *Euphemites sp.*, bivalves indet. [crushed]	8	0	480	0
Shale, dark; many small ironstone nodules and ironstone bands; *Orbiculoidea sp.*, *Plicochonetes sp.*, *Rugosochonetes sp.* [in position of growth], *Euphemites urii*, *Retispira sp.*, '*Dentalium*' *sp.*, *Posidonia corrugata*, *?Schizodus sp.*, orthocone nautiloids . .	8	0	488	0
Shale, dark; *Fenestella sp.*, *Plicochonetes sp.*, productoid indet. [costate], *Rugosochonetes sp.*, *Nuculopsis gibbosa*, *Posidonia corrugata*, *Retispira* cf. *decussata*, orthocone nautiloids, palaeoniscid scales	0	6	488	6
Shale, dark grey; ironstone nodules; plant remains [rare], *Euphemites sp.* [rare], bivalves indet. [rare] . .	7	0	495	6
Mudstone, dark; ironstone nodules; *Fenestella sp.*, *Lingula squamiformis* [in position of growth], *Euphemites* aff. *pentonensis*, *E. urii*, *Retispira decussata*, *Posidonia corrugata*, orthocone nautiloids, fish remains . .	5	6	501	0
Mudstone, dark; pyritic; much ironstone; trepostomatous bryozoa [incrusting], *Eomarginifera* cf. *setosa*, *?Plicochonetes sp.*, *Euphemites urii*, *Glabrocingulum atomarium*, *Retispira decussata*, *R. striata*, *?Posidonia sp.*, orthocone nautiloids, goniatites [fragments and juv.] including *?Girtyoceras sp.*, trilobite indet. [thoracic segment], *Psammodus sp.* . .	5	9	506	9
Shale, grey; sandy bands with many small ironstone nodules; plant remains indet. [abundant fragments] . .	1	0	507	9
Sandstone, dark grey fine-grained; carbonaceous fragments . .	0	3	508	0

	Thickness ft	Thickness in	Depth ft	Depth in
Shales, grey; sandy bands sporadically micaceous; ironstone nodules; plant remains indet., *Euphemites sp.* [rare], bivalves indet.	10	3	518	3
Mudstone, black slightly pyritic; ironstone nodules; productoid indet., *Euphemites urii*, pleurotomariids indet., *Retispira decussata*, *Straparollus* (*Euomphalus*) *pentangulatus*, *Posidonia corrugata*, orthocone nautiloids, *Pseudorthoceras sp.*	4	3	522	6
Mudstone, black pyritic; large ironstone nodules; *Cyathaxonia cornu*, zaphrentoids, *Buxtonia scabricula*, *Crurithyris urii*, *Dictyoclostus sp.*, *Eomarginifera* cf. *setosa*, *Phricodothyris paricosta*, *Productus concinnus ?*, *Rugosochonetes sp.*, *Spirifer bisulcatus*, *Euphemites urii*, *Pseudozygopleura rugifera*, *Retispira decussata*, *R. striata*, *Straparollus* aff. *pentangulatus*, *Koninckopecten scoticus*, *Palaeoneilo luciniformis*, *Polidevcia* cf. *attenuata*, ?*Schizodus sp.*. orthocone nautiloids	4	0	526	6
Mudstone, dark; ironstone nodules; plant remains including *Cordaites sp.*, *Cyathaxonia cornu*, *Hexaphyllia* aff. *mirabilis*, *Pleurodictyum dechenianum*, *Fenestella sp.*, *Buxtonia scabricula*, *Chonetipustula* cf. *plicata*, *Chonetipustula sp.*, *Crurithyris* cf. *magnispina*, *Crurithyris urii*, *Eomarginifera setosa*, *Megachonetes papilionaceus*?, *Orbiculoidea sp.*, *Phricodothyris sp.*, *Plicochonetes buchianus*, productoids indet. [costate], *Rugosochonetes sp.*, *Schizophoria sp.*, *Spirifer bisulcatus*, *Spiriferellina sp.*, spiriferoids indet. [smooth], *Euphemites urii*, *Lepetopsis* cf. *conoideus*, *Retispira decussata*, *Straparollus* (*Euomphalus*) *pentangulatus*, *Aviculopecten sp.*, *Palaeoneilo undulata*, *Polidevcia attenuata*, '*Pseudamusium*' *auriculatus*, *Catastroboceras quadratum sp.*, orthocone nautiloids, *Weberides mucronatus*, crinoid columnals [up to 10 mm diam.]	4	6	531	0
Ironstone	0	6	531	6
Mudstone, dark; ironstone nodules; *Cladochonus bacillarius*, *Cyathaxonia cornu*, *Pleurodictyum dechenianum*, *Claviradix ashi*, *Diploporaria sp.*, *Fenestella sp.*, *Penniretepora sp.*, trepostomatous bryozoa, *Brachythyris sp.*, ?*Buxtonia scabricula*, *Chonetipustula plicata*, *Composita ambigua*, *Crurithyris magnispina*, *C. urii*, *Dielasma sp.*, *Dictyoclostus sulcatus*, *Eomarginifera setosa*, *Lingula mytilloides*, *Megachonetes papilionacea*, *M.* cf. *siblyi*, *Orbiculoidea nitida*, *Overtonia fimbriata*, *Phricodothyris sp.*, *Plicochonetes buchianus*, *Productus* cf. *carbonarius*, *P. concinnus*, *Rugosochonetes sp.*, *Spirifer bisulcatus*, *Stenoscisma sp.*, *Tornquistia polita*, *Euphemites urii*, *Pseudozygopleura sp.*, *Straparollus* (*Euomphalus*) aff. *pentangulatus*, *Aviculopecten* cf. *murchisoni*, *A.* cf. *nobilis*, *Koninckopecten scoticus*, *Palaeoneilo undulata*, *Posidonia corrugata*, *Streblochondria anisota*, ?*Sulcatopinna sp.*, orthocone nautiloids, *Weberides mucronatus*, ostracods, crinoid columnals [up to 3 mm diam.], *Platycrinites spiniger*, ? fish teeth and scales ..	5	9	537	3
Limestone, grey argillaceous; *Crurithyris urii*, crinoid columnals [small]	0	3	537	6
Mudstone, dark; a few small ironstone nodules; traces of				

	Thickness ft in	Depth ft in
pyrite; chonetoid indet., *Chonetipustula sp.*, *Eomarginifera setosa*, *Lingula sp.*, *Plicochonetes* cf. *crassistrius*, *Productus productus*, *Straparollus sp.*, trilobites indet., ostracods [abundant]	1 0	538 6
Shales, light and dark banded at top with greenish and buff patches; zaphrentoids indet., *Eomarginifera setosa*, productoids indet. [costate], *Plicochonetes sp.*, *Spirifer bisulcatus*, orthocone nautiloids, ostracods, crinoid columnals [small, in bands]	1 6	540 0
THREE-YARD LIMESTONE (to 554 ft 3 in)		
Limestone, dark grey crinoidal (2 *ft core lost*); *Calcisphaera sp.*, *Girvanella sp.*, *Spongiostroma sp.*, endothyrids, bryozoa indet., productoids indet [costate], crinoid columnals [small]	6 0	546 0
Shale, dark; thin coal scares; endothyrids, textulariids, '*Chonetes*' *sp.*, '*Dentalium*' *sp.*, ostracods	1 0	547 0
Shale, dark; *Fenestella sp.*, trepostomatous bryozoa, productoids indet. [costate], *Spirifer bisulcatus*	0 6	547 6
Mudstone, dark; much pyrite; *Fenestella sp.*, *Lophophyllidium sp.*, *Eomarginifera setosa*, *Spirifer sp.*, *Portlockiella parallela*	2 0	549 6
Shale, calcareous; a little pyrite; grades down into muddy limestone; *Aphralysia carbonaria*, *Calcisphaera sp.*, *Girvanella sp.*, endothyrids, tetrataxiids, textulariids, *Fenestella sp.*, *Diploporaria sp.*, worm tubes [incrusting], *Eomarginifera setosa*, *Spirifer sp.*, *Tornquistia polita*, '*Dentalium*' *sp.*, crinoid columnals [up to 5 mm diam.], *Archaeocidaris sp.* [spines]	1 0	550 6
Limestone, dark muddy; productoids indet. [costate], crinoid columnals	2 0	552 6
Coal, bright	0 7	553 1
Seatearth-sandstone, rooty; grades down to dark shale	0 4	553 5
Limestone, shaly; *Pleuropugnoides pleurodon*, *Productus concinnus*?, crinoid columnals [up to 6 mm diam.]	0 10	554 3
Shale, dark pyritic	0 3	554 6
Sandstone, grey calcareous; brachiopods indet.	2 0	556 6
Sandstone, fine-grained; grey beds	6 6	563 0
Mudstone, dark pyritic	1 3	564 3
Sandstone and shale, finely interbedded	1 3	565 6
Limestone, grey sandy	1 0	566 6
Sandstone, calcareous dark partings; *Productus* cf. *concinnus*, crinoid columnals, *Archaeocidaris sp.*, annelid burrows [vertical, U-shaped]	1 8	568 2
Shale, dark; sandy bands	1 10	570 0
Ganister, fine-grained; a few shaly partings near top (2 *ft core lost*) ?[*Fault*]	15 0	585 0
Shale; pyrite-rich at base	0 6	585 6
FIVE-YARD LIMESTONE (to bottom of hole)		
Limestone, grey; shaly at top with pyrite; highly fractured; vertical joints infilled with calcite; *Draffania sp.*, *Girvanella sp.*, endothyrids, *Howchinia bradyana*, tetrataxiids, bryozoans indet. [fragments], crinoid columnals [up to 5 mm diam.], *Archaeocidaris sp.* [spines]	6 6	592 0
Bore drilled further but core not raised; believed to be still in limestone	2 0	594 0

References

HULL, J. H. 1968. The Namurian stages of north-eastern England. *Proc. Yorks. geol. Soc.*, **36**, 297–308.

JOHNSON, G. A. L., HODGE, B. L. and FAIRBAIRN, R. A. 1962. The base of the Namurian and of the Millstone Grit in north-eastern England. *Proc. Yorks. geol. Soc.*, **33**, 314–62.

MILLS, D. A. C., and HULL, J. H. 1976. The Geology of the country around Barnard Castle. *Mem. geol. Surv. Gt Br.*

RAYNER, D. H. 1953. The Lower Carboniferous rocks in the north of England: a review. *Proc. Yorks. geol. Soc.*, **28**, 231–315.

READING, H. G. 1957. The stratigraphy and structure of the Cotherstone syncline. *Q. Jnl geol. Soc. Lond.*, **113**, 27–56.

WELLS, A. J. 1957. The stratigraphy and structure of the Middleton Tyas–Sleightholme Anticline, North Yorkshire. *Proc. Geol. Ass.*, **68**, 231–54.

Appendix 4

LIST OF GEOLOGICAL SURVEY PHOTOGRAPHS

COPIES of these photographs are deposited in the libraries of the Institute of Geological Sciences at Exhibition Road, London SW7 2DE and Ring Road Halton, Leeds LS15 8TQ. All may be supplied as black and white prints or lantern slides at a standard tariff; those marked C as colour prints and those marked T as 2 × 2-in colour transparencies.

CARBONIFEROUS LIMESTONE SERIES

L342 T	Four-Fathom Limestone, Browson Bank Quarry. (Plate IIA).
L343 T	Siliceous concretions; base of Four-Fathom Limestone, Browson Bank Quarry. (Plate IIB).
L344 CT	Base of Four-Fathom Limestone, Browson Bank Quarry.
L345 T	Coral band; base of Four-Fathom Limestone, Browson Bank Quarry.

MILLSTONE GRIT SERIES

L311–2 CT	Top Crag Limestone, Crag Bridge, Deepdale Beck. (Plate IVA).
L314–5	Faraday House Shell-Bed, Eggleston Burn near Eggleston. (Plate IIIB).
L321, 392–3 CT	Barnard Castle; the castle on Crag Limestones and associated measures. (Plate I, Frontispiece).
L324 CT	Transgressive sandstone overlying Rookhope Shell-Beds Limestone, Cat Castle Quarry.
L327	Great Limestone in old quarry near Hulands, Bowes.
L328 CT	Working face (1963) in Great Limestone, Lamb Hill Quarry, Bowes. (Plate IIIA).
L329–330 CT	Top of Great Limestone dipping into River Tees near confluence with River Greta, Mortham Tower.
L331 CT	Basement or First Grit overlying shales and thin limestones with *Reticuloceras stubblefieldi*, White House.
L334 CT	Botany Limestone, comprising interbedded limestones and shales, Hedrick Grange Quarry, Kinninvie.
L336 CT	Massive siliceous sandstone, Dunn House Quarry, Staindrop. (Plate IVB).
L337, 340	As above.
L341	Massive siliceous sandstone, Old Quarry, Hollin Hall East Farm, near Gainford.
L390 CT	Massive false-bedded quartzo-feldspathic sandstone ('Second Grit'), old quarry near Houghton le Side.
L399–401 CT	Forcett Quarry, East Layton; small fault in quarry face throwing Great Limestone against underlying sandstones and shales. Copper mineralization occurs as sulphide nodules at the base of the limestone.

Coal Measures

L349	View north-west over River Gaunless valley from Cockfield Fell towards Butterknowle and Copley. Woodland on skyline.
L350	View west over River Gaunless Valley from Cockfield Fell towards Copley and Woodland Fell.
L351 T	Massive sandstone overlying Busty Coal, old quarry on Cockfield Fell. (Plate VIIIA).
L384	Middle Coal Measures between Five-Quarter and Main Coals, Eldon Hill Quarry.
L385, 402–8 CT	View north and east from Brusselton Hill towards Bishop Auckland and Shildon. (Plate VIIIB).

Permian

L356	Middle Magnesian Limestone. Reef-type limestone overlying oolitic dolomite, old quarry, High Coniscliffe.
L397–8 CT	Middle Magnesian Limestone. Algal macrostructures. North bank of River Tees, High Coniscliffe. (Plate XVB).
L357	As above.
L358	Lower Magnesian Limestone. Rubbly dolomite and dolomitic limestone. (Shap Granite Boulder). North bank of River Tees, Piercebridge.
L359	Lower Magnesian Limestone. Bedded dolomitic limestone, nearly ¼ mile ENE of Piercebridge.
L361 CT	Lower Magnesian Limestone. Bedded dolomite and dolomitic limestone. Old quarry north-west of Denton.
L362	Lower Magnesian Limestone. Bedded dolomite and dolomitic limestone. Summerhouse Quarry, Summerhouse. (Plate XVA).
L363 CT	Lower Magnesian Limestone. Bedded dolomite and dolomitic limestone, south bank of River Tees, Piercebridge.
L364 T	Lower Magnesian Limestone. Bedded dolomite and dolomitic limestone, unconformably overlying Namurian sandstone and shale. Cutting on east side of Durham A1(M) motorway, near Cleasby, Darlington (1964).
L365–6, 368 T	As above; closer views of Permian/Carboniferous unconformity (1964).
L367	As above. (Plate XIVA).
L374 T	Lower Magnesian Limestone, Marl Slate and Basal Permian Sandstone and Breccia unconformably overlying Lower Coal Measures sandstone. Thickley Quarry at East Thickley, near Shildon.
L375 T	As above. Hand indicates unconformity. (Plate XIII B).
L376 CT	As L374. Closer view.
L377 T	Close-up of Permian–Carboniferous unconformity. Hammer head on unconformity. Thickley Quarry at East Thickley, near Shildon.
L378 T	Lower Magnesian Limestone (disturbed bedding at top), Marl Slate and Basal Permian Sandstone and Breccia unconformably overlying Middle Coal Measures sandstone. Eldon Hill Quarry.
L380	As above. (Plate XIIIA).
L379, L381–2	Folding in Lower Magnesian Limestone. Eldon Hill Quarry.
L383	Lower Magnesian Limestone and Marl Slate unconformably overlying Middle Coal Measures. Hammer head on unconformity. Eldon Hill Quarry.
L386	Lower Magnesian Limestone. Rubbly bedded dolomitic limestone overlain by breccia. Old quarry west of Shackleton Beacon, Heighington.

L387 CT	Lower Magnesian Limestone. Brecciated calcitic dolomite. Shackleton Beacon near Heighington. (Plate XIVB).
L388	Lower Magnesian Limestone. Bedded dolomitic limestone. Old quarry near Southfield House, south of Shildon.
L389	Lower Magnesian Limestone. Steeply dipping massive calcitic dolomite, underlain by thinly bedded dolomite. Old quarry near Elm Grange Farm, Heighington.
L391	View of drift-covered Lower and Middle Magnesian Limestone country looking east and south-east from near Elm Grange, Heighington. Cleveland Hills in background.

Igneous Rocks

L347	Cleveland Dyke. South wall of Haggerleases Quarry near Butterknowle. (Plate XVIB).
L348 T	Old quarry in Cleveland Dyke looking west-north-west across Cockfield Fell. (Plate XVIA).
L352	Cleveland Dyke. South face of western pit, Bolam Quarry (1963). Westphalian sediments form floor of quarry. (Plate XVIIA).
L353 T	Cleveland Dyke. North face of western pit, Bolam Quarry. Tholeiitic dolerite split into two leaves separated by Westphalian sediments. (Plate XVIIB).
L354	As L352, showing Lower Coal Measures sediments underlying tholeiitic dolerite.

Pleistocene and Recent

L316	Landslip of Boulder clay, Eggleston Burn. (Plate XIXA).
L317 CT	Glacial Drainage Channels—flat bottomed in foreground, overflow type on sky line. From near Hill Top, Eggleston-Stanhope Road.
L318 CT	Glacial Drainage Channels on skyline cutting through Hett Dyke. From near Hill Top, Eggleston-Stanhope Road. (Plate XIXB).
L319 CT	Interbedded fluvio-glacial gravels and false-bedded sands. Sayer's Old Quarry, near Eggleston.
L332 CT	Glacial Drainage Channels cutting Namurian sandstones (well-marked dip slopes). Billy Lane Farm, Kinninvie.
L333 T	Romaldkirk Moraine, south side. From Cotherstone–Romaldkirk road.
L335 T	View east-north-east over River Tees Terraces; Primrose Hill Farm, near Winston. (Plate XXB).
L338	View south-west down River Tees valley from near Winston.
L339 T	Knoll of Glacial Sand and Gravel, Hill House, near Headlam.
L360 T	View overlooking high terrace of River Tees south-east of Piercebridge. Bank feature of boulder clay overlying Lower Magnesian Limestone in background.
L369	Boulder Clay bluff, Manfield Scar; River Tees near Manfield.
L370 CT	View south-west towards rock-floored drumlin at Micklow, near Aldbrough.
L372–373 CT	Contorted glacial sand and gravel. Stapleton Manor, Stapleton.
L371	As above. (Plate XXA).

Scenery

L385, 402–408 CT	Panorama looking north and east from Brusselton Hill towards Bishop Auckland and Shildon. (L405, Plate VIIIB).
L394 CT	Barnard Castle from Startforth.
L395–6 CT	The River Tees and Barnard Castle from Startforth.

The following photographs, listed above under stratigraphical headings are also of scenic interest:

Upper Carboniferous (Namurian): L311–2, L392, L393.

Upper Carboniferous (Westphalian): L349, L350, L385, L402–08.

Permian: L391.

Igneous: L348.

Pleistocene and Recent: L317, L318, L335, L338, L360.

INDEX

Fossil names qualified by aff. or cf. are indexed under the unqualified name. BH=Borehole.

Printed for Her Majesty's Stationery Office
by McCorquodale Ltd., London.
HM 7609 Dd. 289668 K12 7/76 McC 569